Advanced Materials for a Sustainable Environment

This book summarizes recent and critical aspects of advanced materials for environmental protection and remediation. It explores the various development aspects related to environmental remediation, including design and development of novel and highly efficient materials, aimed at environmental sustainability. Synthesis of advanced materials with desirable physicochemical properties and applications is covered as well. Distributed across 13 chapters, the major topics covered include sensing and elimination of contaminants and hazardous materials via advanced materials along with hydrogen energy, biofuels, and CO_2 capture technology.

- Discusses the development in design of synthesis process and materials with sustainable approach.
- Covers removal of biotic and abiotic wastes from the aqueous systems.
- Includes hydrogen energy and biofuels under green energy production.
- Explores removal of environmental (soil and air) contaminants with nanomaterials.
- Reviews advanced materials for environmental remediation in both liquid and gas phases.

Emerging Materials and Technologies

Series Editor:
Boris I. Kharissov

The *Emerging Materials and Technologies* series is devoted to highlighting publications centered on emerging advanced materials and novel technologies. Attention is paid to those newly discovered or applied materials with potential to solve pressing societal problems and improve quality of life, corresponding to environmental protection, medicine, communications, energy, transportation, advanced manufacturing, and related areas.

The series takes into account that, under present strong demands for energy, material, and cost savings, as well as heavy contamination problems and worldwide pandemic conditions, the area of emerging materials and related scalable technologies is a highly interdisciplinary field, with the need for researchers, professionals, and academics across the spectrum of engineering and technological disciplines. The main objective of this book series is to attract more attention to these materials and technologies and invite conversation among the international R&D community.

For more information about this series, please visit: www.routledge.com/Emerging-Materials-and-Technologies/book-series/CRCEMT

Advanced Materials for a Sustainable Environment
Development Strategies and Applications

Edited by
Naveen Kumar and Peter Ramashadi Makgwane

CRC Press
Taylor & Francis Group
Boca Raton London New York

CRC Press is an imprint of the
Taylor & Francis Group, an **informa** business

Designed cover image: © Shutterstock

First edition published 2023
by CRC Press
6000 Broken Sound Parkway NW, Suite 300, Boca Raton, FL 33487-2742

and by CRC Press
4 Park Square, Milton Park, Abingdon, Oxon, OX14 4RN

CRC Press is an imprint of Taylor & Francis Group, LLC

© 2023 selection and editorial matter, Naveen Kumar and Peter Ramashadi Makgwane; individual chapters, the contributors

ISBN: 9781032073057 (hbk)
ISBN: 9781032073064 (pbk)
ISBN: 9781003206385 (ebk)

DOI: 10.1201/9781003206385

Typeset in Times
by codeMantra

Contents

Preface

The growing increase in the generation of new complex pollutants matrixes is due to rapid extension of industrial activities and high rate of urbanization, transport emissions, newly developing high-technology agricultural practices, and emergence of pharmaceutical health-threatening wastes disposal, which presents an alarming concern on a healthy sustainable environment. The application of advanced materials in the spheres of environmental remediation has been overwhelming in the past decade. As a result, there is a high expectation of developing new innovative remedial solutions to these evolving environmental pollution issues and developing greener processes' conversion of renewable materials into environmentally friendlier products. This book systematically discusses the recent trends in the design and development of advanced materials ranging from metal oxides, various carbons, and hybrids nanomaterials from fundamental structured layout, synthesis, and applications to develop sustainable and green environment processes. It consists of 13 chapters written by experts in advanced materials design and applications in various green processes, which can be categorized into three themes. The first theme introduces the broader overview of advanced materials and their applications in diversified processes, including a dedicated chapter on green approach methods to production technology focusing on process designing and manufacturing aspects. The second theme focuses on advanced nanomaterials for sensing and removal of organic and inorganic heavy metals pollutants, including antibacterial self-disinfections. The final theme discusses advanced materials and renewable process in carbon dioxide capture technology and conversion, including new developments in hydrogen and biofuels production for sustainable green energy and the environment. This book appeals to a broad readership, including professional industrial, academic researchers, and graduate students who are interested in advanced materials design towards sustainable environmental applications.

Editors

Naveen Kumar has 18 years of research experience and is currently working as an Associate Professor in Department of the Chemistry, Maharshi Dayanand University, Rohtak (INDIA), and holding various academic positions in the institute. He is actively engaged in the field of material chemistry, out of which photocatalysis and environmental chemistry are major areas of the research. He has published more than 60 articles in the journals of high repute. He also worked at Universitat Politècnica de València, Valencia, Spain, in an international research project entitled as "Development of a New Generation CIGS Based Solar Cells" [NANICIS-269279] in 2013 and 2014. He is also working as a reviewer in various reputed journals. He was a guest editor for MPDI Catalyst (2021) in nanocatalysis and photocatalysis. He has presented research papers at national and international conferences and delivered invited lectures in various institutes.

Peter Ramashadi Makgwane is a Principal Scientist at the Council for Scientific and Industrial Research (CSIR), Pretoria (South Africa), and Extraordinary Professor in Chemistry at the University of the Western Cape (South Africa). He completed his MSc Chemistry from University of Pretoria (South Africa) in 2006, and PhD Chemistry from Nelson Mandela University (South Africa) in 2010. He was a visiting scholar at Polish Academy of Sciences Institute of Physical Chemical (PAS-IPC) catalysis research group, Warsaw (Poland), in 2017 and National Research Centre (Cairo, Egypt) in 2016. He was an invited guest editor for *Journal of Nanotechnology and Nanoscience* (2014) and MDPI Catalyst (2021) in nanocatalysis and photocatalysis themed issues. Professor Makgwane has published a number of research articles in heterogeneous catalysis, nanocatalysis, and photocatalysis focused on chemicals conversion, environmental remediation, and semiconductor gas chemical sensing. His main research interest is in applying catalysis for green and sustainable processes in chemicals, energy, and environment.

Contributors

David Abad-Correa
Applied Engineering
Centro Nacional del Hidrógeno
Puertollano, Spain

Reda M. Abdelhameed
Applied Organic Chemistry Department,
    National Research Centre
Chemical Industries Research Institute
Giza, Egypt

Dana Susan Abraham
Department of Chemistry
Central University of Kerala
Periya, India

Md. Ahmaruzzaman
Department of Chemistry
National Institute of Technology
Silchar, India

Hanan B. Ahmed
Chemistry Department, Faculty of Science
Helwan University
Cairo, Egypt

Francisco Manuel Baena-Moreno
Chemical and Environmental Engineering
    Department, School of Engineering
University of Seville
Seville, Spain

Margandan Bhagiyalakshmi
Department of Chemistry
Central University of Kerala
Periya, India

Mário J. F. Calvete
Department of Chemistry
University of Coimbra
Coimbra, Portugal

Rui M. B. Carrilho
Department of Chemistry
University of Coimbra
Coimbra, Portugal

P. Chinnamuthu
Department of Electronics and Communication
    Engineering
NIT Nagaland
Chumukedima, India

Bijit Choudhuri
Department of Electronics and Communication
    Engineering
NIT Silchar
Silchar, India

Barbara Souza Damasceno
Laboratório de Plasmas e Processos (LPP)
Instituto Tecnológico de Aeronáutica (ITA)
São José dos Campos, Brazil

Jnyanashree Darabdhara
Department of Chemistry
National Institute of Technology
Silchar, India

Pooja Devi
Department of Chemistry
Guru Jambheshwar University of Science &
    Technology
Hisar, India

Seema Devi
Department of Chemistry
Guru Jambheshwar University of Science &
    Technology
Hisar, India

Lucas D. Dias
São Carlos Institute of Physics
University of São Paulo
São Carlos, Brazil

Hossam E. Emam
Department of Pretreatment and Finishing of
    Cellulosic based Textiles, National Research
    Centre
Textile Research and Technology Institute
Giza, Egypt

Luz Marina Gallego-Fernández
Chemical and Environmental Engineering
    Department, School of Engineering
University of Seville
Seville, Spain

Armstrong Godoy-Jr
Departamento Acadêmico de Física (DAFIS)
Universidade Tecnológica Federal do Paraná
    (UTFPR)
Londrina, 86036-370

Rashi Gusain
Department of Chemical Sciences
University of Johannesburg
Doornfontein, South Africa
Centre for Nanostructures and Advanced
    Materials, DSI-CSIR Nanotechnology
    Innovation Centre
Council for Scientific and Industrial Research
Pretoria, South Africa

Isabela Machado Horta
Laboratório de Plasmas e Processos (LPP)
Instituto Tecnológico de Aeronáutica (ITA)
São José dos Campos, Brazil

Jitender Jindal
R.P.S Degree College
Balana, Mahendergarh

Jyoti Kataria
Department of Chemistry
Guru Jambheshwar University of Science &
    Technology
Hisar, India

Naveen Kumar
Department of Chemistry
Maharshi Dayanand University
Rohtak, India

Neeraj Kumar
Department of Chemical Sciences
University of Johannesburg
Doornfontein, South Africa
Centre for Nanostructures and Advanced
    Materials, DSI-CSIR Nanotechnology
    Innovation Centre
Council for Scientific and Industrial Research
Pretoria, South Africa

Douglas Marcel Gonçalves Leite
Laboratório de Plasmas e Processos (LPP)
Instituto Tecnológico de Aeronáutica (ITA)
São José dos Campos, Brazil

Peter R. Makgwane
Centre for Nanostructures and Advanced
    Materials (CeNAM)
Council for Scientifc and Industrial
    Research (CSIR)
Pretoria, South Africa

Mabuatsela Virginia Maphoru
Department of Chemistry
Tshwane University of Technology
Pretoria, South Africa

André Luis de Jesus Pereira
Laboratório de Plasmas e Processos (LPP)
Instituto Tecnológico de Aeronáutica (ITA)
São José dos Campos, Brazil

Mariette M. Pereira
Department of Chemistry
University of Coimbra
Coimbra, Portugal

Sreejarani Kesavan Pillai
Centre for nanostructures and advanced
    materials
Council for Scientific and Industrial Research
Pretoria, South Africa

Parul Raturi
Department of Physics
Omkaranand Sarswati Government Degree
    College
Devpryag, India

Suprakas Sinha Ray
Department of Chemical Sciences
University of Johannesburg
Doornfontein, South Africa
Centre for Nanostructures and Advanced
    Materials, DSI-CSIR Nanotechnology
    Innovation Centre
Council for Scientific and Industrial Research
Pretoria, South Africa

Fábio M. S. Rodrigues
Department of Chemistry
University of Coimbra
Coimbra, Portugal

Akbar Samadi
Institute for Frontier Materials (IFM)
Deakin University
Geelong, Australia

Argemiro Soares da Silva Sobrinho
Laboratório de Plasmas e Processos (LPP)
Instituto Tecnológico de Aeronáutica (ITA)
São José dos Campos, Brazil

Muhammad Tahir
Chemical and Petroleum Engineering
    Department
UAE University
Al Ain, UAE

Sehar Tasleem
School of Chemical and Energy Engineering,
    Faculty of Engineering
Universiti Teknologi Malaysia
Johor Bahru, Malaysia

Fernando Vega
Chemical and Environmental Engineering
    Department, School of Engineering
University of Seville
Seville, Spain

Mari Vinoba
Kuwait Institute for Scientific Research
Kuwait City, Kuwait

Shuaifei Zhao
Institute for Frontier Materials (IFM)
Deakin University
Geelong, Australia

1 Advanced Materials towards Environmental Protection
Attributes and Progress

Naveen Kumar
Maharshi Dayanand University

Peter R. Makgwane
Council for Scientific and Industrial Research (CSIR)

Jitender Jindal
R.P.S Degree College

CONTENTS

DOI: 10.1201/9781003206385-1

1.1 INTRODUCTION

Environmental protection awareness and technology advancement are greatly needed in recent times because of the huge amounts of pollutants that are released in the environment as a result of human and industrial development. Although nature has its own capability for the remediation of these pollutants, the increase in population drastically spoils the natural resources. It decreases the capability to tackle with the environmental problems. Secondly, various synthetic pollutants have long-lasting life existence once disposed; hence, their timely remediation is appreciable before penetration in living beings. All the components in the atmosphere, soil, air, and water are polluted but the water quality is most drastically decreased. Thus, for a sustainable environment, water pollution management, an updating research direction, is a great area of concern. Water contamination is caused by various types of discharges such as heavy metal ions, dyes, pesticides, pharmaceutical residues, and their elimination from reaching the natural resources is now in the focus all over the world [1].

Material researchers are continuously working to design materials with desired properties, which are highly needed to explore a treatment technology in the interest of water, soil, and air decontamination. The initial steps in environmental protection are the proper detection, qualitative and quantitative analyses, and in the subsequent steps, the treatment of pollutants by different modes. In both detection and treatment methods, the regular development is achieved by the researchers. In addition, environmental protection is not only limited to the monitoring and remediation of pollution-causing agents but also searching the ways that reduce the pollution by their origin and utilization, e.g., development of the technologies in the energy sectors, which can minimize pollution primarily (e.g., production of green and renewable energy resources). In the energy sector, the development of photovoltaic technology and hydrogen energy production are two main themes that have a great scope and are considerable ways for a sustainable environment and protection.

In conclusion, both the steps, namely, (i) proper detection, monitoring, and decontamination of the environment and (ii) development and implementation of the technologies in green energy production are associated for the establishment of pollution-free environment. For the consideration and achievement of the above-mentioned ways, the common target is the development of the efficient and cost-effective materials for the development of the technology to be implemented for environmental protection directly or indirectly.

Recently, a variety of hybrid materials are developed, which exhibit unique structural, optical, morphological, and conducting properties for potential applications in the streams for the environmental protection. In this chapter, various modified and advanced materials like doped, nanocomposites, polymer-assisted materials, ionic liquid (IL)-associated materials, metal organic frameworks (MOFs), various zeolites, and clay-based nanomaterials are discussed. In the subsequent section, the various applications employed in the environmental protection by using these advanced materials have also been explored.

1.2 NEED FOR THE ADVANCED MATERIALS

Although nature has its own capability to decontaminate the natural resources but at higher concentrations or in dealing with synthetic chemicals, the support of technology is required. In the subsequent sections, the prominent aspects are elaborated, which define the intense desire for the development of advanced materials for environmental remediation.

1.2.1 EMERGING POLLUTANTS IN ATMOSPHERE

In the present section, the emerging pollutants are reviewed, which seeded for the desire to develop advanced materials for their remedial measures.

1.2.1.1   Pharmaceutical Residues

Pharmaceutical residues are released from household utilization, hospitals, and livestock farming. These residue wastes are bioactive in nature, so their release in water is highly hazardous for the biotic components. In the pharmaceutical residues, the major area is covered by antibiotics, which causes the antibiotic resistance when enters into living beings through the food chain. In the treatment techniques, the processes that are highly efficient in the removal of wastes are ozonation, biofilter with granulated activated carbon (GAC), ozonation and biofilter, and ultrafiltration [2]. The pharmaceuticals like paracetamol and ibuprofen are removed efficiently from the water but the antibiotics removal is difficult, and the drugs like sulfamethazine or carbamazepine can be removed to the maximum level of up to 23% [3]. Pharmaceutical drugs are transformed into metabolites in organism and in the environment fully or partially. The metabolites that are released from the living organisms through the process of excretion also elevate these contaminants. So, the determination of the metabolites, toxicity, and their remediation require a great research attention. The drugs and chemicals that are maximum in use are various antibiotics, analgesics, anti-inflammatory agents, and triclosan, a chemical found in products of household purposes like toothpastes, soups, and deodorant [4].

1.2.1.2   Endocrine-Disrupting Chemicals (EDC)

EDC are the chemicals that interfere with the hormonal system of the body. These chemicals are found in air, water, soil, cosmetic care products, and food sources. Some common EDCs are lead, cadmium, DDT, atrazine, bisphenol A, phthalates, triclosan, polychlorinated biphenyls, perfluorochemicals, etc. [5]. These chemicals are a serious threat to human beings because they are capable of causing neurological disorders and reducing stress toleration, indirectly are linked to diabetes, have an effect on reproductive health by restriction in sex hormones, and weaken the immune system [6]. A long exposure to multiple EDC can cause more adverse effects [7].

1.2.1.3   Dyes and Dye-Containing Hazardous Substances

Dyes and dye-containing substances are primarily released by textile and leather industries and pollute the environment globally. Dyes as organic compounds are highly soluble in water and thus difficult to remove by conventional techniques. Other than textile dyes, food dyes are also a significant pollutant released by dye-manufacturing plants. Most of the dyes have mutagenic and carcinogenic properties. The widely used food dyes are Brilliant Blue, Indigo carmine, Citrus Red, Fast green, Allura Red, tetrazine, etc. In most of the cosmetic products, dyes or pigments are added in different concentrations. Also, hair dye industries are accounting for 80% of dyes used in the cosmetic sector [8]. Most of these dyes have the benzene moiety, which itself is highly hazardous to health.

1.2.1.4   Polycyclic Aromatic Hydrocarbon (PAH)-Based Emerging Contaminants

PAHs are highly toxic, persistent chemicals, which are released in the environment by certain anthropogenic activity. These compounds have jointed aromatic rings like anthracene, benzofluoranthene, cyclopenta pyrene, dibenzoanthracene, benzoperylene, benzopyrene, chrysene, dibenzopyrene, indenopyrene, and 5-methylchrysene [9]. These chemicals are highly carcinogenic in nature. PAHs are classified as low molecular weight (2–3 aromatic rings) and high molecular weight (4 or more aromatic rings) compound. The primary cause of the PAHs is the incomplete combustion of the organic matter. Accumulation of these substances is accelerated in the soil because of their high hydrophobicity and low solubility in water [10]. Further, this pollution is transferred to the waterbodies, plants, and foods. In drinking water, PAH concentration is found to be in a range of 1–11 ng/L [11]. The world largest used beverages like tea and coffee are also contaminated with PAHs. PAHs like benzopyrene, pyrene, phenanthrene, and naphthalene are found in newspaper ink also. It is very difficult to eliminate these compounds completely from the environment. PHAs in waterbodies cause high toxicity to biotic components. In animals, PHAs causes genotoxicity, cardiotoxicity, neurotoxicity, and behavioral alterations [12]. Therefore, these PHAs are highly hazardous chemicals whose concentration is increasing day by day.

1.2.1.5   Biocide Contaminants

Biocide is a term that mainly includes pesticides and antimicrobials. Pesticides are the chemicals mainly used in the agriculture and highly polluted the environment, especially the waterbodies, which drastically affect the health of living beings. Pesticides are considered mutagens that alter DNA and cause hormonal disturbance in the body. In monocropping areas, there is excessive use of pesticides and biocides. Organochlorine- and organophosphate-type pesticides are stable and highly persistent. These types of pesticides interfere with the central nervous system and re-regulate the enzyme processing [13]. Biocides are used mainly in the medical-associated fields such as disinfectant, germicides, and antibacterial, and these biocides cause a significant risk to health at higher concentration [14].

1.2.1.6   Gaseous and Volatile Pollutants

Volatile organic compounds (VOCs) (like formaldehydes, benzene, toluene, butadiene, styrene, tri- and tetra-chloroethylene) [15] and pollutant gases (like nitrogen oxides, sulfur oxides, and hydrogen sulfides) are released in the atmosphere by industries, automobiles, and chemical processes [16]. These gaseous pollutants have short- or long-term adverse effects on the health of living beings. The prime sources of the VOCs are paint, varnishes, adhesives, pressed wood products, cleaners, and disinfectant. Thermal drying of sludges also releases a variety of VOCs as a result of breakdown of the organic matter by heating. Inorganic compounds, such as SO_2 and NH_3, which are poisonous and corrosive gases, are also a component in the stream of VOCs released in thermal drying of wastes [17].

In the cosmetic and pharmaceutical products, benzyl alcohol is the commonly found chemical, and upon photoreaction, this release reactive oxygen species (ROS), which are responsible for the formation of volatile organic compounds in the atmosphere [18]. Further, ultrafine aerosols are also dangerous for human health [19].

Besides eye and throat irritation, these VOCs cause damage to the organs like kidney and liver. The aromatic hydrocarbons like benzene and toluene are carcinogenic also. To detect these pollutants and for their remediation, the advanced materials with better efficiency are always in demand.

1.2.2   Emerging Demand of Renewable and Clean Energy Resources

With the gradual increase in population and industrialization, there is a demand for high energy for sustainable development; hence, renewable energy sources are extremely valued. In renewable energies, solar energy and biomass energy are considered great options and are available at very low cost, and moreover, it is a green source of energy. The solar energy is utilized by solar cells. A variety of solar cells are employed with advanced materials used as anode, cathode, and electrolyte. Second, the solid electrolyte in the solar cells also impacts the efficiency significantly. So, regular inventions are needed for the materials' design to be utilized in photovoltaic cells to achieve better results over the present technology at the commercial level.

Biomass energy is contributing about 30% to the renewable source of energy worldwide. The utilization of biomass energy not only increases the income of the farmers but is a fantastic way to meet out the agricultural wastes. In the biomass conversion to fuels, the great demand is for the advanced active catalyst materials, which can convert efficiently the biomass to methanol, ethanol, methane, and hydrogen. So, effective catalysts are greatly needed to meet the demand of biofuels and their utilization at the commercial level.

1.3   DESIGN AND ENGINEERING OF ADVANCED MATERIALS

Advanced materials are associated with some unique and key characteristics in the context of structural, optical, surface, and redox properties. Other than these characteristics, chemical stability, thermal stability, and multi-utility of the materials also contribute significantly towards the advancement of the materials for diverse applications. Applications of materials are characteristic

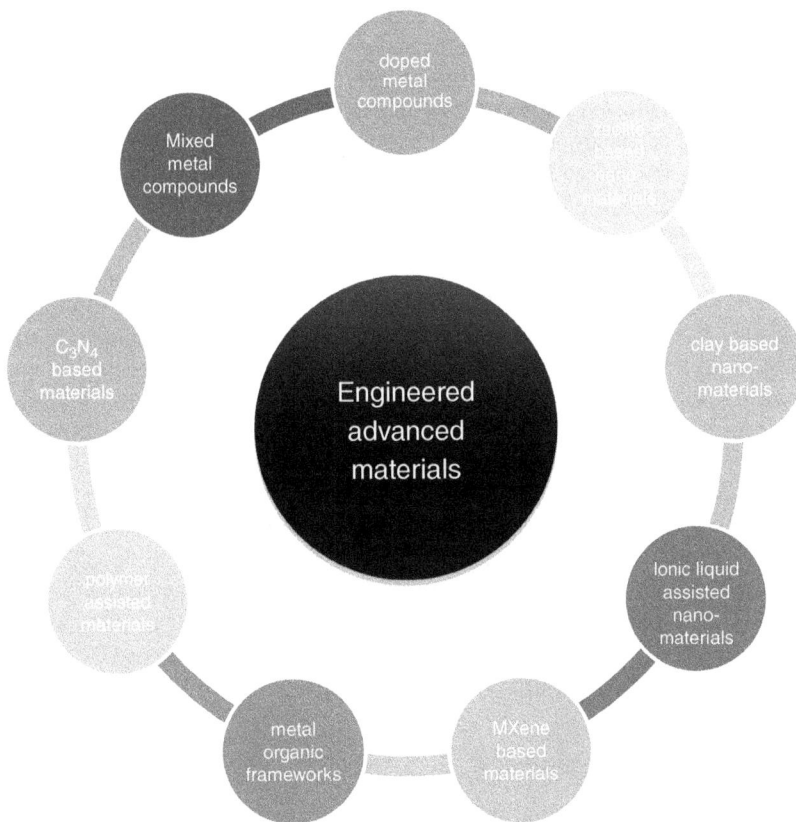

FIGURE 1.1 Engineered advanced materials for sustainable environment.

dependent, so keeping the applications in mind, the researchers are regularly practicing on the materials design and properties optimization to achieve a breakthrough in reference to a particular application. As an example, the photocatalytic activated materials should be accompanied by advancement in the optical and electronic properties amenable to improved absorption, suitable band gap alignment, and large lifetime of charge carriers. Further, surface properties are highly associated with surface-based reactions. The porous materials with high surface area are highly applicable in the field of catalysis by the availability of more number of active sites. The advanced characteristics in redox properties modulations are very much crucial in the sensing technology.

Researchers have a keen interest to develop the materials with multi-utility characteristics. Therefore, in place of single compounds, mixed or hybrid materials are preferred so that the maximum desired features may be accumulated in a single material's composites. In view of this, the next section overviews a variety of materials (Figure 1.1) employed in diverse applications like catalysis, energy production, energy storage, sensing, and environmental decontamination.

1.3.1 Doped Metal Compounds

Doping in metal compounds provides a flexible way to tune to the structural properties of the nano-materials. Doping can enhance the important properties such as electrical conductivity, electronic and optical, significantly. In this regard, a foreign species is inserted in the base lattice to induce the creation of defect state, which significantly modifies the optical and electronic properties. It can be particularly useful to modulate the energy band gap. Charge separation and transfer is improved by the transfer of charge carriers to the effective defect trapping sites. Additionally, electronic transitions from the defect states to the conduction band or from the valance band to the defect states are allowed under sub-band gap irradiation. In the dopants, metals are the prime choice but a number of reports are there, which adopted the non-metal doping and exhibited the enhanced optical properties [20]. The presence of metal ion dopant in the photocatalyst matrix significantly improves the interfacial electron transfer, which reduces charge carrier recombination rate; therefore, photoreactivity of the materials increases. Metal dopants like Cu^{2+}, Mn^{2+}, Co^{2+}, Ni^{2+}, and rare earth play an important role in altering the electronic structure and modulation possibilities of the host material [21]. Other than these commonly employed metals, some other metal ions such as Sb (V) [22], Rh (III) [23], Ga (III) [24], and Sn (IV) [25] have also been explored with improved properties. Cu-, Ag-, and Ni-doped montmorillonite-TiO_2 composites were reported with enhanced photocatalytic activity, and in case of Cu and Ag, the twice efficiency is observed in comparison with P25 TiO_2 [26]. Co-doping of metals is also beneficial as the doping metals may compensate the charge, which results in better visible light response [27]. W and Mo are also reported as dopant in the lattice like TiO_2 and not only Mo^{6+} and W^{6+} ions but also Mo^{4+}, Mo^{5+}, W^{4+}, W^{5+}, as well as Ti^{3+} ions, which participated in photocatalytic enhancing activity [28,29].

Non-metal anion doping has also been widely applied to improve various structural, surface, and optical properties. Most widely doped non-metals are B, C, S, and N [30–32]. The photocatalytic efficiency might be enhanced by extending the lifetime of the generated photoexcited species [33]. N-doped TiO_2 exhibited better optical and photocatalytic activities, which are attributed to the increase in charge carrier efficiency [34]. Further, B-, N-, P-, and I-doped TiO_2 are found with advanced photocatalytic activity towards bisphenol A [35].

1.3.2 MIXED METAL COMPOUNDS

Metal compound mixing is an easy and effective way to improve the structural, optical, and surface properties of the materials. The presence of other materials may alter the growth properties and also direct the morphology of the materials, which causes hierarchy in the materials shape and size, and also alter the surface properties. ZnO and TiO_2 are the most widely used materials in photocatalysis, photovoltaics, and semiconductor-based applications but practical utility exploration is limited by their absorption in ultraviolet region only and short lifespan of charge carriers [36,37]. The rapid recombination of photogenerated electron and hole pairs deteriorates the performance of a photocatalyst. Hence, in order to overcome these barriers, efficient transportation and separation of charge carriers in photocatalyst need to be emphasized by the introduction of different compounds.

The potential semiconductor materials such as ZnO, TiO_2, SiO_2, MoO_3, CdS, MgO, WO_3, SnO_2, Cu_2O, and In_2O_3 are used individually as well as in a mixed state for various applications [38–46]. The design and introduction of the materials is also based on the other properties that need to be enhanced, but in the photoactive materials, two common interests are (i) the production of visibly active materials and (ii) the increase in lifespan of charge carriers. So with these two common interests and other desired properties, the materials are designed and optimized.

In photocatalyst materials, one of the most important desires is to recover the material for its re-use. One of the most effective ways is to make smart magnetically active materials so that the photocatalyst materials may be recovered by magnet after photocatalysis cycle. For this feature, the materials frequently added are Fe_2O_3 [47], Fe_3O_4 [48], $CoFe_2O_4$ [49], $ZnFe_2O_4$ [50], Mn-Zn ferrite [51], Ni-Zn ferrite [52], and $BiFeO_3$ [53].

Surface decoration of the mixed or nanocomposite materials with metals is also a fantastic way to improve the optical properties. In the surface-decorated nanomaterials, the most frequently used metals are Ag, Pd, Au, Pt, Cu, Bi, and Ni [54]. Simultaneous decoration of multiple metals also exhibited a pronounced effect towards improvement in the properties. Ag and Bi dual nanoparticle modification on BiOBr illustrates the increase in photocatalytic activity [55]. In most of the cases, the enhancement in the charge carrier response is attributed to the surface plasmon resonance (SPR).

1.3.3 CARBON NITRIDE (C_3N_4)-BASED MATERIALS

C_3N_4 is designated as a remarkable material with unique properties and used alone as well as in mixed form with a diverse nature of compounds. Due to a suitable band gap, high chemical stability, and versatile characteristics, C_3N_4 is highly suitable for the formation of hybrid structure. Naturally, g-C_3N_4 exists in the form of s-triazine [C_3N_3] and tri-s-triazine [C_6N_7] [56]. The most widely used sources to prepare C_3N_4 are melamine [57], urea [58], thiourea [59], and cyanamide [60]. C_3N_4 usage is slightly restricted only due to its low absorption and fewer number of active sites for the catalytic activity due to lesser surface area. To encounter these problems, it is used to design the hybrid materials and known to enhance its photocatalytic activity along with the materials in combination [61]. 2D g-C_3N_4 forms a heterostructure photocatalyst with remarkable catalytic properties. C_3N_4 is used to couple with a variety of metal semiconductors such as ZnO [62,63], TiO_2, $BaTiO_3$ [64], $CuFe_2O_4$ [65], $CaSnO_3$ [66], Bi_2MoO_6 [67], $ZnIn_2S_4$ [68], MoS_2 [69], $BiPO_4$ [70], and BiOBr [71]. A potential photoreduction activity is exhibited by the C_3N_4 quantum dots synchronized with Au quantum dots over CeO_2/Fe_3O_4 photocatalysts [58]. Various heterojunction schemes are investigated and proposed, such as Type I and Type II, Z-scheme, S-scheme, and p-n junctions, and on the ground of various photocatalysis reactions on the C_3N_4 heterojunction catalysis, ASS-Z scheme is a profound scheme for enhanced photocatalysis [72]. Doping of non-metals like P and S in g-C_3N_4 also enhanced the photochemical activity [73]. Doping of noble metal (Au, Pt, Ag, and Pd) is accompanied by improved photocatalytic activity. Au act as a mediator for charge flow and lead to an increase in charge carrier ability. For example, in g-CNS/Au/CdS, Au boosts the catalytic activity and effectively degrades the dye with 100% efficiency in 5 minutes [73].

Electrons are transmitted very quickly in the two-dimensional C_3N_4, so an increase in electrical conductivity is observed when C_3N_4 is coupled with other compounds, as described in the above paragraph. The common characteristic that is improved in the composites is the charge carrier ability due to the formation of heterojunctions. C_3N_4-coupled compounds exhibited good photoactivity towards hydrogen production [69,74] and photocatalytic reduction of CO_2 [70].

1.3.4 POLYMER-ASSISTED MATERIALS

Porous polymeric materials have proved to be better candidates for adsorption to assist photocatalytic materials. The important characteristic feature other than absorption and conduction is the surface area; the materials with higher surface area provide more number of active sites. Polymers are considered to be electrical insulators before the invention of conducting polymers. Conducting polymers possess unique electrical and optical properties similar to those of inorganic semiconductors [75]. A conjugated carbon chain consists of alternating single and double bonds, where the highly delocalized, polarized, and electron-dense "π" bonds are responsible for its electrical and optical behaviors. Typical conducting polymers includes polyacetylene (PA), polyaniline (PANI), polypyrrole (PPy), polythiophene (PTH), poly(para-phenylene) (PPP), poly(phenylenevinylene) (PPV), and polyfuran (PF) [76].

The mechanical, electrical, and optical properties of conducting polymers are highly associated with side chains and attached dopants [77]. Conducting polymers possess both localized and delocalized states, and the delocalization of pi bonds leads to the generation of charge carriers like polarons, solitons [78]. The conductivity of conducting polymers increases with an increase in dopant concentration [79,80]. Conducting polymers with advanced conducting characteristics made a significance contribution in the field of energy and environment. The assistance of conducting polymers with metal oxides, metal sulfides, etc. enhances their applied characteristics [81,82]. The binary or ternary nanocomposite materials containing conducting polymers as one of the components exhibited the potential activities in the field of photocatalysis decontamination of the contaminants.

Generally, conducting polymers have low electrical conductivity and optical properties in their pristine state; however, doping with suitable materials can give them excellent properties [76,83–85].

Polyacetylene exhibits a conductivity in the range of 10^5 S/cm but when the doping level increases, its conductivity rises drastically to 10^2–10^3 S/cm [86], and depending upon the dopant material, its properties like electrochemical or optical and mechanical may be tuned. PEDOT@ZnO@GQDs, a ternary nanocomposite, composed of sensitized ZnO nanorods, graphene quantum dots (GQDs), and conducting polymer PEDOT, exhibited a good photocatalytic efficiency for the degradation of dye [87]. PEDOT and GQDs both inhibited the re-joining of electron hole (e–/h+) because both are hole transport materials. The positive charge on GQDs is stabilized by the negatively charged PEDOT polymer through the intermolecular charge transfer.

1.3.5 Metal Organic Frameworks (MOFs)

MOFs are the compounds containing the metal ion co-ordinated with the organic ligands, and the special feature found in these compounds is the high porosity (up to 93%) and surface area (up to 10,000 m^2/g) [88]. These characteristic properties contributes to the adsorption properties of MOFs towards H_2 and CO_2 photoreduction. The applications of MOFs are not limited and extended to catalysis and supercapacitor applications. A variety of metals and ligands with different valences are responsible for the different structures and hence properties of the MOFs. There is a diverse range of MOFs applicable in diverse applications. MOFs are crystalline materials and exhibit the band gap between 1.0 and 5.5 eV. MOFs are employed as photocatalytic materials in the applications such as water decontamination, CO_2 photoreduction, and H_2 production. All these particular applications are directed to handle the environmental issues.

MOFs are also reported as template to carry out the synthesis of nanomaterials with high porosity [89]. To increase the thermal and chemical stability, these metal organic frameworks are coupled with various other compounds such as nanoparticles of metals, metal oxides, metal sulfides, carbon materials, and polymer [90]. MOF structures are tailorable and hence possess a structural diversity. This will enable the MOFs for easy tuning of the properties like absorption of light, selectivity of the materials, and immobilization of the applied materials. Further, charge transfer pathways of MOFs or assisted materials make them suitable for photocatalytic applications. Some of the widely adopted inorganic cations in MOFs are Al^{3+}, Fe^{3+}, Zr^{4+}, Cr^{3+}, Ag^+, Cu^{2+}, Ni^{2+}, and Zn^{2+}. The usual organic ligands are carboxylic acids, imidazole, and various pyridine-based moieties. The type of ligand decides that the MOFs will be porous or non-porous, and if porous, it will be meso- or microporous. Generally, MOFs are classified as Types I, II, and III. Type I are MOFs containing M-Oxo clusters and exhibit an efficient charge transfer. These are called pristine MOFs. Pristine MOFs are having high band gap so the absorption part is poor. In Type II MOFs, the dye-sensitized MOFs are considered and these MOFs exhibited good stability and high separation of charge carriers. In Type III MOFs, pristine MOFs are associated with photo redox species like metal, metal oxides, or other compounds.

1.3.6 Mxene-Based Materials

Mxene is a recent approach in the materials, but its activity is extremely fast and implicated in various applications such as environment, energy, catalysis, and sensing. Basically, Mxenes represent a formula $M_{n+1}X_nT_x$, where M represents a transition metal, X represents carbon and nitrogen, and T_x represents the associated functionalized groups like –OH, –F, –O [91]. These surface functionalities significantly influence the surface interaction properties; therefore, they are highly decisive in adsorption and separation efficiency. Further, Mxene-based hybrid materials have more competitive features and have been successfully applied in photocatalysis, adsorption, and electrochemical sensing. High hydrophilicity of Mxene-based materials extends their pollutants adsorption capacities. In the functionalities, –O and –OH are more stable functionalities in comparison with –F. Mxene adsorption is so developed that it can be compared with carbon materials. The most widely used transition metals are Ti, V, and Mo [92] for Mxenes.

Mxene materials effectively adsorb the heavy metal ions. $Ti_3C_2(OH/ONa)_xF_{2-x}$ Mxenes materials exhibit a large adsorption capacity for the Pb(II) (140 mg/g) [93]. The authors also reported the first principal calculation on adsorption by $Ti_3C_2(OH/ONa)_xF_{2-x}$ and submitted that adsorption significantly affected by the structural position of –OH; the –OH vertical to the titanium shows better adsorption for Pb(II). Metal oxides/Mxene materials are reported as photocatalysts with advanced characteristics in comparison with individual photocatalysts. The most tried MO with Mxenes is TiO_2, and TiO_2/Ti_3C_2 nanocomposite materials are better than TiO_2 and Ti_3C_2 alone [94,95]. The Mxene-associated materials are not limited to the adsorption and photocatalysis of organic moieties or inorganic metal ions but also explored their utilities up to radionuclide adsorption and gaseous contaminant adsorption [96]. Mxene materials are also employed as transducer materials in biosensors. Mxenes also have integration properties with the enzymes. In different sensing principles, ratiometric fluorescence, colorimetry, SPR, SERS, and ECL [97], photoluminescent sensors are the most effective ones. In this line, MQDs/MNSs expressed very good activity towards heavy metals (Zn^{2+}, Cu^{2+}, and Fe^{3+}) with high sensitivity. Not only sensitivity, but also selectivity is the beauty of these compounds. Colorimetric sensing by the Mxene-based materials needs the attention due to simplicity of the analysis. These materials are based on the SPR-based sensing. In these types, further investigations are highly required, especially the biocompatible characteristics of the Mxene-based materials. So, regarding this, steps should be moved for implications towards more stable, less toxic Mxene materials with desired characteristics in various highlighted fields like photocatalysis, sensing, and adsorption applications towards environmental contaminants.

1.3.7 IONIC LIQUID (IL)-ASSISTED NANOMATERIALS

IL is now a much known terminology in the material science. ILs are the compounds that exist in ion pairs containing large organic cations in combination with the variety of anions. Recently, the liquids become very popular in the field of materials and are applicable in diversified fields directly or indirectly. First implication of the ILs is the replacement of the volatile organic solvents in the synthesis processes, and by this reason, the ILs are designated as green solvents. ILs are prominently applied in synthesis, energy, sensory, and as electrolyte materials. Recently, these ILs are also employed in biomedicine and drug delivery. ILs are also applicable in perovskite solar cells; the hydrophobic part in these liquids causes moisture resistance and thus protects the perovskite materials thin layers from erosion. As electrolyte also, the ILs perform better than the solid electrolytes and also eliminate the burning problems of liquid electrolytes. This enhances the performance of electrochemical devices [98].

Polyionic liquids (PILs), a macromolecular framework, may be applied as porous materials in industrial application. This is a new approach, and mainly ionic porous organic polymers are in focus. Ionic functionalities on polymer compounds improve the surface interaction properties, which increase the CO_2 capture capacity [99] and solid-phase extraction [100]. PILs are easily processable into thin films as compared to other materials. ILs are also employed for the removal of heavy metal ions from the environment, and the mechanism is different from that of the conventional techniques due to the presence of heterogeneous nature of IL by the presence of both hydrophilic and hydrophobic parts. Further, it is observed that the adsorption of metal ions like Ni^{2+}, Cu^{2+}, and Pb^{2+} on ILs is not affected by the presence of other ions, and this characteristic is highly useful to treat industrial effluents containing a variety of species [101]. IL chitosan with graphene nanocomposites is also reported with good efficiency for the removal of Cr^{6+} [102]. Pb^{2+} and As^{3+} are also effectively adsorbed by 1-butyl-3-methylimidazolium chloride [BMIM]Cl-associated SiO_2 nanocomposites. ILs are not only applicable for the metal ions but may be employed to remove the organic contaminants like dyes, pesticides, and phenolic compounds from the water. Hydrophobic part of the ILs attracts the organic moieties. Therefore, these green solvents in biphasic aqueous mixtures have the capability to remove the organic and inorganic contaminants. Similar reports were there, which showed the removal of pesticides, pharmaceuticals, and hormones [103,104].

Further, ILs are also highly applicable in the extraction process, and the extraction capabilities are varied depending upon the alkyl chain [105,106]. The extraction capabilities increase with an increase in alkyl chain in the ILs. The ILs also recover easily so do not cause secondary pollution and reduce the processing cost. NaOH recovers the maximum portion (97%) of the ILs by stripping of the contaminants. Recently, advanced oxidation processes are also employed for the mineralization of the organic moieties and recovery of the ILs [107,108].

1.3.8 CLAY-BASED NANOMATERIALS

Basically, clays are the aluminosilicates in which silica-oxygen tetrahedral are arranged in the form of layer with the apical oxygen shared with octahedral, which consist of Al^{3+}, Fe^{3+}, and Mg^{2+} surrounded by six oxygen atoms. Only 2/3 of the octahedral sites are occupied by the ions and other empty sites are prone for the substitution of the metal ion. The layer pattern signifies a particular clay material and the properties exhibited, as a consequence. Further, in natural and synthetic processes, the material's variation is very high. The modified clay materials showed an enhancement in properties, particularly in surface characteristics and electronic properties. In the modification processes, the prominent one is the production of clay-based nanocomposites.

Zero-valent iron (ZVI) clay composites also exhibited effective properties towards the water decontamination [109]. Low prices and non-toxicity are the beauty of these materials. ZVI-modified clays exhibited outstanding activities. Here, synergism takes place between ZVI and the clay part. The clay part adsorbed contaminants effectively and then removed by the ZVI. To make the inorganic clays effectively for the organic pollutant adsorption and removal, the interlayer space in clays can be modified with alkyl chains, functional groups, and other organic moieties. Biopolymers are also found in the suitable combination for the development of clay nanocomposites for the decontamination of organic moieties [110,111]. Polypeptides and polysaccharides are the best candidates for the above type of nanocomposites due to their easy availability and less cost. Further, the properties of both clay and biopolymers improved by this combination. The specificity and wettability of these nanocomposites are better than those of their individual components [112,113]. The new hybrid materials came up with high surface area and adsorption properties. Some of the prominent polysaccharides employed for biopolymer-clay nanocomposites are chitosan [114,115], cellulose [116,117], starch [118], alginate [119,120], xanthan gum, and cyclodextrin [121,122].

In the smectite clays, the interlayer space is expandable, and thus, the space is very much suitable for the modifications to improve the selective adsorption. Cationic surfactants are found as a suitable candidate to make the organophilic clay-based assemblies [123]. To achieve the desired characteristics for a particular application, a variety of combinations of clays with surfactant is possible. Phospholipids are found as a good surfactant of biological origin to make the organophilic clay-based materials [124,125]. The surfactants arrange themselves parallel to the silicate layer. Pillared clays are also well-known materials for the adsorption, the structures of these clays are suitable for organic modifications, and the organic cations are intercalated in the interlayer space. The most common pillared clay is montmorillonite [126,127]. Cationic dyes are also observed for the modification of clays. These modified clays exhibited enhanced photocatalytic activities, and the adsorption properties varied with the structure of the dye [128,129].

1.3.9 ZEOLITE-BASED NANOMATERIALS

Zeolite has been very well-known term for a long time. Structurally, these aluminosilicates consist of SiO_4-Al_2O_3 tetrahedra associated with mono- or divalent cations (Na^+, K^+, Ca^{2+}) for charge balance. Some of the natural zeolites are quartz, feldspars, and phyllosilicates. These acquire hierarchal characteristics by the variable aluminosilicate structural arrangement. A variety of synthetic zeolites with good adsorption capacity for heavy metal ions are reported. The Si/Al ratio in the zeolites highly influence the efficiency towards the adsorption of micropollutants [130]. Peoples

TABLE 1.1

Zeolite-Based Materials for Pollutant Decontamination

Pollutant Type	Zeolite and Zeolite-Based Materials
CO_2	Zeolite A [132], zeolite 13X [133], zeolite A+P1 [134]
SO_2	Mix zeolite X and A [135]
Benzene, toluene, o-xylene m-xylene, p-xylene	Zeolite X [136]
Se(IV)	Zeolite from low-calcium fly ash [137]
Ni^{2+}	Modified natural zeolite 3A-type ($Mg(OH)_2$ [138] , zeolite from fly ash X-type [139]
Pb^{2+}	Natural zeolite, modified zeolite (0.1 H_2SO_4)
	Modified zeolite (0.2 H_2SO_4)
	Modified zeolite (0.5 H_2SO_4)
	Modified zeolite (1.0 H_2SO_4) [140]

working in this area always have a desire to bring about efficient and low-cost zeolite-based materials and their application towards environmental decontamination.

International Zeolite Association reported 248 forms of zeolite frameworks. FAU (zeolite-X and -Y), MFI (ZSM-5), MOR (mordenite), BEA (beta), and HEU (clinoptilolite) are the widely used photocatalysts for environmental decontamination [131]. Zeolites are applicable for both aqueous and gaseous contaminants. Some of the recently applied zeolites used for a variety of decontaminates are given in Table 1.1.

Zeolites may act as a support for a variety of other materials such as metal oxides, carbon materials to modify the properties of these materials towards the adsorption and degradation of contaminants. Iron oxide-modified zeolites are found with excellent photodegradation ability towards organic molecule degradation like nitroanilines [141]. IL-modified zeolites with efficient decontamination properties are also reported [142]. Passive NO_x adsorbents (PNAs) are found excellent to control the emissions from the diesel vehicles. Well-reported PNAs are oxides of metals like Al, Ti, Ce, Cu, etc. and zeolite-based nanomaterials of noble metals like Pd, Pt, and Ag [143]. Potassium tungsten phosphate/ZSM-5 nanocomposite reported recently expressed the excellent catalytic activity for the production of ethane and pesticide decontamination [144]. A variety of TiO_2/zeolite composites such as TiO_2-beta zeolite/M(M=Ag, Pd, Au) [145,146], g-C_3N_4-TiO_2/zeolite [147], and N-doped TiO_2/$ZnFe_2O_4$/zeolite [148] materials are involved in the decontamination of diverse pollutants (dyes, pesticides, antibiotics, etc.). In the same trend, metal sulfides/zeolites also exhibited the good features for decontamination of pollutants and the metal sulfides primarily employed are CdS, ZnS, MoS_2, SnS, and CuS [149,150]. Other zeolite-based materials, which are well explored and exhibit the good adsorption and catalyst properties, are metal-doped zeolites, halide/zeolites, perovskite zeolites, etc. [131].

1.4 ADVANCED MATERIALS APPLICATION TOWARDS SUSTAINABLE ENVIRONMENT

Technology is greatly needed to clean and preserve the natural resources to acquire a sustainable environment. Natural resources like water, air, and soil require regular monitoring and remediation of health-hazardous pollutants in the environment. In the above sections, the advanced materials are highlighted, which are employed to monitor the environmental index, decontaminate the environment, or generate green energy. The striking methodologies are discussed in the following sections, where the advanced materials elaborated above are utilized to achieve a sustainable environment.

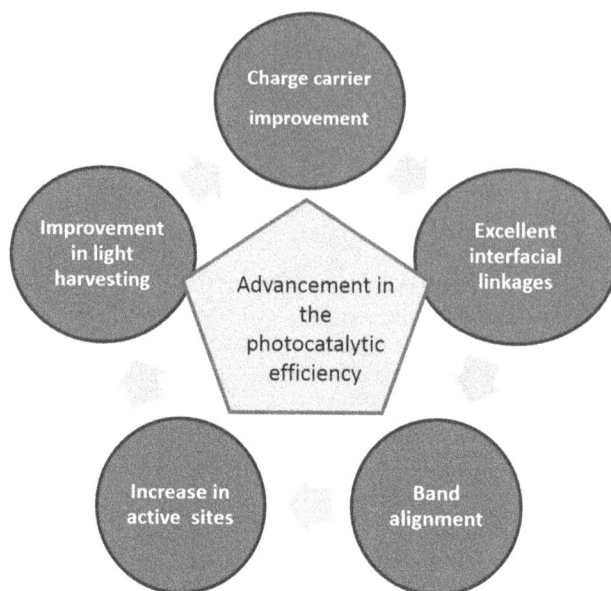

FIGURE 1.2 Desired features for an efficient catalyst.

1.4.1 PHOTOCATALYTIC DECONTAMINATION

Photocatalysis is a simple but effective methodology for the decontamination of water, having less procedural costs. In this process, the most important part is the photocatalyst and its characteristics. In general, semiconductor materials act as photocatalysts. This is considered under heterogeneous photocatalysis and also known as the "advanced oxidation process". In the semiconductor materials, one of the prime characteristics is the band gap of the materials. Based on the band gap of the materials, the excitation source can be selected. Broadly speaking, the photocatalyst materials are UV or visible active. A variety of advanced materials, as discussed in Section 1.3, are now reported with an efficient photocatalytic performance. Photocatalytic reactor assembly is required to carry out the oxidation procedure comprising a photocatalytic reactor vessel, which is normally made up of glass vessel, a light source for photoexcitation, an oxygen source, and a detector for the analysis that depends on the decontaminants. UV visible spectrophotometer or GC-MS techniques are mostly employed for the determination of concentration during photocatalytic reactions. In brief, the photocatalysis involves the excitation of the photocatalytic material by suitable light source, which creates the charge carriers (electron and holes), which in turn generate the various ROS. These ROS participate in the reaction and mineralize the contaminants. The important factors other than absorption of light, which acts as a deciding factor in the efficiency of photocatalyst, are the concentration and lifetime of charge carriers (electrons and holes). Most of the above-discussed advanced materials are associated with the formation of heterojunctions, which increase the lifetime of charge carriers. To be an efficient photocatalytic material, prime characteristics need to be developed, which are shown in Figure 1.2.

1.4.2 HYDROGEN (H_2) PRODUCTION AND STORAGE

H_2 energy seems a hot topic of discussion because researchers are considering the hydrogen as fuel to solve the future energy crises. H_2 is a clean and highly efficient source of energy. Hydrogen can be obtained by a number of processes such as water splitting, ethanol, and biomass materials. Among different methodologies, the photo-electrochemical production of H_2 is the unique one [151,152]. In the hydrogen production, a variety of binary, ternary, and quaternary nanocomposites and modified materials were employed as photocatalyst as discussed in Section 1.3. All these advanced materials are eligible to work as photocatalyst for the hydrogen production with different efficiency. An optimum catalyst is required, which effectively absorbs radiation, and possesses narrow band gap and low recombination of charge carriers. Further low cost, reusability, and chemical stability are some of the secondary important characteristics that should be associated with the photocatalytic materials. The second step after the production of hydrogen is the storage. Over the conventional gaseous and cryogenic storage methods, solid storage systems are efficient and adsorb the hydrogen reversibly. The metal hydrides, metal oxides, and other porous nanocomposites are the efficient ones for the storage of hydrogen. Electrochemical storage of hydrogen is very efficient, in which atomic hydrogen is adsorbed and produced by the electrochemical breakdown of aqueous solutions.

1.4.3 CHEMICAL SENSING

The qualitative and quantitative analysis is a critical step before we think for the remediation process of the environmental contaminants. Recently, sensing techniques are highly developed and employed in various sectors, particularly to detect and quantify the environmental contaminants. Recently, sensors with immense accuracy and fast response were developed [153,154]. Along with accuracy and fast response, selective response is extremely important. Researchers are regularly working in the development sensor technology for all states of pollutants: solid, liquid, or gas [155]. The common characteristics that needed to establish a good sensing technology are high sensitivity, good selectivity, reliable, large detection range, and good reproducibility.

In the atmosphere, a huge amount of pollutant gases are released due to industrial development. Nitrogen oxides, sulfur oxides, ammonia, and hydrogen sulfide are commonly found toxic gases, which are dangerous to living beings after an optimum concentration. Therefore, the materials that exhibited the properties to detect and capture the gaseous pollutants are of great need [156]. The other volatile compounds released in the atmosphere are aromatic hydrocarbons. Hybrid materials show a remarkable performance in this technology featured by unique conductivity, optical, electrical, and surface properties [157]. Out of sensing techniques, electrochemical sensing technique is a reliable technique in terms of accuracy, fast response, and cost. An electrochemical sensor basically contains a transducer which reacts with analyte and passes an electric signal to the electrical circuit for producing the results. In the electrochemical sensors, the most important is the working electrode. For better resolution of the redox peaks, the modified electrodes are highly preferable [158,159]. The electrodes can be modified with a variety of materials that amplify both the selectivity and sensitivity of the electrode. The materials discussed in Section 1.3 are applicable for the modification of the electrodes. Materials chosen are also associated with the redox potential. Graphene materials are associated with the sensing of materials at lower potential. Further, reduced graphene enhanced the electrochemical activity [160]. Further, porous electroactive materials are more desirable because of the presence of more number of active sites. Other applied materials in the electrochemical sensing are carbon nanotubes and conducting polymer-assisted materials. Both of these well followed the desired properties to be better electrochemical sensor. Noble metal-modified carbonaceous materials also perform better in the electrochemical sensing but high cost of these materials limits their uses [161].

1.4.4 ADSORPTION

Adsorption is a simple and effective methodology for the decontamination of water and air, which is easy to operate at low cost. The prime requirement is the availability of suitable and efficient adsorbent material. A variety of modified materials discussed above (Section 1.3) act as effective adsorbents, especially the porous materials. Adsorption of the pollutants (both inorganic and organic) is highly associated with the surface structures and modifications. Electrostatic forces are the prime one for an effective adsorption characteristic. To make the adsorbent surfaces active for electrostatic attractions, general classes of materials are functionalized with a wide variety of diversified species (inorganic or organic). Other implicative factors are surface area and porosity of the base materials. Materials with suitable pore size and porosity are effectively employed for the adsorption of atmospheric contaminants.

Modified carbon materials are also considered with good adsorption efficiencies [162]. The first modification in the carbon is the activation or functionalization, which makes the surface of carbon materials prone to the adsorption of organic and inorganic contaminants [163,164]. Carbon adsorbents from agriculture waste are the effective ones on the adsorption characteristics and cost. Pesticide and herbicide molecules are also decontaminated by the adsorption process.

In the contaminants, aromatic hydrocarbons both in liquid phase and in gas phase are highly carcinogenic for the biotic components. For the adsorptive removal of the aromatic hydrocarbons, the adsorbent should contain a hydrophobic part that attracts the aromatic rings. In the organically modified clay materials, the hydrophobicity that varies with cation is used for the modification. Further, the structure of the cations also influences the amount of adsorbate, and the order of adsorption of alkyl benzenes also varies [165,166]. Aromatic pollutants are effectively removed by the zeolite. A number of zeolite-based materials are discussed in Section 1.3.9.

1.4.5 LITHIUM-ION BATTERIES

Lithium-ion batteries are now making a remarkable place in the rechargeable battery technology. In the materials requirement, both cathodic and anodic materials are equally important. As anode, carbon is found an efficient electrode. $C/LiCoO_2$ lithium-ion batteries are commercialized. As cathodic material, olivine-structured $LiMPO_4$ where (M = Fe, Mn, Co, Ni) are employed [167]. The materials with different M have their own pros and cons. For example, $LiFePO_4$ has slightly lower energy density than $LiCoO_2$ but the low cost and environmentally favorable $LiFePO_4$ will also be a reliable cathodic material in lithium-ion battery technology. Further, material synthesis and processing conditions affect the materials' efficiency significantly. The electrochemical performance is increased by 60% after heat treatment at 500°C. The discharge potential of $LiMnPO_4$ is about 0.5 V vs. Li/Li^+, which is higher than that of $LiFePO_4$ (3.45 V). $LiCoPO_4$ and $LiNiPO_4$ both also have high potential of 4.8 and 5.1 V vs. Li/Li+, respectively [168]. The operating voltages like those mentioned above cause side reactions. Other popular class for cathodic materials is orthosilicate, Li_2MSiO_4 (M = Fe, Mn, Co), where Li_2FeSiO_4 is more suitable as Si and Fe materials are available at low cost.

Carbon, pure tin, pure silicon, various different hierarchal SnO_2, SnO_2/C nanocomposites, and various tin-based intermetallic compounds belongs to anodic materials [169]. Sn- and Si-based materials have a great potential due to their high energy density and stable life cycles.

1.5   CONCLUSION AND FUTURE REMARKS

In recent years, materials have been very advanced in all fields. From simple metallic or non-metallic state, now we are at bimetallic to tetrametallic nanocomposite materials. Further, hybrid materials came with highly superior characteristics. For environmental remediation, a variety of materials have been invented, and further, there is a lot of scopes. The materials may be tried with blends of unique materials, which may improve the properties, which may be useful in certain specific applications. In the adsorbent materials, there is still a lot of scope in the modification of natural materials and also in the synthetic compounds for progress in porous features. Recently, development in the computational studies of the materials directed the scientists for the prediction of possible investigations for the development of the materials. In the photocatalytic materials, the visible active materials needed more experimentation so that efficient materials may be achieved, which can decontaminate the systems upon just exposure to sunlight. Self-cleaning materials will be the future ceramic materials, which will be applied to exposed areas of the constructions. In the energy sector, biofuels and rechargeable batteries have a large scopes. In the conversion of biomass to fuels, efficient catalysts are the topic of keen interest. In the battery technology, efficiency and cost are both the prime topics to work. In the near future, achieving the goal towards alternative forms of transportation, such as plug-in hybrid electric vehicles (PHEVs) and all electric vehicles (EVs), relies on either the discovery of novel Li-ion battery materials, which have a very high energy density, or the novel multi-electron transfer system. The development of porous materials with surface modifications is also of great interest in the catalysis of different chemical reactions. In the detection and monitoring, sensor-based technology needs more development so that highly sensitive sensors may be developed at low cost. In the polymeric materials also, there is a lot of scope because polymers are the great support for the immobilization of various active materials. Recently developed materials such as MOFs and Mxenes and their associates still have a lot of scope in their structural variability and its diversified utility. Further, readily formed Mxenes may be tried in a lot of combinations with other metal compounds, polymers, and carbon materials.

REFERENCES

[1] L. Jiang, Y. Wang, C. Feng, Application of photocatalytic technology in environmental safety, *Procedia Engineering*. 45 (2012) 993–997. https://doi.org/10.1016/j.proeng.2012.08.271.

[2] C. Baresel, M. Ek, H. Ejhed, A.-S. Allard, J. Magnér, L. Dahlgren, K. Westling, C. Wahlberg, U. Fortkamp, S. Söhr, M. Harding, J. Fång, J. Karlsson, Sustainable treatment systems for removal of pharmaceutical residues and other priority persistent substances, *Water Science and Technology*. 79 (2019) 537–543. https://doi.org/10.2166/wst.2019.080.

[3] N. Ratola, A. Cincinelli, A. Alves, A. Katsoyiannis, Occurrence of organic microcontaminants in the wastewater treatment process. A mini review, *Journal of Hazardous Materials*. 239–240 (2012) 1–18. https://doi.org/10.1016/j.jhazmat.2012.05.040.

[4] J. Rogowska, M. Cieszynska-Semenowicz, W. Ratajczyk, L. Wolska, Micropollutants in treated wastewater, *Ambio*. 49 (2020) 487–503. https://doi.org/10.1007/s13280-019-01219-5.

[5] T.T. Schug, A. Janesick, B. Blumberg, J.J. Heindel, Endocrine disrupting chemicals and disease susceptibility, *The Journal of Steroid Biochemistry and Molecular Biology*. 127 (2011) 204–215. https://doi.org/10.1016/j.jsbmb.2011.08.007.

[6] E. Diamanti-Kandarakis, J.-P. Bourguignon, L.C. Giudice, R. Hauser, G.S. Prins, A.M. Soto, R.T. Zoeller, A.C. Gore, Endocrine-disrupting chemicals: An endocrine society scientific statement, *Endocrine Reviews*. 30 (2009) 293–342. https://doi.org/10.1210/er.2009-0002.

[7] R. Lauretta, A. Sansone, M. Sansone, F. Romanelli, M. Appetecchia, Endocrine disrupting chemicals: Effects on endocrine glands, *Frontiers in Endocrinology*. 10 (2019) 178. https://doi.org/10.3389/fendo.2019.00178.

[8] L.D. Ardila-Leal, R.A. Poutou-Piñales, A.M. Pedroza-Rodríguez, B.E. Quevedo-Hidalgo, A brief history of colour, the environmental impact of synthetic dyes and removal by using laccases, *Molecules*. 26 (2021) 3813. https://doi.org/10.3390/molecules26133813.

[9] Z. Zelinkova, T. Wenzl, The occurrence of 16 EPA PAHs in food – A review, polycyclic, *Aromatic Compounds*. 35 (2015) 248–284. https://doi.org/10.1080/10406638.2014.918550.

[10] A.B. Patel, S. Shaikh, K.R. Jain, C. Desai, D. Madamwar, Polycyclic aromatic hydrocarbons: Sources, toxicity, and remediation approaches, *Frontiers in Microbiology*. 11 (2020) 562813. https://doi.org/10.3389/fmicb.2020.562813.

[11] K. Skupińska, I. Misiewicz, T. Kasprzycka-Guttman, Polycyclic aromatic hydrocarbons: Physicochemical properties, environmental appearance and impact on living organisms, *Acta Poloniae Pharmaceutica*. 61 (2004) 233–240.

[12] K. Bhuyan, A. Giri, Polycyclic aromatic hydrocarbon compounds as emerging water pollutants: Toxicological aspects of phenanthrene on aquatic animals, in: V. Shikuku (Ed.), *Advances in Environmental Engineering and Green Technologies*, IGI Global, 2020: pp. 45–67. https://doi.org/10.4018/978-1-7998-1871-7.ch004.

[13] M.A. Hassaan, A. El Nemr, Pesticides pollution: Classifications, human health impact, extraction and treatment techniques, *The Egyptian Journal of Aquatic Research*. 46 (2020) 207–220. https://doi.org/10.1016/j.ejar.2020.08.007.

[14] J.-H. Kim, M.-Y. Hwang, Y. Kim, A potential health risk to occupational user from exposure to biocidal active chemicals, *International Journal of Environmental Research and Public Health*. 17 (2020) 8770. https://doi.org/10.3390/ijerph17238770.

[15] X. Cong, J. Zhang, Y. Pu, A novel living environment exposure matrix of the common organic air pollutants for exposure assessment, *Ecotoxicology and Environmental Safety*. 215 (2021) 112118. https://doi.org/10.1016/j.ecoenv.2021.112118.

[16] T.-M. Chen, W.G. Kuschner, J. Gokhale, S. Shofer, Outdoor air pollution: Nitrogen dioxide, sulfur dioxide, and carbon monoxide health effects, *The American Journal of the Medical Sciences*. 333 (2007) 249–256. https://doi.org/10.1097/MAJ.0b013e31803b900f.

[17] S. Xue, W. Ding, L. Li, J. Ma, F. Chai, J. Liu, Emission, dispersion, and potential risk of volatile organic and odorous compounds in the exhaust gas from two sludge thermal drying processes, *Waste Management*. 138 (2022) 116–124. https://doi.org/10.1016/j.wasman.2021.11.040.

[18] J. Lin, H. Zhao, H. Cao, Y. Zhao, C. Chen, Photoinduced release of odorous volatile organic compounds from aqueous pollutants: The role of reactive oxygen species in increasing risk during cross-media transformation, *Science of the Total Environment*. 822 (2022) 153397. https://doi.org/10.1016/j.scitotenv.2022.153397.

[19] P. Azimi, D. Zhao, C. Pouzet, N.E. Crain, B. Stephens, Emissions of ultrafine particles and volatile organic compounds from commercially available desktop three-dimensional printers with multiple filaments, *Environmental Science & Technology*. 50 (2016) 1260–1268. https://doi.org/10.1021/acs.est.5b04983.

[20] R. Marschall, L. Wang, Non-metal doping of transition metal oxides for visible-light photocatalysis, *Catalysis Today*. 225 (2014) 111–135. https://doi.org/10.1016/j.cattod.2013.10.088.

[21] K. Wu, Q. Zhao, L. Chen, Z. Liu, B. Ruan, M. Wu, Effect of transition-metal ion doping on electrocatalytic activities of graphene/polyaniline-M^{2+} (Mn^{2+}, Co^{2+}, Ni^{2+}, and Cu^{2+}) composite materials as Pt-free counter electrode in dye-sensitized solar cells, *Polymer-Plastics Technology and Materials*. 58 (2019) 40–46. https://doi.org/10.1080/03602559.2018.1455864.

[22] J. Moon, H. Takagi, Y. Fujishiro, M. Awano, Preparation and characterization of the Sb-doped TiO_2 photocatalysts, *Journal of Materials Science*. 36 (2001) 949–955. https://doi.org/10.1023/A:1004819706292.

[23] F.E. Oropeza, R.G. Egdell, Control of valence states in Rh-doped TiO_2 by Sb co-doping: A study by high resolution X-ray photoemission spectroscopy, *Chemical Physics Letters*. 515 (2011) 249–253. https://doi.org/10.1016/j.cplett.2011.09.017.

[24] M. Nolan, First-principles prediction of new photocatalyst materials with visible-light absorption and improved charge separation: Surface modification of rutile TiO_2 with nanoclusters of MgO and Ga_2O_3, *ACS Applied Materials & Interfaces*. 4 (2012) 5863–5871. https://doi.org/10.1021/am301516c.

[25] S. Selvinsimpson, P. Gnanamozhi, V. Pandiyan, M. Govindasamy, M.A. Habila, N. AlMasoud, Y. Chen, Synergetic effect of Sn doped ZnO nanoparticles synthesized via ultrasonication technique and its photocatalytic and antibacterial activity, *Environmental Research*. 197 (2021) 111115. https://doi.org/10.1016/j.envres.2021.111115.

[26] J. Ménesi, R. Kékesi, L. Kőrösi, V. Zöllmer, A. Richardt, I. Dékány, The effect of transition metal doping on the photooxidation process of titania-clay composites, *International Journal of Photoenergy*. 2008 (2008) 1–9. https://doi.org/10.1155/2008/846304.

[27] A. Kudo, R. Niishiro, A. Iwase, H. Kato, Effects of doping of metal cations on morphology, activity, and visible light response of photocatalysts, *Chemical Physics*. 339 (2007) 104–110. https://doi.org/10.1016/j.chemphys.2007.07.024.

[28] O. Avilés-García, J. Espino-Valencia, R. Romero, J.L. Rico-Cerda, M. Arroyo-Albiter, R. Natividad, W and Mo doped TiO_2: Synthesis, characterization and photocatalytic activity, *Fuel*. 198 (2017) 31–41. https://doi.org/10.1016/j.fuel.2016.10.005.

[29] A. Khlyustova, N. Sirotkin, T. Kusova, A. Kraev, V. Titov, A. Agafonov, Doped TiO_2: The effect of doping elements on photocatalytic activity, *Materials Advances*. 1 (2020) 1193–1201. https://doi.org/10.1039/D0MA00171F.

[30] S. In, A. Orlov, R. Berg, F. García, S. Pedrosa-Jimenez, M.S. Tikhov, D.S. Wright, R.M. Lambert, Effective visible light-activated B-doped and B, N-codoped TiO_2 photocatalysts, *Journal of the American Chemical Society*. 129 (2007) 13790–13791. https://doi.org/10.1021/ja0749237.

[31] H. Irie, Y. Watanabe, K. Hashimoto, Carbon-doped anatase TiO_2 powders as a visible-light sensitive photocatalyst, *Chemistry Letters*. 32 (2003) 772–773. https://doi.org/10.1246/cl.2003.772.

[32] T. Umebayashi, T. Yamaki, S. Tanaka, K. Asai, Visible light-induced degradation of methylene blue on S-doped TiO_2, *Chemistry Letters*. 32 (2003) 330–331. https://doi.org/10.1246/cl.2003.330.

[33] J. Chen, G. Wu, T. Wang, X. Li, M. Li, Y. Sang, H. Liu, Carrier step-by-step transport initiated by precise defect distribution engineering for efficient photocatalytic hydrogen generation, *ACS Applied Materials & Interfaces*. 9 (2017) 4634–4642. https://doi.org/10.1021/acsami.6b14700.

[34] H. Li, P. Zhang, S. Yin, Y. Wang, Q. Dong, C. Guo, T. Sato, Effect of transition metal elements addition on the properties of nitrogen-doped TiO_2 photocatalysts, *Journal of Physics: Conference Series*. 339 (2012) 012013. https://doi.org/10.1088/1742-6596/339/1/012013.

[35] C.-Y. Kuo, H.-K. Jheng, S.-E. Syu, Effect of non-metal doping on the photocatalytic activity of titanium dioxide on the photodegradation of aqueous bisphenol A, *Environmental Technology*. 42 (2021) 1603–1611. https://doi.org/10.1080/09593330.2019.1674930.

[36] V. Kumari, N. Kumar, S. Yadav, A. Mittal, S. Sharma, Novel mixed metal oxide ($ZnO.La_2O_3.CeO_2$) synthesized via hydrothermal and solution combustion process – A comparative study and their photocatalytic properties, *Materials Today: Proceedings*. 19 (2019) 650–657. https://doi.org/10.1016/j.matpr.2019.07.748.

[37] A. Mittal, S. Sharma, V. Kumari, S. Yadav, N.S. Chauhan, N. Kumar, Highly efficient, visible active TiO_2/CdS/ZnS photocatalyst, study of activity in an ultra low energy consumption LED based photo reactor, *Journal of Materials Science: Materials in Electronics*. 30 (2019) 17933–17946. https://doi.org/10.1007/s10854-019-02147-6.

[38] V. Kumari, S. Yadav, A. Mittal, K. Kumari, B. Mari, N. Kumar, Surface plasmon response of Pd deposited ZnO/CuO nanostructures with enhanced photocatalytic efficacy towards the degradation of organic pollutants, *Inorganic Chemistry Communications*. 121 (2020) 108241. https://doi.org/10.1016/j.inoche.2020.108241.

[39] S. Sharma, N. Kumar, B. Mari, N.S. Chauhan, A. Mittal, S. Maken, K. Kumari, Solution combustion synthesized TiO_2/Bi_2O_3/CuO nano-composites and their photocatalytic activity using visible LEDs assisted photoreactor, *Inorganic Chemistry Communications*. 125 (2021) 108418. https://doi.org/10.1016/j.inoche.2020.108418.

[40] C. Anderson, A.J. Bard, An improved photocatalyst of TiO_2/SiO_2 prepared by a sol-gel synthesis, *The Journal of Physical Chemistry*. 99 (1995) 9882–9885. https://doi.org/10.1021/j100024a033.

[41] J. Wang, S. Dong, C. Yu, X. Han, J. Guo, J. Sun, An efficient MoO_3 catalyst for in-practical degradation of dye wastewater under room conditions, *Catalysis Communications*. 92 (2017) 100–104. https://doi.org/10.1016/j.catcom.2017.01.013.

[42] D. Jing, L. Guo, A Novel method for the preparation of a highly stable and active cds photocatalyst with a special surface nanostructure, *The Journal of Physical Chemistry B*. 110 (2006) 11139–11145. https://doi.org/10.1021/jp060905k.

[43] A. Islam, S.H. Teo, M.R. Awual, Y.H. Taufiq-Yap, Improving the hydrogen production from water over MgO promoted Ni–Si/CNTs photocatalyst, *Journal of Cleaner Production*. 238 (2019) 117887. https://doi.org/10.1016/j.jclepro.2019.117887.

[44] J. Jin, J. Yu, D. Guo, C. Cui, W. Ho, A hierarchical Z-scheme CdS-WO_3 photocatalyst with enhanced CO_2 reduction activity, *Small*. 11 (2015) 5262–5271. https://doi.org/10.1002/smll.201500926.

[45] C. Xu, L. Cao, G. Su, W. Liu, H. Liu, Y. Yu, X. Qu, Preparation of ZnO/$Cu2O$ compound photocatalyst and application in treating organic dyes, *Journal of Hazardous Materials*. 176 (2010) 807–813. https://doi.org/10.1016/j.jhazmat.2009.11.106.

[46] J. He, P. Lyu, B. Jiang, S. Chang, H. Du, J. Zhu, H. Li, A novel amorphous alloy photocatalyst (NiB/In_2O_3) composite for sunlight-induced CO_2 hydrogenation to HCOOH, *Applied Catalysis B: Environmental*. 298 (2021) 120603. https://doi.org/10.1016/j.apcatb.2021.120603.

[47] A. Dehno Khalaji, P. Machek, M. Jarosova, α-Fe_2O_3 nanoparticles: Synthesis characterization magnetic properties and photocatalytic degradation of methyl orange, *Advanced Journal of Chemistry-Section A*. 4 (2021). https://doi.org/10.22034/ajca.2021.292396.1268.

[48] V. Alfredo Reyes Villegas, J. Isaías De León Ramírez, E. Hernandez Guevara, S. Perez Sicairos, L. Angelica Hurtado Ayala, B. Landeros Sanchez, Synthesis and characterization of magnetite nanoparticles for photocatalysis of nitrobenzene, *Journal of Saudi Chemical Society*. 24 (2020) 223–235. https://doi.org/10.1016/j.jscs.2019.12.004.

[49] C. Haw, W. Chiu, S. Abdul Rahman, P. Khiew, S. Radiman, R. Abdul Shukor, M.A.A. Hamid, N. Ghazali, The design of new magnetic-photocatalyst nanocomposites ($CoFe_2O_4$ –TiO_2) as smart nanomaterials for recyclable-photocatalysis applications, *New J. Chem*. 40 (2016) 1124–1136. https://doi.org/10.1039/C5NJ02496J.

[50] I. Ullah, F. Ali, Z. Ali, M. Humayun, Glycol stabilized magnetic nanoparticles for photocatalytic degradation of xylenol orange, *Materials Research Express*. 5 (2018) 055509. https://doi.org/10.1088/2053-1591/aac3b8.

[51] K. Laohhasurayotin, S. Pookboonmee, D. Viboonratanasri, W. Kangwansupamonkon, Preparation of magnetic photocatalyst nanoparticles—TiO_2/SiO_2/Mn–Zn ferrite—and its photocatalytic activity influenced by silica interlayer, *Materials Research Bulletin*. 47 (2012) 1500–1507. https://doi.org/10.1016/j.materresbull.2012.02.030.

[52] R. Liu, H.T. Ou, Synthesis and application of magnetic photocatalyst of Ni-Zn ferrite/TiO$_2$ from IC lead frame scraps, *Journal of Nanotechnology*. 2015 (2015) 1–7. https://doi.org/10.1155/2015/727210.

[53] S.M. Masoudpanah, S.M. Mirkazemi, Structural, magnetic and photocatalytic properties of BiFeO$_3$ nanoparticles, *Journal of Nanostructures*. 7 (2017). https://doi.org/10.22052/jns.2017.03.003.

[54] T. Chankhanittha, N. Komchoo, T. Senasu, J. Piriyanon, S. Youngme, K. Hemavibool, S. Nanan, Silver decorated ZnO photocatalyst for effective removal of reactive red azo dye and ofloxacin antibiotic under solar light irradiation, *Colloids and Surfaces A: Physicochemical and Engineering Aspects*. 626 (2021) 127034. https://doi.org/10.1016/j.colsurfa.2021.127034.

[55] M. Qin, K. Jin, X. Li, R. Wang, Y. Li, H. Wang, Novel highly-active Ag/Bi dual nanoparticles-decorated BiOBr photocatalyst for efficient degradation of ibuprofen, *Environmental Research*. 206 (2022) 112628. https://doi.org/10.1016/j.envres.2021.112628.

[56] M. Raaja Rajeshwari, S. Kokilavani, S. Sudheer Khan, Recent developments in architecturing the g-C$_3$N$_4$ based nanostructured photocatalysts: Synthesis, modifications and applications in water treatment, *Chemosphere*. 291 (2022) 132735. https://doi.org/10.1016/j.chemosphere.2021.132735.

[57] K.N. Van, H.T. Huu, V.N. Nguyen Thi, T.L. Le Thi, D.H. Truong, T.T. Truong, N.N. Dao, V. Vo, D.L. Tran, Y. Vasseghian, Facile construction of S-scheme SnO$_2$/g-C$_3$N$_4$ photocatalyst for improved photoactivity, *Chemosphere*. 289 (2022) 133120. https://doi.org/10.1016/j.chemosphere.2021.133120.

[58] Y. Wei, X. Li, Y. Zhang, Y. Yan, P. Huo, H. Wang, G-C$_3$N$_4$ quantum dots and Au nano particles co-modified CeO$_2$/Fe$_3$O$_4$ micro-flowers photocatalyst for enhanced CO$_2$ photoreduction, *Renewable Energy*. 179 (2021) 756–765. https://doi.org/10.1016/j.renene.2021.07.091.

[59] Y. Hong, E. Liu, J. Shi, X. Lin, L. Sheng, M. Zhang, L. Wang, J. Chen, A direct one-step synthesis of ultrathin g-C$_3$N$_4$ nanosheets from thiourea for boosting solar photocatalytic H$_2$ evolution, *International Journal of Hydrogen Energy*. 44 (2019) 7194–7204. https://doi.org/10.1016/j.ijhydene.2019.01.274.

[60] H. Li, L. Wang, Y. Liu, J. Lei, J. Zhang, Mesoporous graphitic carbon nitride materials: Synthesis and modifications, *Research on Chemical Intermediates*. 42 (2016) 3979–3998. https://doi.org/10.1007/s11164-015-2294-9.

[61] Q. Yang, T. Wang, F. Han, Z. Zheng, B. Xing, B. Li, Bimetal-modified g-C$_3$N$_4$ photocatalyst for promoting hydrogen production coupled with selective oxidation of biomass derivative, *Journal of Alloys and Compounds*. 897 (2022) 163177. https://doi.org/10.1016/j.jallcom.2021.163177.

[62] H. Jung, T.-T. Pham, E.W. Shin, Effect of g-C$_3$N$_4$ precursors on the morphological structures of g-C$_3$N$_4$/ZnO composite photocatalysts, *Journal of Alloys and Compounds*. 788 (2019) 1084–1092. https://doi.org/10.1016/j.jallcom.2019.03.006.

[63] M. Sher, M. Javed, S. Shahid, O. Hakami, M.A. Qamar, S. Iqbal, M.M. AL-Anazy, H.B. Baghdadi, Designing of highly active g-C$_3$N$_4$/Sn doped ZnO heterostructure as a photocatalyst for the disinfection and degradation of the organic pollutants under visible light irradiation, *Journal of Photochemistry and Photobiology A: Chemistry*. 418 (2021) 113393. https://doi.org/10.1016/j.jphotochem.2021.113393.

[64] T. Xian, H. Yang, L.J. Di, J.F. Dai, Enhanced photocatalytic activity of BaTiO$_3$@g-C$_3$N$_4$ for the degradation of methyl orange under simulated sunlight irradiation, *Journal of Alloys and Compounds*. 622 (2015) 1098–1104. https://doi.org/10.1016/j.jallcom.2014.11.051.

[65] Y. Yao, F. Lu, Y. Zhu, F. Wei, X. Liu, C. Lian, S. Wang, Magnetic core–shell CuFe$_2$O$_4$@C$_3$N$_4$ hybrids for visible light photocatalysis of Orange II, *Journal of Hazardous Materials*. 297 (2015) 224–233. https://doi.org/10.1016/j.jhazmat.2015.04.046.

[66] G. Venkatesh, G. Palanisamy, M. Srinivasan, S. Vignesh, N. Elavarasan, T. Pazhanivel, A.M. Al-Enizi, M. Ubaidullah, A. Karim, K.M. Prabu, CaSnO$_3$ coupled g-C$_3$N$_4$ S-scheme heterostructure photocatalyst for efficient pollutant degradation, *Diamond and Related Materials*. 124 (2022) 108873. https://doi.org/10.1016/j.diamond.2022.108873.

[67] Q. Wang, C. Chen, S. Zhu, X. Ni, Z. Li, Acetylene black quantum dots as a bridge for few-layer g-C$_3$N$_4$/MoS$_2$ nanosheet architecture: 0D–2D heterojunction as an efficient visible-light-driven photocatalyst, *Research on Chemical Intermediates*. 45 (2019) 4975–4993. https://doi.org/10.1007/s11164-019-03876-3.

[68] X. Liu, S. Wang, F. Yang, Y. Zhang, L. Yan, K. Li, H. Guo, J. Yan, J. Lin, Construction of Au/g-C$_3$N$_4$/ZnIn$_2$S$_4$ plasma photocatalyst heterojunction composite with 3D hierarchical microarchitecture for visible-light-driven hydrogen production, *International Journal of Hydrogen Energy*. 47 (2022) 2900–2913. https://doi.org/10.1016/j.ijhydene.2021.10.203.

[69] H. Yuan, F. Fang, J. Dong, W. Xia, X. Zeng, W. Shangguan, Enhanced photocatalytic hydrogen production based on laminated MoS$_2$/g-C$_3$N$_4$ photocatalysts, *Colloids and Surfaces A: Physicochemical and Engineering Aspects*. (2022) 128575. https://doi.org/10.1016/j.colsurfa.2022.128575.

[70] X. Wang, Y. Ren, Y. Li, G. Zhang, Fabrication of 1D/2D $BiPO_4$/g-C_3N_4 heterostructured photocatalyst with enhanced photocatalytic efficiency for NO removal, *Chemosphere*. 287 (2022) 132098. https://doi.org/10.1016/j.chemosphere.2021.132098.

[71] J. Chen, X. Liu, M. Que, L. Yang, H. Zheng, Z. Liu, T. Yang, Y. Li, X. Yang, S. Zhu, In situ forming heterointerface in g-C_3N_4/BiOBr photocatalyst for enhancing the photocatalytic activity, *Journal of Physics and Chemistry of Solids*. 163 (2022) 110609. https://doi.org/10.1016/j.jpcs.2022.110609.

[72] S. Sharma, V. Dutta, P. Raizada, A.A.P. Khan, Q.V. Le, V.K. Thakur, J.K. Biswas, R. Selvasembian, P. Singh, Controllable functionalization of g-C_3N_4 mediated all-solid-state (ASS) Z-scheme photocatalysts towards sustainable energy and environmental applications, *Environmental Technology & Innovation*. 24 (2021) 101972. https://doi.org/10.1016/j.eti.2021.101972.

[73] W. Li, C. Feng, S. Dai, J. Yue, F. Hua, H. Hou, Fabrication of sulfur-doped g-C_3N_4 /Au/CdS Z-scheme photocatalyst to improve the photocatalytic performance under visible light, *Applied Catalysis B: Environmental*. 168–169 (2015) 465–471. https://doi.org/10.1016/j.apcatb.2015.01.012.

[74] W. Fang, S. Yao, L. Wang, C. Li, Enhanced photocatalytic overall water splitting via hollow structure Pt/g-C_3N_4/BiOBr photocatalyst with S-scheme heterojunction, *Journal of Alloys and Compounds*. 891 (2022) 162081. https://doi.org/10.1016/j.jallcom.2021.162081.

[75] T. Nezakati, A. Seifalian, A. Tan, A.M. Seifalian, Conductive polymers: Opportunities and challenges in biomedical applications, *Chemical Reviews*. 118 (2018) 6766–6843. https://doi.org/10.1021/acs.chemrev.6b00275.

[76] K. Namsheer, C.S. Rout, Conducting polymers: A comprehensive review on recent advances in synthesis, properties and applications, *RSC Advances*. 11 (2021) 5659–5697. https://doi.org/10.1039/D0RA07800J.

[77] Ch. Seidel, Handbook of Conducting Polymers. Volumes 1 and 2. Hg. von Terje A. Skotheim. ISBN 0-8247-7395-0 und 0-8247-7454-X. New York/Basel: Marcel Dekker Inc. 1986. XVIII+XVII, 1417 S., gcb. $ 150.00., *Acta Polymerica*. 38 (1987) 101–101. https://doi.org/10.1002/actp.1987.010380126.

[78] T.-H. Le, Y. Kim, H. Yoon, Electrical and electrochemical properties of conducting polymers, *Polymers*. 9 (2017) 150. https://doi.org/10.3390/polym9040150.

[79] H.S. Nalwa, ed., *Handbook of Organic Conductive Molecules and Polymers*, Wiley, Chichester; New York, 1997.

[80] B.R. Hsieh, A review of: "'Organic Electroluminescent Materials and Devices', Edited by S. Miyata and H. S. Nalwa, Gordon and Breach Science Publishers, Amsterdam, 1997; ISBN 2-919875-10-8; x+487 pages, 12 color plates; $180.00," *Molecular Crystals and Liquid Crystals Science and Technology. Section A. Molecular Crystals and Liquid Crystals*. 326 (1999) 425–427. https://doi.org/10.1080/10587259908025430.

[81] V.V. Tran, T.T.V. Nu, H.-R. Jung, M. Chang, Advanced photocatalysts based on conducting polymer/metal oxide composites for environmental applications, *Polymers*. 13 (2021) 3031. https://doi.org/10.3390/polym13183031.

[82] J. Zia, F. Fatima, U. Riaz, A comprehensive review on the photocatalytic activity of polythiophene-based nanocomposites against degradation of organic pollutants, *Catalysis Science & Technology*, 11 (2021) 6630–6648. https://doi.org/10.1039/D1CY01129D.

[83] C.R. Martin, Template synthesis of electronically conductive polymer nanostructures, *Accounts of Chemical Research*. 28 (1995) 61–68. https://doi.org/10.1021/ar00050a002.

[84] A.G. MacDiarmid, W.E. Jones, I.D. Norris, J. Gao, A.T. Johnson, N.J. Pinto, J. Hone, B. Han, F.K. Ko, H. Okuzaki, M. Llaguno, Electrostatically-generated nanofibers of electronic polymers, *Synthetic Metals*. 119 (2001) 27–30. https://doi.org/10.1016/S0379-6779(00)00597-X.

[85] C.-S. Park, D. Kim, B. Shin, D. Kim, H.-K. Lee, H.-S. Tae, Conductive polymer synthesis with single-crystallinity via a novel plasma polymerization technique for gas sensor applications, *Materials*. 9 (2016) 812. https://doi.org/10.3390/ma9100812.

[86] J.L. Bredas, G.B. Street, Polarons, bipolarons, and solitons in conducting polymers, *Accounts of Chemical Research*. 18 (1985) 309–315. https://doi.org/10.1021/ar00118a005.

[87] R.J. Tayade, T.S. Natarajan, H.C. Bajaj, Photocatalytic degradation of methylene blue dye using ultraviolet light emitting diodes, *Industrial & Engineering Chemistry Research*. 48 (2009) 10262–10267. https://doi.org/10.1021/ie9012437.

[88] N. Al Amery, H.R. Abid, S. Al-Saadi, S. Wang, S. Liu, Facile directions for synthesis, modification and activation of MOFs, *Materials Today Chemistry*. 17 (2020) 100343. https://doi.org/10.1016/j.mtchem.2020.100343.

[89] B. Liu, H. Shioyama, T. Akita, Q. Xu, Metal-organic framework as a template for porous carbon synthesis, *Journal of the American Chemical Society*. 130 (2008) 5390–5391. https://doi.org/10.1021/ja7106146.

[90] S. Swetha, B. Janani, S.S. Khan, A critical review on the development of metal-organic frameworks for boosting photocatalysis in the fields of energy and environment, *Journal of Cleaner Production*. 333 (2022) 130164. https://doi.org/10.1016/j.jclepro.2021.130164.

[91] M.M. Tunesi, R.A. Soomro, X. Han, Q. Zhu, Y. Wei, B. Xu, Application of MXenes in environmental remediation technologies, *Nano Convergence*. 8 (2021) 5. https://doi.org/10.1186/s40580-021-00255-w.

[92] M. Öper, U. Yorulmaz, C. Sevik, F. Ay, N. Kosku Perkgöz, Controlled CVD growth of ultrathin Mo_2C (MXene) flakes, *Journal of Applied Physics*. 131 (2022) 025304. https://doi.org/10.1063/5.0067970.

[93] J. Guo, Q. Peng, H. Fu, G. Zou, Q. Zhang, Heavy-metal adsorption behavior of two-dimensional alka-lization-intercalated MXene by first-principles calculations, *The Journal of Physical Chemistry C*. 119 (2015) 20923–20930. https://doi.org/10.1021/acs.jpcc.5b05426.

[94] Y. Gao, L. Wang, A. Zhou, Z. Li, J. Chen, H. Bala, Q. Hu, X. Cao, Hydrothermal synthesis of TiO_2/Ti_3C_2 nanocomposites with enhanced photocatalytic activity, *Materials Letters*. 150 (2015) 62–64. https://doi.org/10.1016/j.matlet.2015.02.135.

[95] C. Peng, X. Yang, Y. Li, H. Yu, H. Wang, F. Peng, Hybrids of two-dimensional Ti_3C_2 and TiO_2 expos-ing {001} facets toward enhanced photocatalytic activity, *ACS Applied Materials & Interfaces*. 8 (2016) 6051–6060. https://doi.org/10.1021/acsami.5b11973.

[96] Y. Zhang, L. Wang, N. Zhang, Z. Zhou, Adsorptive environmental applications of MXene nanomateri-als: A review, *RSC Advances*. 8 (2018) 19895–19905. https://doi.org/10.1039/C8RA03077D.

[97] S.K. Bhardwaj, H. Singh, M. Khatri, K.-H. Kim, N. Bhardwaj, Advances in MXenes-based optical biosensors: A review, *Biosensors and Bioelectronics*. 202 (2022) 113995. https://doi.org/10.1016/j.bios.2022.113995.

[98] Y. Pei, Y. Zhang, J. Ma, M. Fan, S. Zhang, J. Wang, Ionic liquids for advanced materials, *Materials Today Nano*. 17 (2022) 100159. https://doi.org/10.1016/j.mtnano.2021.100159.

[99] R.V. Barrulas, M. Zanatta, T. Casimiro, M.C. Corvo, Advanced porous materials from poly(ionic liq-uid)s: Challenges, applications and opportunities, *Chemical Engineering Journal*. 411 (2021) 128528. https://doi.org/10.1016/j.cej.2021.128528.

[100] W. Zhang, Q. Zhao, J. Yuan, Porous polyelectrolytes: The interplay of charge and pores for new func-tionalities, *Angewandte Chemie International Edition*. 57 (2018) 6754–6773. https://doi.org/10.1002/anie.201710272.

[101] R. Lertlapwasin, N. Bhawawet, A. Imyim, S. Fuangswasdi, Ionic liquid extraction of heavy metal ions by 2-aminothiophenol in 1-butyl-3-methylimidazolium hexafluorophosphate and their association constants, *Separation and Purification Technology*. 72 (2010) 70–76. https://doi.org/10.1016/j.seppur.2010.01.004.

[102] L. Li, C. Luo, X. Li, H. Duan, X. Wang, Preparation of magnetic ionic liquid/chitosan/graphene oxide composite and application for water treatment, *International Journal of Biological Macromolecules*. 66 (2014) 172–178. https://doi.org/10.1016/j.ijbiomac.2014.02.031.

[103] G.D. Bozyiğit, M.F. Ayyıldız, D.S. Chormey, N.B. Turan, F. Kapukıran, G.O. Engin, S. Bakırdere, Removal of selected pesticides, alkylphenols, hormones and bisphenol A from domestic wastewater by electrooxidation process, *Water Science and Technology*. 85 (2022) 220–228. https://doi.org/10.2166/wst.2021.635.

[104] W. Liu, J. Quan, Z. Hu, Detection of organophosphorus pesticides in wheat by ionic liquid-based disper-sive liquid-liquid microextraction combined with HPLC, *Journal of Analytical Methods in Chemistry*. 2018 (2018) 1–10. https://doi.org/10.1155/2018/8916393.

[105] E. Calla-Quispe, J. Robles, C. Areche, B. Sepulveda, Are ionic liquids better extracting agents than toxic volatile organic solvents? A combination of ionic liquids, microwave and LC/MS/MS, applied to the Lichen *Stereocaulon glareosum*, *Frontiers in Chemistry*. 8 (2020) 450. https://doi.org/10.3389/fchem.2020.00450.

[106] S.P.M. Ventura, F.A. e Silva, M.V. Quental, D. Mondal, M.G. Freire, J.A.P. Coutinho, ionic-liquid-mediated extraction and separation processes for bioactive compounds: Past, present, and future trends, *Chemical Reviews*. 117 (2017) 6984–7052. https://doi.org/10.1021/acs.chemrev.6b00550.

[107] A. Brinda Lakshmi, A. Balasubramanian, S. Venkatesan, Extraction of phenol and chlorophenols using ionic liquid $[Bmim]^+ [BF_4]^-$ dissolved in tributyl phosphate, *Clean Soil Air Water*. 41 (2013) 349–355. https://doi.org/10.1002/clen.201100632.

[108] R. Goutham, P. Rohit, S.S. Vigneshwar, A. Swetha, J. Arun, K.P. Gopinath, A. Pugazhendhi, Ionic liquids in wastewater treatment: A review on pollutant removal and degradation, recovery of ionic liq-uids, economics and future perspectives, *Journal of Molecular Liquids*. 349 (2022) 118150. https://doi.org/10.1016/j.molliq.2021.118150.

[109] N. Ezzatahmadi, G.A. Ayoko, G.J. Millar, R. Speight, C. Yan, J. Li, S. Li, J. Zhu, Y. Xi, Clay-supported nanoscale zero-valent iron composite materials for the remediation of contaminated aqueous solutions: A review, *Chemical Engineering Journal*. 312 (2017) 336–350. https://doi.org/10.1016/j.cej.2016.11.154.

[110] F. Ali, H. Ullah, Z. Ali, F. Rahim, F. Khan, Z. Ur Rehman, Polymer-clay nanocomposites, preparations and current applications: A review, *Current Nanomaterials*. 1 (2016) 83–95. https://doi.org/10.2174/240 5461501666160625080118.

[111] A. Gil, L. Santamaría, S.A. Korili, M.A. Vicente, L.V. Barbosa, S.D. de Souza, L. Marçal, E.H. de Faria, K.J. Ciuffi, A review of organic-inorganic hybrid clay based adsorbents for contaminants removal: Synthesis, perspectives and applications, *Journal of Environmental Chemical Engineering*. 9 (2021) 105808. https://doi.org/10.1016/j.jece.2021.105808.

[112] A.C.S. Alcântara, M. Darder, Building up functional bionanocomposites from the assembly of clays and biopolymers, *The Chemical Record*. 18 (2018) 696–712. https://doi.org/10.1002/tcr.201700076.

[113] R. Zafar, K.M. Zia, S. Tabasum, F. Jabeen, A. Noreen, M. Zuber, Polysaccharide based bionanocomposites, properties and applications: A review, *International Journal of Biological Macromolecules*. 92 (2016) 1012–1024. https://doi.org/10.1016/j.ijbiomac.2016.07.102.

[114] P. Monvisade, P. Siriphannon, Chitosan intercalated montmorillonite: Preparation, characterization and cationic dye adsorption, *Applied Clay Science*. 42 (2009) 427–431. https://doi.org/10.1016/j.clay.2008.04.013.

[115] M.N.V. Ravi Kumar, A review of chitin and chitosan applications, *Reactive and Functional Polymers*. 46 (2000) 1–27. https://doi.org/10.1016/S1381-5148(00)00038-9.

[116] N. Peng, D. Hu, J. Zeng, Y. Li, L. Liang, C. Chang, Superabsorbent cellulose–clay nanocomposite hydrogels for highly efficient removal of dye in water, *ACS Sustainable Chemistry & Engineering*. 4 (2016) 7217–7224. https://doi.org/10.1021/acssuschemeng.6b02178.

[117] Md.M. Islam, M.N. Khan, S. Biswas, T. Rabia Choudhury, P. Haque, T. U Rashid, M. Mizanur Rahman, Preparation and characterization of bijoypur clay-crystalline cellulose composite for application as an adsorbent, *Advanced Material Science*. 2 (2017). https://doi.org/10.15761/AMS.1000126.

[118] F. Chivrac, E. Pollet, M. Schmutz, L. Avérous, New approach to elaborate exfoliated starch-based nanobiocomposites, *Biomacromolecules*. 9 (2008) 896–900. https://doi.org/10.1021/bm7012668.

[119] A.A. Edathil, P. Pal, F. Banat, Alginate clay hybrid composite adsorbents for the reclamation of industrial lean methyldiethanolamine solutions, *Applied Clay Science*. 156 (2018) 213–223. https://doi.org/10.1016/j.clay.2018.02.015.

[120] A.A. Edathil, P. Pal, F. Banat, Amine contaminants removal using alginate clay hybrid composites and its effect on foaming, *International Journal of Industrial Chemistry*. 10 (2019) 145–158. https://doi.org/10.1007/s40090-019-0180-9.

[121] I.A. Shabtai, Y.G. Mishael, Polycyclodextrin–clay composites: Regenerable dual-site sorbents for bisphenol a removal from treated wastewater, *ACS Applied Materials & Interfaces*. 10 (2018) 27088–27097. https://doi.org/10.1021/acsami.8b09715.

[122] A. Alsbaiee, B.J. Smith, L. Xiao, Y. Ling, D.E. Helbling, W.R. Dichtel, Rapid removal of organic micropollutants from water by a porous β-cyclodextrin polymer, *Nature*. 529 (2016) 190–194. https://doi.org/10.1038/nature16185.

[123] A. (Fern) Phuekphong, K. (Jaa) Imwiset, M. Ogawa, Designing nanoarchitecture for environmental remediation based on the clay minerals as building block, *Journal of Hazardous Materials*. 399 (2020) 122888. https://doi.org/10.1016/j.jhazmat.2020.122888.

[124] B. Wicklein, M. Darder, P. Aranda, E. Ruiz-Hitzky, Bio-organoclays based on phospholipids as immobilization hosts for biological species, *Langmuir*. 26 (2010) 5217–5225. https://doi.org/10.1021/la9036925.

[125] V.G. Marques, M.R.F. Gonçalves, A.G. Osorio, A.B. Tessaro, G.E.H. da Silva, G.F.R. Paganotto, N.L.V. Carreño, Evaluation of the influence of manufacture parameters during organoclay synthesis, *Cerâmica*. 64 (2018) 418–424. https://doi.org/10.1590/0366-69132018643712298.

[126] K. Mogyorósi, A. Farkas, I. Dékány, I. Ilisz, A. Dombi, TiO$_2$-based photocatalytic degradation of 2-chlorophenol adsorbed on hydrophobic clay, *Environmental Science & Technology*. 36 (2002) 3618–3624. https://doi.org/10.1021/es015843k.

[127] D.F. Montaño, H. Casanova, W.I. Cardona, L.F. Giraldo, Functionalization of montmorillonite with ionic liquids based on 1-alkyl-3-methylimidazolium: Effect of anion and length chain, *Materials Chemistry and Physics*. 198 (2017) 386–392. https://doi.org/10.1016/j.matchemphys.2017.06.027.

[128] M. Altamirano, A. Senz, H.E. Gsponer, Luminescence quenching of tris(2,2′-bipyridine) ruthenium(II) by 2,6-dimethylphenol and 4-bromo-2,6-dimethylphenol in sol–gel-processed silicate thin films, *Journal of Colloid and Interface Science*. 270 (2004) 364–370. https://doi.org/10.1016/j.jcis.2003.09.006.

[129] T. Okada, T. Morita, M. Ogawa, Tris(2,2′-bipyridine)ruthenium(II)-clays as adsorbents for phenol and chlorinated phenols from aqueous solution, *Applied Clay Science*. 29 (2005) 45–53. https://doi.org/10.1016/j.clay.2004.09.004.

[130] J. Szerement, A. Szatanik-Kloc, R. Jarosz, T. Bajda, M. Mierzwa-Hersztek, Contemporary applications of natural and synthetic zeolites from fly ash in agriculture and environmental protection, *Journal of Cleaner Production*. 311 (2021) 127461. https://doi.org/10.1016/j.jclepro.2021.127461.

[131] G. Hu, J. Yang, X. Duan, R. Farnood, C. Yang, J. Yang, W. Liu, Q. Liu, Recent developments and challenges in zeolite-based composite photocatalysts for environmental applications, *Chemical Engineering Journal*. 417 (2021) 129209. https://doi.org/10.1016/j.cej.2021.129209.

[132] J.T. Soe, S.-S. Kim, Y.-R. Lee, J.-W. Ahn, W.-S. Ahn, CO_2 capture and Ca^{2+} exchange using zeolite a and 13X prepared from power plant fly ash: CO_2 capture and Ca^{2+} exchange using zeolite A and 13X, *Bulletin of the Korean Chemical Society*. 37 (2016) 490–493. https://doi.org/10.1002/bkcs.10710.

[133] Z. Zhang, Y. Xiao, B. Wang, Q. Sun, H. Liu, Waste is a misplayed resource: Synthesis of zeolites from fly ash for CO_2 capture, *Energy Procedia*. 114 (2017) 2537–2544. https://doi.org/10.1016/j.egypro.2017.08.036.

[134] N. Czuma, I. Casanova, P. Baran, J. Szczurowski, K. Zarębska, CO_2 sorption and regeneration properties of fly ash zeolites synthesized with the use of differentiated methods, *Scientific Reports*. 10 (2020) 1825. https://doi.org/10.1038/s41598-020-58591-6.

[135] N. Czuma, K. Zarębska, P. Baran, Analysis of the influence of fusion synthesis parameters on the SO2 sorption properties of zeolites produced out of fly ash, in: *E3S Web of Conferences*, Vol. 10, 2016: p. 00010. https://doi.org/10.1051/e3sconf/20161000010.

[136] T. Zhu, X. Zhang, Y. Han, T. Liu, B. Wang, Z. Zhang, Preparation of zeolite X by the aluminum residue from coal fly ash for the adsorption of volatile organic compounds, *Frontiers in Chemistry*. 7 (2019) 341. https://doi.org/10.3389/fchem.2019.00341.

[137] X. Zhang, X. Li, F. Zhang, S. Peng, S.H. Tumrani, X. Ji, Adsorption of Se(IV) in aqueous solution by zeolites synthesized from fly ashes with different compositions, *Journal of Water Reuse and Desalination*. 9 (2019) 506–519. https://doi.org/10.2166/wrd.2019.036.

[138] H. Pahlavanzadeh, M. Motamedi, Adsorption of nickel, Ni(II), in aqueous solution by modified zeolite as a cation-exchange adsorbent, *Journal of Chemical & Engineering Data*. 65 (2020) 185–197. https://doi.org/10.1021/acs.jced.9b00868.

[139] Y. Zhang, J. Dong, F. Guo, Z. Shao, J. Wu, Zeolite synthesized from coal fly ash produced by a gasification process for Ni^{2+} removal from water, *Minerals*. 8 (2018) 116. https://doi.org/10.3390/min8030116.

[140] M. Abatal, A.V. Córdova Quiroz, M.T. Olguín, A.R. Vázquez-Olmos, J. Vargas, F. Anguebes-Franseschi, G. Giácoman-Vallejos, Sorption of Pb(II) from aqueous solutions by acid-modified clinoptilolite-rich tuffs with different Si/Al ratios, *Applied Sciences*. 9 (2019) 2415. https://doi.org/10.3390/app9122415.

[141] V. Arumugam, K.G. Moodley, A. Dass, R.M. Gengan, D. Ali, S. Alarifi, M. Chandrasekaran, Y. Gao, Ionic liquid covered iron-oxide magnetic nanoparticles decorated zeolite nanocomposite for excellent catalytic reduction and degradation of environmental toxic organic pollutants and dyes, *Journal of Molecular Liquids*. 342 (2021) 117492. https://doi.org/10.1016/j.molliq.2021.117492.

[142] J. Yao, H. Wang, Zeolitic imidazolate framework composite membranes and thin films: Synthesis and applications, *Chemical Society Reviews*. 43 (2014) 4470–4493. https://doi.org/10.1039/C3CS60480B.

[143] J. Li, X. Meng, F.-S. Xiao, Zeolites for control of NO emissions: Opportunities and challenges, *Chem Catalysis*. 2 (2022) 253–261. https://doi.org/10.1016/j.checat.2021.11.011.

[144] A. Jevremović, B. Nedić Vasiljević, A. Popa, S. Uskoković-Marković, L. Ignjatović, D. Bajuk-Bogdanović, M. Milojević-Rakić, The environmental impact of potassium tungstophosphate/ZSM-5 zeolite: Insight into catalysis and adsorption processes, *Microporous and Mesoporous Materials*. 315 (2021) 110925. https://doi.org/10.1016/j.micromeso.2021.110925.

[145] I.H.A.E. Maksod, A. Al-Shehri, S. Bawaked, M. Mokhtar, K. Narasimharao, Structural and photocatalytic properties of precious metals modified TiO_2-BEA zeolite composites, *Molecular Catalysis*. 441 (2017) 140–149. https://doi.org/10.1016/j.mcat.2017.08.012.

[146] L.N.Q. Tu, N.V.H. Nhan, N. Van Dung, N.T. An, N.Q. Long, Enhanced photocatalytic performance and moisture tolerance of nano-sized Me/TiO_2–zeolite Y (Me=Au, Pd) for gaseous toluene removal: Activity and mechanistic investigation, *Journal of Nanoparticle Research*. 21 (2019) 194. https://doi.org/10.1007/s11051-019-4642-y.

[147] S.-H. Liu, W.-X. Lin, A simple method to prepare g-C_3N_4-TiO_2/waste zeolites as visible-light-responsive photocatalytic coatings for degradation of indoor formaldehyde, *Journal of Hazardous Materials*. 368 (2019) 468–476. https://doi.org/10.1016/j.jhazmat.2019.01.082.

[148] M. Aram, M. Farhadian, A.R. Solaimany Nazar, S. Tangestaninejad, P. Eskandari, B.-H. Jeon, Metronidazole and cephalexin degradation by using of urea/TiO_2/$ZnFe_2O_4$/clinoptiloite catalyst under visible-light irradiation and ozone injection, *Journal of Molecular Liquids*. 304 (2020) 112764. https://doi.org/10.1016/j.molliq.2020.112764.

[149] M. Wang, H. Yao, L. Zhang, X. Zhou, Synthesis of highly-efficient photocatalyst for visible- light-driven hydrogen evolution by recycling of heavy metal ions in wastewater, *Journal of Hazardous Materials.* 383 (2020) 121149. https://doi.org/10.1016/j.jhazmat.2019.121149.

[150] W. Zhang, X. Xiao, L. Zheng, C. Wan, Fabrication of TiO_2/MoS_2@zeolite photocatalyst and its photocatalytic activity for degradation of methyl orange under visible light, *Applied Surface Science.* 358 (2015) 468–478. https://doi.org/10.1016/j.apsusc.2015.08.054.

[151] M. Ahmed, I. Dincer, A review on photoelectrochemical hydrogen production systems: Challenges and future directions, *International Journal of Hydrogen Energy.* 44 (2019) 2474–2507. https://doi.org/10.1016/j.ijhydene.2018.12.037.

[152] A. Gondolini, N. Sangiorgi, A. Sangiorgi, A. Sanson, Photoelectrochemical hydrogen production by screen-printed copper oxide electrodes, *Energies.* 14 (2021) 2942. https://doi.org/10.3390/en14102942.

[153] T.-D. Nguyen, J.S. Lee, Recent development of flexible tactile sensors and their applications, *Sensors.* 22 (2021) 50. https://doi.org/10.3390/s22010050.

[154] U. Yaqoob, M.I. Younis, Chemical gas sensors: Recent developments, challenges, and the potential of machine learning—a review, *Sensors.* 21 (2021) 2877. https://doi.org/10.3390/s21082877.

[155] F. Zhang, K. Yang, Z. Pei, Y. Wu, S. Sang, Q. Zhang, H. Jiao, A highly accurate flexible sensor system for human blood pressure and heart rate monitoring based on graphene/sponge, *RSC Advances.* 12 (2022) 2391–2398. https://doi.org/10.1039/D1RA08608A.

[156] Z. Wang, C. Gao, S. Hou, H. Yang, Z. Shao, S. Xu, H. Ye, A DFT study of As doped WSe_2: A NO_2 sensing material with ultra-high selectivity in the atmospheric environment, *Materials Today Communications.* 28 (2021) 102654. https://doi.org/10.1016/j.mtcomm.2021.102654.

[157] A. Shakeel, K. Rizwan, U. Farooq, S. Iqbal, A.A. Altaf, Advanced polymeric/inorganic nanohybrids: An integrated platform for gas sensing applications, *Chemosphere.* 294 (2022) 133772. https://doi.org/10.1016/j.chemosphere.2022.133772.

[158] Y. Yu, M. Pan, J. Peng, D. Hu, Y. Hao, Z. Qian, A review on recent advances in hydrogen peroxide electrochemical sensors for applications in cell detection, *Chinese Chemical Letters.* (2022) S1001841722001590. https://doi.org/10.1016/j.cclet.2022.02.045.

[159] P. Zhao, X. Zhang, J. Hao, Melamine-imprinted electrochemical sensor of graphene/ionic liquid composites and its use for the detection of melamine in dairy products, *ChemPhysMater.* (2021) S2772571521000140. https://doi.org/10.1016/j.chphma.2021.11.003.

[160] N. Baig, A. Waheed, M. Sajid, I. Khan, A.-N. Kawde, M. Sohail, Porous graphene-based electrodes: Advances in electrochemical sensing of environmental contaminants, *Trends in Environmental Analytical Chemistry.* 30 (2021) e00120. https://doi.org/10.1016/j.teac.2021.e00120.

[161] L. Wang, N. Wang, J. Wen, Y. Jia, S. Pan, H. Xiong, Y. Tang, J. Wang, X. Yang, Y. Sun, Y. Chen, P. Wan, Ultrasensitive sensing of environmental nitroaromatic contaminants on nanocomposite of Prussian blue analogues cubes grown on glucose-derived porous carbon, *Chemical Engineering Journal.* 397 (2020) 125450. https://doi.org/10.1016/j.cej.2020.125450.

[162] M.M. Sabzehmeidani, S. Mahnaee, M. Ghaedi, H. Heidari, V.A.L. Roy, Carbon based materials: A review of adsorbents for inorganic and organic compounds, *Materials Advances.* 2 (2021) 598–627. https://doi.org/10.1039/D0MA00087F.

[163] U. Kamran, Y.-J. Heo, J.W. Lee, S.-J. Park, Functionalized carbon materials for electronic devices: A review, *Micromachines.* 10 (2019) 234. https://doi.org/10.3390/mi10040234.

[164] E. García-Bordejé, E. Pires, J.M. Fraile, Carbon materials functionalized with sulfonic groups as acid catalysts, in: *Emerging Carbon Materials for Catalysis*, Elsevier, 2021: pp. 255–298. https://doi.org/10.1016/B978-0-12-817561-3.00008-1.

[165] L. Kaluđerović, Z.P. Tomić, R. Đurović-Pejčev, L. Životić, Adsorption behaviour of clomazone on inorganic and organically modified natural montmorillonite from Bogovina (Serbia), *Clay Minerals.* 55 (2020) 342–350. https://doi.org/10.1180/clm.2021.3.

[166] A. Nennemann, Y. Mishael, S. Nir, B. Rubin, T. Polubesova, F. Bergaya, H. van Damme, G. Lagaly, Clay-based formulations of metolachlor with reduced leaching, *Applied Clay Science.* 18 (2001) 265–275. https://doi.org/10.1016/S0169-1317(01)00032-1.

[167] J. Chen, Recent progress in advanced materials for lithium ion batteries, *Materials.* 6 (2013) 156–183. https://doi.org/10.3390/ma6010156.

[168] N. Tolganbek, Y. Yerkinbekova, S. Kalybekkyzy, Z. Bakenov, A. Mentbayeva, Current state of high voltage olivine structured $LiMPO4$ cathode materials for energy storage applications: A review, *Journal of Alloys and Compounds.* 882 (2021) 160774. https://doi.org/10.1016/j.jallcom.2021.160774.

[169] X. Jiang, Y. Chen, X. Meng, W. Cao, C. Liu, Q. Huang, N. Naik, V. Murugadoss, M. Huang, Z. Guo, The impact of electrode with carbon materials on safety performance of lithium-ion batteries: A review, *Carbon.* 191 (2022) 448–470. https://doi.org/10.1016/j.carbon.2022.02.011.

2 Green Approaches to Catalytic Processes under Alternative Reaction Media

Mariette M. Pereira, Mário J. F. Calvete,
Fábio M. S. Rodrigues, and Rui M. B. Carrilho
University of Coimbra

Lucas D. Dias
University of São Paulo

CONTENTS

2.1 INTRODUCTION

The worldwide uncontrolled industrial development in the 20th century contributed to massive environmental problems/disasters, which culminated in the organization of the Pollution Prevention Act of 1990 [1], where industrials and academics declared that, instead of treating and disposing chemical waste, U.S. national policy should eliminate pollution by improving design (including cost-effective changes in processes and products). This was the birth of the "Green Chemistry philosophy" [2], initially set out in 1998 in the pioneering book *Green Chemistry: Theory and Practice* [3], by Paul Anastas and John Warner, where the 12 principles of Green Chemistry were described for the first time (Figure 2.1).

Several preventive actions have been taken worldwide since then, with the aim of decreasing environmental problems and defining future strategies for transforming Green Chemistry into a real tool to enhance sustainability in its economic, social, and environmental aspects [4]. This chapter is strongly engaged with the 12 Principles of Green Chemistry philosophy, as they apply to the development of highly sustainable chemical processes for the preparation of fine chemicals [5–7]. Among these processes, we highlight the use of catalytic processes instead of stoichiometric reagent-based processes (Principle 9) and the use of alternative solvents (Principle 5). This philosophy points ways in which chemistry can help to increase the sustainability of our planet, through (i) reducing energy consumption; (ii) replacing toxic stoichiometric reagents; (iii) increasing selectivity, thus minimizing waste; (iv) lowering consumption of toxic solvents; (v) reutilizing catalysts to minimize waste; and (vi) using alternative/benign solvents.

DOI: 10.1201/9781003206385-2

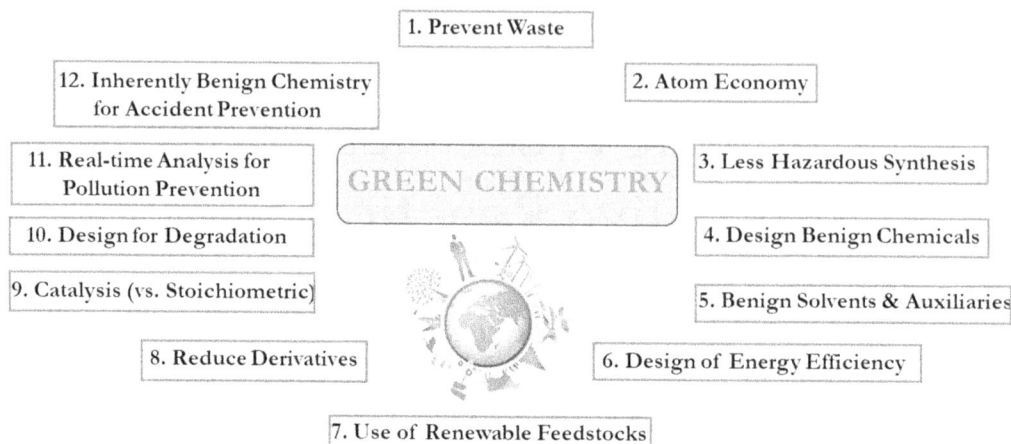

FIGURE 2.1   The 12 principles of Green Chemistry.

2.2   CATALYSIS IN THE GREEN CHEMISTRY CONTEXT

Catalysis can be regarded as a "foundational pillar" of Green Chemistry [8], significantly contributing to the success of a sustainable chemical industry in the 20th century [9]. The relevance of catalysis in our days is clearly demonstrated by the huge number of publications over the last decade (Figure 2.2).

The term "catalysis" was first proposed by Jöns Jakob Berzelius (1779–1848) [10], who wrote about "the property of exerting on other bodies an action which is very different from chemical affinity. By means of this action, they produce decomposition in bodies, and form new compounds into the composition of which they do not enter..." [11]. In addition, according to IUPAC [12], it can be defined as "a substance that increases the rate of a reaction without modifying the overall standard Gibbs energy change in the reaction; the process is called catalysis". Therefore, catalysis is based on the utilization of a substance, the "catalyst", that reduces energy input and frequently simplifies the final purification processes, due to higher selectivity (Figure 2.3) [13]. These features, in combination with the use of renewable feedstock such as biomass, CO_2 and terpenes, transform catalysis into a key tool for decreasing costs and waste, making a major contribution to the overall sustainability of planet Earth [14].

The relevance of this field is also clearly demonstrated by the large number of Chemistry Nobel Prizes (10) awarded by the Royal Swedish Academy of Sciences to scientists who have made relevant contributions to enzymatic or chemical catalysis, which have greatly enhanced both industrial and economic development [15,16].

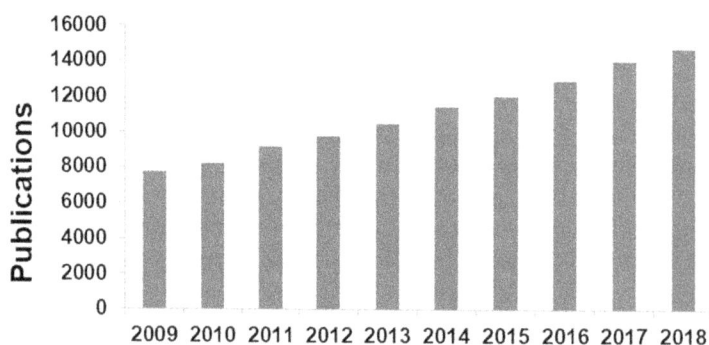

FIGURE 2.2   Publications on "catalysis" from 2010 to 2020. Search carried out on Web of Science® search engine using "catalysis" as keyword.

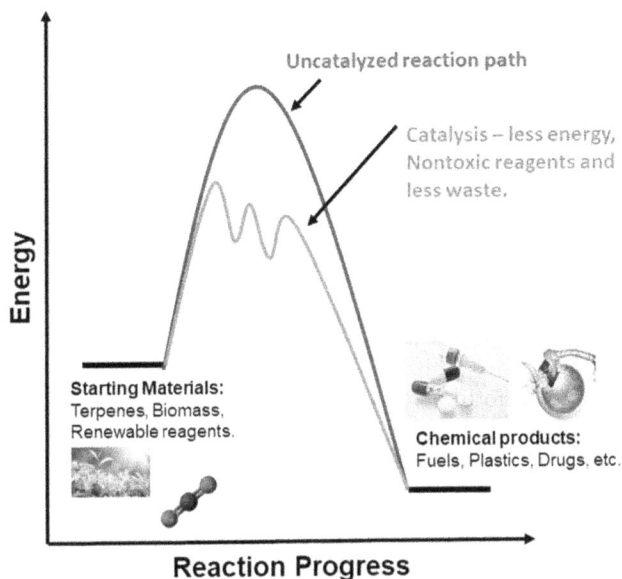

FIGURE 2.3 General potential energy diagram showing the effect of a catalyst in a hypothetical reaction.

In 1909, Wilhelm Ostwald was awarded with The Nobel Prize in Chemistry for his studies on the rates of numerous chemical reactions, including reactions that occur in the presence of bases or acids. As result, his research revealed that a chemical entity – a catalyst – can affect a chemical reaction's rate, but is not present in its end-products. This discovery is considered a landmark for the catalysis field since from then on, it was possible to understand what was occurring in many industrial and biochemical processes [16].

In 1912, Victor Grignard and Paul Sabatier were awarded the Nobel Prize for Chemistry, for the discovery of stable organomagnesium compounds, normally called "Grignard reagents". These organometallic complexes, although mainly used as stoichiometric reagents, have made a great contribution to the development of organic/medicinal chemistry. Furthermore, the results of P. Sabatier may be considered as the first example of the introduction of metals as hydrogenating catalysts of organic compounds [16].

Some years later, in 1918, Fritz Haber received his Nobel Prize in Chemistry for the catalytic synthesis of ammonia from nitrogen (N_2) and hydrogen (H_2). This discovery was essential for the development of fertilizers and explosives. When nitrogen and hydrogen gases pass through an apparatus at a controlled temperature, pressure, and flow rate, and in the presence of an iron catalyst, ammonia is formed in an energy-efficient process. The annual world production of synthetic nitrogen fertilizer is currently more than 100 million tons. The food base of half of the current world population is based on the Haber–Bosch process [16].

In 1963, Karl Ziegler and Giulio Natta shared the Nobel Prize in Chemistry for their studies on the development of catalytic processes able to promote controlled polymerization at room temperature and atmospheric pressure. Ziegler developed aluminum catalysts for creating molecular chains, in which aluminum's electrons are grouped so that active molecules are drawn to them and wedge themselves in between the molecular chain and the aluminum atom. Particularly, his polyethylene did not consist of ethylene units randomly attached in a branched pattern, as polyethylene made by the old process, but consisted of very ordered, very long, straight-chain molecules. Natta also discovered a catalyst that formed molecular chains with their parts oriented in certain directions. He used catalytic systems to make a variety of polymers, such as synthetic rubber molecularly identical with natural rubber. Thus, this discovery has paved the way to produce rubbery and textile-like materials [16].

Then, in 1973, The Nobel Prize in Chemistry was divided equally between Ernst Otto Fischer and Geoffrey Wilkinson for their pioneering studies on organometallic chemistry, which was not considered an obvious work; it was a prize in "chemistry for chemists". Independently, Fischer and Wilkinson conceptualized compounds that contained bonds between metal atoms and a whole molecule, arranged in a way, which has not been considered as possible or probable earlier, called sandwich compounds, e.g., ferrocene. They did not prepare the first sandwich but they were the first to grasp the odd nature of the compound and its conceptual importance. In this sense, the Fischer and Wilkinson's studies had much importance for the practical applications, in particular on the development of homogeneous catalysis of the reactions of unsaturated hydrocarbons that is also of considerable industrial importance [16].

In 1975, John Warcup Cornforth and Vladimir Prelog were awarded with the Nobel Prize in Chemistry regarding their contribution on studies of stereochemistry of enzyme-catalyzed reactions and stereochemistry of organic molecules and reactions, respectively. Cornforth investigated the use of enzymes as catalysts in several synthetic transformations; moreover, he studied how an enzyme replaces a certain hydrogen atom in a reaction molecule. However, Prelog investigated the connection between the atoms and organic molecules and how they react yielding other molecules. Prelog also studied organic molecules that occur in two different reversed forms and their reactivity in a reaction. Based on Cornforth and Prelog's evidences and results, from then on it was possible to promote the synthesis of an important family of compounds applied on Pharmaceutical Industry, including natural products [16].

The Nobel Prize for Chemistry in 2001 was shared by William S. Knowles and Ryoji Noyori to highlight the relevance of their work on chiral-catalyzed hydrogenation reactions, and K. Barry Sharpless for his work on chiral-catalyzed oxidation reactions. Sharpless developed a broad set of catalytic oxidation reactions, in particular, chiral epoxidation, asymmetric dihydroxylation, and oxyamination. Sharpless's synthetic oxidation methodologies allowed the manufacture of highly efficient antibiotics, anti-inflammatory drugs, and heart medicines, in addition to many chiral agricultural chemicals [16].

In 2005, the Royal Swedish Academy of Sciences awarded Chauvin, Grubbs, and Schrock for notable work in olefin metathesis, which allows promoting the redistribution of fragments of olefins by the scission and regeneration of carbon-carbon double bond, yielding few sub-products and hazardous wastes, which commented their work as "a great step forward for Green Chemistry". This methodology has been applied in several industrial processes, including the synthesis of new drugs [16].

In 2010, Heck, Negishi, and Suzuki were awarded with the Nobel Prize in Chemistry, at catalysis field regarding the development of palladium-catalyzed cross-coupling reactions. This versatile tool allows creating sophisticated carbon-based molecules being this specific bond the basis of life. Nowadays, the palladium-catalyzed cross-coupling reaction has been used in research worldwide, as well as in the commercial production of pharmaceutical compounds [16].

More recently, the Swedish Academy awarded the Chemistry Nobel Prize 2018 to the scientists Frances H. Arnold, George P. Smith, and Gregory P. Winter. In Arnold's work, she performed the well-designed chemical modification of enzymes, particularly those of the cytochrome P-450 family, to produce powerful biocatalysts for the production of fine chemicals. The utilization of Arnold's enzymes enables the more environmentally friendly manufacture of various chemical substances such as pharmaceuticals, as well as the production of renewable fuels for a greener transport sector [16].

Finally, the most recent Chemistry Nobel Prize, in 2021, was granted to Benjamin List and David W.C. MacMillan, on the design of new precise tools for molecule construction, using organocatalysis, a methodology that avoids the use of polluting metal complexes. Independently, they developed a third type of catalysis, named as "asymmetric organocatalysis", based on the

ONE-POT SEQUENTIAL REACTIONS

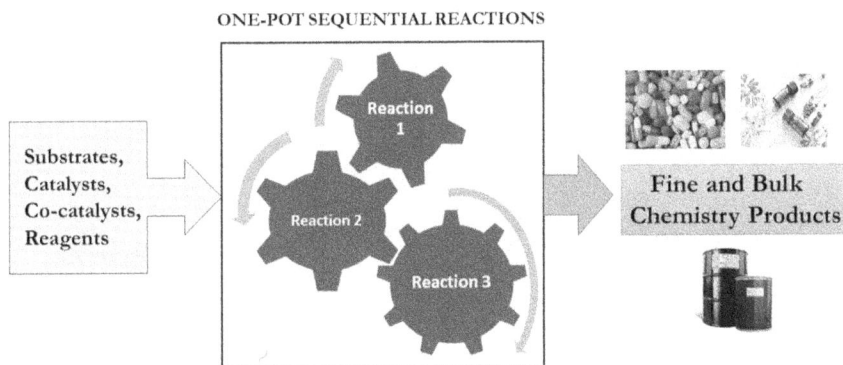

FIGURE 2.4 General procedure for one-pot sequential reactions.

utilization of small organic molecules as catalysts, considered both environmentally friendly and cheap to produce. Their work had high impact on the research for new pharmaceuticals, using greener approaches [16].

The Nobel Prize Lectures of all these scientists have highlighted the relevance of pursuing studies on the development of catalytic reactions to obtain more sustainable chemical industrial processes, preferably using renewable raw materials such as terpenes, CO_2, and O_2 and alternative/benign solvents. However, many environmental challenges, not to mention problems with industrial development, still lie ahead. The solutions are likely to be given by multidisciplinary studies at the interface between chemistry, engineering, physics, and biology [16].

As discussed above, the development of multidisciplinary teams has made an enormous contribution towards the design of more sustainable chemical processes. As an example, in the near future, the development of alternative modern synthetic chemistry is likely to require the development of sequential catalytic reactions, or bifunctional and/or synergic catalysts, preferentially immobilized onto solid supports. The use of sequential reactions is a great challenge, but helps to achieve Green Chemistry goals since they reach significant waste/solvent minimization, avoiding the need for intermediate purification processes, and are, in general, highly atom economic (Principles of Green Chemistry 1, 2, 4, and 5). These one-pot sequential processes are classified as cascade [17], domino [18,19], and tandem reactions [20,21] according to the type of reaction and mechanism involved (Figure 2.4).

Another challenge in using the Green Chemistry principles is the combination of the best features of heterogeneous (high stability and easy recovering) and homogeneous (high selectivity) catalysts via immobilization of the latter onto solid supports [22]. In the literature, several papers describe the immobilization of synthetic or enzymatic homogeneous catalysts onto a variety of solid supports such as polymers [23,24], silica [25,26], mesoporous materials [27,28], carbon materials [29,30], and magnetic nanoparticles (MNPs) [31,32] (Figure 2.5).

2.3 ALTERNATIVE SOLVENTS AND REACTION MEDIA

According to IUPAC, solvent is a class of chemical molecules that are in "liquid or solid phase containing more than one substance, when for convenience one (or more) substance, which is called the solvent, is treated differently from the other substances, which are called solutes" [33], being used in multiple applications such as organic synthesis of pharmaceuticals, pesticides, polymers and commodities, liquid-liquid or solid-liquid extractions, chromatography, paints and coatings, cleaners, inks, etc [34]. The term "solvent" derives from Latin and means "to loosen", which is exactly the opposite principle of Green Chemistry philosophy and explains the motivation for finding alternative solvents. In addition, Sheldon mentioned that approximately 85% of the total mass of chemicals

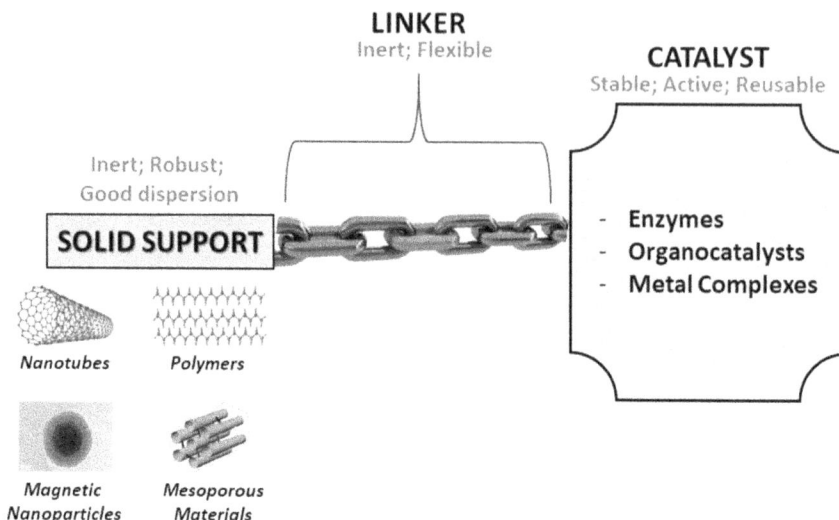

FIGURE 2.5 General approach for immobilization of a catalyst.

involved in pharmaceutical manufacture comprises solvents, and recovery efficiencies are typically between 50% and 80% [35]. Therefore, according to Green Chemistry and Sustainable Technology, the choice of the appropriate solvents requires a paradigm shift. They should be non-toxic, non-hazardous, non-corrosive, non-flammable, and preferentially coming from renewable resources [36]. Furthermore, there is also a strong increase in stringent environmental legislation in Europe [37] and America [38–40].

Many industrial companies have developed guides for appropriate solvent selection and replacement. Solvents are not only used as a reaction medium but also in extractions and cleaning. So, careful reaction *design* that contemplates how solvents can be ideally reused or recovered may have a noteworthy impact on the processes economics and sustainability [41]. Solvents have been assessed in a systematic way considering three main aspects:

   i. Worker safety – including carcinogenicity, mutagenicity, reprotoxicity, skin absorption/ sensitization, and toxicity;
  ii. Process safety – including flammability, potential for high emissions through high vapor pressure, static charge, potential for peroxide formation and odor issues;
 iii. Environmental and regulatory considerations – including ecotoxicity and ground water contamination, potential environment, health and safety (EHS) regulatory restrictions, ozone depletion potential, photo-reactive potential.

The list of Table 2.1 covers the solvents commonly used in medicinal chemistry. For those solvents in the red/undesirable category, solvent replacement is suggested, according to Table 2.2.

Another difficulty associated with solvent use is the energy consumption required for residual solvent removal from final products, which is generally achieved by evaporation or distillation [42]. Moreover, volatile organic solvents (VOCs) are not only a major problem for atmospheric pollution but also a serious worker health problem [43].

All chemical synthetic processes, involving catalysts, in batch or in continuous flow, require the contact between the reactants and the catalyst. For many years, it was believed that a solvent was always required, usually organic, but scientific advances aiming to achieve greener and more sustainable solutions led to several examples of solventless reactions [44], systems involving mechanochemistry [45] or cases where after heating (conventional or microwave) [46], one of the reagents

TABLE 2.1
Pfizer Solvent Selection Guide for Medicinal Chemistry

Preferred	Usable	Undesirable
Water	Cyclohexane	Pentane
Acetone	Heptane	Hexane(s)
Ethanol	Toluene	Di-isopropyl ether
2-Propanol	Methylcyclohexane	Diethyl ether
1-Propanol	t-Butyl methyl ether	Dichloromethane
Ethyl acetate	Isooctane	Dichloroethane
Isopropyl acetate	Acetonitrile	Chloroform
Methanol	2-Methyltetrahydrofuran	Dimethyl formamide
Methyl ethyl ketone	Tetrahydrofuran	N-Methylpyrrolidinone
1-Butanol	Xylenes	Pyridine
t-Butanol	Dimethyl sulfoxide	Dimethyl acetate
	Acetic acid	Dioxane
	Ethylene glycol	Dimethoxyethane
		Benzene
		Carbon tetrachloride

TABLE 2.2
Solvent Replacement Table

Undesirable Solvents	Alternative
Pentane	Heptane
Hexane(s)	Heptane
Di-isopropyl ether or diethyl ether	2-MeTHF or t-butyl methyl ether
Dioxane or dimethoxyethane	2-MeTHF or t-butyl methyl ether
Chloroform, dichloroethane, or carbon Tetrachloride	Dichloromethane
Dimethyl formamide, dimethyl acetamide, or N-methylpyrrolidone	Acetonitrile
Pyridine	Triethylamine (if pyridine used as base)
Dichloromethane (extractions)	Ethyl acetate, MTBE, toluene, 2-MeTHF
Dichloromethane (chromatography)	Ethyl acetate/heptane
Benzene	Toluene

Notes: Dichloromethane is the recommended alternative to other chlorinated solvents (i.e., least bad); dipolar aprotic solvents (*N,N*-dimethyl formamide, *N,N*-dimethyl acetamide, and *N*-methylpyrrolidone) have no satisfactory alternative (acetonitrile is a relatively poor substitute), and replacements are an identified research priority.

would melt, dissolving the remaining ones and the catalyst. In this case, the reaction occurs in the liquid state but without the addition of any additional solvent, as demonstrated by Pereira on the synthesis of bacteriochlorin from porphyrins [47]. This is a big step for reducing waste, although the transposition to a large-scale industrial process is still a problem due to security issues, which is also an important goal of sustainability. In addition, organic solvents are, in general, also required in the work-up and purification processes. Regarding this point, the use of catalysts is very important, since they will improve the selectivity for products and, having less sub-products, chromatography may be avoided and ideally, the target compounds can be purified just by recrystallization [48].

Once again, solvents are required and, to overcome this problem, two industrial changes are necessary: (i) implementation of solvent reutilization process and (ii) change practices and use renewable solvents. Finally, the assessment for the quality of the products is also needed, particularly in pharmaceutical industry, where the use of HPLC-based purifications inputs large amount of organic solvents (Figure 2.6).

The Brundtland Commission of the United Nations presented a report [49], where the problem of solvent usage is a key point to meet in the future a sustainable development, "development that meets the needs of the present without compromising the ability of future generations to meet their own needs." Accordingly, the development of chemical processes should be designed in line with the Green Chemistry principles, and measured by the metrics developed so far (atom economy, E-factor, mass efficiency, mass intensity, life cycle assessment, etc.) [5,50,51]. Regarding the multiple aspects required to improve sustainability of chemical processes, we highlight the replacement of classic VOCs, which are considered as the major contributors for industrial waste streams. Therefore, the use of alternative "green and sustainable solvents", defined by Charles L. Liotta as "A green and sustainable solvent can be defined as a solvent that addresses environmental issues, contributes to the optimization of the overall process, and is cost-effective" is imperative [52]. Figure 2.7 summarizes the alternative solvents that we can envisage to optimize the sustainability of a chemical process.

In a classroom context, the analysis of Figure 2.7 leads students, in general, to answer that the ideal greener industrial processes would be solventless reactions or reactions using water as solvent. However, when asked about the real effect of solvents in the rate and selectivity of some reactions (S_N1 and S_N2 as examples) and about the consequences for the planet of a hypothetical

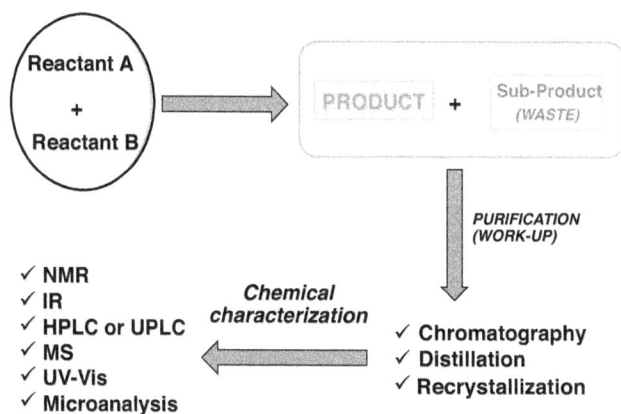

FIGURE 2.6 General process for synthesis, purification, and characterization of a new compound.

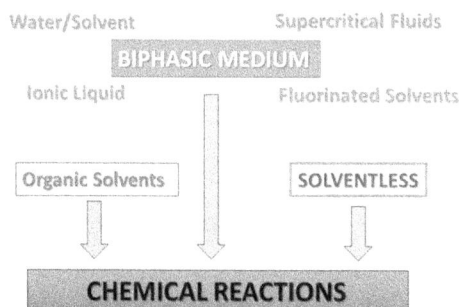

FIGURE 2.7 Alternative reaction media overview.

massive industrial change of solvents to water, the answer switches straightaway to an assumption that this would not be a universal solution, as water would immediately become scarce, compromising the future of multiple communities in the world regarding the need of potable water [53] (Pereira et al., 2019).

Therefore, among other aspects, the search for alternative solvents and reaction processes always should deal with less consumption, less waste (higher selectivity), and higher yields. In the last few decades, Academia and Industry started to analyze the sustainability of the chemical processes regarding the effect of solvents and catalysts in the reaction rates (low energy consumption, 6th Principle of Green Chemistry), product yield and selectivity (Ninth Principle of Green Chemistry), and recycling strategies for catalysts and solvents [54,55]. To that respect, Figure 2.7 shows alternative solvent solutions for implementing a chemical process under a less toxic approach. In the last decades, the use of biphasic solutions based on water: ionic liquid (IL); organic solvent: fluorinated solvent; or organic solvent: CO_2 supercritical are just examples of the scientific ideas to improve the sustainability of a chemical process.

In the next sections, we present selected examples of industries that changed their practices toward the use of catalytic processes in alternative solvents aiming the enhancement of the sustainability of their chemical processes.

2.3.1 Water as Solvent for Catalytic Industrial Processes

Regarding Green Chemistry approaches, water may be considered as an ideal renewable alternative solvent. It is the most abundant molecule on the planet; it is benign, non-flammable, presents no toxicity, and is liquid, at room temperature. Moreover, it is an excellent solvent capable of dissolving many ionic compounds and polar covalent molecules. Humans have used water since pre-historical ages and, for centuries, it was the only solvent available for scientists as organic volatile solvents were made available in large variety and quantities only after the industrial revolution of 19th century [56,57].

Since the water molecule is not linear and the oxygen atom has a higher electronegativity than hydrogen atoms, a permanent dipole is formed, giving to the water molecule its permanent dipolar moment where the oxygen atom carries a slight negative charge, whereas the hydrogen atoms are slightly positive. The high polarity of water allows the efficient solvation of ions. In the ion dissolution process, the dipolar moment of the water molecules is aligned with the ionic species to minimize the energy and each ion dissolved is solvated with several water molecules (Figure 2.8a and b). The structural characteristics of the water molecule make it both good hydrogen bond donor and bond acceptor with high dielectric constant. This property reduces the strength of the electrostatic forces between dissolved ions and allows them to separate and move freely in solution.

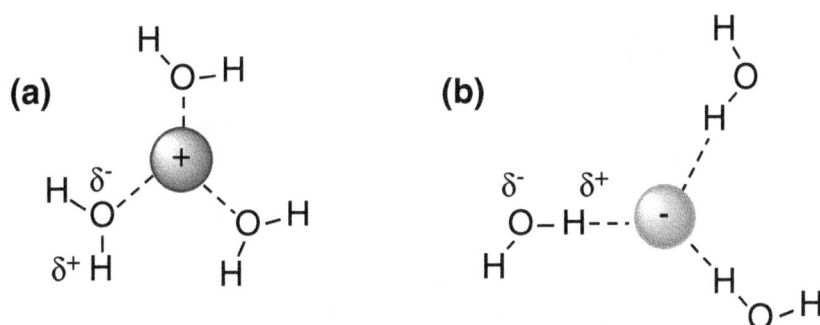

FIGURE 2.8   (a) Cation water solvation, (b) anion water solvation.

TABLE 2.3

Chemical and Physical Properties of Water at 0.1 MPa [56,61,62]

Property	Value
Melting point	0°C
Boiling point	100°C
Critical temperature and pressure	374°C, 22.1 MPa
Density (4°C)	1.00 g/cm^3
Specific heat capacity	4.19 J/g/K
Dielectric constant	78.30
Dipole moment	5.9×10^{-30} C m
Relative permittivity (ε_r)	78.5
α	1.23
β	0.49
π^*	1.14
Donor number	1.46
Acceptor number	54.8

The maximum density value is obtained at 4°C (above its melting point). Ice water is less dense due to the rigid hexagonal and open structure where water molecules exhibit a tetrahedral coordination due to its possible four hydrogen bonds. When ice melts, some of the hydrogen bonds break and the tridimensional structure breaks down to a denser phase. As the temperature of water is raised, its density and polarity decrease, reaching a supercritical state at temperatures above 374°C and pressure over 22.1 MPa [58]. At this point, or even near this point, water starts to assume many of the organic solvents' properties. These effects are attributed to the reduction of the hydrogen bonds at very high temperatures. To explore these water characteristics, some studies have been made, using superheated, subcritical, and supercritical water as solvent [59–61].

Regarding the potential use of water as solvent in an industrial chemical process, the most relevant chemical–physical properties are summarized in Table 2.3. We highlight its high dipole moment (5.9×10^{-30} C m), high polarizability π^*(1.14), and relative high permittivity ε_r (78.5), which are good parameters to measure solute–solvent interactions. The high hydrogen bond donating α (1.23) and accepting β (0.49) parameters are also relevant to explain the use of water as solvent for ionic and polar species. In addition, water also has a high specific heat transfer value that provides a safer handle of exothermic reactions due to the absorption of energy during the chemical process [62].

The above-mentioned properties make water an attractive alternative solvent for use in biphasic catalytic systems, as it readily separates from common organic solvents due to its high polarity and density. There are many solvents that, in combination with water, form biphasic systems, for example, with fluorinated solvents, some ILs, and the great majority of volatile organic solvents [56,63]. Generally, in biphasic systems, the catalyst is dissolved in one phase and the reactants and/or products in another phase, which allows catalyst recovery and reuse by simple phase separation. Ideally, the catalyst solution remains in the reactor without the need of further treatment for the next operation. These biphasic water/organic solvent systems have been one of the most widely used alternative sustainable processes for industrial applications. Water is cheaper than organic solvents, and the reuse of catalysts allowed the development of sustainable and economically attractive catalytic homogenous processes. However, regarding Green Chemistry approaches, some of these processes require the use of phase-transfer agents, which increase the final waste [63,64].

Additionally, the development of metal complexes containing water-soluble ligands is one of the most common strategies to perform catalytic transformations under aqueous biphasic conditions [65].

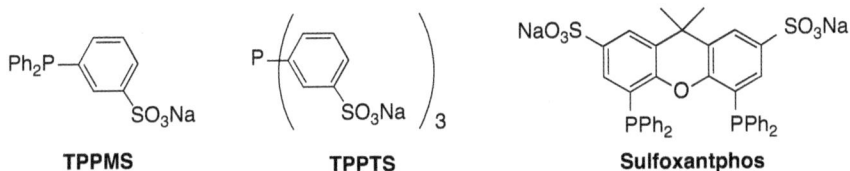

FIGURE 2.9 Structure of popular water-soluble phosphines, where TPPMS = diphenyl(*m*-sulfonatophenyl) phosphine sodium salt, TPPTS = tris(*m*-sulfonatophenyl)phosphine sodium salt, and sulfoxantphos = sulfonated xantphos sodium salt derivative.

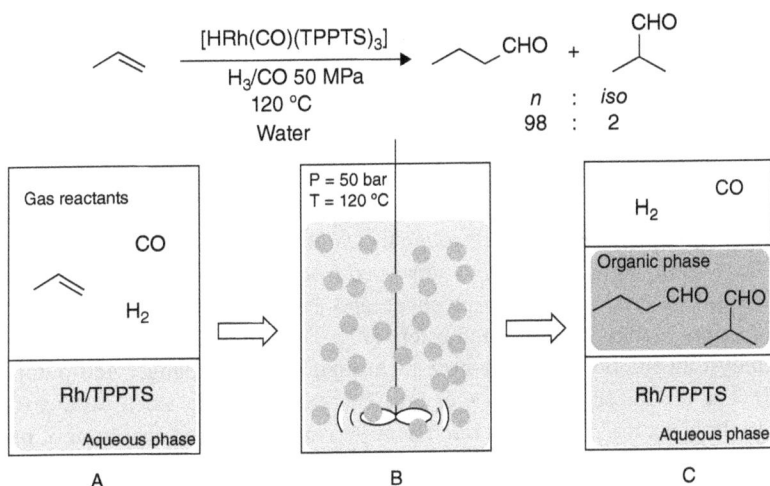

FIGURE 2.10 Simplified OXEA propene hydroformylation system.

For this purpose, several water-soluble phosphine-type ligands have been synthesized (Figure 2.9) [66]. These ligands are usually obtained through the introduction of sulfonic groups in their structures. For instance, the reaction of triphenylphosphine with fuming sulfuric acid yields sulfonated derivatives with $-SO_3^-$ groups in meta positions, which are usually isolated as alkali metal salts (Na^+ or K^+) [66,67].

One of the most emblematic applications of water as solvent in industrial catalytic biphasic processes, and also a milestone in Green Chemistry, is the large-scale biphasic Ruhrchemie-Rhône-Poulenc (now Hoechst Celanese) propene hydroformylation process. This aqueous two-phase process was started at Rhône-Poulenc, in 1975 [68], followed by process development at Ruhrchemie AG, leading to the implementation of the first oxo-plant using a two-phase process in 1984 [69]. The high industrial interest of this sustainable approach led several academic groups to study the synthesis of water-soluble triphenylphosphine trisulfonate (a.k.a. TPPTS) [70–75]. This Ruhrchemie-Rhône-Poulenc oxo-process (now called OXEA process) is the first industrial biphasic hydroformylation system in which the catalyst is present in the aqueous phase (Figure 2.10).

In this system, the Rh/TPPTS is present in the aqueous phase, while the reactants (propene, CO, and H_2) are fed in the gas state (Figure 2.10a). To put both phases in contact and allow the catalytic transformation, the reaction should be carried out under vigorous stirring at, typically, 120°C and 5 MPa (Figure 2.10b) [76]. After the reaction completion, phase separation occurs; butanal is recovered by simple decantation, followed by distillation of the newly formed organic phase, while the catalyst remains in the aqueous phase (Figure 2.10c) of the reactor.

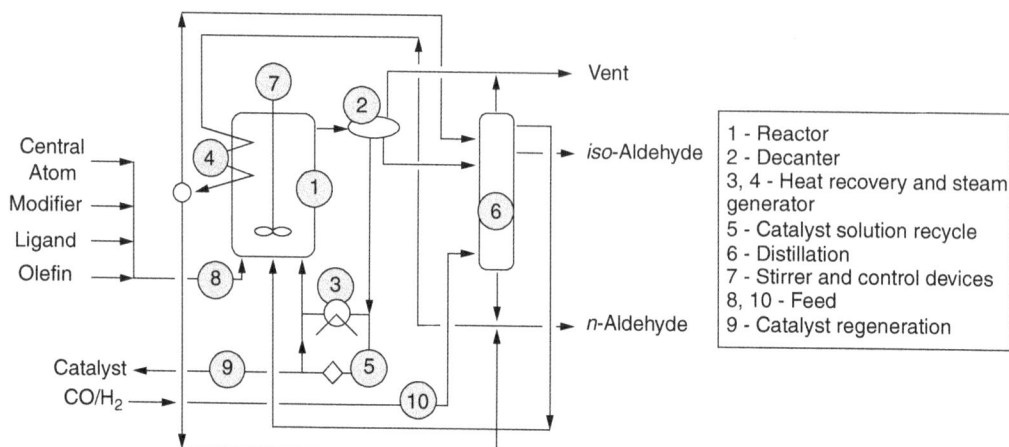

FIGURE 2.11   Detailed schematics of the OXEA propene hydroformylation biphasic industrial process.

The industrial plant scheme is expectably more complex, but quite straightforward to understand (Figure 2.11) [77]. The process is performed in a tank reactor (1) where the olefin and the syngas are added from the bottom of the reactor through the catalyst phase (aqueous phase) under intensive stirring (7). The resulting aldehyde crude is separated from the aqueous phase through decantation of the top phase (2). A mixture of butyraldehyde and isobutyraldehyde in 98:2 ratio is obtained with few by-products such as alcohols and esters [77]. Then, the aqueous solution containing the Rh/TPPTS catalyst (at the bottom) is re-heated through a heat exchanger and pumped back into the reactor (5). The olefin excess and syngas are separated from the aldehyde phase in a stripper and fed back to the reactor (3). The generated heat (4) is used for the generation of process steam, which is used for subsequent distillation (6) of the crude aldehyde phase to separate butyraldehyde (bp = 75°C) from isobutyraldehyde (bp = 63°C).

Regarding Green Chemistry principles, we may consider this biphasic propene hydroformylation process as outstanding, since it includes the inherent biphasic catalysis advantages (prevent waste, atom economy, reduction of derivatives, reuse of catalysts) along with the use of water as alternative solvent (prevent waste, less hazardous synthesis, benign solvents). Additionally, there is a great overall energy conservation throughout the process (design for energy efficiency). The plant operates under milder reaction conditions when compared to conventional methods (pressure and temperature), and produces very little amounts of waste once the catalyst is reused, and the loss of Rh is very low (parts per billion range). The process has nearly 100% atomic economy and is highly chemo- and regioselective, providing around 800,000 tons of n-butanal a year. This aldehyde is a key starting material for conversion into 2-ethylhexanol, used as plasticizer, diesel fuels, and lubricant additives [78]. However, problems associated with contact between catalyst (water phase) and reagents (organic phase) limited the application of this system to short-chain alkenes. So, it is still a great challenge to find appropriate solvents/phase-transfer agents to enlarge the process to longer-chain alkenes. Nevertheless, this system remains as one of the best achievements for the application of alternative solvent systems in real industrial homogeneous catalysis.

Boosted by such case of success, a number of other processes have been scaled up and are used at the industrial level. Rhodia (former Rhône-Poulenc) developed a biphasic aqueous system for the production of vitamin E precursors *via* biphasic catalytic C-C coupling process, using again Rh/TPPTS as water-soluble catalyst [79–81]. In this process, the myrcene and acetylacetone are activated by the Rh catalyst to give C_{13} geranylacetone ester; the reaction proceeds in a biphasic liquid/liquid system where conversion is regulated by stirring time (Figure 2.12a). The products are removed from the reactor through decantation of the organic phase, and the catalyst remains in the aqueous phase in a similar way of the previous example.

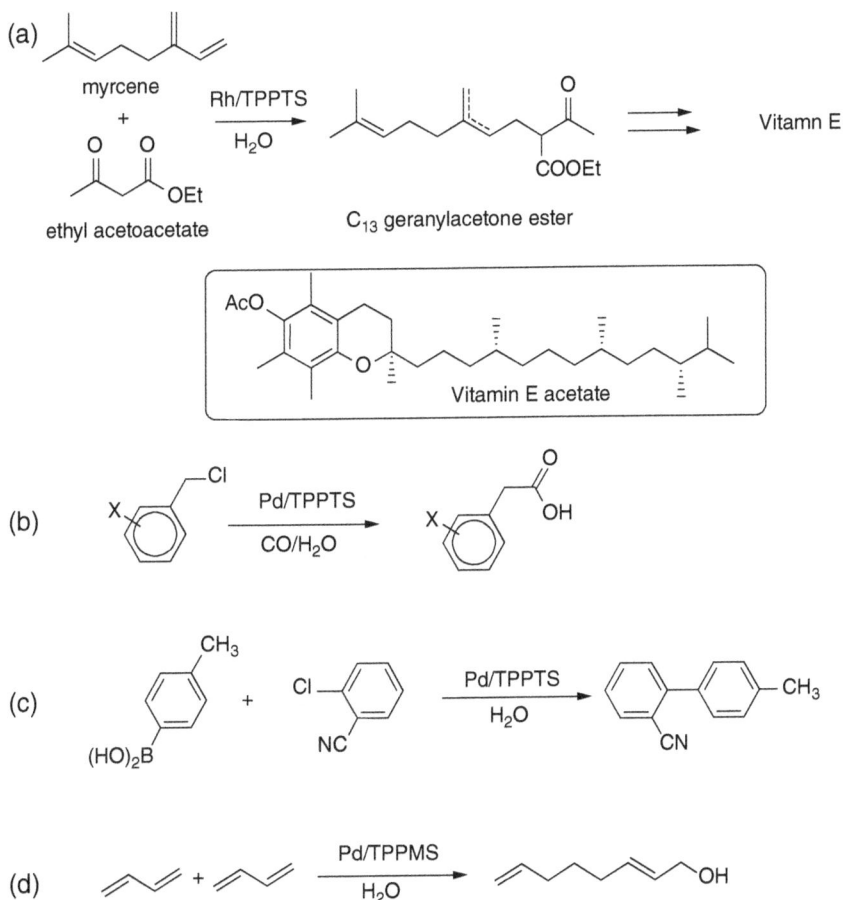

FIGURE 2.12 Industrial reactions carried out under biphasic conditions.

Hoechst Celanese (at the time still as Ruhrchemie-Rhône-Poulenc) developed a process for preparing phenylacetic acid and derivatives [82,83], using Pd/TPPTS as water-soluble catalyst for the carbonylation of benzyl chloride (Figure 2.12b). In this process, the benzyl halide is dissolved in an organic solvent and the Pd/TPPTS in an alkaline aqueous solution, with pressure and temperature up to 3 MPa CO and 90°C, respectively. Moreover, to neutralize the strong acid liberated during the carbonylation reaction (HX, X=Cl, Br, I), an excess of base is necessary, allowing also the separation of the product phenylacetic acid sodium salt in the aqueous phase. Thus, the product can be easily isolated by acidification of the water phase and extraction of the neutralized phenylacetic acid with an organic solvent. In this alternative process, a significant decrease in by-products and a solid kilogram cost reduction were achieved.

Another relevant example of application of water as alternative solvent, in industrial processes, is the use of Pd/TPPTS as catalyst for Suzuki coupling in a biphasic system, developed by Clariant AG, where a set of biphenyl derivatives (angiotensin (II) inhibitors) has been prepared (Figure 2.12c) [84].

Kuraray Corporation also uses a continuous biphasic aqueous catalytic system (water and a selected organic solvent – alcohols, ethers, sulfoxides, and others) to promote the hydrodimerization of butadiene, catalyzed by palladium/monosulfonated triphenylphosphine (Pd/TPPMS) to produce octa-2,7-dien-1-ol (Figure 2.12d) [85]. To avoid the formation of polymers as secondary products, the control of butadiene feed is required. Finally, the octa-2,7-dien-1-ol product is extracted and

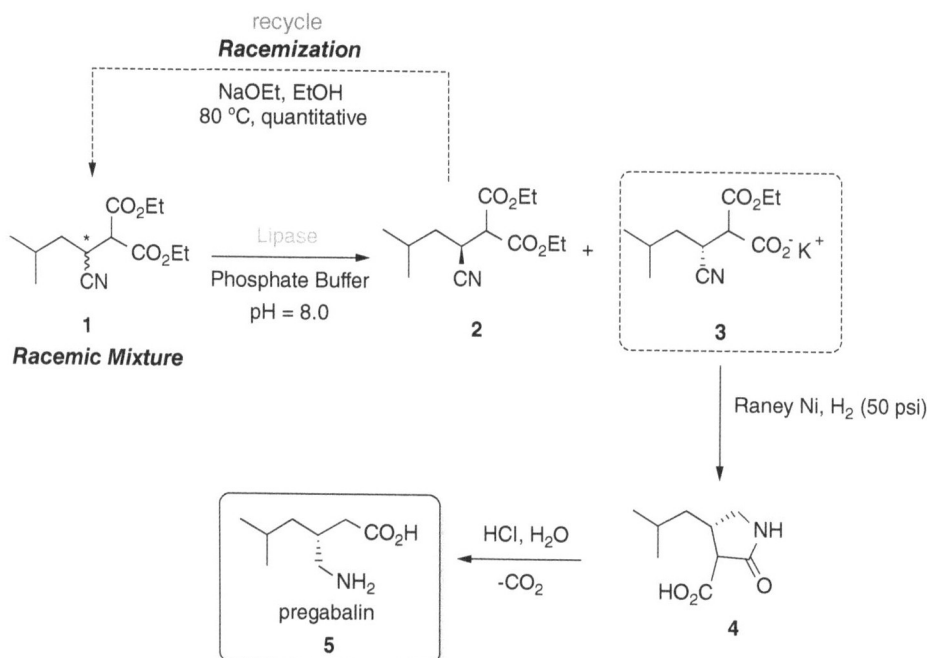

FIGURE 2.13   Water-based synthesis of pregabalin catalyzed by lipase.

isolated by distillation (any unreacted butadiene is recovered and fed back to the reactor). The resulting aqueous layer containing the water-soluble catalyst is then reintroduced in the reactor.

The above examples clearly demonstrate the relevance of the development of water-soluble catalysts to enhance the application of alternative aqueous biphasic systems in catalytic industrial processes. Other pertinent cases where water plays a key role as alternative solvent in industry are in enzymatic kinetic resolution processes [86] and in biocatalytic reactions, for the production of commodities, pharmaceuticals, and food additives [87]. Such relevance is well evidenced by the nearly two hundred processes that have been implemented by industry or that are in the edge of being commercialized [88].

As a selected example, we highlight the synthesis of pregabalin 5 ((S)-(+)-3-aminomethyl-5-methyl-hexanoic acid (Figure 2.13), marketed under the trade name Lyrica® as an anticonvulsant agent (exhibiting affinity to the human $\alpha_2\delta$ calcium channel). Its first synthetic process involved more than ten steps, using expensive and toxic reagents and solvents [89]. Alternatively, pregabalin was prepared in a lipase (extracted from the microorganism *Thermomyces lanuginosus*)-mediated water biocatalytic process, which was able to promote the enzymatic kinetic resolution of the racemic cyanodiester 1 (Figure 2.13) [90–93].

This biocatalytic process, carried out through enzymatic resolution of the racemic mixture of 3-cyano-2-ethoxycarbonyl-5-methyl-hexanoic acid ethyl ester, is performed in aqueous potassium phosphate buffer medium (pH=8.0) for 24 hours. The lipase is able to enantioselectively hydrolyze the ester of (S)-enantiomer (water soluble), which is then used as starting material for the subsequent steps. The unreacted (R)-enantiomer diester was treated with a strong base (NaOH), which, by abstraction of the acidic α-proton of the malonate moiety, induces racemization and regeneration of racemic substrate. The aqueous solution containing the (S)-enantiomer was concentrated under vacuum and submitted to hydrogenation/cyclization reaction using heterogeneous Raney/nickel (50% aqueous solution) and H_2 (50 psi) for 20 hours. The Raney Ni catalyst was recovered by filtration; the pH of the previously used solution was adjusted to 3.0 using 37% HCl, followed by extraction with ethyl acetate. Finally, the solution was concentrated under vacuum, affording the

(S)-4-isobutyl-2-oxo-pyrrolidine-3-carboxylic acid in 40%–42% isolated yield and *ee* >97%. Lastly, its acidic hydrolysis (HCl 37%) yields the pregabalin in 80%–85% and *ee* >99.5% (Figure 2.13). It should be noted that the recycling/racemization step of the unreacted diester is clearly in line with Green Chemistry philosophy, reducing the amount of sub-product by 50%, thereby improving the E-factor of the overall process (the E-factor is the ratio of the mass of waste per mass of product: E-factor=kg of waste/kg of product) [94].

This process is a definite example of how traditional chemical resolution can be substituted by a biocatalytic water process avoiding the use of toxic and volatile organic solvents. It should be noted that, in typical classic processes for the production of active pharmaceutical ingredients (API), 80% of the waste are solvents [95].

Another example on the application of biocatalysis in aqueous medium, applied by pharmaceutical industry, is the synthesis of atorvastatin calcium, the first drug having annual sales over ten billion dollars, being considered *"the goose that laid the golden eggs"* of Pfizer company for many years.

The atorvastatin calcium is a potent hypolipidemic and/or hypocholesterolemic agent [96]. The key step for the synthesis of atorvastatin is the preparation of chiral synthon 10 (Figure 2.14). The original synthesis includes six steps – most of them are carried out under strict conditions. To overcome the use of toxic/expensive reagents and polluting solvents, a greener approach was later developed, using a microbial deoxyribose-5-phosphate aldolase (DERA) enzyme [97]. This enzyme is able to catalyze the sequential aldol condensation between benzyl chloroformate (CBz) N-protected aminopropylaldehyde and acetaldehyde, yielding the corresponding lactol with excellent *ee* (98%) and *d.e.* (97%). This intermediate was then converted through catalytic dehydrogenation using Pt/C as catalyst, air or oxygen as terminal oxidant, and at pH 8.0, into the corresponding carboxylic acid, followed by protection, esterification, and hydrogenation, which affords the useful atorvastatin intermediate A. Finally, the API atorvastatin is obtained by Paal-Knorr condensation of A with diketone A, followed by deprotection, lactonization, and finally, lactone ring opening (Figure 2.14). This alternative chemoenzymatic synthetic route of atorvastatin is significantly shortened, allowing the reduction of hundreds of metric tons of raw materials and solvents each year, concomitantly with a significant reduction in energy consumption.

In sum, despite the apparent attractiveness regarding the replacement of classic volatile organic solvents by water, there are still several issues to overcome, when considering the overall sustainability of water-based catalytic industrial processes. These concerns are as follows: (i) most of the industrial substrates/catalysts are not water soluble and, consequently, catalyst's structural modification is required (involving the use of less benign materials/solvents); (ii) water has a high boiling point and consequent product isolation/purification needs distillation (high energy consumption); and (iii) the hypothetical massive consumption of water by industries, all over the world, would cause a calamitous potable water shortage.

2.3.2 Ionic Liquids in Industry

ILs are organic salts composed by ions with melting point below 100°C. The first reported IL was described in 1914, by Paul Walden (ethylammonium nitrate) [98]. There are five main IL categories, based on their organic cation: (i) N-alkylpyridinium, (ii) 1,3-dialkylimidazolium, (iii) N, N-dialkylpyrrolidinium, (iv) tetraalkylammonium, and (v) tetraalkylphosphonium (Figure 2.15). Moreover, there are nine main IL types, based on their organic anion: (i) hexafluorophosphate $\left(PF_6^-\right)$, (ii) bis(trifluoromethylsulfonyl)imide $\left(N\left(SO_2CF_3\right)_2^-\right)$, (iii) tetrafluoroborate $\left(BF_4^-\right)$, (iv) trifluoromethylsulfonate $\left(CF_3SO_2^-\right)$, (v) ethanoate $\left(CH_3CO_2^-\right)$, (vi) methanesulfonate $\left(CH_3SO_2^-\right)$, (vii) nitrate NO_3^-, and (viii) halide (e.g., Cl⁻) (Figure 2.15) [99]. Although the presented IL families are the most widely used as alternative solvents [100,101], as electrolytes [102] and as catalysts [103–106], there is still a wide room of possible anion/cation combination to synthesize ILs. This is well illustrated by Plechkova and Seddon sentence: "If there are one million possible simple systems, then there are one billion (10^{12}) binary combinations of these, and one trillion (10^{18}) ternary systems possible!!" [107].

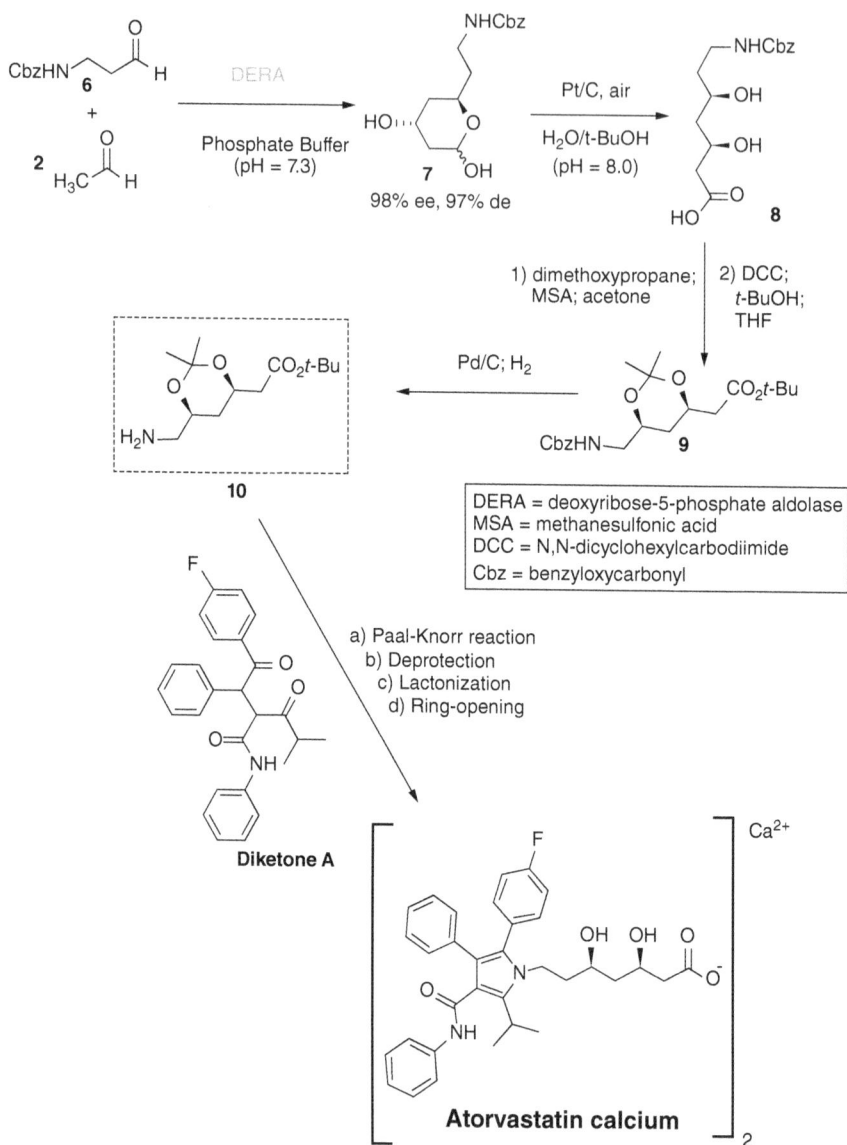

FIGURE 2.14 Intermediate synthesis of atorvastatin in aqueous medium catalyzed by deoxyribose-5-phosphate aldolase enzyme.

Their relevant physical–chemical properties [108–114] make them particularly suitable to be used as substitutes for classic volatile organic solvents in catalysis, namely:

- High solvating ability enhances the solubility of organometal catalysts, polar organic molecules, and gases;
- Absence of vapor pressure minimizes the exposure to potential toxic vapor;
- Low melting point and high thermal stability (decomposition > 350°C);
- Easy modulation of physical and chemical properties by fine tuning of their structures;
- Non-flammability, thus safer to use;
- Easily recyclable through physical separations;
- Polar nature, but non-coordinating species.

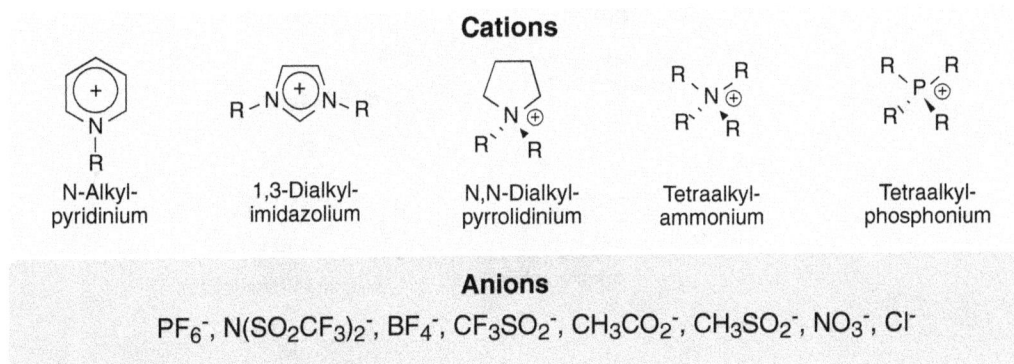

Cations

N-Alkyl-pyridinium   1,3-Dialkyl-imidazolium   N,N-Dialkyl-pyrrolidinium   Tetraalkyl-ammonium   Tetraalkyl-phosphonium

Anions

PF_6^-, $N(SO_2CF_3)_2^-$, BF_4^-, $CF_3SO_2^-$, $CH_3CO_2^-$, $CH_3SO_2^-$, NO_3^-, Cl^-

FIGURE 2.15 Most common cation/anion combination of IL.

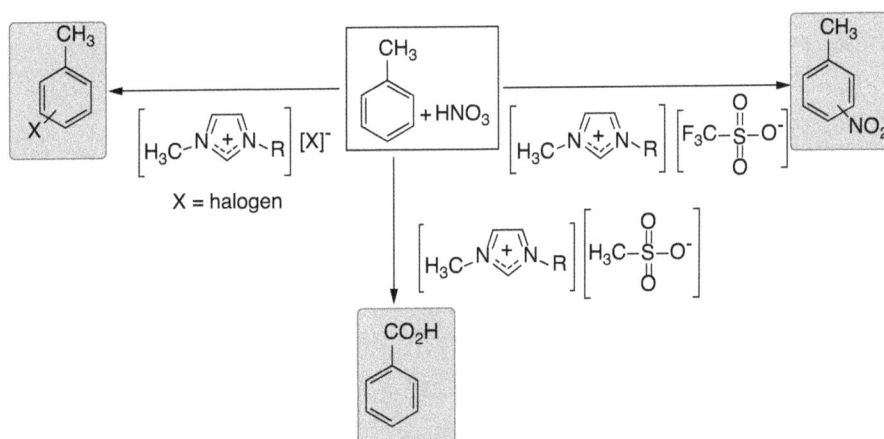

FIGURE 2.16 Effect of the IL structure on product selectivity: reaction of toluene with nitric acid, performed in different ionic liquids, always reaching yields over 99%.

Therefore, ILs can be considered excellent candidates for industrial alternative solvent applications [115]. However, it should be noted that the structure/physicochemical properties of ILs may also control the reaction outcome. A notable example is the reaction between toluene and nitric acid that gives three different products, depending on the cationic and anionic combination of ILs [116]. The use of imidazolium triflate (IL) gives nitrated products, imidazolium methanesulfonate IL promotes the oxidation of toluene to benzoic acid, while imidazolium halogenated ILs, acting as reagent/catalyst/solvent, give mono-halogenated toluene derivatives (Figure 2.16). Moreover, imidazolium methanesulfonate and triflate ILs hold the high advantage of easy recovery and reusability by simple distillation or solvent/solvent extraction.

One of the first applications of ILs in industrial processes was developed by the Eastman Chemical Company, using the thermally stable trioctyl(octyldecyl) phosphonium iodide as solvent for the isomerization of 3,4-epoxybut-1-ene to 2,5-dihydrofuran (Figure 2.17a). This process was implemented in a 1,400 metric tons/year plant that operated in the period 1994–2004 [107,117]. Furthermore, the high solubility of the Sn catalyst in this IL allowed its recovery through the simple liquid-liquid extraction.

Degussa developed a biphasic process for polydimethylsiloxane modification by hydrosilylation, in which $H_2[PtCl_6]$ was combined with 1-butyl-4-methylpyridinium tetrafluoroborate IL, allowing its easy reuse concomitantly with the separation/isolation of the insoluble polymeric product (Figure 2.17b) [118].

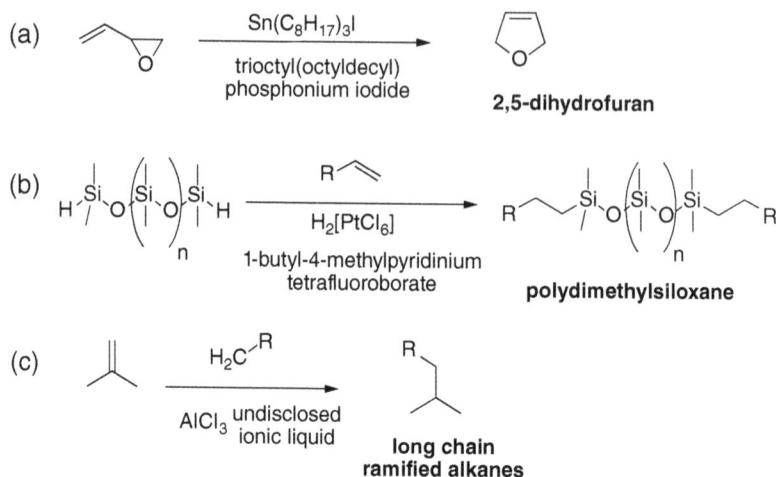

FIGURE 2.17   Selected industrial-catalyzed reactions involving the use of ionic liquids as solvents.

In 2006, PetroChina announced a process, called "ionikylation", in which an undisclosed IL containing the Lewis acid $AlCl_3$ is used as solvent and catalyst for the alkylation of isobutene (Figure 2.17c). After a pilot plant evaluation, the process was transferred to their China sulfuric acid-based alkylation unit (65,000 ton/year) [119]. The process upgrade not only increased the overall yield, when compared to the sulfuric acid-catalyzed reaction, but also increased the process capacity to 91,000 ton/year. Remarkably, in 2019, this was the largest commercial use of ILs, with PetroChina having recently enlarged their capacity to 300,000 ton/year production.

Another emblematic case of industrial utilization of ILs refers to the Difasol process, developed by the IFP (Institut Français du Pétrole), for the dimerization of alkenes in an IL factory plant and marketed by Axens, a subsidiary of IFP. The original process, Dimersol process, consisted in the direct butenes dimerization, without any addition of solvent, catalyzed by a cationic nickel complex of the general form $[PR_3NiCH_2R'][AlCl_4]$, developed by Nobel laureate Yves Chauvin at IFP (Figure 2.18) [120]. Generally, the dimeric products are further hydroformylated, reduced to alcohols, and then converted into higher-value molecules.

Conversely, the Difasol aqueous-organic biphasic process, further developed by Chauvin and Olivier-Bourbigou, used a chloroaluminate(III) IL (2:1 molar ratio $AlCl_3$:tetramethylammonium chloride) as solvent for the nickel-catalyzed dimerization reactions, but with a much higher catalyst activity and dimer selectivity [121], which allowed the use of a smaller reactor (Figure 2.18.). This improved biphasic system enables the easy product separation (organic phase) from the IL phase containing the catalyst. The catalyst remains active and selective despite the presence of the IL.

Besides the above-mentioned industrial catalytic examples, the widespread use of ILs has not been implemented since they are difficult to obtain in highly pure form, they are quite expensive to synthesize, a few are water and/or oxygen sensitive, and they may be leached into the product phase. An additional problem for the industrial application of ILs as solvents relates to the moderate solubility of gas molecules (CO, CO_2, O_2, or H_2) used in many catalytic industrial processes [122].

2.3.3   CARBON DIOXIDE AND OTHER SUPERCRITICAL FLUIDS IN INDUSTRIAL CATALYSIS

A supercritical fluid (SCF), first reported by LaTour in 1822 [123], is any chemical substance that exists at pressures and temperatures above its critical values, where distinct liquid and gas phases do not exist. They can flow out through solids like a gas or dissolve materials like a liquid [124]. From the analysis of a typical phase diagram for a pure substance (Figure 2.19), the critical point is

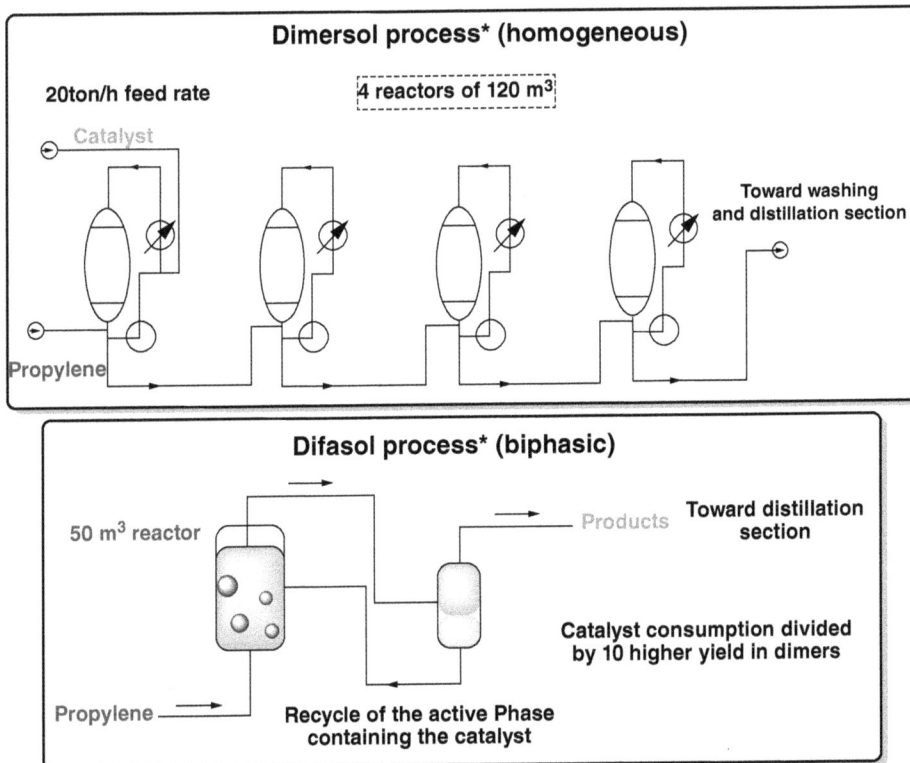

FIGURE 2.18 Schematics of Dimersol and Difasol butane dimerization processes.

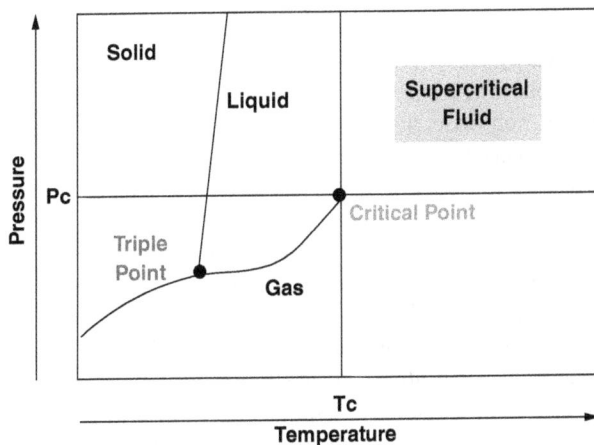

FIGURE 2.19 Typical phase diagram for a pure component.

the highest temperature and pressure for which a substance exists in equilibrium between vapor and liquid phases. Thus, in a closed system, the rise of both temperature and pressure causes a decrease in the liquid density, as result of thermal expansion, and an increase in the gas density. So, the densities of both phases are identical at the critical point, becoming indistinguishable beyond this point, creating a SCF [125,126].

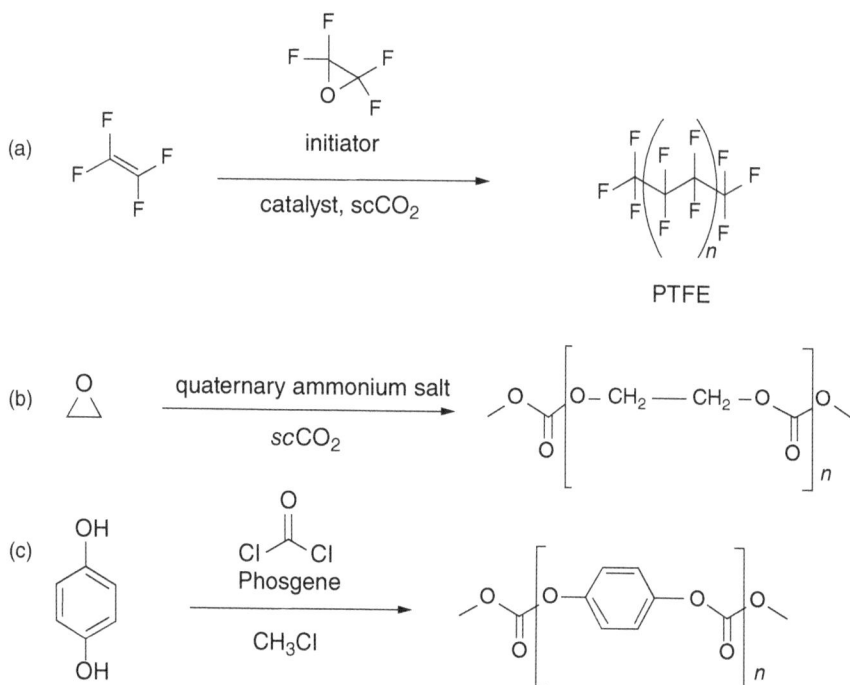

FIGURE 2.20   (a) Polytetrafluoroethylene synthesis, (b) alternative polycarbonate synthesis using scCO$_2$, (c) polycarbonate conventional synthesis using phosgene.

The unique physical–chemical properties of SCFs make them excellent candidates for the replacement of classic organic solvents, particularly for use in biphasic catalytic reactions [127]. They have been applied in several industrial processes, including synthetic and extraction procedures, as, for instance, the extraction of caffeine from coffee, using supercritical carbon dioxide (scCO$_2$), first described by Zosel [128]. This is the most applied SCF as alternative solvent, reaching the supercritical phase at temperature above 31°C and pressure above ~7.3 MPa [129–131]. Moreover, it is inexpensive, non-toxic, non-volatile, non-flammable, and relatively inert and not less relevant; it is a renewable material.

Several catalytic processes using scCO$_2$ as solvent have also been industrially developed – being the polymerization processes the mostly used (Figure 2.20). For instance, DuPont, inventor of the well-known polytetrafluoroethylene (PTFE – TEFLON™) polymers, replaced in 2000 the classic toxic and environmentally polluting solvents (chlorofluorocarbons) for the polymerization process by scCO$_2$, diminishing typical side reactivity [132–134]. Later, in 2016, Chemours company, a DuPont's subsidiary, has become the biggest world producer of fluoropolymers, reaching 1.3 billion dollars revenue, using scCO$_2$ as alternative solvent (Figure 2.20a).

Another significant example for the use of scCO$_2$ at the industrial level is the polycarbonate production from alkene oxides, implemented by Asahi Kasei Corporation [135], which manages to deliver 50,000 ton/year. This alternative process (Figure 2.20b) avoids the use of highly hazardous conventional phosgene-based process (Figure 2.20c), along with the reduction of CO$_2$ emissions.

Moreover, scCO$_2$ has been also widely used as alternative solvent for organometallic-based catalytic reactions at the industrial level, including hydrogenation, oxidation, and metathesis. For instance, Thomas Swan and Company patented (2002) the continuous hydrogenation of several organic compounds using scCO$_2$ as solvent in a 1,000 ton/year plant [136,137]. Hoffmann-La Roche patented the synthesis of (D, L)-α-tocopherol, a pharmaceutical synthon of vitamin E,

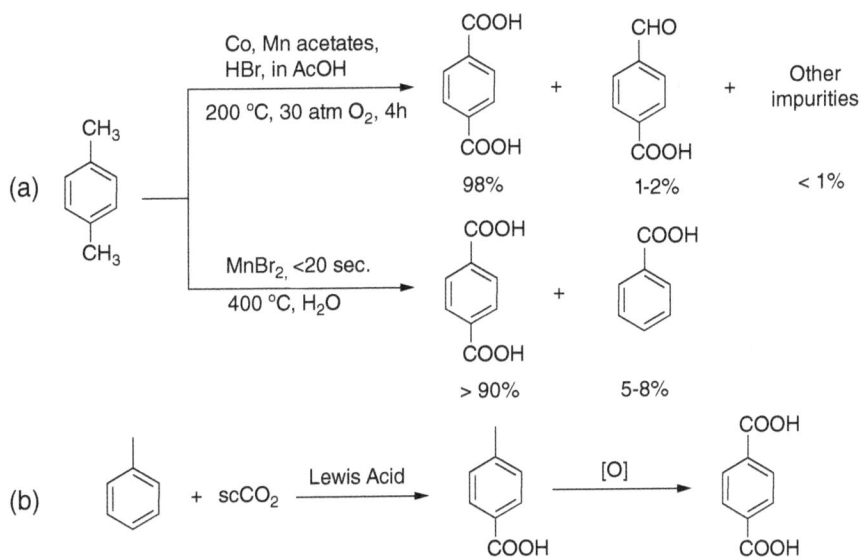

FIGURE 2.21 (a) Amoco's terephthalic acid synthesis, (b) alternative Samsung's terephthalic acid synthesis using scCO$_2$ as solvent and reactant.

by coupling of trimethyl hydroquinone with isophytol, using zinc chloride as catalyst, in scCO$_2$ [138].

In addition, the industrial carboxylation of toluene using scCO$_2$ for the production of terephthalic acid, patented by Samsung Petrochemical [139], in substitution of the previous Amoco process is also a relevant example [140]. While Amoco relied on the use of toxic manganese and cobalt salts in the oxidation of *p*-xylene (Figure 2.21a), Samsung Petrochemical implemented the carboxylation of toluene, using scCO$_2$, to give *p*-toluic acid, which was further oxidized to terephthalic acid. The inventors showed that a Lewis acid mixture of aluminum chloride and aluminum powder could activate CO$_2$ at supercritical conditions to form *p*-toluic acid from toluene, with excellent yield and selectivity (Figure 2.21b).

Boehringer Ingelheim patented the use of scCO$_2$ simultaneously as solvent and protecting group, in the continuous olefin ring-closing metathesis (RCM) for the preparation of the insect repellent epilachnene. As secondary amines tend to poison the metathesis ruthenium catalyst used, the author overcame this problem by using scCO$_2$ as *in situ* amine-protecting group. Then, the ruthenium-catalyzed RCM reaction could be efficiently performed, followed by carbamic acid deprotection (Figure 2.22) [141].

The use of SCFs in the synthesis of polyamides, particularly Nylon 6, was patented by DeSimone (North Carolina, University) [142], but was never translated to industry. The conventional manufacture of caprolactam, the precursor for Nylon 6, consists of two steps: reaction of cyclohexanone with hydroxylamine followed by Beckman rearrangement. The amount of ammonium sulfate produced as a by-product (5 kg/kg caprolactam) is considered very large; consequently, this process has a large E-factor (Figure 2.23a). An alternative synthetic route (Figure 2.23b) was patented by Invista, having a lower E-factor as it involves the formation of the lactam from 6-aminocapronitrile, followed by hydrolysis in scH$_2$O in less than 2 minutes [143].

The use of SCFs, namely, scCO$_2$, still withstands several challenges. The fabrication of large-scale apparatuses designed to function at quite high pressures implies a significant energy consumption. Moreover, additives and complex modifiers are currently needed to increase the solubility of the substrates in scCO$_2$ – being some of them based on volatile organic solvents.

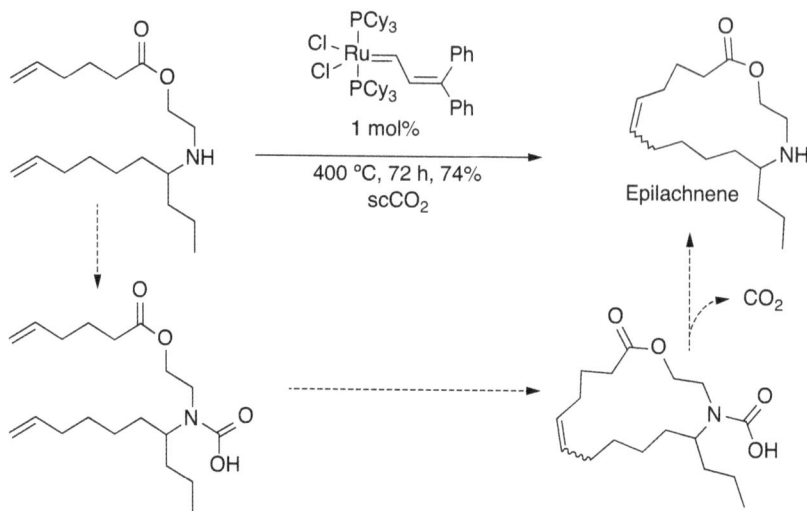

FIGURE 2.22 Metathesis under supercritical CO₂. No protection of secondary amine is necessary.

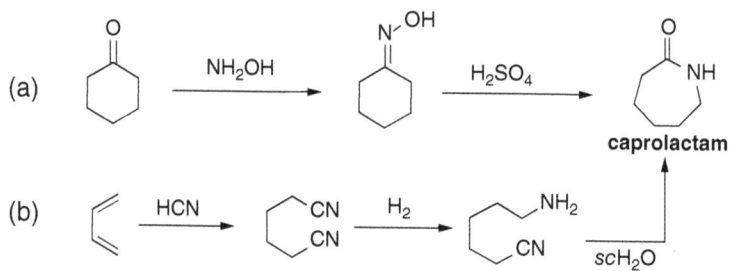

FIGURE 2.23 (a) Conventional caprolactam synthesis, (b) alternative caprolactam synthesis using water in supercritical conditions.

2.3.4 Renewable Solvents in Industry

The evolution from traditional linear economy to circular economy is mandatory, where recovery, regeneration, recycling, and use of renewable resources must be key features in all chemical processes [144]. So, the chemical industry is also considering the use of renewable materials and the ones obtained from waste valorization, as relevant sources for replacing the classic organic volatile solvents [145]. The primary sources of renewable solvents arise from plant biomass, including (D)-limonene extracted from citrus waste [146], γ-valerolactone extracted from lignocellulosic biomass [147], and 2-methyltetrahydrofuran (2-MeTHF) extracted from corn and sugar cane [148] (Figure 2.24).

To analyze the potential of bio-derived solvents for replacing n-hexane, some of the physical properties of n-hexane and (D)-limonene are presented in Table 2.4. The advantages of the use of (D)-limonene versus n-hexane are evident: it has lower vapor pressure and higher flash point, preventing solvent evaporation and fire risks. Moreover, the significantly higher boiling point of (D)-limonene widens the industrial reaction range, while its similar surface tension and water solubility allow its direct replacement in homogeneous or biphasic reaction media [149].

These relevant physical properties induced the industry to promote the replacement of toxic solvents (methyl ethyl ketone, xylene, CFCs, and some halogenated hydrocarbons) in manufacturing

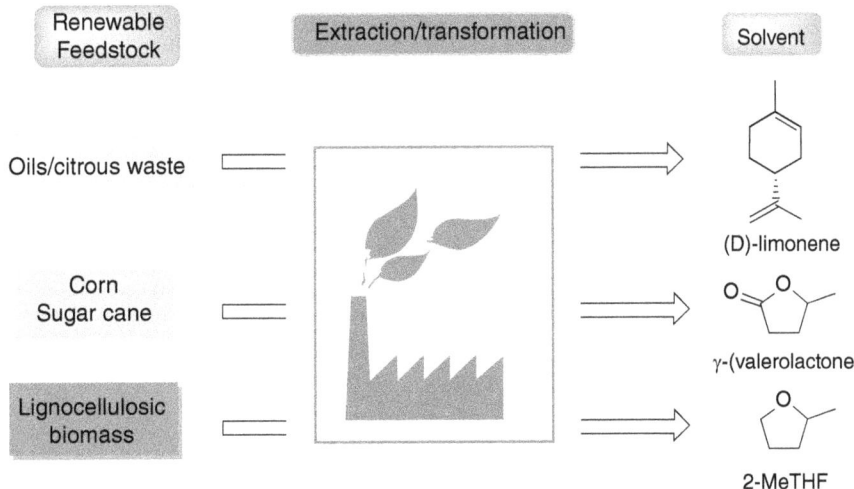

FIGURE 2.24   Structure and renewable source of (D)-limonene, γ-valerolactone, 2-MeTHF.

TABLE 2.4
Physical Properties of *n*-Hexane *vs* (D)-Limonene [149]

Property	*n*-Hexane	(D)-Limonene
Boiling point (°C)	69	176
Flash point (°C)	−22	48
Vapor pressure[a] (MPa)	2	0.2
Density[a] (g/mL)	0.655	0.841
Viscosity[a] (cP)	0.30	0.90
Surface tension[a] (dyn/cm)	17.9	25.8
Solubility in water (g/L)	0.011	0.020

[a] at 25°C.

and/or cleaning processes. Additionally, γ-valerolactone, a derivative from biomass, received particular attention as industrial additive for cleaning processes and fuel production.

Nevertheless, 2-MeTHF is clearly the mostly used renewable solvent with robust applications at industrial level, including pharmaceuticals, as depicted by tens of thousands patents registered so far for its use as alternative reaction solvent. Below, we include a few selected examples. 2-MeTHF has been used as solvent for the synthesis of Grignard reagents containing benzyl and allyl magnesium halides (Figure 2.25a), by Chemetall [150] and Pfizer [151].

Takeda pharmaceutical patented the substitution of THF by 2-MeTHF for the preparation of fused imidazole compounds, through Reformatsky's reaction. They described as additional advantage, i.e., the high solubility of the catalyst $ZnBr_2$ in 2-MeTHF (Figure 2.25b) [152].

Given the moderate boiling point, stability against nucleophiles, and low water solubility, 2-MeTHF is often used for dichloromethane replacement, with particular interest in biphasic reactions. In this context, Pfizer patented a two-phase reaction to promote the deprotection of *tert*-butyloxycarbonyl group (Figure 2.25c) [153,154].

Other examples of protection-deprotection biphasic reactions using 2-MeTHF have been patented by Vertex Pharmaceuticals and Lonza, including the synthesis of ataxia telangiectasia and

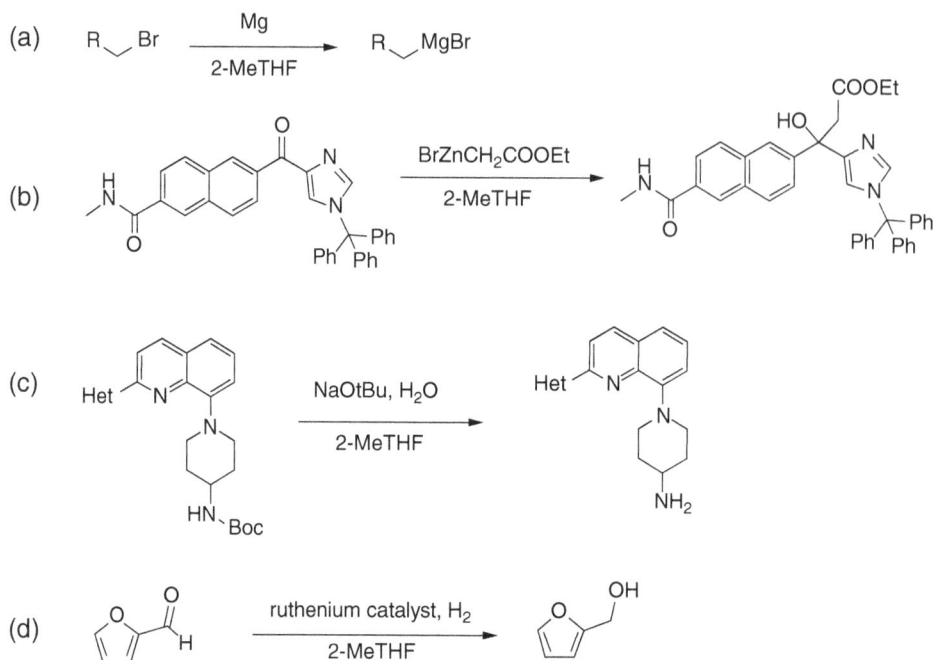

FIGURE 2.25 Industrial processes using 2-MeTHF as solvent (a) Grignard synthesis, (b) Takeda's imidazole synthesis, (c) NH$_2$-protecting group removal, (d) selective aldehyde hydrogenation.

rad3 (ATR)-related kinase inhibitors (aminopyrazine-isoxazoles) [155] and difluoromethylornithine derivatives, respectively [156]. Additionally, 2-MeTHF was also used as solvent for organometallic-catalyzed reactions, namely, hydrogenation for the large-scale preparation of allyl-derived alcohols, using a ruthenium-based catalyst and hydrogen gas (Figure 2.25d) [157], Sonogashira [158], Suzuki-Miyaura [159–161], Ni-catalyzed coupling reactions [162], and lithium-based reactions [163,164].

Several attempts are being made to replace most of the classic VOCs by alternative bio-based solvents. However, several of these new solvents are still VOCs and toxic, still having a long way from being considered as perfect green solvents. Moreover, research is still needed to promote the real implementation of most of these green solvents obtained from renewable sources, in order to translate their use to industrial processes. Moreover, we consider that the use of bio-based solvents is only an alternatively sustainable option, if they are obtained from waste, and not from intensive agricultural processes, which can damage related ecosystems.

2.4 SOLVENTLESS REACTIONS

"The best solvent is no solvent". Who does not remember this sentence, stated by Sheldon? [94] In accordance, many industrial reactions are performed in the gas phase or in liquid or solid phase, without any solvent added [165]. Likewise, many materials can be prepared without added solvents *via* solvent-free synthetic methodologies [166].

Without addition of solvent seems to be the main point, as the term "without solvent" has evolved through the years [130] (Kerton, 2013b). The general and common definition for reactions carried without solvent is "solvent-free reactions", but the term "solventless reactions" has been frequently used as well [130,167,168].

These "solvent-free reactions" can be divided into four classes: (i) liquid-liquid reactions (one reagent dissolves the other) [169]; (ii) solid–solid synthesis, in which two solids interact directly and form a third solid product without the intervention of a liquid or vapor phase (co-crystallization)

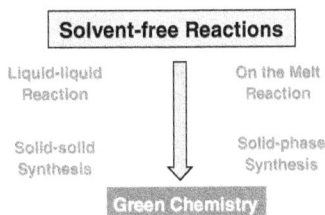

FIGURE 2.26 Relationship between Green Chemistry and solvent-free reactions.

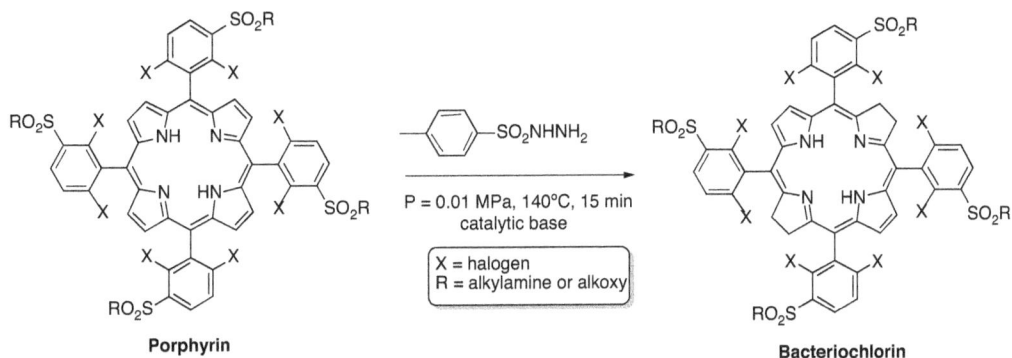

FIGURE 2.27 Solvent-free synthesis of bacteriochlorins.

[170]; (iii) on the melt reactions (any system in which pure solid reagents react together, in the absence of any solvent in which one melts and dissolves the other) [47]; and (iv) solid-phase synthesis (reaction of molecules from a liquid phase with a solid substrate, e.g., polymer-supported syntheses) [171] (Figure 2.26).

As selected examples, industrial-catalyzed transesterification reactions have been largely used in solvent-free liquid-liquid conditions – being, in general, reactants simultaneously used as solvents and reagents. Products such as esterified glycerol derivatives [172,173], phenolic ester derivatives [174], ester citrates [175], fatty acid esters [176], carbamates [177], and some polymers [178] have been prepared using this methodology.

We highlight the solvent-free industrial synthesis of redaporfin, a photosensitizer in phase I/II for head and neck cancers, specifically designed by Luzitin SA (www.luzitin.pt), for application in photodynamic therapy of cancer [179], where stable halogenated sulfonamide bacteriochlorins have been prepared via "on the melt" reduction of porphyrins with p-toluene sulfonylhydrazide (melts above 100°C and dissolves the porphyrin), in the total absence of solvents catalyzed or not by bases [47]. It should also be noticed that this solvent-free synthetic methodology gave higher yields of isolated o-halogenated phenyl-bacteriochlorins (Figure 2.27), when compared with the previously reported methods, using solvents [180].

The broad structural features of these stable NIR photosensitizers containing halogens and either sulfonamide or sulfoether groups [47,181] clearly show the great versatility of this solvent-free method to prepare large amounts of amphiphilic bacteriochlorins, compounds with multiple applications, which combine high singlet–triplet intersystem crossings, high bioavailability, and high absorption at the therapeutic window.

Ideally, we can say that industrial catalytic processes should have all to benefit from using solvent-free chemistry regarding environmental and economic issues. However, the solvent-free reactions still have some challenges to overcome regarding scale-up processes due to security issues.

ACKNOWLEDGMENTS

L.D. D. thanks Fundação de Amparo à Pesquisa do Estado de São Paulo (FAPESP) for his Postdoc Grant 2019/13569-8. F.M.S.R. thanks Fundação para a Ciência e Tecnologia (FCT-Portugal) for his scholarship, under the FCT-PhD program CATSUS (PD/BD/114340/2016). Thanks are also due to FCT (Fundação para a Ciência e Tecnologia – Portugal) and FEDER (European Regional Development Fund) for funding the project DUALPDI - POCI-01–0145-FEDER-027996.

REFERENCES

[1] U.S. Environmental Protection Agency, *Summary of the Pollution Prevention Act. United States Environmental Protection Agency-1990.* Available at https://www.epa.gov/laws-regulations/summary-pollution-prevention-act.

[2] P.T. Anastas, *Origins and Early History of Green Chemistry*, 1st Ed., World Scientific, Singapore, 2018.

[3] P.T. Anastas, J.C. Warner, *Green Chemistry: Theory and Practice*, 1st Ed., Oxford University Press, Oxford, 1998.

[4] J.H. Clark, *Green Chemistry and Environmentally Friendly Technologies*, 1st Ed., Blackwell Science Publishers, Oxford, 2005.

[5] M.M. Pereira, M.J.F. Calvete, *Sustainable Synthesis of Pharmaceuticals: Using Transition Metal Complexes as Catalysts*, RSC Books, Cambridge, 2018.

[6] E.L. Scott, J.P.M. Sanders, A. Steinbuchel. Perspectives on chemicals from renewable resources, In *Sustainable Biotechnology*, (O.V. Singh, S.P. Harvey, Eds.) Springer, New York, 2009, pp. 195–210.

[7] M.M. Yung, Catalytic conversion of biomass to fuels and chemicals, *Top. Catal.*, 59 (2016) pp. 1–1.

[8] P.T. Anastas, M.M. Kirchhoff, T.C. Williamson, Catalysis as a foundational pillar of green chemistry, *Appl. Catal. A*, 221 (2001) pp. 3–13.

[9] J.L. Casci, C.M. Lok, M.D. Shannon, Fischer-Tropsch catalysis: The basis for an emerging industry with origins in the early 20th century, *Catal. Today*, 145 (2009) pp. 38–44.

[10] B. Lindstrom, and L.J. Pettersson, A brief history of catalysis, *Cattech*, 7 (2003) pp. 130–138.

[11] J.J. Berzelius, Katalytische kraft, *Edinburgh New Philos. J.*, 21 (1836) p. 223.

[12] K.J. Laidler, A glossary of terms used in chemical kinetics, including reaction dynamics (IUPAC Recommendations 1996), *Pure Appl. Chem.*, 68 (1996) pp. 149–192.

[13] J. Wisniak, The history of catalysis. From the beginning to Nobel Prizes, *Educ. Química*, 21 (2010) pp. 60–69.

[14] K.M. Nicholas, *Selective Catalysis for Renewable Feedstocks and Chemicals*, 1st Ed., Springer, Heidelberg, 2014.

[15] A.M. Echavarren, Nobel Prize Awarded for Catalysis, *Chemcatchem*, 2 (2010) pp. 1331–1332.

[16] NobelPrizeWebsite, *The Nobel Prize in Chemistry*. Avaliable at https://www.nobelprize.org/prizes/chemistry/.

[17] L.Q. Lu, J.R. Chen, W.J. Xiao, Development of cascade reactions for the concise construction of diverse heterocyclic architectures, *Acc. Chem. Res.* 45 (2012) pp. 1278–1293.

[18] A. Grossmann, D. Enders, N-Heterocyclic carbene catalyzed domino reactions, *Angew. Chem. Int. Edit.*, 51 (2012) pp. 314–325.

[19] L.F. Tietze, T. Kinzel, C.C. Brazel, The domino multicomponent allylation reaction for the stereoselective synthesis of homoallylic alcohols, *Acc. Chem. Res.*, 42 (2009) pp. 367–378.

[20] M.M. Hussain, P.J. Walsh, Tandem reactions for streamlining synthesis: Enantio- and diastereoselective one-pot generation of functionalized epoxy alcohols, *Acc. Chem. Res.*, 41 (2008) pp. 883–893.

[21] J.C. Wasilke, S.J. Obrey, R.T. Baker, G.C. Bazan, Concurrent tandem catalysis, *Chem. Rev.*, 105 (2005) pp. 1001–1020.

[22] E. Rafiee, S Eavani, Heterogenization of heteropoly compounds: A review of their structure and synthesis, *RSC Adv.*, 6 (2016) pp. 46433–46466.

[23] I. Aranaz, N. Acosta, A. Heras, Enzymatic D-p-hydrophenyl glycine synthesis using chitin and chitosan as supports for biocatalyst immobilization, *Biocatal. Biotransfor.*, 36 (2018) pp. 89–101.

[24] J.C. Pessoa, M.R. Maurya, Vanadium complexes supported on organic polymers as sustainable systems for catalytic oxidations, *Inorg. Chim. Acta*, 455 (2017) pp. 415–428.

[25] A. Kumar, G.D. Park, S.K.S. Patel, S. Kondaveeti, S. Otari, M.Z. Anwar, V.C. Kalia, Y. Singh, S.C. Kim, B.K. Cho, J.H. Sohn, D.R. Kim, Y.C. Kang, J.K. Lee, SiO_2 microparticles with carbon nanotube-derived mesopores as an efficient support for enzyme immobilization, *Chem. Eng. J.*, 359 (2019) pp. 1252–1264.

[26] R. Zhong, A.C. Lindhorst, F.J. Groche, F.E. Kuhn, Immobilization of N-heterocyclic carbene compounds: A synthetic perspective. *Chem. Rev.*, 117 (2017) pp. 1970–2058.

[27] J.S. Chang, J.S. Hwang, S.E. Park, Preparation and application of nanocatalysts via surface functionalization of mesoporous materials, *Res. Chem. Intermediat.*, 29 (2003) pp. 921–938.

[28] A.C.B. Neves, M.J.F. Calvete, T.M.V.D.P. Melo, M.M. Pereira, Immobilized catalysts for hydroformylation reactions: A versatile tool for aldehyde synthesis, *Eur. J. Org. Chem.*, (2012) pp. 6309–6320.

[29] M. Adeel, M. Bilal, T. Rasheed, A. Sharma, H.M.N. Iqbal, Graphene and graphene oxide: Functionalization and nano-bio-catalytic system for enzyme immobilization and biotechnological perspective, *Int. J. Biol. Macromol.*, 120 (2018) pp. 1430–1440.

[30] A. Soozanipour, A. Taheri-Kafrani, Enzyme immobilization on functionalized graphene oxide nanosheets: Efficient and robust biocatalysts, In *Enzyme Nanoarchitectures: Enzymes Armored with Graphene*, (C.V. Kumar, Ed.), 1st Ed., Elsevier Academic Press Inc., San Diego, 2018, pp. 371–403.

[31] M.B. Gawande, P.S. Branco, R.S. Varma, Nano-magnetite (Fe_3O_4) as a support for recyclable catalysts in the development of sustainable methodologies, *Chem. Soc. Rev.*, 42 (2013) pp. 3371–3393.

[32] L.M. Rossi, N.J.S. Costa, F.P. Silva, R. Wojcieszak, Magnetic nanomaterials in catalysis: Advanced catalysts for magnetic separation and beyond, *Green Chem.*, 16 (2014) pp. 2906–2933.

[33] A.D. McNaught, A. Wilkinson, *IUPAC. Compendium of Chemical Terminology*, 1st Ed., Blackwell Scientific Publications, Oxford, 1997.

[34] G. Wypych, *Handbook of Solvents Volume 1: Properties*, 1st Ed., ChemTec Publishing, Ontario, 2014.

[35] R.A. Sheldon, Green solvents for sustainable organic synthesis: State of the art, *Green Chem.*, 7 (2005) pp. 267–278.

[36] P.G. Jessop, Searching for green solvents, *Green Chem.*, 13 (2011) pp. 1391–1398. doi: 10.1039/c0gc00797h.

[37] D. Prat, A. Wells, J. Hayler, H. Sneddon, C.R. McElroy, S. Abou-Shehada, P.J. Dunn, CHEM21 selection guide of classical- and less classical-solvents, *Green Chem.*, 18 (2016) pp. 288–296.

[38] Cornell Law School, Legal Information Institute, *40 CFR Part 59 – National volatile organic compound emission standards for consumer and commercial products.* Available at https://www.law.cornell.edu/cfr/text/40/part-59

[39] U.S. Environmental Protection Agency, *Guide: How to Determine if Solvents That Can No Longer Be Used in the Workplace Are Hazardous Waste-2016.* Available at https://www.epa.gov/hwgenerators/guide-how-determine-if-solvents-can-no-longer-be-used-workplace-are-hazardous-waste.

[40] U.S. Environmental Protection Agency, *Clean Air Act Guidelines and Standards for Solvent Use and Surface Coating Industry-2019.* Available at https://www.epa.gov/stationary-sources-air-pollution/clean-air-act-guidelines-and-standards-solvent-use-and-surface.

[41] K. Alfonsi, J. Colberg, P.J. Dunn, T. Fevig, S. Jennings, T.A. Johnson, H.P. Kleine, C. Knight, M.A. Nagy, D.A. Perry, M. Stefaniak, Green chemistry tools to influence a medicinal chemistry and research chemistry based organization, *Green Chem.*, 10 (2008) pp. 31–36.

[42] L. Cseri, M. Razali, P. Pogany, G. Szekely, Organic solvents in sustainable synthesis and engineering, In *Green Chemistry: An Inclusive Approach* (B. Török, T. Dransfield, Eds.) Elsevier, B. V., Amsterdam, 2018, pp. 513–553.

[43] M. Ikeda, Public-health problems of organic solvents, *Toxicol. Lett.*, 64 (1992) pp. 191–201.

[44] K. Tanaka, F. Toda, Solvent-free organic synthesis, *Chem. Rev.*, 100 (2000) pp. 1025–1074.

[45] J.L. Howard, Q. Cao, D.L. Browne, Mechanochemistry as an emerging tool for molecular synthesis: What can it offer?, *Chem. Sci.*, 9 (2018) pp. 3080–3094.

[46] C.A. Henriques, S.M.A. Pinto, M. Pineiro, J. Canotilho, M.E.S. Eusebio, M.M. Pereira, M.J.F. Calvete, Solventless metallation of low melting porphyrins synthesized by the water/microwave method, *RSC Adv.*, 5 (2015) pp. 64916–64924.

[47] M.M. Pereira, A.R. Abreu, N.P.F. Goncalves, M.J.F. Calvete, A.V.C. Simoes, C.J.P. Monteiro, L.G. Arnaut, M.E. Eusebio, J. Canotilho, An insight into solvent-free diimide porphyrin reduction: A versatile approach for meso-aryl hydroporphyrin synthesis, *Green Chem.*, 14 (2012) pp. 1666–1672.

[48] P.T. Anastas, E.S. Beach, Green Chemistry: The emergence of a transformative framework, *Green Chem. Lett. Rev.*, 1 (2007) pp. 9–24.

[49] G.H. Brundtland, *Our Common Future, From One Earth to One World: An Overview by the World Commission on Environment and Development, 1987.* Available at https://sustainabledevelopment.un.org/content/documents/5987our-common-future.pdf.

[50] D.J. Constable, *Green Chemistry Metrics, Part 11: Handbook of Green Chemistry*, 1st Ed., Wiley-VCH, Weinheim, 2018.

[51] R.A. Sheldon, Metrics of green chemistry and sustainability: Past, present, and future, *ACS Sustain. Chem. Eng.*, 6 (2018) pp. 32–48.

[52] P. Pollet, E.A. Davey, E.E. Ureña-Benavides, C.A. Eckert, C.L. Liotta, Solvents for sustainable chemical processes, *Green Chem.*, 16 (2014) pp. 1034–1055.

[54] P. Anastas, N. Eghbali, Green Chemistry: Principles and practice, *Chem. Soc. Rev.*, 39 (2010) pp. 301–312.

[55] G. Náray-Szabó, L.T. Mika, Conservative evolution and industrial metabolism in Green Chemistry, *Green Chem.*, 20 (2018) pp. 2171–2191.

[56] D. Adams, P. Dyson, S. Tavener, *Chemistry in Alternative Reaction Media*, 1st Ed., John Wiley & Sons, Chichester, 2003.

[57] S. Mithen, The domestication of water: Water management in the ancient world and its prehistoric origins in the Jordan Valley, *Philos. T. R. Soc. A*, 368 (2010) pp. 5249–5274.

[58] W Leitner, P.G. Jessop, *Green Solvents: Supercritical Solvents – Volume 4*, 1st Ed., Wiley-VCH, New York, 2013.

[59] C.A. Henriques, S.M.A. Pinto, G.L.B. Aquino, M. Pineiro, M.J.F. Calvete, M.M. Pereira, Ecofriendly porphyrin synthesis by using water under microwave irradiation, *Chemsuschem*, 7 (2014) pp. 2821–2824.

[60] N.S. Kus, Organic reactions in subcritical and supercritical water, *Tetrahedron*, 68 (2012) pp. 949–958.

[61] Y. Marcus, Extraction by subcritical and supercritical water, methanol, ethanol and their mixtures, *Separations*, 5 (2018) p. 4.

[62] K. Hartonen, M.-L. Riekkola, Water as the first choice green solvent, In *The Application of Green Solvents in Separation Processes*, (F. Pena-Pereira, M. Tobiszewski, Eds.), 1st Ed., Elsevier B. V., Amsterdam, 2017, pp. 19–55.

[63] D. Vogt, Other biphasic concepts, In *Aqueous-Phase Organometallic Catalysis: Concepts and Applications*, (B. Cornils, W.A. Herrmann, Eds.), 1st Ed., Wiley-VCH, Weinheim, 2004.

[64] F. Joó, Biphasic catalysis-homogeneous, In *Encyclopedia of Catalysis*, (I.T. Horvath, Ed.), 1st Ed., Wiley-VCH, Weinheim, 2010.

[65] K.H. Shaughnessy, Hydrophilic ligands and their application in aqueous-phase metal-catalyzed reactions, *Chem. Rev.*, 109 (2009) pp. 643–710.

[66] L.C. Matsinha, S. Siangwata, G.S. Smith, B.C.E. Makhubela, Aqueous biphasic hydroformylation of olefins: From classical phosphine-containing systems to emerging strategies based on water-soluble nonphosphine ligands, *Catal. Rev.*, 61 (2019) pp. 111–133.

[67] F. Joó, Aqueous biphasic hydrogenations, *Acc. Chem. Res.*, 35 (2002) pp. 738–745.

[68] E.G. Kuntz, *Procede d'hydroformylation des olefins*, French Patent FR2314910 (1975); FR2349562 (1976); FR2366237, to Rhône-Poulenc Recherche (1975).

[69] B. Cornils, J. Hibbel, G. Kessen, W. Konkol, B. Lieder, E. Wiebus, H. Kalbfel, H. Bach, *Process for the hydroformylation of olefins*, U.S. Patent US4533755, to Ruhrchemie Aktiengesellschaft (1985).

[70] B. Cornils, E.G. Kuntz, Introducing Tppts and related ligands for industrial biphasic processes, *J. Organomet. Chem.*, 502 (1995) pp. 177–186.

[71] W.A. Herrmann, C.W. Kohlpaintner, Water-soluble ligands, metal-complexes, and catalysts - synergism of homogeneous and heterogeneous catalysis. *Angew. Chem. Int. Ed.*, 32 (1993) pp. 1524–1544.

[72] W.A. Herrmann, F.E. Kuhn, Basic aqueous chemistry, In *Aqueous-Phase Organometallic Catalysis*, (B. Cornils, W.A. Herrmann, Eds.), 1st Ed., Wiley-VCH, Weinheim, 2004, pp. 44–56.

[73] E.G. Kuntz, Homogeneous catalysis in water, *Chemtech*, 17 (1987) pp. 570–575.

[74] E.G. Kuntz, C.P.E. Lyon, Hydrosoluble ligands for a new technology, In *Aqueous Organometallic Chemistry and Catalysis*, NATO ASI Series, (I.T. Horváth, F. Joó, Eds.), Springer, Dordrecht, 1995, vol. 5, pp. 177–181.

[75] E. Wiebus, B. Cornils, Biphasic systems: Water-organic, In *Catalyst Separation, Recovery and Recycling*, (D.J. Cole-Hamilton, R.P. Tooze, Eds.), 1st Ed., Springer, Dordrecht, 2006.

[76] C.W. Kohlpaintner, R.W. Fischer, B. Cornils, Aqueous biphasic catalysis: Ruhrchemie/Rhone-Poulenc oxo process, *Appl. Catal. A*, 221 (2001) pp. 219–225.

[77] E. Wiebus, B. Cornils, Industrial-scale oxo synthesis with an immobilized catalyst, *Chem, Ing. Tech.*, 66 (1994) pp. 916–923.

[78] R.A. Sheldon, I.W.C.E. Arends, U. Hanefeld, *Green Chemistry and Catalysis*, 1st Ed., Wiley-VCH, Weinheim, 2007.

[79] J.-M. Grosselin, C. Mercier, Procédé de préparation de la vitamine A, French Patent FR8716629, to Rhône-Poulenc Santé (1987).

[80] C. Mercier, P. Chabardes, Organometallic chemistry in industrial vitamin-A and vitamin-E synthesis, *Pure Appl. Chem.*, 66 (1994) pp. 1509–1518.

[81] G. Mignani, D. Morel, Y. Colleuille, C. Mercier, A novel C-10 terpene synthon -2-methyl-6-methylene-1,3e, 7-octatriene, *Tetrahedron Lett.*, 27 (1986) pp. 2591–2594.

[82] C.W. Kohlpaintner, M. Beller, Palladium-catalyzed carbonylation of benzyl chlorides to phenylacetic acids - A new two-phase process, *J. Mol. Catal. A*, 116 (1997) pp. 259–267.

[83] R.A. Sheldon, L. Maat, G. Papadogianakis, Process for preparing arylacetic acid and arylpropionic acid derivatives, U.S. Patent US5536874, to Hoechst Celanese (1996).

[84] S. Haber, H.J. Kleiner, Verfahren zur Kreuzkopplung von Boronsäuren mit Halogenverbindungen, German Patent DE19527118A1, to Hoechst AG (Clariant AG) (1995).

[85] Y. Tokitoh, N. Yoshimura, *Process for continuous production of octa-2,7-dien-1-ol*, U.S. Patent US5057631, to Kuraray Corporation (1991).

[86] R.A. Sheldon, J.M. Woodley, Role of biocatalysis in sustainable chemistry, *Chem. Rev.*, 118 (2018) pp. 801–838.

[87] N.Q. Ran, L.S. Zhao, Z.M. Chen, J.H. Tao, Recent applications of biocatalysis in developing green chemistry for chemical synthesis at the industrial scale, *Green Chem.*, 10 (2008) pp. 361–372.

[88] J.M. Woodley, Integrating protein engineering with process design for biocatalysis, *Philos. T. R. Soc. A*, 376 (2018) p. 20170062.

[89] R.B. Silverman, R. Andruszkiewicz, Gamma amino butyric acid analogs and optical isomers, U.S. Patent 6197819B1, to Northwestern University (2001).

[90] M. Albert, F. Zepeck, A. Berger, W. Riethorst, H. Schwab, D. Luschnig, P. Remler, J. Salchenegger, D. Osl, D. De Souza, *Process for the stereoselective enzymatic hydrolysis of 5-methyl-3-nitromethyl-hexanoic acid ester*, World patent WO2009141362A2, to Sandoz, AG (2009).

[91] S. Debarge, D.T. Erdman, P.M. O'Neill, R. Kumar, M.J. Karmilowicz, Process and intermediates for the preparation of pregabalin, World patent WO2014155291Al, To Pfizer Pharmaceuticals (2014).

[92] S. Hu, C.A. Martinez, J. Tao, W.E. Tully, P.G.T. Kelleher, Y.R. Dumond, Preparation of pregabalin and related compounds, World patent WO2006000904A2, to Warner-Lambert Company Llc (2006).

[93] W.R.L. Notz, R.W. Scott, L. Zhao, J. Tao, Transaminase-based processes for preparation of pregabalin, World patent WO2008127646A2, to Bioverdant, Inc (2008).

[94] R.A. Sheldon, The E factor 25 years on: The rise of green chemistry and sustainability, *Green Chem.*, 19 (2017) pp. 18–43.

[95] C. Jimenez-Gonzalez, A.D. Curzons, D.J.C. Constable, V.L. Cunningham, Cradle-to-gate life cycle inventory and assessment of pharmaceutical compounds, *Int. J. Life Cycle Ass.*, 9 (2004) pp. 114–121.

[96] B.D. Roth, *Trans-6-[2-(3- or 4-carboxamido-substituted pyrrol-1-yl)alkyl]-4-hydroxypyran-2-one inhibitors of cholesterol synthesis*, U.S. patent US4681893A to Warner Lambert CO (1987).

[97] S. Hu, J. Tao, J. Xie, Process for producing atorvastatin, pharmaceutically acceptable salts thereof and intermediates thereof, World patent WO2006134482, To Pfizer, Inc (2006).

[98] P Walden, Molecular weights and electrical conductivity of several fused salts, *Bull. Acad. Imper. Sci.*, 1800 (1914) pp. 405–422.

[99] Y.X. Qiao, W.B. Ma, N. Theyssen, C. Chen, Z.S. Hou, Temperature-responsive ionic liquids: Fundamental behaviors and catalytic application, *Chem. Rev.*, 117 (2017) pp. 6881–6928.

[100] D.M. Freudendahl, S. Santoro, S.A. Shahzad, C. Santi, T. Wirth, Green chemistry with selenium reagents: Development of efficient catalytic reactions, *Angew. Chem. Int. Edit.*, 48 (2009) pp. 8409–8411.

[101] S. Singhal, S. Agarwal, M. Singh, S. Rana, S. Arora, N. Singhal, Ionic liquids: Green catalysts for alkene-isoalkane alkylation, *J. Mol. Liq.*, 285 (2019) pp. 299–313.

[102] M. Díaz, A. Ortiz, I. Ortiz, Progress in the use of ionic liquids as electrolyte membranes in fuel cells, *J. Membr. Sci.*, 469 (2014) pp. 379–396.

[103] W. Ma, H. Yuan, H. Wang, Q. Zhou, K. Kong, D. Li, Y. Yao, Z. Hou, Identifying catalytically active mononuclear peroxoniobate anion of ionic liquids in the epoxidation of olefins, *ACS Catal.*, 8 (2018) pp. 4645–4659

[104] A.K. Chakraborti, S.R. Roy, On catalysis by ionic liquids, *J. Am. Chem. Soc.*, 131 (2009) pp. 6902–6903.

[105] C. Dai, J. Zhang, C. Huang, Z. Lei, Ionic liquids in selective oxidation: Catalysts and solvents, *Chem. Rev.*, 117 (2017) pp. 6929–6983.

[106] R.L. Vekariya, A review of ionic liquids: Applications towards catalytic organic transformations, *J. Mol. Liq.*, 227 (2017) pp. 44–60.

[107] N.V. Plechkova, K.R. Seddon, Applications of ionic liquids in the chemical industry, *Chem. Soc. Rev.*, 37 (2008) pp. 123–150.

[108] E.W. Castner Jr., J.F. Wishart, Spotlight on ionic liquids, *J. Chem. Phys.*, 132 (2010) p. 120901.

[109] D.D. Irge, Ionic liquids: A review on greener chemistry applications, quality ionic liquid synthesis and economical viability in a chemical processes, *Am. J. Phys. Chem.*, 5 (2016) pp. 74–79.

[110] D.R. MacFarlane, M. Kar, J.M. Pringle, *Fundamentals of Ionic Liquids, From Chemistry to Applications*, 1st Ed., Wiley-VCH, Weinheim, 2017.

[111] Y. Marcus, *Ionic Liquid Properties, From Molten Salts to RTILs*, 1st Ed., Springer International Publishing, Zurich, 2016.

[112] A. Mohammad, D. Inamuddin, *Green Solvents II, Properties and Applications of Ionic Liquids*, 1st Ed., Springer Science+Business Media, Dordrecht, 2012.

[113] S.P.M. Ventura, F.A.E. Silva, M.V. Quental, D. Mondal, M.G. Freire, J.A.P. Coutinho, Ionic-liquid-mediated extraction and separation processes for bioactive compounds: Past, present, and future trends, *Chem. Rev.*, 117 (2017) pp. 6984–7052.

[114] J.S. Wilkes, Properties of ionic liquid solvents for catalysis, *J. Mol. Catal. A*, 214 (2004) pp. 11–17.

[115] K. Dong, X. Liu, H. Dong, X. Zhang, S. Zhang, Multiscale studies on ionic liquids, *Chem. Rev.*, 117 (2017) pp. 6636–6695.

[116] M.J. Earle, S.P. Katdare, K.R. Seddon, Paradigm confirmed: The first use of ionic liquids to dramatically influence the outcome of chemical reactions, *Org. Lett.*, 6 (2004) pp. 707–710.

[117] G.W. Phillips, S.N. Falling, J.R. Monnier, *Continuous process for the manufacture of 2,5-dihydrofurans from γ, δ-epoxybutenes*, U.S. patent US5315019, To Eastman Chemical Co (1994).

[118] K Hell, U Hesse, B Weyershausen, Process for the production of organically modified Polysiloxanes by using ionic liquids, European patent EP1382630B1, to Evonik Goldschmidt GmbH (2004).

[119] Z.C. Liu, R. Zhang, C.M. Xu, R.G. Xia, Ionic liquid alkylation process produces high-quality gasoline, *Oil Gas J.*, 104 (2006) pp. 52–56.

[120] Y. Chauvin, J.F. Gaillard, D.V. Quang, J.W. Andrews, IFP dimersol process for dimerization of C3 and C4 olefinic cuts, *Chem. Ind.*, 9 (1974) 375–378.

[121] K.D. Hope, D.W. Twomey, M.S. Driver, D. Stern, J. Collins, T. Harris, Method for manufacturing high viscosity polyalphaolefins using ionic liquid catalysts, U.S. patent US20050113621A1, to Chevron Phillips Chemical Co LP (2005).

[122] Z. Lei, C. Dai, B. Chen, Gas solubility in ionic liquids, *Chem. Rev.*, 114 (2014) pp. 1289–1326.

[123] C. C. de la Tour, Exposé de quelques résultats obtenu par l'action combinée de la chaleur et de la compression sur certains liquides, tels que l'eau, l'alcool, l'éther sulfurique et l'essence de pétrole rectifiée, *Ann. Chim. Phys.*, 21 (1822) pp. 127–132.

[124] Z. Knez, E. Markocic, M. Leitge, M. Primozic, M.K. Hrncic, M. Skerget, Industrial applications of supercritical fluids: A review, *Energy*, 77 (2014) pp. 235–243.

[125] W.H. Hauthal, Advances with supercritical fluids, *Chemosphere*, 43 (2001) pp. 123–135.

[126] O. Kajimoto, Solvation in supercritical fluids: Its effects on energy transfer and chemical reactions. *Chem. Rev.*, 99 (1999) pp. 355–390.

[127] M. Perrut, Supercritical fluid applications: Industrial developments and economic issues, *Ind. Eng. Chem. Res.*, 39 (2000) pp. 4531–4535.

[128] K. Zosel, Separation with supercritical gases: Practical applications, *Angew. Chemie Int. Edit.*, 17 (1978) pp. 702–709.

[129] F. Kerton, Industrial applications of green solvents, *RSC Green Chem. Ser.*, (2013) pp. 285–304.

[130] F. Kerton, 'Solvent-free' chemistry, *RSC Green Chem. Ser.*, (2013) pp. 51–81.

[131] F. Kerton, Supercritical fluids, *RSC Green Chem. Ser.*, (2013) pp. 115–148.

[132] P.D. Brothers, Polymerization of fluoromonomers in carbon dioxide, U.S. patent US6103844A, to Chemours Co FC LLC (2000).

[133] E.F. Debrabander, P.D. Brothers, Polymerization of fluoropolymers in carbon dioxide, U.S. patent US6051682A, to Chemours Co FC LLC (2000).

[134] R.C. Wheland, P.D. Brothers, *Copolymers of maleic anhydride or acid and fluorinated olefins*, U.S. patent US6107423A, to Chemours Co FC LLC (2000).

[135] S. Fukuoka, M. Kawamura, K. Komiya, M. Tojo, H. Hachiya, K. Hasegawa, M. Aminaka, H. Okamoto, I. Fukawa, S. Konno, A novel non-phosgene polycarbonate production process using by-product CO_2 as starting material, *Green Chem.* 5 (2003) pp. 497–507.

[136] M. Poliakoff, T.M. Swan, T. Tacke, M.G. Hitzler, S.K. Ross, S. Wieland, Supercritical hydrogenation, U.S. patent US6156933, to Thomas Swan and Co Ltd (2000).

[137] S.K. Ross, N.J. Meehan, M. Poliakoff, D.D.N. Cater, Supercritical hydrogenation, European patent EP1373166A2, to Thomas Swan and Co Ltd (2004).

[138] W. Bonrath, S. Wang, Process for manufacturing d, l-alpha-tocopherol, European patent EP1000940A1, to Hoffmann-La Roche AG (1999).

[139] J.S. Lee, M. Hronec, K.H. Lee, J.W. Kwak, Y.H. Chu, Environmentally benign and simplified method for preparation of aromatic dicarboxylic acid, U.S. patent US20100087676A1, to Samsung Petrochemical Co Ltd (2007).

[140] R.A.F. Tomas, J.C.M. Bordado, J.F.P. Gomes, p-Xylene oxidation to terephthalic acid: A literature review oriented toward process optimization and development, *Chem. Rev.*, 113(10) (2013), pp. 7421–7469.

[141] W. Leitner, N. Theyssen, Z. Hou, K.W. Kottsieper, M. Solinas, D. Giunta, Process for continuous ring-closing metathesis in compressed carbon dioxide, U.S. patent US7482501B2, to Boehringer Ingelheim International GmbH (2005).

[142] J. DeSimone, R. Givens, Y.X. Ni, *Synthesis of polyamides in liquid and supercritical CO₂*, U.S. patent US6025459A, to North Carolina University (2000).

[143] M. Poliakoff, P. Hamley, E.G. Cepeda, G.R. Aird, D.A.S. Coote, C. Yan, W.B. Thomas, I. Pearson, Preparation of lactams, World patent WO2006078403A1, to Invista Technologies S.A R.L (2006).

[144] J.H. Clark, T.J. Farmer, L. Herrero-Davila, J. Sherwood, Circular economy design considerations for research and process development in the chemical sciences, *Green Chem.*, 18 (2016) pp. 3914–3934.

[145] R.A. Sheldon, Green chemistry and resource efficiency: Towards a green economy, *Green Chem.*, 18 (2016) pp. 3180–3183.

[146] R. Ciriminna, M. Lomeli-Rodriguez, P.D. Cara, J.A. Lopez-Sanchez, M. Pagliaro, Limonene: A versatile chemical of the bioeconomy, *Chem. Commun.*, 50 (2014) pp. 15288–15296.

[147] D.M. Alonso, S.G. Wettstein, J.A. Dumesic, Gamma-valerolactone, a sustainable platform molecule derived from lignocellulosic biomass, *Green Chem.*, 15 (2013) pp. 584–595.

[148] F.V. Pace, P. Hoyos, L. Castoldi, P.D. de Maria, A.R. Alcantara, 2-Methyltetrahydrofuran (2-MeTHF): A biomass-derived solvent with broad application in organic chemistry, *Chemsuschem*, 5 (2012) pp. 1369–1379.

[149] Z. Li, K.H. Smith, G.W. Stevens, The use of environmentally sustainable bio-derived solvents in solvent extraction applications—A review, *Chin. J. Chem. Eng.*, 24 (2016) pp. 215–220.

[150] S. Haber, D. Hauk, U. Wietelmann, D. Dawidowski, P. Rittmeyer, J. Roeder, Organomagnesium synthesis agent, World patent WO2007099173A1, to Chemetall GmbH (2007).

[151] J.H. Tatlock, I.J. Mcalpine, M.B. Tran-Dube, E.Y. Rui, M.J. Wythes, R.A. Kumpf, M.A. Mctigue, Substituted nucleoside derivatives useful as anticancer agents, World patent WO2016135582A1, To Pfizer, Inc (2016).

[152] J. Kawakami, K. Nakamoto, S. Nuwa, S. Handa, S. Miki, Process for producing fused imidazole compound, Reformatsky reagent in stable form, and process for producing the same, U.S. patent US7662974B2, to Takeda Pharmaceutical Co Ltd (2010).

[153] D. Ripin, M. Vetelino, L. Wei, Processes for the preparation of substituted bicyclic derivatives, U.S. patent US20050026940A1, to Pfizer Inc (2005).

[154] N. Tom, D. Ripin, M Castaldi, Processes for the preparation of benzoimidazole derivatives, U.S. patent US7183414B2, to Pfizer Inc (2007).

[155] J.-D. Charrier, J. Studley, F.Y.T.M. Pierard, S.J. Durrant, B.J. Littler, R.M. Hughes, D.A. Siesel, P. Angell, A. Urbina, Y. Shi, Processes for making compounds useful as inhibitors of *ATR* kinase, World patent WO2013049726A3, to Vertex Pharmaceuticals Inc (2013).

[156] B. Gutmann, M. Bersier, P. Hanselmann, C.O. Kappe, M. Koeckinger, C. Hone, *Method for preparation of difluoromethylornithine*, World patent WO2019057951A1, to Lonza, Inc (2019).

[157] L. Browne, D. Grainger, *Hydrogenation of a compound comprising an α, β-unsaturated carbonyl group*, World patent WO2019043415A1, to Johnson Matthey Public Limited Company (2019).

[158] D.D. Wirth, C.M. Yates, *Antifungal compounds and processes for making*, World patent WO2017049196A1, to Viamet Pharmaceuticals, Inc (2017).

[159] M. Bio, E. Fang, J.E. Milne, S. Wiedeman, A. Wilsily, *Method for the preparation of (1,2,4)-triazolo(4,3-a)pyridines*, World patent WO2014210042A3, to Amgen Inc (2015).

[160] I. Cerna, R. Vlasakova, R. Krulis, *Method of preparing abiraterone acetate of high purity applicable on industrial scale*, World patent WO2016004910A1, to Zentiva, K.S (2016).

[161] C.F. Gelin, T.P. Lebold, B.T. Shireman, J.M. Ziff, *Substituted 2-azabicycles and their use as orexin receptor modulators*, World patent WO2014165070A1, to Janssen Pharmaceutica Nv (2014).

[162] J. Miller, J. Penney, *Process for preparing alkynyl-substituted aromatic and heterocyclic compounds*, U.S. patent US7105707B2, to PharmaCore Inc (2006).

[163] E. Aktoudianakis, G. Chin, B..K Corkey, J. Du, K. Elbel, R.H. Jiang, T. Kobayashi, R. Martinez, S.E. Metobo, M. Mish, S. Shevick, D. Sperandio, H. Yang, J. Zablocki, *Benzimidazolone derivatives as bromodomain inhibitors*, World patent WO2014160873A1, to Gilead Sciences, Inc (2014).

[164] M. Hintze, J. Wen, *Process for preparing methyllithium*, U.S. patent US6861011, to Rockwood Lithium Inc (2003).

[165] G.-M. Côme, *Gas-Phase Thermal Reactions: Chemical Engineering Kinetics*, 1st Ed., Springer Science+Business Media, Dordrecht, (2001).

[166] A.R. West, *Solid State Chemistry and Its Applications*, 2nd Ed, John Wiley and Sons Ltd, Hoboken, 2014.

[167] K. Tanaka, *Solvent-Free Organic Synthesis*, 2nd Ed, Wiley-VCH, Weinheim, 2009.

[168] T. Welton, All solutions have a solvent, *Green Chem.*, 8 (2006) pp. 13–13. doi: 10.1039/b514869n.

[169] D. Margetić, and V. Štrukil, *Mechanochemical Organic Synthesis*, 1st Ed., Elsevier, Amsterdam, 2016.

[170] A.O.L. Evora, E.R.A. Castro, T.M.R. Maria, M.T.S. Rosado, M.R. Silva, A.M. Beja, J. Canotilho, M.E.S. Eusebio, Pyrazinamide-diflunisal: A new dual-drug co-crystal, *Cryst. Growth Des.*, 11 (2011) pp. 4780–4788.

[171] L. Ferrazzano, D. Corbisiero, G. Martelli, A. Tolomelli, A. Viola, A. Ricci, W. Cabri, Green solvent mixtures for solid-phase peptide synthesis: A dimethylformamide-free highly efficient synthesis of pharmaceutical-grade peptides. *ACS Sust. Chem. Eng.*, 715 (2019) pp. 12867–12877.

[172] C.F. Cooper, *Preparation of esterified propoxylated glycerin by transesterification*, U.S. patent US5175323A, to Arco Chemical Technology, L.P (1992).

[173] T.A. Pelloso, A.D. Roden, G.L. Boldt, *Synthesis of acetoglyceride fats*, World patent WO1994018290A1, to Nabisco Inc (1994).

[174] V.A. Majerczak, M.S. Gibson, R.J. Orlando, *Solventless synthesis of hydrophilic phenol ester derivatives*, World patent WO2002040447A2, to The Procter & Gamble Company (2002).

[175] T.E. Enright, P. Salehi, K. Morimitsu, *Solventless reaction process*, U.S. patent US20120277462A1, to Xerox (2012).

[176] P.J. Corrigan, *Synthesis of polyol fatty acid polyesters*, U.S. patent US6620952B1, to The Procter & Gamble Company (2003).

[177] S. Wershofen, S. Klein, A. Vidal-Ferran, E. Reixach, F.X. Rius-Ruiz, *Process for preparing aromatic carbamates*, European patent EP2408737B1, Bayer Intellectual Property GmbH (2015).

[178] T. Estrin, *Solventless method for preparation of carboxylic polymers*, U.S. patent US7026410B2, to Henkel Corp (2006).

[179] L.G.S.A. Moreira, M.M. Pereira, S.J.F.S. Simões, S.P.M. Simões, K. Urbanska, G. Stochel, *Process for preparing chlorins and their pharmaceutical uses*, World patent WO2010047611A9, to Universidade De Coimbra, Bluepharma - Industria Farmacêutica, S.A (2010).

[180] H.W. Whitlock Jr, R. Hanauer, M.Y. Oester, B.K. Bower, Diimide reduction of porphyrins, *J. Am. Chem. Soc.*, 91 (1969) pp. 7485–7489.

[181] A.V.C. Simoes, A. Adamowicz, J.M. Dabrowski, M.J.F. Calvete, A.R. Abreu, G. Stochel, L.G. Arnaut, M.M. Pereira, Amphiphilic meso(sulfonate ester fluoroaryl)porphyrins: Refining the substituents of porphyrin derivatives for phototherapy and diagnostics, *Tetrahedron*, 68 (2012) pp. 8767–8772.

3 Sensing of Environmental Contaminants Using Advanced Nanomaterial

Parul Raturi
Omkaranand Sarswati Government Degree College

Bijit Choudhuri
NIT Silchar

P. Chinnamuthu
NIT Nagaland

CONTENTS

3.1 INTRODUCTION

The growing complexity in human living standards and corresponding industrial requirements are influencing our environment in a significant manner. The cost of the treatment to encounter all the pollution-induced ailments and reduction in working efficiency are creating serious impediment to the growth of human society. So, emphasis is given to sustainable goals on good health and well-being, clean water and sanitation, affordable and clean energy, etc. [1]. Consequently, substantial effort is paid to develop highly efficient and eco-friendly sensors to assess the environment. Recent development of novel efficient nanofabrication and characterization facilities has opened the chances to synthesize new functional materials suitable for environment sensors [2]. Nanostructures are materials with dimensions in 1–100 nm range, and they can exhibit exciting characteristics and offer distinct advantages over their bulk counterparts [3,4]. As a result, significant research effort is concentrated to develop nanostructured devices for humidity sensing, pH sensing, biosensing, heavy metal detection, etc. [5–8]. Detection and monitoring of the humidity and pH of the soil is essential for the smart agricultural monitoring.

DOI: 10.1201/9781003206385-3

Moreover, heavy metals are originated through the natural environmental contamination sources such as earth's crust and other human activities such as smelting operations, mining, industrial waste, agricultural and domestic waste, etc. [9,10]. Few other environmental sources for the heavy metal ion discharge include atmospheric accumulation of these metals, soil erosion, leaching, metal corrosion, and many more [11–15]. Metal oxides are the versatile class of materials, which has the ability to be operated under the harsh environmental condition due to stability and mechanical robustness. Moreover, they are widely used by the researchers for the environmental sensing applications due to their relatively low cost and exquisite sensing abilities. Metal oxides became materials of interest for every scientific community ranging from material science to engineering from physics to chemistry. Further, it was observed that by using the nanostructures of the metal oxide, their sensing abilities can be enhanced [16,17]. Recent advancements in the nanotechnology offered a new prospective for the synthesis of metal oxide materials with favorable physical and chemical properties and morphologies to attain high sensitivity and selectivity for environmental monitoring [17–19]. In order to meet the criteria for the environmental monitoring, metal oxide nanostructures are proven as active materials for the sensing due to their superior characteristics such as high crystallinity, low cost, ease of synthesis, and remarkable physical/chemical properties. Real-time monitoring of different aspects of the environmental impact is now a matter of interest for the scientific community. The scope of this chapter includes the use of the nanostructures for the environmental sensing – mainly, pH sensor, humidity, and heavy metal ion detection by using the metal oxide nanostructures are focused.

3.2 pH SENSOR

pH level of a solution stands for potential of hydrogen or power of hydrogen. Basically, the pH scale is associated with the concentration of H+ in a solution. The solution becomes acidic when $0 < pH$ level < 7. For $pH = 7$, the solution becomes neutral, and for $pH > 7$, the solution becomes basic. pH-level determination is essential to determine the content of acid in soil, water, etc. to perform agriculture. Many researchers are working to develop highly sensitive and fast-responsive pH sensors with oxide nanostructure using cost-effective growth procedure.

Kao et al. demonstrated multianalyte electrolyte–insulator–semiconductor (EIS) sensor for pH sensing [20]. A ZnO thin film was deposited on n-Si substrate using radiofrequency (RF) reactive sputtering followed by 300-nm-thick aluminum back contact deposition. The XRD analysis revealed that better crystalline properties in ZnO can be achieved with post-rapid thermal annealing at 600°C. The device was fabricated using standard photolithography process followed by silver gel connection on printed circuit board (PCB). The ZnO membrane-based EIS sensor with post-annealing at 600°C exhibited a best sensitivity of 42.45 mV/pH. The device showed reduced hysteresis voltage and higher sensitivity. The schematic diagram of the device is shown in Figure 3.1. Maiolo et al. demonstrated an extended gate thin-film transistor (EGTFT)-based pH sensor using low-temperature polycrystalline silicon (LTPS) process [21]. Nanoporous ZnO membrane was used as the active material in the device, and EGTFT device was prepared on a flexible polymide (PI) substrate. The leakage current was <10 pA and I_{on}/I_{off} was >10^6. The device showed a sensitivity of 59 mV/pH slope in threshold voltage characteristic under the pH variation in 1–9 range. This superior performance was attributed to the presence of the surface states at ZnO-electrolyte interface, which can hold the charges from acid/base solution leading to the accumulation of surface charge and corresponding potential.

Zhuiykov et al. reported the impact of morphological parameter variation on the properties of planar electrochemical pH sensors [22]. Screen-printed sub-micron Cu_2O-doped RuO_2 was used as sensing electrode of the sensor. The performance of the sensor was related to the thickness of the sensing electrode. The sensing time was improved to ~25 seconds for 5-μm-thick sensing electrode as compared to ~80–120 seconds for 2-μm-thick sensing electrode. Manjakkal and coworkers reported the sensing mechanism of the RuO_2-Ta_2O_5 thick-film conductometric pH sensors [23].

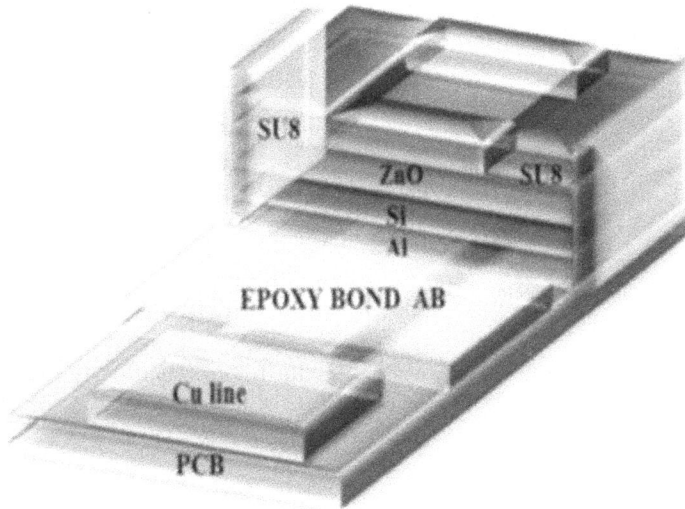

FIGURE 3.1 EIS with ZnO membrane structure. (Reused from Haur Kao et al., 2014) [20].

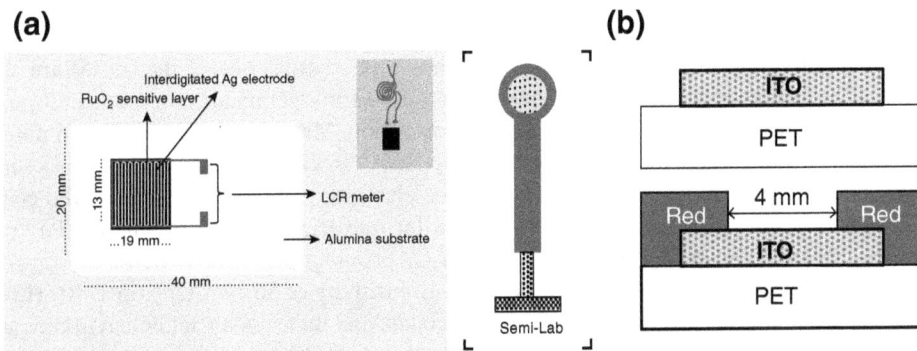

FIGURE 3.2 (a) Schematic view of RuO$_2$-based conductometric pH sensor. Inset: Image of the fabricated thick-film pH sensor. (Reused from Manjakkal et al. 2015 [26]). (b) Schematic and cross-section of the ITO/PET electrode. (Reused from Lue et al., 2012 [27]).

Interdigitated pH sensor was prepared using screen-printing method on alumina substrate. From electrochemical impedance spectroscopy, it can be observed that the electrical properties of the device such as resistance and capacitance were affected by the change in pH value at low frequencies. Xu et al. reported pH sensing by RuO$_2$ nanoparticles (NPs) embellished vertically aligned carbon nanotubes (CNT) [24]. MWCNTs were grown on Ta substrate followed by RuO$_2$ NP deposition by magnetron sputtering. This solid-state sensor showed high sensitivity of −55 mV/pH for pH from 2 to 12. The sensor exhibited a low response time (<40 seconds). Zhao et al. investigated the pH sensing by anodization of titanium substrate electrodes into TiO$_2$ nanotubes. The sensor was examined under a wide range of variation in pH level (from 2 to 12) [25]. The pH sensors exhibited a response similar to the ideal Nernst response and showed a high sensitivity of 54.5 mV/pH. The response time of the sensor was less than 30 seconds. However, the response of the detector degraded with exposure to annealing. Manjakkal et al. fabricated a RuO$_2$ thick-film-based interdigitated pH sensor [26]. The schematic view and image of the sensor is shown in Figure 3.2a. Ag/Pd thick-film paste was used to screen-print the interdigitated electrode on top of the alumina substrate

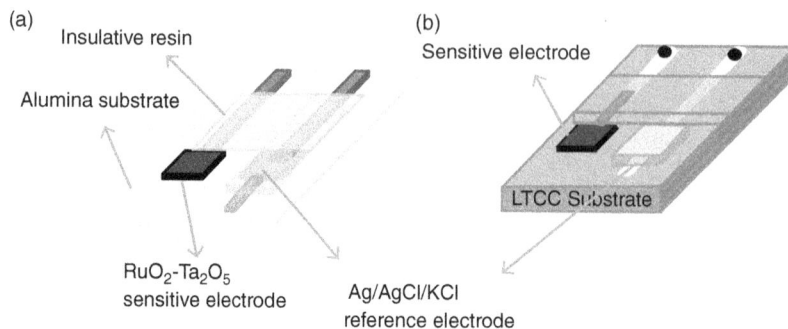

FIGURE 3.3 Schematic representation of potentiometric RuO_2–Ta_2O_5 sensors: (I) on an alumina substrate (II) on LTCC substrate. (Reused from Manjakkal et al., 2016 [28]).

followed by sintering. Subsequently, the active layer of RuO_2 was screen-printed on top of the electrode. Electrochemical impedance spectroscopy revealed that when the pH level changed from 3 to 7, the adsorption resistance changed from 70 to 275 Ω. In parallel, the relaxation time also changed from 0.05 to 0.16 second.

Lue et al. investigated the pH-sensing properties of extended gate field-effect transistors (EGFET) on flexible polyethylene terephthalate (PET) substrate [27]. RF-sputtered indium tin oxide (ITO) coating was used as the electrode of the sensor on PET substrate. The definition of ITO was carried out by a photolithography system. The schematic and cross-section of the sensor are shown in Figure 3.2b. The device exhibited a pH sensitivity of 50.1 mV/pH under room temperature. The sensors showed a lifetime of >55 days with reliable operation. Manjakkal et al. fabricated potentiometric pH sensor and investigated its properties [28]. RuO_2–Ta_2O_5 thick film was used as a sensing electrode, and Ag/AgCl/KCl was used as a reference electrode. Low-temperature cofired ceramic (LTCC) substrate was used alongside with alumina for preparing the sensor. The Ag/Pd contact was screen-printed followed by sensitive RuO_2–Ta_2O_5 printing. The schematic of the sensors is shown in Figure 3.3. The sensor showed an excellent sensitivity of 56.19 mV/pH in 1–10 pH range. The response time in the basic solution was >15 seconds, and under in an acidic environment, the response time reduced to >8 seconds. The sensor exhibited 54±1 mV/pH sensitivity even after 1 year of measurement and showed good reproducibility of result.

Lale and coworkers developed a potentiometric pH sensor using electrochemical microcell silicon-compatible technology [29]. Pt-Pt-Ag microdevices on silicon wafer were fabricated using evaporation and lift-off method and IrO_2 was deposited on Pt-Pt-Ag electrode using the electrode-position method. Similarly, tungsten was deposited on an oxidized silicon substrate followed by an oxygen plasma process to oxidize tungsten into its oxide WO_3. Subsequently, Ti/Pt and Ag patterning was made. The W/WO_3 showed a sensitivity of 55 mV/pH, while Pt/IrO_2 exhibited 60 mV/pH sensitivity. The hysteresis for Pt/IrO_2 and W/WO_3 was +5±0.5 mV and −50±2 mV, respectively. Lin et al. developed indium-gallium-zinc oxide (IGZO) NP-decorated silicon nanowire (SiNW)-based EGFET for pH sensing [30]. Ag-assisted electroless chemical etching process was employed to obtain SiNW. The IGZO NPs were deposited on SiNW using the RF sputtering process. The IGZO NP/SiNW structure showed the highest pH sensitivity of 50 mV/pH, which was ~1.4 times better than the pH sensitivity shown by the pristine SiNW sensor (36 mV/pH). This enhancement in sensitivity was due to the adsorption of additional H^+ ions by IGZO coating. Chin et al. fabricated and characterized electrolyte-ion-sensitive membrane oxide semiconductor (EIOS)-based pH sensors [31]. At first, silicon wafer was oxidized into SiO_2 by dry thermal oxidation. Subsequently, Ta film was deposited using RF sputtering system in the Ar atmosphere followed by oxidation into Ta_2O_5 in the O_2 environment. The sensor showed a maximum sensitivity of 56.17 mV/pH toward H^+. After 1 year's operation, the sensitivity reduced from (−56.19±2) to (−49.59±2) mV/pH. Manjakkal and

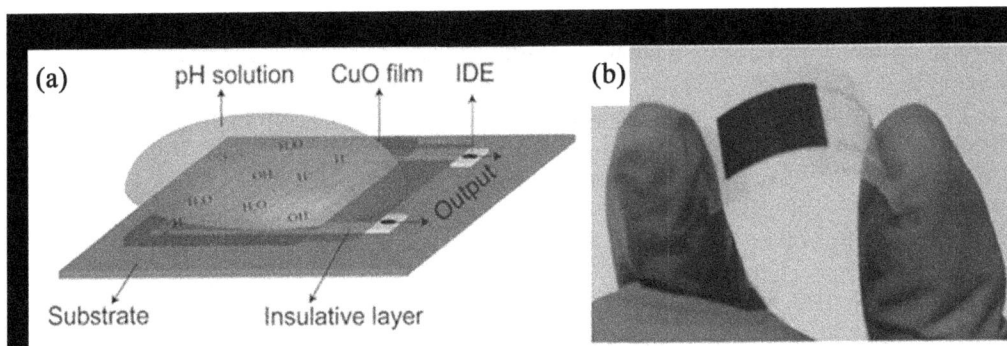

FIGURE 3.4 (a) Schematic representation of CuO-based IDE pH sensor and (b) image of the flexible CuO-based pH sensor. (Reused from Manjakkal et al., 2018 [32]).

co-workers reported pH sensing by CuO nanorod (NR)- and nanoflower (NF)-based printed flexible sensors [32]. They used hydrothermal synthesis methods to prepare CuO NR and NF. Silver paste was used to define interdigitated electrodes on alumina and PET substrates. Afterward, the CuO NR and NF paste was screen-printed on electrodes to define the sensitive area. The schematic of the sensor is shown in Figure 3.4. A sensitivity of 0.64 F/pH was observed in the pH range of 5–8.5.

Hussain et al. reported the pH sensing by Co_3O_4 nanostructure grown by the hydrothermal method [33]. Co_3O_4 nanostructure was grown on p-Si substrate using the hydrothermal synthesis method. The sensor exhibited a sensitivity of −58.45 mV/pH, which was very close to the theoretical approximation of −59.1 mV/pH at 298 K. This similarity was due to the high surface-to-volume ratio of Co_3O_4 nanostructure. The sensor exhibited a response time of 53 seconds. Young et al. investigated the pH response of ZnO nanostructure [34]. At first, a seed ZnO layer was grown using RF magnetron sputtering on a glass substrate followed by hydrothermal synthesis of one-dimensional (1-D) ZnO NR. The ZnO NR sensor exhibited a responsivity and linearity of 44.56 mV/pH and 0.983, respectively.

3.3 HUMIDITY SENSOR

Jeong et al. developed TiO_2 NF-based flexible resistive-type humidity sensor [35]. Ag patterning was made on polyimide (PI) substrate using roll-to-roll gravure printing method to deposit the contact, and the TiO_2 NF was used as the sensing material. When the relative humidity (RH) changed from 20% to 95%, resistance changed in the $\sim10^3–10^4$ scale. The maximum sensitivity of the sensor was observed to be 485.7 RH%$^{-1}$. This high sensitivity was attributed to the high surface-to-volume ratio of NF structure. The response and recovery times were in the order of hundreds of seconds. The sensors also offered resistance to mechanical stress, thus maintaining their properties. Steele and co-workers developed a capacitive RH sensor using TiO_2 nanostructure/interdigitated electrode substrates [36]. Countersunk interdigitated electrode substrates were used for electrodes. Sensitive TiO_2 nanostructure was grown using glancing angle deposition incorporated in an electron beam evaporation system. When the RH was changed from 2% to 95%, per unit area capacitance changed from ~1 to ~800 nF/cm^2. With an increase in film thickness from 280 nm to 8.5 μm, the response time varied from 64 to 1,440 ms. Zhang et al. developed tin dioxide/reduced graphene oxide (rGO) nanocomposite on flexible substrate and investigated humidity-sensing properties [37]. Cu/Ni-interdigitated electrodes were deposited on a flexible PI substrate using a sputtering system and photolithography technique. SnO_2/rGO nanocomposite was prepared using hydrothermal synthesis and hydrothermal reduction method, respectively. The capacitance response of the device showed a ~550-fold increase from 246.53 to 138,267 pF with a change in RH level from 11% to 97%. The maximum sensitivity was calculated to be 1,604.89 pF/%RH. Park and coworkers reported the preparation of rGO/MoS$_2$ hybrid composites and application in humidity sensing [38]. Interdigitated electrode definition was done by standard photolithography process followed by Pt/Ti evaporation using an electron beam evaporator. rGO/MoS$_2$ solution was dropped on the Pt/SiO$_2$/Si and Pt/IED/PET substrate. The response of the sensor increased from 9.03% to 23.85% at 10%–90% RH. The minimum theoretical detection limit was calculated to be 0.01783% RH. This improvement in sensing was due to oxygen-containing functional groups on the rGO surface and dangling bonds at the MoS$_2$ edge. Hsu N-F, et al. synthesized dandelion-like nanostructures of ZnO on Si substrate using hydrothermal synthesis method [39]. Comb-shaped Pt-interdigitated electrodes were prepared using photolithography. The conductance increased ~360 times with a change in RH from 11% to 95%. The sensitivity increased from 10^2 to 10^5 with the synthesis temperature variation from 400°C to 700°C. Liang and co-workers reported the hydrothermal synthesis of ZnO NRs on oxidized Si wafers [40]. Magnetron-sputtered ZnO layer served as the seed layer for subsequent hydrothermal synthesis of ZnO. A 15-nm TiO$_2$ thin film was deposited using atomic-layer deposition (ALD) system to form ZnO-TiO$_2$ core-shell NR. The sensor exhibited a response of 39 at 85% RH exposure. For 33%–95% RH levels, response and recovery times were 10–40 and 5–20 seconds, respectively. Zhang et al. reported highly sensitive humidity sensors using NaNbO$_3$ nanofibers [41]. Pt/Ti-interdigitated electrodes were patterned on an alumina substrate using lift-off photolithography and sputtering techniques. Far-field electrospinning process was used to assemble NaNbO$_3$ nanofibers on the substrate. With a change in RH level from 20% to 80%, the maximum sensitivity was found to be 10^5. The response time of the humidification process was less than 3 seconds. This high sensitivity was attributed to electric field-induced proton transfer between H$_3$O$^+$ due to the physisorption of water molecules. Sun et al. developed high sensitivity capacity humidity sensors using Zn$_{1-x}$Ni$_x$O nanostructures, where x was Ni doping concentration [42]. Zn$_{1-x}$Ni$_x$O nanowires (NWs) were grown using hydrothermal synthesis methods. The interdigitated electrode was prepared by Ti/Au patterning using photolithography and DC magnetron sputtering. With an increase in RH from 11% to 95%, the capacitance sensitivity of the 5% Ni-doped ZnO humidity sensor exhibited more than four-order variation in magnitude. The response and recovery times were improved to 27 and 2 seconds, respectively. This improvement in performance was attributed to Ni ion and oxygen vacancy in the non-stoichiometric oxide. Wang and coworkers investigated humidity-sensing properties of CuO/rGO nanostructures [43]. Modified Hummer's method was used to prepare graphene oxide from graphite powder. Microwave-assisted hydrothermal method was used to synthesize urchin-like CuO/rGO nanocomposite. This composite was mixed with ethanol to coat a ceramic substrate attached with Ag-Pd-interdigitated electrodes. The maximum response was 22,700 when RH varied in high range (75%–98%). The response and recovery times for the composite were 2 and 17 seconds, respectively. Under high humidity conditions, the hole barrier at CuO/rGO Schottky junction reduced greatly, thus decreasing the impedance greatly. This phenomenon resulted in very high response from the sensor under high humidity conditions.

3.4 HEAVY METAL SENSOR

The metallic elements having density greater than 5 g/cm^3 are termed as heavy metals. There are various examples of metals such as cadmium (Cd), zinc (Zn), mercury (Hg), chromium (Cr), and arsenic (As), which are very toxic even if present in low concentration as these metals cannot be destroyed or degraded [44]. Corresponding ions of heavy metals can be accumulated in animals and plants on exposure to polluted environment, resulting in bioaccumulation [45]. Moreover, in comparison with the adults, children are most widely affected by the heavy metals. Organ failure or neurotoxicity may be due to the heavy metal exposure beyond the permissible limit. Among the children, lead consumption may result in problems in learning neuropsychological development and impaired growth. Few heavy metals such as Cd and Cr are carcinogenic [46]. However, detection and removal of the heavy metal ions become difficult.

TABLE 3.1

The List of Sources, Permissible Limit, and Consequences of Heavy Metal

Heavy Metals	Limit Value (mg/L)	Toxicity	Sources
As	0.01	Causes skin damage, cancer, neurobehavior sickness	Coloring agent in textile, wall paper, and toy-making industry
Cd	0.003	Kidney damage, cancer, and obstructive lung disease	Electroplating industry, battery
Cr	0.05	Diarrhea, skin and mucous membrane irritation, bronchopulmonary effects and systemic effects involving kidney, liver, gastrointestinal tract, and circulatory system	Electroplating, leather tanning, and textile industry
Cu	2	Liver damage, insomnia	Sewage effluent, fertilizers, and pesticide
Ni	0.07	Nausea, chronic asthma, and cancer	Electroplating, printing, silver refineries, battery manufacturing industry
Zn	-	Depression, stomach cramps, nausea, and vomiting	Electroplating, smelting, and ore processing, as well as acid mine drainage, effluents from chemical processes and discharge of untreated domestic sewage
Pb	2	Effect circulatory and nervous system, impaired growth in children, and induced learning disabilities	Plastic, paint, pipe, steel, lead acid batteries, and gasoline
Hg	0.006	Rheumatoid arthritis, effect circulatory, and nervous system	Thermometer, electrical properties

It is evident from Table 3.1 that industrial activities are primary source of the heavy metals that might be due to improper and insufficient monitoring and treatment of industrial wastewater. Real-time monitoring of the heavy metals can be achieved by using the portable sensors. Substantial improvement in such sensors regarding their selectivity, sensitivity, and detection capabilities can be achieved by the incorporation of the metal oxide nanostructures. Hence, appropriate monitoring and control of the heavy metals released into the environment can be targeted. In order to detect heavy metals, selectively bind the heavy metal to enzymes or protein. In biosensors, metal oxide nanostructures can be integrated with protein, enzymes, or small molecules for the detection of the heavy metal ions. Portable sensor might be proven useful for the *in situ* detection of heavy metals at the site of discharge. For the determination of the concentration of heavy metals in the industrial wastewater, one of the most acceptable methods is grab sampling and further laboratory analysis of the sample.

For the sensing of the heavy metal ions in the polluted water sources, various biosensors such as electrochemical and optical sensors were also developed. Due to defined 3-D structures and precise detection abilities, aptamer-based biosensors are drawing interest and becoming popular very fast. Aptamers are basically oligonucleotide peptide molecules of size less than 25 kDa [47]. These aptameters have high specificity and affinity to the target molecules [48]. Among various classes of the biosensors used for the heavy metal ion sensing, EC biosensors are found to be superior due to their rapid detection ability, high sensitivity, size miniaturization, low cost, and ease of handling [49]. This class of the biosensors is constructed by two or three EC cells in order to transfer EC signal corresponding to a biological event [50]. A typical EC biosensor will perform amperometric/voltammetric detection (measurable current), potentiometric detection (measurable potential/charge accumulation), conductometric detection (measurable conductive properties) between electrodes, and impedimetric detection (measurable impedance) [49–52]. The detection limits achieved by the researchers by using the aptamer EC sensors are shown in the table. Furthermore, optical sensors are the sensors that can capture a variety of the signals, including ultraviolet, visible, and infrared, resulting from different biological/chemical/physical reaction, and these signals are further transformed into different form of energy. Depending upon the light source, optical biosensors are classified as fluorometric, bioluminescence, chemiluminescence, and colorimetric sensors [53]. Nanomolar-range detection limit could be achieved by using these optical sensors when used for the detection of the heavy metal [54,55]. In order to measure ionic in- and ex-fluxes at the membrane, field-effect transistors were introduced [56]. Though these electrical biosensors were not difficult to instruct, many of these sensors were used for the detection of heavy metal, and most of them were used for the detection of the DNA, viruses, and protein. By using single-walled CNT, aptamer-based FETs were successfully constructed to achieve the detection limit of picomolar range [57]. Moreover, silicon NW-based FETs emerged as the promising candidates in the field of biosensors [58]. These biosensors have various advantages such as high sensitivity, selectivity, and ability of onsite real-time detection; however, along with several advantages, these sensors suffered from the disadvantages of loss in their sensitivity in case of the samples having high ionic concentration [59]. Loss in the sensitivity, when used for the high ionic concentration samples, might be attributed to the shortened debye length [60]. In order to address these limitations, various strategies were proposed by the researchers to make these sensors efficient for the sensitive detection of high ion concentration. Alternative strategies to use different measurement modes and nanoelectronics sensing element configurations were proposed – one of them was the use of memristive biosensors. Preliminary concept of the memristive was followed by the various researchers, and several designs of the devices, which can be used as the resistive membrane when employed with metal oxides, were explored. Titanium dioxide is the most widely used metal oxide as the memresistance due to its functional properties and robustness. However, many more transition metal oxides were also used. A typical memristor configuration Pt/TiO$_2$/conducting glass is schematically shown in Figure 3.5.

The design proposed by the researcher as shown in Figure 3.5 is expected to act as high-performance device for the detection of the heavy metals once the surface modification is carried out to increase the sensitivity of TiO$_2$ layers to heavy metals. Nonetheless, numerous attempts were made to develop portable sensors for heavy metal detection in industrial wastewater. Use of the metal oxide-based nanostructures for mini-portable sensors is of great importance for *in situ* monitoring of the industrial wastewater. Introducing metal oxide nanostructures into the sensor design not only opens up a new prospective for the *in situ* monitoring of the heavy metals in the industrial wastewater; however, integration of the biomolecules with the metal oxide nanostructures enhances selectivity to the heavy metal ions.

The presence of the heavy metals in the industrial discharge became the worldwide issue that must be resolved to keep the environment safe. The heavy metal discharged into industrial wastewater not only impacts the aquatic life but also severely affects the health of the surrounding local communities. However, various sensors are described already for the real-time monitoring of the heavy metals, but only detection does not completely solve the issue; regular treatments of the wastewater laden with hazardous heavy metals at industrial facilities must be needed to restrict the presence of heavy metals in industrial wastewater before discharging it into the environment.

Conventional methods used for the treatment of wastewater involve multiple steps. Initially, in order to separate grease and solids from water, primary treatment is done, which is then followed by biological treatment in order to remove the dissolved organic contaminants; in addition to this small fraction of the phosphorus, nitrogen and heavy metals are also eliminated; this process is termed as secondary treatment. Under secondary treatment process, generally dissolved heavy metals are separated out. The advantages and disadvantages of the various techniques for eliminating the heavy metals are summarized in Table 3.2. In many cases, nature of the targeted wastewater is not simple in these cases; in order to deal with the complexity of the wastewater, combination of

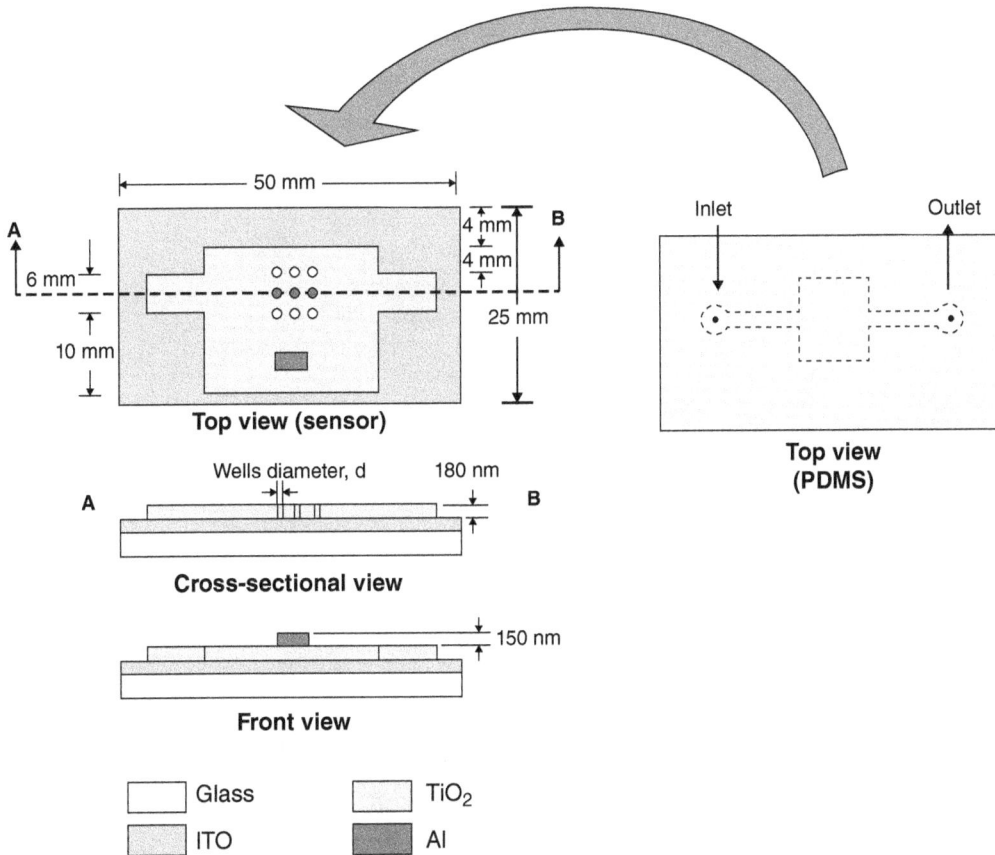

FIGURE 3.5 Design for the memristor biosensor structure for metal ion detection. (Reused from Alias et al., 2020 [61]).

the different methods is utilized for the removal of the heavy metals. Metal oxides exhibit favorable properties for the adsorption of the ions on their surface. Moreover, some oxides show semiconducting behavior, which enables the adsorption of the heavy metal ions and their subsequent conversion into their useful counterparts. On the basis of the research carried out in the literature, it is observed that some metal oxides such as zirconia oxide and titanium di oxide can be used as adsorbent as well as photocatalyst, due to which some targeted heavy ions can not only be adsorbed on their surfaces but also be further reduced under the irradiation of light. On the other hand, few metal oxides such as maghemite and Fe_2O_3 can be used as an adsorbent only; they cannot perform a photocatalytic activity [62–64]. There are some traditional oxides such as kaolinite, bentonite, and montmorillonite, which can be used effectively for the remediation of wastewater.

3.4.1 ADSORPTION METHODS

The adsorption process is simple, cost-effective, and easier compared to other process used for the wastewater containing heavy metals [66,67]. Adsorption process can be classified into two categories, namely, physical and chemical adsorption. In case of the physical adsorption, adsorbates are attached on the adsorbent surface due to van der Waals interaction. This kind of adsorption does not affect the electronic structure of the molecule of atom significantly. van der Waals force gets originated through the interaction between permanent, transient, and induced dipoles. Commercial adsorbents usually undergo physical adsorption. Physical adsorption occurs only

TABLE 3.2

Comparison Chart between Methods Used for Heavy Metals Removal

Method	Advantages	Disadvantages	Reference
Ion exchange	i. Reduction of biological sludge/chemicals ii. Sorbent's recycling iii. High efficiency	i. Difference of ions affinity in resin and ions to be removed is needed ii. High operational cost	[65]
Adsorption	i. High concentration of heavy metals can be removed ii. Abundance of sorbents iii. Cost-effective	i. Waste by-products ii. Less selective iii. Efficiency dependent on the waste by-products	[66]
Precipitation	i. Simple and cost-effective ii. Limited to low heavy ion concentration	i. Complex mixture of metals can be removed ii. Secondary pollution due to the use of chemicals	[67]
Photocatalysis	i. Cost-effective ii. Nanostructured metal oxides or modified metal oxides can be used to achieve nearly complete removal by using this method	i. Less efficiency with high concentration of heavy metals ii. Not applicable for all heavy metals iii. pH dependent iv. Difficult to treat complex metal mixtures Only possible with minute/low concentration heavy metal ions	[68–70]
Electrochemical	i. Fast ii. Less sludge production	i. Expensive ii. Chemicals used can result in secondary pollution	[71]
Membrane Filtration	i. Simple and highly efficient ii. Water and salts can be reused	i. Membrane fouling (high maintenance cost) ii. Sludge generation iii. Membrane size dependent	[70,71]

under the conditions favorable for the multilayer absorption. On the other hand, in case of chemical adsorption, chemical reaction takes place between the adsorbate and the adsorbent, resulting in the formation of new bond between the adsorbate and the substrate surface. The monolayer can be formed by the adsorbate, which can be further utilized for the catalytic reaction. Adsorption process involves the multiple steps. First, pollutants are transported to the adsorbent's surface from the bulk solution followed by the adsorption of the pollutants on the active sites of the adsorbent [45]. Further transportation within the sorbent's particles takes place. Pore size distribution, surface area, functional groups present on the surface, and adsorbent's polarity are the factors that greatly affect the effectiveness of the adsorption process [67]. Various studies were carried out by the researchers for the removal of the heavy metal ions; few of them are listed in Table 3.3.

3.4.1.1 Adsorption Isotherms and Kinetics

The modeling of experimental data from the adsorption processes is a very important means of predicting the mechanisms of various adsorption systems. The adsorption behavior of nanomaterials can be illustrated by using the few models introduced below.

Adsorption Isotherm: In order to analyze the capability of the adsorbents, adsorption isotherms are useful. Under the condition of the adsorption equilibrium, the relationship between the equilibrium concentration of the adsorbate at constant temperature and adsorbate amount on the adsorbents is termed as adsorption. A variety of the models were used to determine the adsorption isotherm, such as Freundlich isotherm, Langmuir, sips models, etc. Freundlich and Langmuir models are the extensively used models.

TABLE 3.3
Adsorption Rate of Different Materials on Specific Heavy Metal

Adsorbent	Adsorption Rate (mg/g)					Reference
	Cd	Cu	Hg	Ni	Zn	
Natural clay		44.84				[72]
Chitosan					75	[73]
Bentonite	11.20					[74]
Bentonite				92.59		[75]
Bentonite					68.49	[76]
Kaolinite		10.78				[77]
Kaolinite				2.10		[78]
Kaolinite					4.95	[79]
Montmorillonite	6.30					[80]
Montmorillonite					154.60	[81]
Zeolite	10.87	23.25			12.85	[82]

a. Langmuir Model

According to the assumption of Langmuir model, uniform adsorption takes place on the adsorbent's active sites and saturation occurs once the adsorptive sites get occupied by the adsorbates, which means no further adsorption will take place on these active sites [83]. According to the assumption of this model, all the active adsorption sites will have equal binding energy and binding of only single adsorbate can take place at each active site. Langmuir model can be expressed in the form of the following linear equation [84]:

$$\frac{1}{q_e} = \frac{1}{bq_mC_e} + \frac{C_e}{q_m} \tag{3.1}$$

where C_e (mg/L) is termed as equilibrium concentration and q_e is the adsorbed amount at equilibrium. q_m (mg/g) and b are the constants for the maximum single-layer capacity and adsorption energy, respectively.

The favorability of the adsorption process can be determined by the equation given by:

$$R_L = \frac{1}{1 + bC_0} \tag{3.2}$$

where C_0 is the highest initial concentration of the adsorbate and b (L/mg) is termed as Langmuir constant. If the value of the $R_L > 1$, then it represents the unfavorable adsorption; if the value of the $R_L = 1$, then it indicates the linear adsorption. However, $0 < R_L < 1$ corresponds to the favorable adsorption, and $R_L = 0$ corresponds to the irreversibility.

b. Freundlich Isotherm

The Freundlich isotherm is used to explain non-ideal sorption [85]. This model is used for the heterogeneous surface having an exponential distribution of the active sites. In contrast to Langmuir isotherm, multilayer adsorption is the basis of the Freundlich model and its linear form can be expressed as follows [86]:

$$\ln q_e = \ln K_f + \frac{1}{n} \ln C_e \tag{3.3}$$

where K_f (mg/g) is known as the Freundlich constant for the adsorption capacity of adsorbent, and $1/n$ indicates the adsorption intensity.

c. Sips Model

This model is the hybrid model of the Freundlich and Langmuir isotherm [87]. This model is used for the prediction of the heterogeneous adsorption systems at low concentration of the adsorbate. It predicts the multilayer adsorption capacity of the Freundlich isotherm model [88]. However, at high adsorbate concentration, it reduces to the Langmuir isotherm [89]. The expression for the sips model is given below [90]:

$$q_e = \frac{K_s\, C_e^{\beta s}}{1 - a_s\, C_e^{\beta s}} \tag{3.4}$$

where K_s is termed as the isotherm model constant (L/g), a_s is the isotherm exponent constant (L/g), and β_s is the sip's isotherm exponent.

The expression for the linearized form of the equation is given by [90]:

$$\beta_s\, \ln C_e = -\ln\left(\frac{K_s}{q_e}\right) + \ln\left(a_s\right) \tag{3.5}$$

Parameters associated with this model are dependent on temperature, concentration, and pH [91,92]. Moreover, for linear regression and non-linear regression, isotherm constants are different [93].

3.4.1.2 Adsorption Kinetics

For the designing of the adsorption systems, adsorption kinetics plays the significant role. Adsorption kinetics is used to describe the relationship between the adsorbate amount adsorbed on the surface of the adsorbent (q_t) and contact time (t). Pseudo-first-order and pseudo-second-order models are the most commonly used kinetic models. Depending upon the adsorption capacity, pseudo-first-order model is used to describe adsorption for the liquid-solid phase systems [94]. However, the pseudo-second-order model is used for the solid-phase systems. General expression for the pseudo-first-order model is given by [95]:

$$\frac{dq_t}{dt} = K\,(q_e - q_t) \tag{3.6}$$

The integral form of the above equation can also be written as:

$$log\,(q_e - q_t) = logq_e - \frac{k_1}{2.303}t$$

where q_e is the adsorbent's (mg/g) adsorption capacity, q_t is the adsorption capacity (mg/g), t is the time of contact, and k_1 and K are the rate constants.

The following equation can be used to express pseudo-second-order model [96]:

$$\frac{dq_t}{dt} = K\,(q_e - q_t)^2 \tag{3.7}$$

The integral form of the above equation is given as follows:

$$\frac{t}{q_t} = \frac{1}{k_2\, q_e^2} + \frac{1}{q_e}\,t \tag{3.8}$$

where q_t is the adsorption capacity (mg/g) at the contact time of t, q_e is the adsorbent's equilibrium adsorption capacity, and k_2 is the rate constant for the pseudo-second-order kinetic model.

3.4.2 ELECTROCHEMICAL DETECTION

Due to high sensitivity of detection and low instrumentation cost, electrochemical detection technique is the most preferred technique for the accurate detection of the heavy metal ions. Anodic stripping voltammetry is most widely used among the available EC techniques for sensing the heavy metal ions. This anodic stripping voltammetry is used in combination of another technique usually combined with pulse voltammetry such as square wave or differential pulse termed as differential pulse ASV and square wave ASV, respectively. Moreover, three electrodes, namely, working electrode, counter electrode, and reference electrode, are used in these techniques. In this method, the produced current at the working electrode is measured as the result of applied voltage. The choice of the materials for working electrode in the working cell plays a significant role as all the redox reactions take place at the interface of working electrode and analyte. Mercury was widely used by the researchers as working electrode due to its superior sensing properties; however, due to its toxicity and less environmental-friendly characteristics, it was replaced by bismuth due to its less toxic and environmental-friendly nature [97–101]. Silver and gold were also explored for the use of these materials for sensing of heavy metal ions. Metal oxide nanomaterials, especially hematite, were recently used for the modification of the working electrode [102]. Further, for the improvement of the selectivity and detection limit of the working electrode for heavy metal ion, other oxides such as tin oxide and zinc oxide were used, as given in Table 3.4 [103]. It is evident from the table that SWASV (square wave anodic stripping voltammetry) is the mostly widely used method for the detection of the heavy metals due to the fast scan rate, in comparison with pulse voltammetry [108]. Modification of the working electrode with nanostructured materials such as nanoflakes and NPs is shown in Figure 3.6 [104]. These sensors are helpful in the real-time monitoring of the heavy metals at the point of discharge.

TABLE 3.4

Selectivity and Detection Limit of the Working Electrode for Heavy Metal Ion Other Oxides

Electrode	Analyte	Limit of Detection	Sensitivity ($\mu A/\mu M$)	Technique	Reference
SnO_2/Rgo/GCE	Cd(II)	0.1015 nM	18.4	SWAS	[102]
	Cu(II)	0.2269 nM	14.98		
	Hg(II)	0.2789 Nm	28.2		
	Pb(II)	0.1839 nM	18.6		
GO/[Ru(bpy)$_3$]$^{2+}$/Au	Cd(II)	2.80 Nm	17.51	DPV	[104,105]
	As(III)	2.30 Nm	23.6		
	Hg(II)	1.60 Nm	1.60		
	Pb(II)	1.41 nM	1.41		
ZnO/Rgo/GCE	Cu(II)	0.03 μM	–	SWAS	[106]
	Hg(II)	0.06 μM			
	Pb(II)	0.03 μM			
	Cd(II)	0.04 μM			
Bi/Fe$_2$O$_3$ NPs/G/GCE	Zn(II)	0.11 μg/L	–	DPSAV	[107]
	Pb(II)	0.07 μg/L			
	Cd(II)	0.08 μg/L			
AuNPs/SPCE	Cu(II)	1.4 ng/L	3.52	SWAS	[108]
	Pb(II)	2.1 ng/L	5.94		

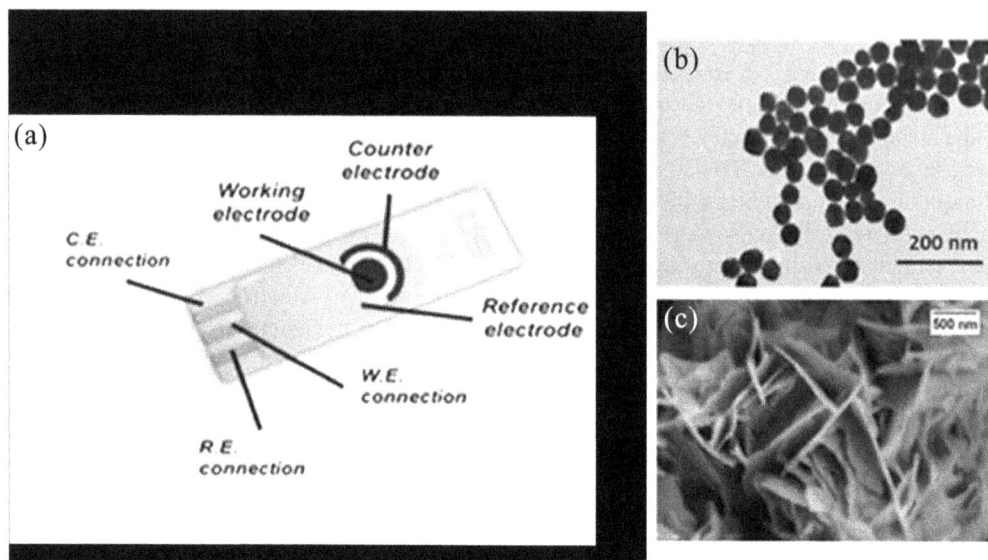

FIGURE 3.6 Typical electrochemical electrode: (a) commercial screen-printed electrode, (b) nanoparticles for working electrode modification, and (c) oxide nanoflakes for working electrode modification. (Reused from Alias et al., 2020 [61]).

3.4.3 PHOTOCATALYST

A variety of the interpretations and the definitions of the term "photocatalysis" have been used as well as proposed by the researchers. Photocatalysis is a chemical reaction occurring as a result of the photoabsorption of the photocatalyst, which does not go any significant change during and after the reaction. In other words, photocatalyst acts catalytically without undergoing any kind of change in its structure or composition under the photoirradiation. This definition of the photocatalysis is consistent with most of the definitions proposed by the researchers and explains the phenomenon with a general approach. There are various applications of the photocatalysis that are explored by the researchers; however, in the present chapter, we will be mainly focusing on the application of the photocatalysis in the field of treatment of heavy metal and organic contaminants present in the environment. For the treatment of heavy metal ions as well as organic pollutants, photocatalysis is one of the effective, simple, environmental-friendly cost-effective processes. Under this process, generally either reduction or oxidation of the target pollutant is carried out. Wide band gap metal oxide semiconductors such as TiO_2, ZnO, ZrO_2, CuO, and Nb_2O_5 were widely used by the researchers for the photocatalytic activity due to their electronic structure [63,109–116]. These semiconductor metal oxides having a wide band gap provide a favorable light absorption for the generation of electron/hole pair. When appropriate light energy is supplied to the metal oxide semiconductor, electron/hole pair generation takes place in the conduction band and valance band, respectively. In order to absorb the pollutants (organic or heavy metal ions), electrons are transferred from the conduction band, and thus, reduction takes place. However, this reduction process is always accompanied by the oxidation in which holes present in the valance band are used for the oxidation of the water. Though, it is not possible to use photocatalysis for the removal of all types of heavy metal ions. TiO_2 was used by the researchers for the photocatalytic removal of a series of heavy metal ions Cd (II), Cr (VI), Cu (II), and Pb (II) [117].

Among these heavy metal ions, the maximum removal of Cr (VI) was observed; however, the minimum removal of Cd (II) was reported. Further study has shown that the TiO_2 nanostructures were successfully used for the removal of up to 99.8% Cr (VI) from the contaminated wastewater.

The removal of the Cr (VI) involves first the reduction of Cr (VI) into Cr (III), which is easier to remove via precipitation from the treated water. It was also observed that the reduction was greatly dependent on the pH [113]. The pH optimum for the removal of Cr (VI) to attain the maximum removal was 2 [114]. However, the maximum performance of photocatalytic removal was observed for Cu (III) at pH 2–3.5 [115] and Hg at pH 4–4.1 [117]. In few reports, in order to enhance the photocatalytic performance, hole scavengers are added to increase the rate of photocatalytic reaction. Recombination rate is suppressed as the result of addition of scavenger, leading to improved photoreduction [118].

3.4.4 SUMMARY

The human society health standard is very closely related to the environment around its inhabitants. The pollution level of the environment is a very critical parameter, and the recent surge in pandemic cases demonstrated its deadly effect on patients with comorbidity issues related to deteriorated environment. So, it is of utmost importance to assess the environment quality and take an immediate measure to improve its condition. This chapter discussed about the different types of sensors associated with the sensing of environmental parameters and their relationship with different contaminants in the atmosphere. The preparation method and performance of different evaluation tools like pH sensors, humidity sensors, biosensors, heavy metal detection, and remedial are elaborated. The recent extensive investigation of IoT compatible environment sensors and the cloud network can lead to a very efficient monitoring and improvement of different environment parameters. However, there are some challenges such as reading stabilization issues, ability to recover from condensation, repeatability, speed, robustness in outdoor application, compatibility, etc. The present research attention is concentrated at these research gaps, and realization of these goals will help us to have a better world to live with.

REFERENCES

[1] U. N. SDG, *Sustainable Development Goals*, United Nations, New York City 2018.

[2] J. K. Patra and K.-H. Baek, "Green nanobiotechnology: factors affecting synthesis and characterization techniques," *J. Nanomater.*, vol. 2014, p. 219, 2014.

[3] J. Jeevanandam, A. Barhoum, Y. S. Chan, A. Dufresne, and M. K. Danquah, "Review on nanoparticles and nanostructured materials: history, sources, toxicity and regulations," *Beilstein J. Nanotechnol.*, vol. 9, no. 1, pp. 1050–1074, 2018.

[4] M. A. Azmi and K. F. Shad, "Role of nanostructure molecules in enhancing the bioavailability of oral drugs," in *Nanostructures for Novel Therapy*, edited by Denisa Ficai and Alexandru Mihai Grumezescu, Elsevier, 2017, pp. 375–407.

[5] F. Ejehi, R. Mohammadpour, E. Asadian, P. Sasanpour, S. Fardindoost, and O. Akhavan, "Graphene oxide papers in nanogenerators for self-powered humidity sensing by finger tapping," *Sci. Rep.*, vol. 10, no. 1, pp. 1–11, 2020.

[6] M. T. Ghoneim et al., "Recent progress in electrochemical pH-sensing materials and configurations for biomedical applications," *Chem. Rev.*, vol. 119, no. 8, pp. 5248–5297, 2019.

[7] R. Batool, A. Rhouati, M. H. Nawaz, A. Hayat, and J. L. Marty, "A review of the construction of nanohybrids for electrochemical biosensing of glucose," *Biosensors*, vol. 9, no. 1, p. 46, 2019.

[8] F. Fu and Q. Wang, "Removal of heavy metal ions from wastewaters: a review," *J. Environ. Manage.*, vol. 92, no. 3, pp. 407–418, 2011.

[9] J. O. Nriagu, "A global assessment of natural sources of atmospheric trace metals," *Nature*, vol. 338, no. 6210, pp. 47–49, 1989, doi: 10.1038/338047a0.

[10] J. E. Fergusson, *The Heavy Elements: Chemistry, Environmental Impact and Health Effects*, (No. 628.53 F4). 1990.

[11] J. H. Duffus, "'Heavy metals'—a meaningless term," *Chem. Int. IUPAC*, vol. 23, no. 6, pp. 163–167, 2001.

[12] Z. L. He, X. E. Yang, and P. J. Stoffella, "Trace elements in agroecosystems and impacts on the environment," *J. Trace Elem. Med. Biol.*, vol. 19, no. 2–3, pp. 125–140, 2005.

[13] N. Herawati et al., "Cadmium, copper, and zinc levels in rice and soil of Japan, Indonesia, and China by soil type.," *Bull. Environ. Contam. Toxicol.*, vol. 64, no. 1, pp. 33–39, 2000.

[14] S. Shallari, C. Schwartz, A. Hasko, and J.-L. Morel, "Heavy metals in soils and plants of serpentine and industrial sites of Albania," *Sci. Total Environ.*, vol. 209, no. 2–3, pp. 133–142, 1998.

[15] M. Li, H. Gou, I. Al-Ogaidi, and N. Wu, "Nanostructured Sensors for Detection of Heavy Metals: A Review," *ACS Sustainable Chem. Eng.*, vol. 1, no. 2–3, pp. 713–723, 2013.

[16] H. Bin Wu, J. S. Chen, H. H. Hng, and X. W. D. Lou, "Nanostructured metal oxide-based materials as advanced anodes for lithium-ion batteries," *Nanoscale*, vol. 4, no. 8, pp. 2526–2542, 2012.

[17] R. S. Devan, R. A. Patil, J.-H. Lin, and Y.-R. Ma, "One-dimensional metal-oxide nanostructures: recent developments in synthesis, characterization, and applications," *Adv. Funct. Mater.*, vol. 22, no. 16, pp. 3326–3370, 2012, doi: 10.1002/adfm.201201008

[18] A. Kolmakov and M. Moskovits, "Chemical sensing and catalysis by one-dimensional metal-oxide nanostructures," *Annu. Rev. Mater. Res.*, vol. 34, pp. 151–180, 2004.

[19] T. Guo, M.-S. Yao, Y.-H. Lin, and C.-W. Nan, "A comprehensive review on synthesis methods for transition-metal oxide nanostructures," *CrystEngComm*, vol. 17, no. 19, pp. 3551–3585, 2015.

[20] C. Haur Kao et al., "Multianalyte biosensor based on pH-sensitive ZnO electrolyte—insulator—semiconductor structures," *J. Appl. Phys.*, vol. 115, no. 18, p. 184701, 2014.

[21] L. Maiolo et al., "Flexible pH sensors based on polysilicon thin film transistors and ZnO nanowalls," *Appl. Phys. Lett.*, vol. 105, no. 9, p. 93501, 2014.

[22] S. Zhuiykov, E. Kats, K. Kalantar-zadeh, M. Breedon, and N. Miura, "Influence of thickness of submicron Cu_2O-doped RuO_2 electrode on sensing performance of planar electrochemical pH sensors," *Mater. Lett.*, vol. 75, pp. 165–168, 2012.

[23] L. Manjakkal, K. Cvejin, J. Kulawik, K. Zaraska, R. P. Socha, and D. Szwagierczak, "X-ray photoelectron spectroscopic and electrochemical impedance spectroscopic analysis of RuO_2-Ta_2O_5 thick film pH sensors," *Anal. Chim. Acta*, vol. 931, pp. 47–56, 2016.

[24] B. Xu and W.-D. Zhang, "Modification of vertically aligned carbon nanotubes with RuO_2 for a solid-state pH sensor," *Electrochim. Acta*, vol. 55, no. 8, pp. 2859–2864, 2010.

[25] R. Zhao, M. Xu, J. Wang, and G. Chen, "A pH sensor based on the TiO_2 nanotube array modified Ti electrode," *Electrochim. Acta*, vol. 55, no. 20, pp. 5647–5651, 2010.

[26] L. Manjakkal, E. Djurdjic, K. Cvejin, J. Kulawik, K. Zaraska, and D. Szwagierczak, "Electrochemical impedance spectroscopic analysis of RuO_2 based thick film pH sensors," *Electrochim. Acta*, vol. 168, pp. 246–255, 2015.

[27] C.-E. Lue et al., "pH sensing reliability of flexible ITO/PET electrodes on EGFETs prepared by a roll-to-roll process," *Microelectron. Reliab.*, vol. 52, no. 8, pp. 1651–1654, 2012.

[28] L. Manjakkal, K. Zaraska, K. Cvejin, J. Kulawik, and D. Szwagierczak, "Potentiometric RuO_2—Ta_2O_5 pH sensors fabricated using thick film and LTCC technologies," *Talanta*, vol. 147, pp. 233–240, 2016.

[29] A. Lale, A. Tsopela, A. Civélas, L. Salvagnac, J. Launay, and P. Temple-Boyer, "Integration of tungsten layers for the mass fabrication of WO_3-based pH-sensitive potentiometric microsensors," *Sens. Actuators B Chem.*, vol. 206, pp. 152–158, 2015.

[30] J.-C. Lin, B.-R. Huang, and Y.-K. Yang, "IGZO nanoparticle-modified silicon nanowires as extended-gate field-effect transistor pH sensors," *Sens. Actuators B Chem.*, vol. 184, pp. 27–32, 2013.

[31] M. Chen, Y. Jin, X. Qu, Q. Jin, and J. Zhao, "Electrochemical impedance spectroscopy study of Ta_2O_5 based EIOS pH sensors in acid environment," *Sens. Actuators B Chem.*, vol. 192, pp. 399–405, 2014.

[32] L. Manjakkal, B. Sakthivel, N. Gopalakrishnan, and R. Dahiya, "Printed flexible electrochemical pH sensors based on CuO nanorods," *Sens. Actuators B Chem.*, vol. 263, pp. 50–58, 2018.

[33] M. Hussain, Z. H. Ibupoto, M. A. Abbasi, O. Nur, and M. Willander, "Effect of anions on the morphology of Co_3O_4 nanostructures grown by hydrothermal method and their pH sensing application," *J. Electroanal. Chem.*, vol. 717, pp. 78–82, 2014.

[34] S.-J. Young, L.-T. Lai, and W.-L. Tang, "Improving the performance of pH sensors with one-dimensional ZnO nanostructures," *IEEE Sens. J.*, vol. 19, no. 23, pp. 10972–10976, 2019.

[35] H. Jeong, Y. Noh, and D. Lee, "Highly stable and sensitive resistive flexible humidity sensors by means of roll-to-roll printed electrodes and flower-like TiO_2 nanostructures," *Ceram. Int.*, vol. 45, no. 1, pp. 985–992, 2019.

[36] J. J. Steele, M. T. Taschuk, and M. J. Brett, "Response time of nanostructured relative humidity sensors," *Sens. Actuators B Chem.*, vol. 140, no. 2, pp. 610–615, 2009.

[37] D. Zhang, H. Chang, P. Li, R. Liu, and Q. Xue, "Fabrication and characterization of an ultrasensitive humidity sensor based on metal oxide/graphene hybrid nanocomposite," *Sensors Actuators B Chem.*, vol. 225, pp. 233–240, 2016.

[38] S. Y. Park et al., "Room temperature humidity sensors based on rGO/MoS$_2$ hybrid composites synthesized by hydrothermal method," *Sensors Actuators B Chem.*, vol. 258, pp. 775–782, 2018.

[39] N.-F. Hsu, M. Chang, and K.-T. Hsu, "Rapid synthesis of ZnO dandelion-like nanostructures and their applications in humidity sensing and photocatalysis," *Mater. Sci. Semicond. Process.*, vol. 21, pp. 200–205, 2014.

[40] Y.-C. Liang, W.-K. Liao, and S.-L. Liu, "Performance enhancement of humidity sensors made from oxide heterostructure nanorods via microstructural modifications," *RSC Adv.*, vol. 4, no. 92, pp. 50866–50872, 2014.

[41] Y. Zhang et al., "Fast and highly sensitive humidity sensors based on NaNbO$_3$ nanofibers," *RSC Adv.*, vol. 5, no. 26, pp. 20453–20458, 2015.

[42] N. Sun et al., "High sensitivity capacitive humidity sensors based on Zn$_{1-x}$ Ni$_x$O nanostructures and plausible sensing mechanism," *J. Mater. Sci. Mater. Electron.*, vol. 30, no. 2, pp. 1724–1738, 2019.

[43] Z. Wang et al., "Humidity-sensing properties of urchinlike CuO nanostructures modified by reduced graphene oxide," *ACS Appl. Mater. Interfaces*, vol. 6, no. 6, pp. 3888–3895, 2014.

[44] S. Mishra et al., "Heavy metal contamination: an alarming threat to environment and human health," in *Environmental Biotechnology: For Sustainable Future*, Springer, Singapore, 2019, pp. 103–125.

[45] R. C. Sobti, N. K. Arora, and R. Kothari, *Environmental Biotechnology: For Sustainable Future*. Springer, Singapore, 2018.

[46] M. A. Barakat, "New trends in removing heavy metals from industrial wastewater," *Arab. J. Chem.*, vol. 4, no. 4, pp. 361–377, 2011.

[47] G. Mayer, "The chemical biology of aptamers," *Angew. Chemie Int. Ed.*, vol. 48, no. 15, pp. 2672–2689, 2009.

[48] A. D. Ellington and J. W. Szostak, "In vitro selection of RNA molecules that bind specific ligands," *Nature*, vol. 346, no. 6287, pp. 818–822, 1990.

[49] A. Chaubey and B. Malhotra, "Mediated biosensors," *Biosens. Bioelectron.*, vol. 17, no. 6–7, pp. 441–456, 2002.

[50] Z. Zhu and H. S. Zhou, "Nanoparticles (NPs) for biosensing applications: current aspects and prospects," in *The World Scientific Encyclopedia of Nanomedicine and Bioengineering I: Volume 1: Noble Metal Nanoparticles for Biomedical Applications*, Edited by Yu Cheng, Jia Huang, Yarong Liu, Pin Wang, Bingbo Zhang, World Scientific, Singapore, 2017, pp. 177–209.

[51] A. Guiseppi-Elie and L. Lingerfelt, "Impedimetric detection of DNA hybridization: towards near-patient DNA diagnostics," in *Immobilisation of DNA on Chips I*, Edited by Christine Wittmann, Springer, Berlin, Heidelberg, 2005, pp. 161–186.

[52] A. Bonanni, M. J. Esplandiu, M. I. Pividori, S. Alegret, and M. Del Valle, "Impedimetric genosensors for the detection of DNA hybridization," *Anal. Bioanal. Chem.*, vol. 385, no. 7, pp. 1195–1201, 2006.

[53] R. Sharma, K. V Ragavan, M. S. Thakur, and K. Raghavarao, "Recent advances in nanoparticle based aptasensors for food contaminants," *Biosens. Bioelectron.*, vol. 74, pp. 612–627, 2015.

[54] Y. Luan, A. Lu, J. Chen, H. Fu, and L. Xu, "A label-free aptamer-based fluorescent assay for cadmium detection," *Appl. Sci.*, vol. 6, no. 12, p. 432, 2016.

[55] Y. Xiang and Y. Lu, "DNA as sensors and imaging agents for metal ions," *Inorg. Chem.*, vol. 53, no. 4, pp. 1925–1942, 2014.

[56] J. Wei, H. Yang, H. Sun, Z. Lin, and S. Xia, "A fully CMOS-integrated pH-ISFET interface circuit," in *2005 6th International Conference on ASIC*, 2005, vol. 1, pp. 365–367.

[57] E. S. Forzani et al., "Tuning the chemical selectivity of SWNT-FETs for detection of heavy-metal ions," *Small*, vol. 2, no. 11, pp. 1283–1291, 2006.

[58] C. Wang et al., "A label-free and portable graphene FET aptasensor for children blood lead detection," *Sci. Rep.*, vol. 6, no. 1, pp. 1–8, 2016.

[59] K.-I. Chen, B.-R. Li, and Y.-T. Chen, "Silicon nanowire field-effect transistor-based biosensors for biomedical diagnosis and cellular recording investigation," *Nano Today*, vol. 6, no. 2, pp. 131–154, 2011.

[60] E. Stern, R. Wagner, F. J. Sigworth, R. Breaker, T. M. Fahmy, and M. A. Reed, "Importance of the debye screening length on nanowire field effect transistor sensors," *Nano Lett.*, vol. 7, no. 11, pp. 3405–3409, 2007.

[61] N. Alias et al., "Metal oxide for heavy metal detection and removal," in *Metal Oxide Powder Technologies*, edited by, Yarub Al-Douri, Elsevier, 2020, pp. 299–332.

[62] Z. Lockman, *1-Dimensional Metal Oxide Nanostructures: Growth, Properties, and Devices*, edited by Z. Lockman, CRC Press, Boca Raton, 2018.

[63] N. Bashirom, T. W. Kian, G. Kawamura, A. Matsuda, K. A. Razak, and Z. Lockman, "Sunlight activated anodic freestanding ZrO$_2$ nanotube arrays for Cr (VI) photoreduction," *Nanotechnology*, vol. 29, no. 37, p. 375701, 2018.

[64] F. Budiman, N. Bashirom, W. K. Tan, K. A. Razak, A. Matsuda, and Z. Lockman, "Rapid nanosheets and nanowires formation by thermal oxidation of iron in water vapour and their applications as Cr (VI) adsorbent," *Appl. Surf. Sci.*, vol. 380, pp. 172–177, 2016.

[65] C. Balan, I. Volf, and D. Bilba, "Chromium (VI) removal from aqueous solutions by purolite base anion-exchange resins with gel structure," *Chem. Ind. Chem. Eng. Q.*, vol. 19, no. 4, pp. 615–628, 2013.

[66] F. Budiman, T. W. Kian, K. A. Razak, A. Matsuda, and Z. Lockman, "The assessment of Cr (VI) removal by iron oxide nanosheets and nanowires synthesized by thermal oxidation of iron in water vapour," *Procedia Chem.*, vol. 19, pp. 586–593, 2016.

[67] A. Azimi, A. Azari, M. Rezakazemi, and M. Ansarpour, "Removal of heavy metals from industrial wastewaters: a review," *ChemBioEng Rev.*, vol. 4, no. 1, pp. 37–59, 2017.

[68] C. E. Barrera-D\'\iaz, V. Lugo-Lugo, and B. Bilyeu, "A review of chemical, electrochemical and biological methods for aqueous Cr (VI) reduction," *J. Hazard. Mater.*, vol. 223, pp. 1–12, 2012.

[69] K. Gandha et al., "Mesoporous iron oxide nanowires: synthesis, magnetic and photocatalytic properties," *RSC Adv.*, vol. 6, no. 93, pp. 90537–90546, 2016.

[70] N. Bashirom, K. A. Razak, and Z. Lockman, "Synthesis of freestanding amorphous $ZrO2$ nanotubes by anodization and their application in photoreduction of Cr (VI) under visible light," *Surf. Coatings Technol.*, vol. 320, pp. 371–376, 2017.

[71] E. Kaprara, K. Simeonidis, A. I. Zouboulis, and M. Mitrakas, "Evaluation of current treatment technologies for Cr (VI) removal from water sources at sub-ppb levels," in *Proceedings of the 13th International Conference on Environmental Science and Technology, Athens, Greece*, 2013, pp. 5–7.

[72] Y. C. Sharma, "Thermodynamics of removal of cadmium by adsorption on an indigenous clay," *Chem. Eng. J.*, vol. 145, no. 1, pp. 64–68, 2008.

[73] J. R. Evans, W. G. Davids, J. D. MacRae, and A. Amirbahman, "Kinetics of cadmium uptake by chitosan-based crab shells," *Water Res.*, vol. 36, no. 13, pp. 3219–3226, 2002.

[74] B. Sathyanarayana and K. Seshaiah, "Kinetics and equilibrium studies on the sorption of manganese (II) and nickel (II) onto kaolinite and bentonite," *E-Journal Chem.*, vol. 8, no. 1, pp. 373–385, 2011.

[75] T. K. Sen and D. Gomez, "Adsorption of zinc (Zn^{2+}) from aqueous solution on natural bentonite," *Desalination*, vol. 267, no. 2–3, pp. 286–294, 2011.

[76] P. Liu, H. Sehaqui, P. Tingaut, A. Wichser, K. Oksman, and A. P. Mathew, "Cellulose and chitin nanomaterials for capturing silver ions (Ag^+) from water via surface adsorption," *Cellulose*, vol. 21, no. 1, pp. 449–461, 2014.

[77] T. Sen and M. Sarali, "Adsorption of cadmium metal ion (Cd^{2+}) from its aqueous solution by aluminium oxide and kaolin: a kinetic and equilibrium study," *J. Environ. Res. Dev.*, vol. 3, no. 1, pp. 220–227, 2008.

[78] S. Sen Gupta and K. G. Bhattacharyya, "Immobilization of Pb (II), Cd (II) and Ni (II) ions on kaolinite and montmorillonite surfaces from aqueous medium," *J. Environ. Manage.*, vol. 87, no. 1, pp. 46–58, 2008.

[79] A. Et and S. Shahmohammadi-Kalalagh, "Isotherm and kinetic studies on adsorption of Pb, Zn and Cu by kaolinite," *Casp. J. Environ. Sci.*, vol. 9, no. 2, pp. 243–255, 2011.

[80] S. M. Dal Bosco et al., "Removal of Mn (II) and Cd (II) from wastewaters by natural and modified clays," *Adsorption*, vol. 12, no. 2, pp. 133–146, 2006.

[81] L. de Pablo, M. L. Chávez, and M. Abatal, "Adsorption of heavy metals in acid to alkaline environments by montmorillonite and Ca-montmorillonite," *Chem. Eng. J.*, vol. 171, no. 3, pp. 1276–1286, 2011.

[82] Y. S. Ok, J. E. Yang, Y.-S. Zhang, S.-J. Kim, and D.-Y. Chung, "Heavy metal adsorption by a formulated zeolite-Portland cement mixture," *J. Hazard. Mater.*, vol. 147, no. 1–2, pp. 91–96, 2007.

[83] W. Shen et al., "Adsorption of Cu (II) and Pb (II) onto diethylenetriamine-bacterial cellulose," *Carbohydr. Polym.*, vol. 75, no. 1, pp. 110–114, 2009.

[84] Y.-S. Ho, W.-T. Chiu, and C.-C. Wang, "Regression analysis for the sorption isotherms of basic dyes on sugarcane dust," *Bioresour. Technol.*, vol. 96, no. 11, pp. 1285–1291, 2005.

[85] S. V Ramanaiah, S. V. Mohan, and P. N. Sarma, "Adsorptive removal of fluoride from aqueous phase using waste fungus (*Pleurotus ostreatus* 1804) biosorbent: kinetics evaluation," *Ecol. Eng.*, vol. 31, no. 1, pp. 47–56, 2007.

[86] H. M. F. Freundlich et al., "Over the adsorption in solution," *J. Phys. Chem.*, vol. 57, no. 385471, pp. 1100–1107, 1906.

[87] E. Repo, J. K. Warchol, T. A. Kurniawan, and M. E. T. Sillanpää, "Adsorption of Co (II) and Ni (II) by EDTA-and/or DTPA-modified chitosan: kinetic and equilibrium modeling," *Chem. Eng. J.*, vol. 161, no. 1–2, pp. 73–82, 2010.

[88] S. Al-Asheh, F. Banat, R. Al-Omari, and Z. Duvnjak, "Predictions of binary sorption isotherms for the sorption of heavy metals by pine bark using single isotherm data," *Chemosphere*, vol. 41, no. 5, pp. 659–665, 2000.

[89] Y. S. Ho, J. F. Porter, and G. McKay, "Equilibrium isotherm studies for the sorption of divalent metal ions onto peat: copper, nickel and lead single component systems," *Water. Air. Soil Pollut.*, vol. 141, no. 1, pp. 1–33, 2002.

[90] E. Repo, J. K. Warchoł, A. Bhatnagar, and M. Sillanpää, "Heavy metals adsorption by novel EDTA-modified chitosan—silica hybrid materials," *J. Colloid Interface Sci.*, vol. 358, no. 1, pp. 261–267, 2011.

[91] T. M. Elmorsi et al., "Equilibrium isotherms and kinetic studies of removal of methylene blue dye by adsorption onto miswak leaves as a natural adsorbent," *J. Environ. Prot.*, vol. 2, no. 6, p. 817, 2011.

[92] C. Chen, "Evaluation of equilibrium sorption isotherm equations," *Open Chem. Eng. J.*, vol. 7, no. 1, pp. 24–44, 2013.

[93] J. Toth, "State equation of the solid-gas interface layers," *Acta Chim. Hung.*, vol. 69, pp. 311–328, 1971.

[94] S. Cavus and G. Gurdag, "Noncompetitive removal of heavy metal ions from aqueous solutions by poly [2-(acrylamido)-2-methyl-1-propanesulfonic acid-co-itaconic acid] hydrogel," *Ind. Eng. Chem. Res.*, vol. 48, no. 5, pp. 2652–2658, 2009.

[95] A. M. Farhan, N. M. Salem, A. H. Al-Dujaili, and A. M. Awwad, "Biosorption studies of Cr (VI) ions from electroplating wastewater by walnut shell powder," *Am. J. Environ. Eng.*, vol. 2, no. 6, pp. 188–195, 2012.

[96] M. Matouq, N. Jildeh, M. Qtaishat, M. Hindiyeh, and M. Q. Al Syouf, "The adsorption kinetics and modeling for heavy metals removal from wastewater by Moringa pods," *J. Environ. Chem. Eng.*, vol. 3, no. 2, pp. 775–784, 2015.

[97] P. B. Tchounwou, C. G. Yedjou, A. K. Patlolla, and D. J. Sutton, "Heavy metal toxicity and the environment," *Mol. Clin. Environ. Toxicol.*, pp. 133–164, 2012.

[98] S. Lee, S. Bong, J. Ha, M. Kwak, S.-K. Park, and Y. Piao, "Electrochemical deposition of bismuth on activated graphene-nafion composite for anodic stripping voltammetric determination of trace heavy metals," *Sens. Actuators B Chem.*, vol. 215, pp. 62–69, 2015.

[99] G. Zhao, Y. Yin, H. Wang, G. Liu, and Z. Wang, "Sensitive stripping voltammetric determination of Cd (II) and Pb (II) by a Bi/multi-walled carbon nanotube-emeraldine base polyaniline-Nafion composite modified glassy carbon electrode," *Electrochim. Acta*, vol. 220, pp. 267–275, 2016.

[100] K. A. Razak et al., "Effect of hydrothermal reaction temperature on properties of bismuth nanoparticles and its properties as modified electrode for Pb sensors," *J. Phys. Conf. Ser.*, vol. 1082, no. 1, p. 12077, 2018.

[101] L. Shi, Y. Li, X. Rong, Y. Wang, and S. Ding, "Facile fabrication of a novel 3D graphene framework/ Bi nanoparticle film for ultrasensitive electrochemical assays of heavy metal ions," *Anal. Chim. Acta*, vol. 968, pp. 21–29, 2017.

[102] Y. Bonfil, M. Brand, and E. Kirowa-Eisner, "Characteristics of subtractive anodic stripping voltammetry of Pb and Cd at silver and gold electrodes," *Anal. Chim. Acta*, vol. 464, no. 1, pp. 99–114, 2002.

[103] S. Lee, J. Oh, D. Kim, and Y. Piao, "A sensitive electrochemical sensor using an iron oxide/graphene composite for the simultaneous detection of heavy metal ions," *Talanta*, vol. 160, pp. 528–536, 2016.

[104] R. Cornelis, *Handbook of Elemental Speciation: Techniques and Methodology.* John Wiley & Sons, Chichester, West Sussex, 2004.

[105] M. B. Gumpu, M. Veerapandian, U. M. Krishnan, and J. B. B. Rayappan, "Simultaneous electrochemical detection of Cd (II), Pb (II), As (III) and Hg (II) ions using ruthenium (II)-textured graphene oxide nanocomposite," *Talanta*, vol. 162, pp. 574–582, 2017.

[106] W. Liu, "Preparation of a zinc oxide-reduced graphene oxide nanocomposite for the determination of cadmium (II), lead (II), copper (II), and mercury (II) in water," *Int. J. Electrochem. Sci*, vol. 12, pp. 5392–5403, 2017.

[107] W. J. Yi, Y. Li, G. Ran, H. Q. Luo, and N. B. Li, "Determination of cadmium (II) by square wave anodic stripping voltammetry using bismuth--antimony film electrode," *Sens. Actuators B Chem.*, vol. 166, pp. 544–548, 2012.

[108] P. Kanyong, S. Rawlinson, and J. Davis, "Gold nanoparticle modified screen-printed carbon arrays for the simultaneous electrochemical analysis of lead and copper in tap water," *Microchim. Acta*, vol. 183, no. 8, pp. 2361–2368, 2016.

[109] N. Bashirom, M. A. Zulkifli, S. Subagja, T. W. Kian, A. Matsuda, and Z. Lockman, "Cr (VI) removal on visible light active TiO_2 nanotube arrays," *AIP Conf. Proc.*, 2018, vol. 1958, no. 1, p. 20025.

[110] Q. Wang, J. Shang, T. Zhu, and F. Zhao, "Efficient photoelectrocatalytic reduction of Cr (VI) using TiO_2 nanotube arrays as the photoanode and a large-area titanium mesh as the photocathode," *J. Mol. Catal. A Chem.*, vol. 335, no. 1–2, pp. 242–247, 2011.

[111] Q. Wu, J. Zhao, G. Qin, C. Wang, X. Tong, and S. Xue, "Photocatalytic reduction of Cr (VI) with TiO_2 film under visible light," *Appl. Catal. B Environ.*, vol. 142, pp. 142–148, 2013.

[112] J. K. Yang, S. M. Lee, M. Farrokhi, O. Giahi, and M. Shirzad Siboni, "Photocatalytic removal of Cr (VI) with illuminated TiO_2," *Desalin. water Treat.*, vol. 46, no. 1–3, pp. 375–380, 2012.

[113] F. Hashemzadeh, A. Gaffarinejad, and R. Rahimi, "Porous p-NiO/n-Nb_2O_5 nanocomposites prepared by an EISA route with enhanced photocatalytic activity in simultaneous Cr (VI) reduction and methyl orange decolorization under visible light irradiation," *J. Hazard. Mater.*, vol. 286, pp. 64–74, 2015.

[114] H.-Y. Lin, H.-C. Yang, and W.-L. Wang, "Synthesis of mesoporous Nb_2O_5 photocatalysts with Pt, Au, Cu and NiO cocatalyst for water splitting," *Catal. Today*, vol. 174, no. 1, pp. 106–113, 2011.

[115] X. Liu, R. Zheng, R. Yuan, L. Peng, Y. Liu, and J. Lin, "Released defective Nb_2O_5 with optimized solar photocatalytic activity," *ECS J. Solid State Sci. Technol.*, vol. 6, no. 9, p. P665, 2017.

[116] M. A. Zulkifli, N. Bashirom, T. W. Kian, G. Kawamura, A. Matsuda, and Z. Lockman, "Rapid Tio_2 nanotubes formation in aged electrolyte and their application as photocatalysts for Cr (VI) reduction under visible light," *IEEE Trans. Nanotechnol.*, vol. 17, no. 6, pp. 1106–1110, 2018.

[117] E. T. Wahyuni, N. H. Aprilita, H. Hatimah, A. M. Wulandari, and M. Mudasir, "Removal of toxic metal ions in water by photocatalytic method," *Am. Chem. Sci. J.*, vol. 5, no. 2, pp. 194–201, 2015.

[118] L. Wang, N. Wang, L. Zhu, H. Yu, and H. Tang, "Photocatalytic reduction of Cr (VI) over different TiO_2 photocatalysts and the effects of dissolved organic species," *J. Hazard. Mater.*, vol. 152, no. 1, pp. 93–99, 2008.

4 Nano-Engineered Hybrid Materials for Decontamination of Hazardous Organics

Isabela Machado Horta, Barbara Souza Damasceno,
Douglas Marcel Gonçalves Leite,
Argemiro Soares da Silva Sobrinho, and
André Luis de Jesus Pereira
Instituto Tecnológico de Aeronáutica (ITA)

Armstrong Godoy-Jr
Universidade Tecnológica Federal do Paraná (UTFPR)

CONTENTS

4.1 INTRODUCTION

From the industrial revolution, there was an important evolution of industrial processes, and, at the same time, there was a rapid population growth and an increase in life expectancy. These factors, together with the increase in the population's purchasing power, increased the demand for food and consumer goods, bringing the need for advances in agriculture and industrial activities. The consequences of this accelerated development brought serious environmental problems such as increased deforestation and contamination of aquatic systems, soil, and air. The wastes generated by industrial and agricultural activities are, in most cases, toxic and harmful to both the population and the environment. The destinations of most of the wastes produced by human activities and industries are rivers and groundwater. In the countryside, the increased development of agriculture has also increased the indiscriminate use of fertilizers and pesticides, which has been another source of contamination in these aquatic systems [1,2]. Another worrying fact is the significant concentration of drugs found in effluent water from water treatment plants [3,4]. The presence of drugs, which are normally eliminated in the urine and feces of humans and animals, persists due to their inability to be degraded by natural processes and is expected to increase in the coming years due to the increase in the use of drugs such as acetaminophen, carbamazepine, diclofenac, and ibuprofen [4,5].

In summary, these hazardous organic contaminants have been widely employed in many industries due to their physicochemical properties. However, they are bio-toxic, stable, and carcinogenic, causing harmful effects to living organisms and destroying the aesthetic quality of water [6–9]. In order to mitigate this problem, many research groups have been dedicated to studying, developing, and improving new physical, biological, and chemical techniques for decontaminating organic pollutants. Among these techniques, it is possible to highlight the processes of adsorption, biodegradation, electrochemical oxidation, filtration, and photocatalysis. All these methods have shown significant advances in recent years. However, to achieve greater maturity and, consequently, enable its application on a large scale, further studies are needed.

The advancement of these techniques is directly related to the development of new materials or composition of materials that are used in the decontamination process. For purposes of improving properties for specific applications, hybrid materials show some great advantages. The studies surrounding them started early in the 1940s, but the term "hybrid material" was only determined in the mid-1980s [10]. Mostly, the definition of a hybrid material states that it is a combination of organic and inorganic parts, or of two or more materials involving a formation of orbitals [10,11]. Nonetheless, a hybrid material has new properties (derived from the precursor materials and the hybrid interfaces synergistically coupled) that are improved and designed for a specific application [12]. That is because the processing, the structure, and the organic/inorganic materials used for the hybrid highly influence the final properties.

The organic and inorganic parts of the hybrid material contribute differently to its properties. While organic materials have advantages such as being cheaper, recyclable, and sustainable, the inorganic part may provide great mechanical resistance; thermal and chemical stabilities; and electrical, magnetic, optical, chemical, and electrochemical properties [10].

The use of hybrid materials is vast, once they can be processed aiming specific properties. Therefore, they have uses in medicine, cosmetics, biotechnology, smart coating, optical and electronic industries, energy generation and storage, aerospace and textile industries, and environmental remediation [12]. Consequently, the studies on hybrid materials synthesis and application are very individual and are constantly being updated. According to the literature data (web of sciences), the field of study on hybrid materials in the wide sense shows a rising tendency, mainly in the last decade.

However, these materials may get weak physical and mechanical stabilities, high toxicity, and diverse particle size and shape, as well as uncontrollable surface functionalization if not well engineered. The hybrid material limitations occur basically during their assembling. Therefore, it is extremely necessary to control the synthesis procedure, which may require additional production steps and resources to obtain a conjugated and synergic coupled material with advanced and specific properties. Thus, different strategies are required to support the technical, economic, manufacturing, or sustainable limitations of the ordinary materials, favoring their use for real and practical applications [12,13].

In this chapter, we will present some of the main works published in the last decade, mainly the late 2010s and early 2020s, involving the nano-engineering of hybrid materials aiming their application of organic pollutants removal.

4.2 NANO-ENGINEERED HYBRID MATERIALS FOR ADSORPTION APPLICATION

Adsorption is a conventional method applied to decontaminate those pollutants favored by its simplicity, cost-effectiveness, low energy consumption, high efficiency, sludge-free, and ease of operation [7,8,14]. Effective adsorption depends mostly on the adsorbent characteristics including high surface area, shape, porosity, and surface charge. Adsorbents such as activated carbon (AC), graphene oxide (GO), silica, and magnetite have shown great adsorption performance for the removal of pollutants from aqueous media. However, they can present low adsorption capacity, lack of selectivity, and have a difficulty to be recovered and reused. Addressing these drawbacks, hybrid materials are potential candidates to establish new functionalities from each individual material and enhance the adsorption performance [6,8,15,16]. Nano-engineered hybrid materials do not generate secondary pollutants, present selectivity towards specific contaminants, and can be reused. They can also combine high adsorption capacity, high removal efficiency, and rapid kinetics [14,17]. Therefore, these nanohybrids are highly interesting to treat hazardous organic pollutants from wastewater via adsorption. Some of them will be introduced in the course of this work.

4.2.1 ACTIVATED CARBON-BASED HYBRID

AC is a carbonaceous material presenting an internal surface area from 400 to 3,000 m²/g. It is very useful for wastewater decontamination by adsorption, but its selectivity needs to be improved. Hence, different carbonization approaches have been applied to synthesize AC and enhance its properties [16].

In a study developed in 2021, Kalombo and co-workers used granular activated carbon (GAC) enhanced by nanoparticles (NPs) to form nanohybrid with tunable properties, the so-called Janus particles (JPs). JP is formed by organic and inorganic chemicals, having surfaces with different chemical compositions and specific parts with properties from the precursors preserved [16]. Polypyrrole (PPy) is an organic polymer, thermally stable, conductor, and facile to synthesize [16,18]. It also exhibits excellent anion-exchange properties and positively charged nitrogen atoms in the polymer chains, facilitating the removal of anionic pollutants from an aqueous solution [14,16]. PPy-decorated GAC Janus nanohybrid combines the stability and specificity to organic contaminants from GAC as well as the large surface area and capability to capture inorganic pollutants from PPy. Then, PPy was electrosprayed on one side of GAC producing Janus-like PPy-grafted GAC nanohybrid with 137 m²/g of surface area and strongly bonded in the interface of the two materials. According to the properties acquired from this novel nano-engineered material, PPy-decorated GAC Janus hybrid may be a potential adsorbent for hazardous pollutant decontamination [16].

4.2.2 GRAPHENE OXIDE-BASED HYBRIDS

GO presents chemical stability, large surface area, and a huge amount of oxygenated functional groups and some hydroxyl, carboxyl, and epoxy groups attached to its surface. The functionalized surface of GO favors the attachment of inorganic and organic compounds via covalent and ionic bonds, which is very interesting in terms of adsorption. Magnetic NPs have been used to load GO, resulting in a magnetic nanohybrid and avoiding secondary residues generation. Spinel-type manganese ferrite (MF) presents high surface area, magnetization, and chemical stability, and lanthanum is a rare earth element with unique physicochemical properties and affinity to strongly bond with anionic hazardous pollutants. Thus, graphene-based spinel ferrite material doped by lanthanum (LMF@GO) can be used as a potential adsorbent combining great advantages from each material [8].

A study reporting the production of LMF@GO nanohybrid was performed through ultrasonication of GO with lanthanum-substituted manganese ferrites (LMF). Different quantities of LMF were introduced into GO structure, resulting in a partial or integral lamination of GO because of the surface complexation phenomena or delamination [8]. Figure 4.1 shows a schematic illustration about the LMFx%@GO production.

The tailored LMF75% with GO presented conjugated properties. The BET surface area of LMF75%@GO was lower than GO and higher than LMF75% (LMF75%@GO: ~86.8 m²/g, GO: ~314.2 m²/g, and LMF75%: ~34.3 m²/g). Moreover, the magnetization response of LMF decreased when combined with GO. Still, the LMF75%@GO incorporated magnetic property, allowing its recovery after the adsorption process (GO: ~10.9 emu/g and, LMF75%: ~7.9 emu/g). From this finding, an enhancement of the LMF75%@GO adsorption capacity towards the perfluorooctanoic acid

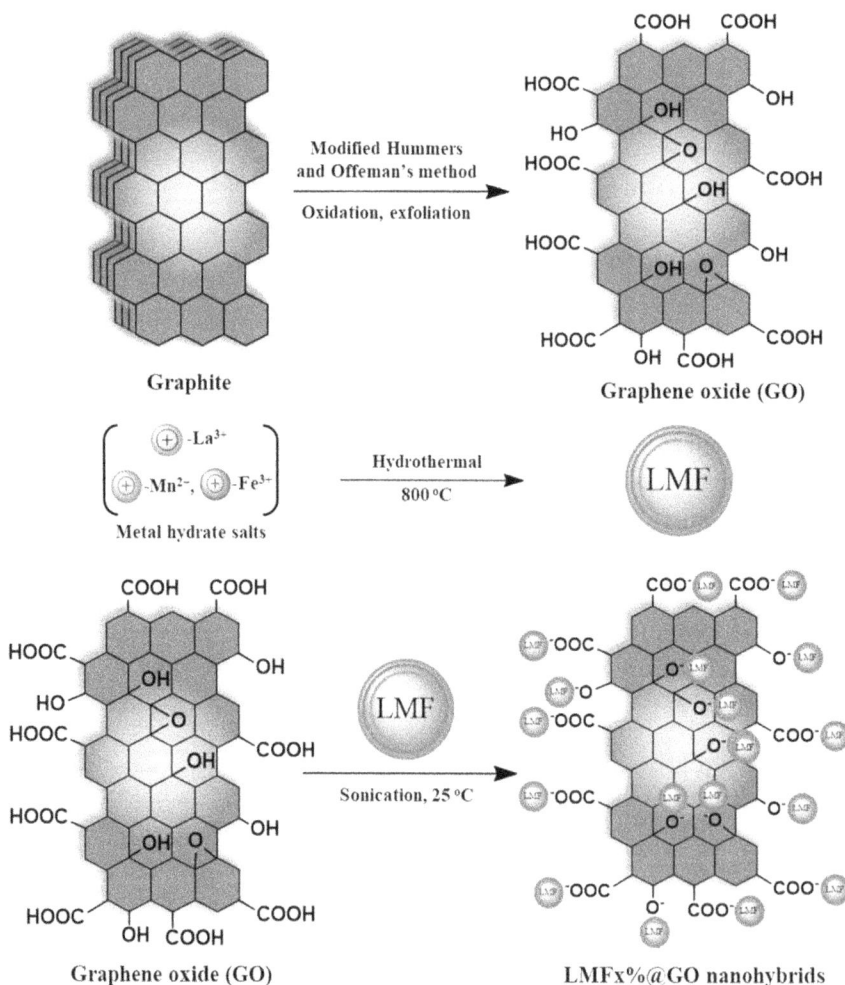

FIGURE 4.1 Schematic route of the LMFx%@GO synthesis. (Reproduced from [8].)

(PFOA) removal from water was found due to the synergistic interactions with anionic species of PFOA. LMF75%@GO exhibited the highest adsorption capacity in comparison with GO and LMF (LMF75%@GO: ~1.6 mg/g, GO: ~0.6 mg/g and, LMF75%: ~1.3 mg/g) [8].

4.2.3 Reduced Graphene Oxide-Based Hybrids

Reduced graphene oxide (rGO) is chemically stable and can be coated with metallic oxide or metal sulfide NPs. Diverse metal NPs (Ni, Pd, Co) exhibit suitable sizes to be introduced among the rGO nanosheets in order to expand the interlayer distance and reduce the rGO agglomeration [7].

A work on the rGO/Fe/Co nanohybrid production used the co-precipitation reaction to join rGO and Fe/Co NPs. The as-prepared hybrid was well dispersed, even so, agglomerations were formed. Nevertheless, it showed ~108.4 m^2/g of surface area and ~101.4 emu/g of magnetic property provided from the rGO and Fe/Co NPs, respectively. This permitted the use, recovery, and reuse of rGO/Fe/Co nanohybrids for methyl blue (MB) dye adsorption from simulated wastewater. The maximum adsorption capacity of rGO/Fe/Co nanohybrids was ~909.1 mg/g and after five cycles of regeneration and reuse, the adsorption efficiency decreased only by ~28%. In comparison with other carbon-based hybrid materials, rGO/Fe/Co hybrid is more practical, functional, and useful adsorbent for water cleaning [7,8,14].

4.2.4 Carbon Nanotube-Based Hybrid

Carbon nanotubes (CNTs) have large specific surface areas, hollow and layered structures, and good chemical stability. However, nanosized CNTs can easily agglomerate and lose mass during the recovery process, limiting their use as adsorbents in wastewater treatment. Then, the combination of CNTs and magnetic materials such as cobalt ferrite ($CoFe_2O_4$) results in a nanohybrid with dual functionalities, including the desirable adsorption performance and effective magnetic separation. Besides, CNT can be chemically modified by polymers with several functional groups to enhance its adsorption capacity and selectivity [14].

Studies on the engineered CNTs-$CoFe_2O_4$@PPy hybrid were developed to investigate the adsorption performance of some organic dyes [14]. CNTs-$CoFe_2O_4$ nanomaterials were synthesized via a solvothermal approach, and further, the obtained material was added to the *in situ* chemical oxidative polymerization of pyrrole, resulting in CNTs-$CoFe_2O_4$@PPy hybrid. The nanohybrid CNTs-$CoFe_2O_4$@PPy with a surface area of ~150.4 m^2/g was used to adsorb MB, methyl orange (MO), and acid fuchsin (AF) dyes, from water. The maximum adsorption capacities of CNTs-$CoFe_2O_4$@PPy towards MB, MO, and AF were ~136.9, ~116.7, and ~135.1 mg/g, respectively.

4.2.5 Silica-Based Hybrid

Clay minerals are silicates formed by silicon-oxygen tetrahedrons and metal (Mg, Al, and Fe)-oxygen octahedrons in their composition. The structure and morphologies of the silicates depend on the tetrahedrons and octahedrons associations. However, after an acid leaching step, some octahedral sheets may be removed, leading to a silicon tetrahedral sheet with higher SiO_2 content. Synthesized silica from natural clay minerals does not change the original shape and exhibits high porosity and surface-active groups. It is noteworthy that acid-leached clay minerals can compose nanohybrid materials with different Si/Al ratios and metal ion content. Regarding Fe(III) ion, it provides magnetically recoverable advantages to the nanohybrid material showing great adsorption interest [19].

In 2020, a work reported the exposition of palygorskite (Pal), montmorillonite (MMT), and nontronite (NNT) to an acid leaching treatment under a hydrothermal condition. Then, the material was used to synthesize layered double hydroxide (LDH) and it was further calcinated to produce magnetic-layered double oxide (LDO) compounds, the so-called Pal-LDO, MMT-LDO, and NNT-LDO.

These hybrid materials showed similar values for the specific surface areas (Pal-LDO: ~100.8 m²/g, MMT-LDO: ~113.7 m²/g, and NNT-LDO: ~105.9 m²/g) and responded to an external magnetic field due to the amount of g-Fe$_2$O$_3$ present. This allows a swift recovery (3 minutes) and reusability of the magnetic adsorbents. Hence, Pal-LDO, MMT-LDO, and NNT-LDO were used as adsorbents, and their adsorption capacities (Q_e) were evaluated towards CR and tetracycline (TC) dyes. Among them, Pal-LDO had the highest Q_e, ~538.4 mg/g for CR and ~626.9 mg/g for TC. After reusing it for four cycles, the maximum Q_e towards CR was ~350.8 mg/L. These results indicate that the new nanohybrids of LDO have optimized structure, organic groups with Mg, Al, and C species on the surface, and more available adsorption sites favoring its adsorption performance [19].

4.2.6 ZEOLITIC IMIDAZOLATE FRAMEWORK-BASED HYBRID

Metal organic framework (MOF) is a type of crystalline material with large surface areas (above 6,000 m²/g) and ordered pore structures. MOF can be engineered in several different structures with versatile properties favoring the adsorption of organic dyes such as MB, MO, and rhodamine B (RhB). Zeolitic imidazolate framework (ZIF) is a type of MOF chemically stable and easy to synthesize, which consists in transition metal cations and imidazole-based ligands. LDH presents low porosity and low surface area. Conversely, it is highly stable metallic node, of low cost, and acts as a platform for the growth of MOFs. ZIF-67 is thermally and chemically stable, presents several pores, and when conjugated with LDH presents controllable features of its size and morphology [17].

In 2020, a study depicted the synthesis of a nano-engineered ZIF-67@LDH material composed by LDH (CoAl-LDH) and ZIF-67 for MO and MB dyes removal from water. Pure ZIF-67, LDH, and ZIF-67@LDH present BET surface area of 593.6, 114.3, and 374.3 m²/g, respectively. Because of the high surface areas, after 100 minutes, the ZIF-67@LDH performance towards removal efficiency was higher than 65% for both dyes and the maximum adsorption capacities were ~57.2 mg/g for MB and ~180.5 mg/g for MO removal, respectively. The great adsorptive response during MO decomposition was due to the hydrogen bonding with nanohybrid material. Besides all these advantages, ZIF-67@LDH was regenerated and reused four times, decreasing their adsorption capacities slightly [17].

4.2.7 NATURAL PLANT SEED FRAMEWORK-BASED HYBRID

AC usually shows better performance in adsorption processes; however, the lack of chemical groups' functionalities impairs the binding efficiency towards organic contaminants. The use of MOFs may increase the adsorption performance when engineered with the AC adsorbent, but it is toxic and still demands more functionalized sites. A way to solve these issues is to use biocompatible cellulosic surface with high functionalities as organic frameworks doped by metal oxide NPs. Black cumin seed (BC) is abundant, biocompatible, of low cost, and nontoxic, and it has many functional groups. Additionally, Fe$_2$O$_3$ is nontoxic metal oxide with limited capacities that can be improved by doping metals. In this way, carbon-based framework materials derived from plants jointly with metal oxides can show an enhancement in chemical, mechanical, and thermal stabilities. To achieve these advantages, binary oxide NPs (Fe$_2$O$_3$–SnO$_2$) were incorporated into the carbon framework of *Nigella sativa* seeds (BC) via co-precipitation as represented in Figure 4.2 [9]. From this, BC and Fe$_2$O$_3$–SnO$_2$ can stay strongly bonded because of the numerous functionalities on the BC surface trapping the hydrated Fe$_2$O$_3$–SnO$_2$ NPs via metal-hydroxyl (M-OH) and BC-hydroxyl (BC-OH) interactions. Thus, Fe$_2$O$_3$–SnO$_2$/BC was used as an adsorbent for water treatment by MB adsorption [9].

The maximum adsorption capacity of Fe$_2$O$_3$–SnO$_2$/BC towards MB dye was ~84.7 mg/g, whereas the removal efficiency of MB was 95% after 15 minutes and 98% after 75 minutes. Additionally, the Fe$_2$O$_3$–SnO$_2$/BC was recovered and reused for seven cycles that is excellent for practical application. After three cycles, the Fe$_2$O$_3$–SnO$_2$/BC removal efficiency decreased from 98% to 89%, and after seven cycles, it reduced from 98% to 35% [9].

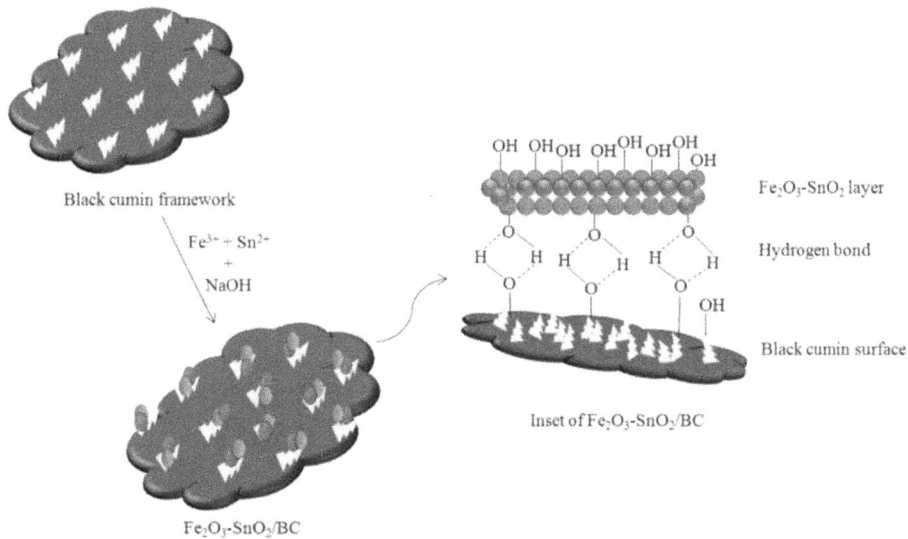

FIGURE 4.2   Illustrative scheme of Fe_2O_3–SnO_2/BC production. (Reproduced from [9].)

4.2.8 BIO-SILICA XEROGEL-BASED HYBRID

Diatomaceous earth (DE) is a sedimentary rock, highly porous, and available, whereas silica xerogel (DX) is a nontoxic, biocompatible, thermally stable, and cheap material. In 2019, a work used DX to change DE surface willing to improve DE surface area. After the surface modification, DE presented $129\,m^2/g$ of surface area. Then, a cross-linker named epichlorohydrin (DXE) was applied to provide amine functionalization on DX via a ring-opening mechanism, resulting in a reduction of the surface area to $105\,m^2/g$ [15]. The material was treated afterwards with hydroxide to obtain amine-activated epichlorohydrin DX (DXEA), decreasing the surface area to $48\,m^2/g$ and was used as adsorbent for the removal of eriochrome black T (EBT), an azo, carcinogenic, and mutagenic dye from water media, as shown in Figure 4.3.

The adsorptive efficiency of DE and DXEA post 5 minutes was about 56.1% and 99.6%, respectively. The satisfactory dye removal efficiency was due to the several active sites available on the surface of DE and DXEA. The high removal efficiency for DXEA is attributable to amine functionality, which results in high driving force and diffusion of EBT molecules. Then, the maximum adsorption capacities of DE and DXEA were ~56.2 and ~62.1 mg/g, respectively. Concerning the selectivity of the adsorbents towards specific dyes, several organic pollutants were mixed with EBT. After the batch adsorption, the EBT dye was removed from the mix, which was possible because the interfering dyes were repelled electrostatically by DXEA, showing selectivity on EBT removal over other organic dyes. Finally, DE and DXEA adsorbents were recovered and reused for five times exhibiting only an ordinary decrease from the first cycle (~99.9%) to the fifth cycle (~82.6%) [15].

4.3   NANO-ENGINEERED HYBRID MATERIALS FOR BIODEGRADATION PROCESS

It is known that the removal of organic pollutants, including organophosphorus compounds (OPs), dyes, and crude oils, from wastewater has been challenging. Many conventional methods are used for environmental remediation; however, low concentration usually remains in the treated water. Biodegradation process using biomaterials (e.g., enzymes and bacterial) is an efficient additional technique to treat organic contaminants at low levels of the pollutant [20–23].

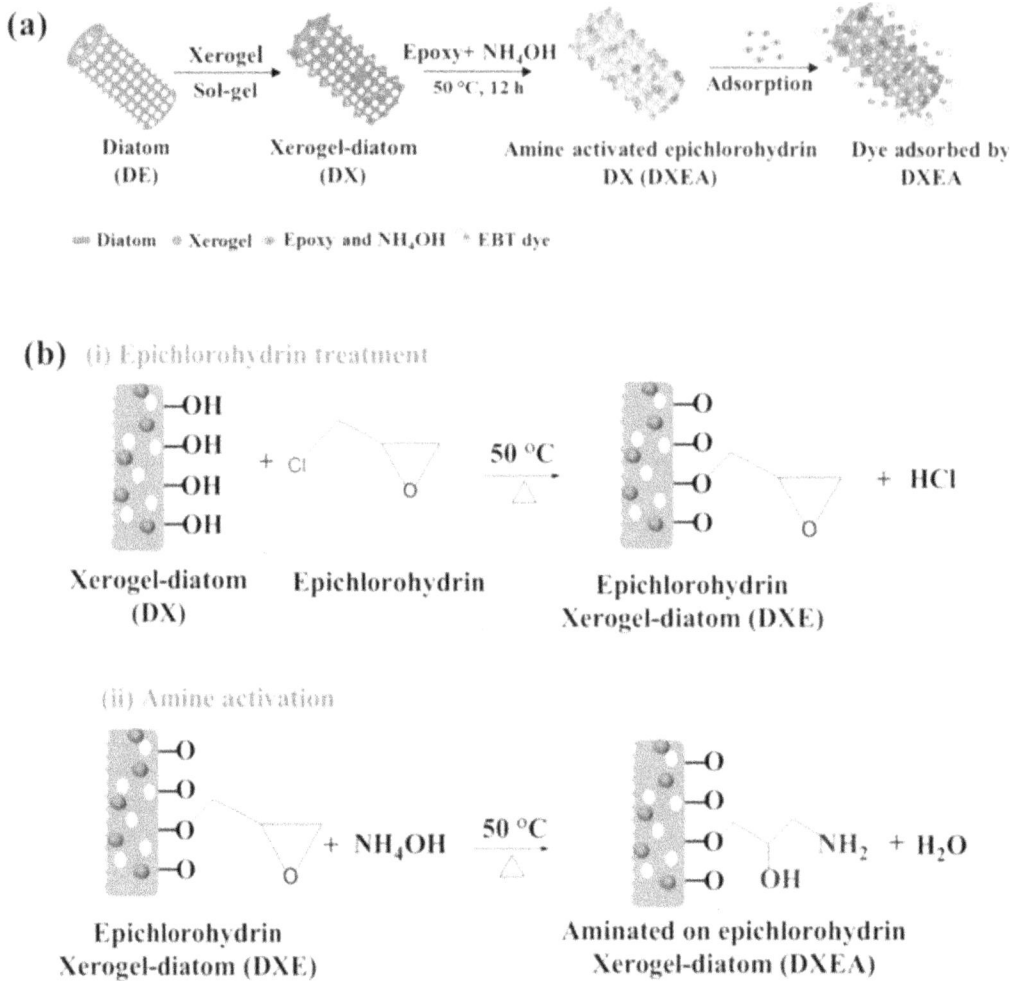

FIGURE 4.3 Illustrative diagram of (a) EBT adsorption and (b) DXEA synthesis mechanism. (Reproduced from [15].)

4.3.1 *Providencia Vermicola*-Based Hybrid

Iron nanoparticle (Fe NP) and activated carbon beads (ACBs) were used as support to grow carbon nanofibers (CNFs) by chemical vapor deposition, resulting in a hybrid material named Fe-CNF/ACB with Fe in the zero-valence state (Figure 4.4a) [20]. Fe-CNF/ACB is a biocompatible catalyst with physical and chemical stabilities. Thus, Fe-CNF/ACB were applied in a hybrid process consisting of catalytic wet air oxidation (cWAO) and microbial degradation for N-(phosphonomethyl)-glycine (PMG) decontamination (Figure 4.4b). PMG, also called glyphosate, is a chemically stable herbicide that causes harmful effects on environment and human health. After applying Fe-CNF/ACB as catalyst in a cWAO procedure to remove PMG, the surface morphology of Fe-CNF/ACB remained the same smooth with spherical-shaped beads as the previous material, revealing a stable characteristic of the material. Even in the presence of bacteria between the CNFs, the material showed to be physically stable, as confirmed by SEM micrographs. Besides, the surface area of Fe-CNF/ACB decreased from ~296 to 280 m^2/g after cWAO and to 212 m^2/g post microbial degradation. This behavior could be explained owing to several active sites available, which were filled

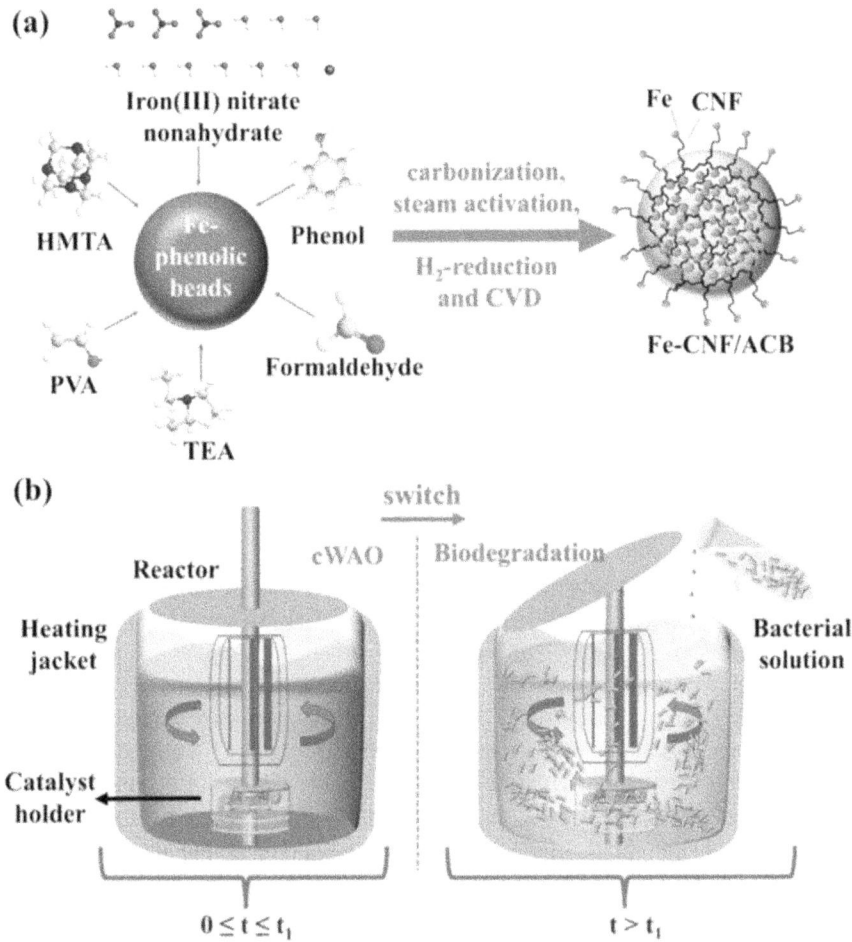

FIGURE 4.4   Illustrative scheme of (a) Fe-CNF/ACB production and (b) PMG removal via cWAO and microbial biodegradation. (Reproduced from [20].)

after the oxidation process, followed by a biofilm formation onto the Fe-CNF/ACB surface post the microbial degradation [20].

Post the oxidation process, a gram-negative bacterium (*Providencia vermicola*) was isolated from wastewater gathered from a fertilizer plant and used in PMG biodegradation in a bacterial solution. Bacterial counts play the most critical role among the PMG-degrading microbes. To degrade PMG (100 mg/L), the initial bacterial counts remained constant at 25 mL/L. By this, about 81% of PMG was biodegraded using the bacterial dose incubated for 1 day, ~91% for 5 days, and completely for 10 days. Additionally, the as-prepared Fe-CNF/ACBs were recovered and reused for five cycles, presenting around 90% of PMG degradation via the hybrid process. After the whole procedure, it was seen that the treated liquid could be safely discharged to water bodies, which infers that Fe-CNF/ACBs are very adequate to degrade and mineralize organic pollutants from wastewater [20].

4.3.2   Phosphotriesterase-Based Hybrid

Phosphotriesterase (PTE) is a biological enzyme from amide hydrolase superfamily that can degrade organophosphorus pesticides (OPs) by hydrolysis breaking the phosphorus ester bond. However, PTE is an instable catalyst, difficult to recover, and expensive. Conversely, immobilized enzyme can

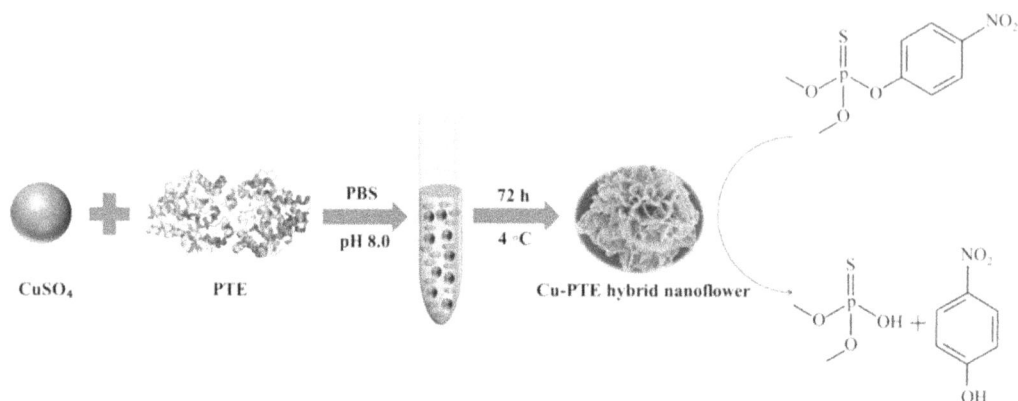

FIGURE 4.5   Illustrative scheme of Cu-PTE hybrid nanoflowers. (Reproduced from [22].)

enhance the specific activity and stability of the PTE. For this fact, organic-inorganic hybrid nano-flowers can be produced via PTE and copper ions (Cu^{2+}) combination, resulting in strong complex Cu-PTE hybrid nanoflowers. The Cu-PTE hybrid nanoflowers were synthesized by mixing PTE with CuSO$_4$ in phosphate buffer solution [22]. Thereafter, the sample was incubated and collected as shown in the scheme presented in Figure 4.5.

PTE hydrolyzed OPs mostly by direct nucleophilic attack of a hydroxide on the phosphorus nuclei. In a work reported in 2021, a toxic OP named "methyl parathion" was used as the substrate to determine the optimal reaction conditions. The PTE and hybrid PTE enzyme catalytic proper-ties were evaluated towards methyl parathion substrate. Cu-PTE hybrid nanoflowers had their catalytic efficiency improved by ~1.8 times in comparison with PTE, which might be related to the flower-like structure owing to its high surface area and proper capture of the enzymes in the nanoflowers [22].

Organic solvents are frequently used to increase the solubility of substrates, but they harm the enzymes' catalytic activities. Because of that, PTE and Cu-PTE hybrid nanoflowers were immersed in n-butanol, n-propanol, isopropanol, and ethyl acetate. It was possible to observe that the hybrid nanoflowers improved the resistance of enzymes to organic solvents, whereas PTE lost all the activ-ity after a certain period. Concerning the reusability, Cu-PTE hybrid nanoflowers maintained 72.3% relative activity after ten consecutive cycles [22].

4.3.3   *BACILLUS LICHENIFORMIS*-BASED HYBRID

Polyurethane foam (PUF) is extensively used for cell immobilization because it is stable, porous, and chemically and physically resistant. However, high porosity is unfavorable due to the cell leak-age process as well as the heating that occurs during PUF polymerization that degrades some bacte-ria. The bacteria *Bacillus licheniformis* can significantly degrade crude oil samples. Thus, a report from 2021 used a hybrid material of PUF, alginate, and microbial cell in aqueous phase to entrap and immobilize the bacterium. Great dispersion and irrelevant cell leakage were the characteristics observed from the strongly bonded bacteria on the hybrid matrix of PUF [21].

The hybrid PUF/alginate/microbial cell removed the heavy crude oil from aqueous phase. The hybrid material showed significantly enhanced removal in each concentration of heavy crude oil. Additionally, the synergistic adsorption and biodegradation capacity permitted the oil removal from wastewaters. It is noteworthy that the maximum monolayer coverage (Q_0) was found out to be 1.25 g/g PUF.

The authors also performed the n-alkanes (C$_9$–C$_{32}$) analysis for crude oil biodegradation. It was observed that light hydrocarbon fractions (C$_9$–C$_{15}$) were the maximum of biodegradation of treated oil fractions. Therefore, *Bacillus licheniformis* entrapped in the PUF hybrid matrix can use a significant number of short-chain alkanes. Besides, the total organic carbon (TOC, %) from recalcitrant clayey soil decreased to 39%–80%. Hence, *Bacillus licheniformis* can considerably reduce TOC from clayey soil using a certain amount of crude oil [21].

4.3.4 LACCASE ENZYME-BASED HYBRID

Immobilized enzymes are reusable, stable, and of low cost. Laccase enzyme is used for textile dye decolorization because it presents low specificity that improves their oxidation. Magnetic NPs are immobilization materials that offer large surface areas, do not interfere with the properties of bio-macromolecules, and have surfaces highly functionalized. A study produced them via co-precipitation route with thiolated chitosan (TCS) and further formed the hybrid named "Fe_3O_4-TCS" [23]. The superparamagnetic Fe_3O_4 NPs (59 emu/g) provided a saturation magnetization of 2.5 emu/g to the Fe_3O_4-TCS hybrid composite. Also, the hybrid can be recovered and reused for wastewater remediation.

The efficiency of decolorization of textile dyes by Fe_3O_4-TCS-laccase was 80% for reactive blue 171 and 43% for acid blue 74. The authors reported that during the first use, the reactive blue 171 dye before being decolorized by the Fe_3O_4-TCS-laccase was first adsorbed on the immobilization support. Also, they saw that the degradation mechanism was mostly due to enzymatic activity. Conversely, an efficient and stable decolorization activity towards acid blue 74 dye was the result of Fe_3O_4-TCS-laccase application, concluding that this material is a potential biodegrader of organic dyes [23].

4.4 NANO-ENGINEERED HYBRID MATERIALS FOR ELECTROCHEMICAL PROCESSES

Electrochemical oxidation processes (EOP), such as anodic oxidation (AO) and electro-Fenton (EF), are versatile and simple processes for water purification, capable of degrading various organic contaminants while generating less waste [24,25]. They show advantages of being environmentally friendly, highly efficient, safe, and easy to use, with a wide variety of electrode materials available [26].

On these electrochemical processes, the electrode material and its properties play an important role on the suitability and efficiency of the treatment, and can be determinant for whether it shows any advantage over different processes or materials. For these electrodes, nano-engineered materials have shown to be an improvement on the catalysis mechanism, thermal and electric conductivities, surface area, and mechanical resistance [25,27] when integrated in anode or cathode. The materials for electrodes (including preparation and characterization techniques used to assess their properties) and their mechanisms for the treatment of wastewater using electrochemical processes have been reported in literature [27–30], mainly classifying by their activity, related to the reactions.

4.4.1 BORON-DOPED DIAMOND

Boron-doped diamond (BDD) is a great material used as anode for EO that shows great electro-catalytic properties and chemical and electrochemical stabilities because of its carbon structure and adjustable conductivity due to the lattice defects created by the boron atoms [25]. It has a wide potential window, mechanical strength, and corrosion resistance, and shows great results on treatment of wastewater, especially concerning organic pollutants, with promising results for industrial-scale treatments [27,28,31].

Studies have compared the use of BDD as anode with different materials (such as Ti/RuO_2, Ti/IrO_2-Ta_2O_5, Pt, RuO_2-based, IrO_2-based, and $Ti/Ru_{0.3}Ti_{0.7}O_2$) for the degradation of hazardous

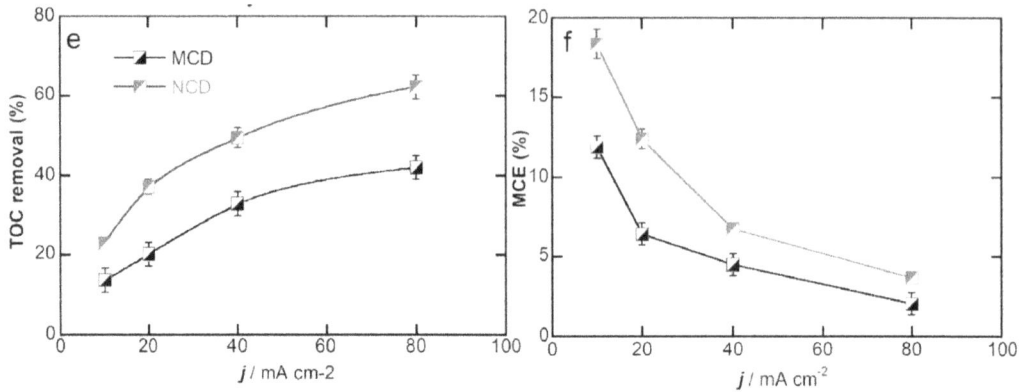

FIGURE 4.6 TOC removal and mineralization current efficiency (MCE) as a function of current density for the NCD and MCD films [25].

organics (phenol with chlorides, naproxen, and clopyralid) [32–34]. They all showed a strong dependency of the electrode material on the efficiency of the process, and a correlation with energy consumption, TOC removal, mineralization, and formation of oxidation products. The BDD anode showed better results of mineralization when used in photoelectron-Fenton process [33,34] and the highest TOC removal [32–34].

A review on the use of BDD electrodes for EO of pesticides (organic pollutants) in water treatment [35] exposes the wide range of use for this material. The discussion around the degradation kinetics and the effect of several process parameters (such as current density, initial pollutant concentration, pH, flowrate, and oxidative species) shows different correlations for the electrode materials with the degradation rate and efficiency, which highlights the complexity and opportunities for these materials on wastewater treatment and removal of organic pollutants. The BDD electrodes have great efficiency but require studies to investigate the degradation pathways and to improve cost. Also, research to understand the generation of oxidants and the unstable hydroxyl radicals – and the generation of more stable oxidants – may increase the pollutant degradation and improve the process.

Therefore, A. J. dos Santos et al. [25] studied the electrocatalytic effects of two different BDD electrodes (microcrystalline diamond [MCD] and nanocrystalline diamond [NCD]) on the EO of an antibiotic (ciprofloxacin) for wastewater treatment. The NCD anode was able to almost totally remove the organic pollutant in 60-minute treatment, in all different several current densities (j) tested (an important factor in electrochemical processes for water treatment), which was an improvement in relation to the MCD, that could only achieve the same performance on a specific j window. The NCD electrode also showed better results in terms of TOC removal and mineralization current efficiency (MCE), as shown in Figure 4.6. In addition, it showed a correlation of pollutant concentration with electrolysis time to completely remove the pollutant, and proved to be viable for electrochemical wastewater treatment.

A study published in 2016 explored the effects of different sp^3/sp^2 carbon ratio on BDD on silicon substrates (p-Si BDD) electrodes for the oxidation of 2,4-dichlorophenoxyacetic acid (2,4-D) on synthetic wastewater [36]. The use of this material and technique was supported by the fact that more conventional modes (such as coagulation or biological treatment) were ineffective for this pollutant, therefore justifying the use of EO with the p-Si BDD electrode. The results showed the importance of the sp^3/sp^2 ratio and the boron content on the electrode performance, in agreement with other studies on the subject [37,38]. The treatment was capable of complete depletion of 2,4-dichlorophenoxyacetic acid, and the evaluation of the sp^3/sp^2 ratio led to a configuration that was more efficient and rapid for mineralization, reducing costs.

A more recent study using BDD anode in AO and EF processes to degrade 2,4-D showed that the EF enhanced the process, with a carbon black-modified graphite-felt cathode (CB-GF), achieving a great MCE and TOC removal with a lower energy consumption [39]. Other organic acids (formic, oxalic, glycolic, glyoxylic, and acetic acids) have been degraded using EO with BDD electrodes, in a study to determine the best conditions and understand the mechanisms involved [40].

4.4.2 Titanium Dioxide Nanotube Arrays

Titanium dioxide nanotube arrays (NTA) have high stability, are of low cost, and have nontoxicity properties that make them a suitable option for wastewater treatment, but show low conductivity. Because of that, there has been great effort to improve such conductivity [27,41]. Mostly, the TiO_2 NTAs are doped with other elements, such as B, Cu, Zn, Pd, Co, Bi [24,42–44], or it can be reduced [27].

The reduction of the TiO_2 NTA into blue-TiO_2 NTA proved it to be a stable electrode with great electrocatalytic performance [27]. It has been produced via electrochemical reduction, thoroughly characterized, and applied for the EO of salicylic acid (SA) [45]. This study by Chang et al. obtained the optimal parameters for the treatment of TiO_2 NTA, and the reduced electrode showed up to be 6.3 times more efficient than Pt electrodes for the oxidation of SA, and remained very stable over eight cycles of EO [45].

In order to increase the structural stability and lifetime for blue-TiO_2 NTA, Yang and Hoffman reported on the activation and deactivation mechanisms and an operational method for the use of blue-TiO_2 NTA in EO processes for wastewater treatment [46]. The electrodes were synthesized by AO and tested for application on wastewater treatment. The electrode was evaluated to have a 16,895-hour lifetime, with a hydroxyl radical production activity that is comparable with BDD electrodes, while operating on a minimum energy consumption. This demonstrates that the blue-TiO_2 NTA is good option as material in terms of cost (including production of the electrode, when compared to BDD), with good efficiency and great potential for studies in terms of improving its stability and use for the degradation of organic pollutants.

Electrodes of Co and Bi-co-doped TiO_2 NTA (Co/Bi/TiO_2 NTA) are a novel material that shows promising results on the elimination of organic pollutants using electrochemical redox process. These electrodes were studied in terms of their efficiency for water treatment and the elimination of meropenem [24,47]. The dopants were used to extend the lifetime of the electrode and enhance OH-species generation. Once the Co/Bi/TiO_2 NTA was investigated in terms of its preparation and viability for electrochemical oxidation treatment for the removal of an organic pollutant in wastewater [47], the electrode was employed as anode and cathode to study oxidation and reduction reactions as a hybrid electrolysis process [24]. Further research should be developed for the improvement of the Co/Bi/TiO_2 NTA.

A first report on pristine TiO_2 NTA as electrode for electrochemical reduction of organic pollutants (nitrobenzene) in 2019 by Ahmadi and Wu [41] studied morphological and operating conditions, reduction mechanism, and kinetics to evaluate it as a promising material for water treatment. They showed that the self-doped TiO_2 NTA had performance improvement with increases in nanotube dimensions (length and diameter) – attributed to the increment on electroactive surface area and electron transferring – on the degradation of nitrobenzene. Also, mineralization was improved with increasing treatment time, and the structure was able to enhance the conductivity. The self-doped TiO_2 NTA showed a good combination of efficiency and stability with low-cost and low-energy consumption.

A study from 2019 produced a composite electrode of TiO_2-NTs/SnO_2-Sb with carbon aerogels (TSS-CA) to degrade three isomers of dihydroxybenzenes that are common pollutants on wastewater of chemical productions [48]. The authors achieved electrodes with high specific surface area and good electrochemical activity, and the performance was attached to the proportion of carbon aerogel (the one with 0.5 wt% was selected as the more appropriate one).

4.4.3 OTHER MATERIALS

Other nanostructured hybrid materials have been used as electrodes for the removal of organic pollutants in wastewater in EO processes, such as nanostructured carbon-based materials (including nanotubes and graphene), nanocomposites, heteroatom-doped graphene, metal oxides modified with carbon, and others [27].

For example, zinc oxide-polyaniline-graphene oxide (ZnO-PANI-GO), a hybrid nanocomposite, was used for EO of a pesticide, as a sensor for detection and degradation of the pollutant [49]. The authors report the synthesis and characterization of the ZnO-PANI-GO electrode and the sensibility for the pesticide compared to other electrodes (such as BDD). The nanocomposite showed a good sensibility and had the lowest limit of detection, and therefore, it is a good option for future research as a hybrid nanomaterial for electrochemical detection and degradation of pesticides.

Many studies cover the use of carbon-based materials, especially rGO, carbon felt (CF) and modified CF, and modified graphite felt (GF) [50–54]. The carbon-based materials show advantages on their adsorption capacity and electroactivity, but having low EO power, they are often modified to achieve efficiency for organic pollutants [55].

In this sense, in 2018, Zhang et al. reported the use of Co-N-doped MoO_2-modified CF cathode to remove ethylenediaminetetraacetic acid (EDTA) using EF process, concluding that the synthesized material is a great option as cathode for EOPs [50]. Yang et al. studied EF process with graphene-modified CF and raw CF cathodes for the degradation of imatinib, and showed that the graphene-modified CF was a better option in terms of mineralization efficiency and toxicity removal [51]. Mi et al. synthesized rGO-Ce/WO_3 nanosheets to modify CF as cathode for the degradation of ciprofloxacin using EF process [52], and obtained great results in terms of degradation, mineralization, and stability. The presence of rGO and Ce was an improvement on the electrode. In 2020, a study by Lai et al. showed the great efficiency and stability of NaOH-activated GF (NaOH-GF) on the degradation of oxytetracycline using EF process [53]. The material was synthesized and characterized to demonstrate the applicability and parameter processes to optimize the use of this electrode on removing non-biodegradable antibiotics.

Wang et al., in 2020, reported a novel electrode of conductive carbon black-modified PbO_2 (C-PbO_2) that was tested against a pristine PbO_2 electrode on the removal of metronidazole (MNZ) from wastewater [54]. They characterized the material in terms of morphology and crystalline structure, and performed electrochemical experiments to understand mechanisms and the effects of different parameter processes, such as electrolyte concentration, pH, current density, and others. The C-PbO_2 had a better electrocatalytic performance and was more stable, due to the presence of carbon black, which improved the current efficiency, the generation of hydroxyl radicals, and the removal of MNZ with oxidation.

Shao et al. developed a 2.5D electrode (which is a two-dimensional anode, a magnet and magnetic granules), which can present combined advantages (like being adsorbent and electrocatalyst) [55]. The electrode consisted of a Ti/Sb-SnO_2 anode magnetically loaded with Fe_3O_4/polyaniline (PANI), resulting in a Ti/Sb-SnO_2/PANI electrode, and is represented in Figure 4.7. They studied the effects of the NPs loading amounts on the structure and electrochemical properties, and the performance on degrading an azo dye (Acid Red G) and a wastewater pollutant (sodium lignosulfonate). The results showed that the different amounts of Fe_3O_4/PANI had different outcomes for the EO performance. In conclusion, the electrode was considered adequate for use in wastewater treatment.

In a study published in 2022, hybrid materials of conductive crystalline polypyrrole (Cryst-PPy), nickel-polypyrrole (Ni-PPy), and copper-polypyrrole (Cu-PPy) were synthesized and tested for the oxidation of phenol using electrochemical process [18]. The authors propose the use of metallic-based NPs of nickel and copper to improve catalytic activity and conductivity. The Cu-PPy electrode had the highest rate for the degradation of phenol and showed to be a new material to be used for EO of phenol in wastewater.

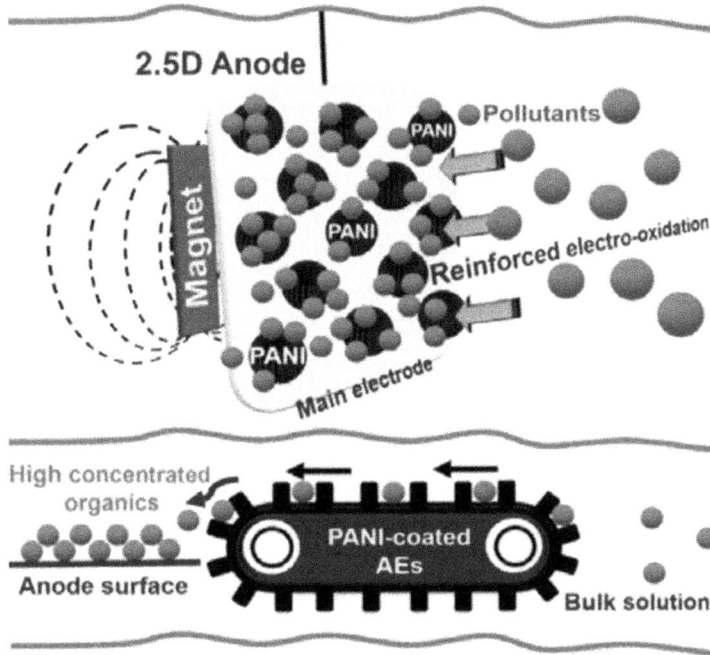

FIGURE 4.7   Representation of the 2.5D anode and its mechanism, showing the main electrode coated with the PANI nanoparticles [55].

4.5   NANO-ENGINEERED HYBRID MATERIALS FOR FILTRATION

The water decontamination by filtration method is, basically, a mechanism that uses a membrane with micropores through which the water to be decontaminated will pass, separating the contaminant from the water [56]. It is essential for the membranes used in the filtration system; some main characteristics are related to their surface and pore structure, aiming at an excellent separation performance, wettability, and antifouling properties [57,58]. Most membrane matrices are made of polymeric materials, and in order to achieve suitable characteristics required for satisfactory filtration performance, many research groups have been trying to incorporate other materials into the membrane matrix [58,59]. These materials own intrinsic characteristics that favor the improvement of the membrane wettability, antifouling characteristic, and hydraulic conductivity, among others [60].

Kusuoro and colleagues produced a polyethersulfone (PES) membrane with incorporated nano-ZnO [61]. The results showed that the nano-ZnO improved the membrane hydrophilicity, and when exposed to UV irradiation, it was observed an enhancement of the liquid flux through the PES-nano-ZnO membrane [61]. In a different approach, they added rGO to the previous structure, forming a nanohybrid PES-rGO/ZnO membrane [57]. The authors realized that the addition of both rGO and ZnO was determinant for the reduction of NPs agglomeration on the membrane surface, increase in wettability, improvement in antifouling properties, and mechanical properties tailoring. Besides, they found a remarkable enhancement of permeate flux and stability.

Tian et al. also used GO in their nanohybrid material [60]. They reported the production of amino-functionalized polypropylene nonwoven/graphene oxide (PP-g-DMAEMA/GO). Figure 4.8 presents the steps of synthesis used by the authors to produce that nanohybrid material.

The authors found that this nanohybrid structure favors the rapid removal of organic pollutants from aqueous solutions, which was possible by a combination of factors. First, the penetrating channels assured water to permeate fluently that was beneficial to the diffusion of the organic pollutants

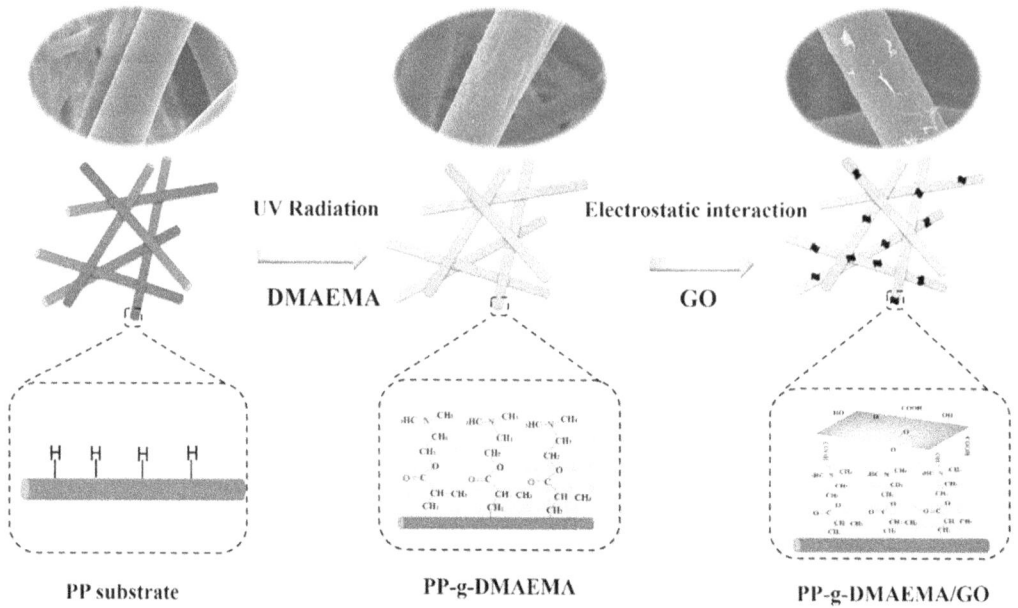

FIGURE 4.8 Illustration of the synthesis of nanohybrid PP-g-DMAEMA/GO. (Reproduced from [60].)

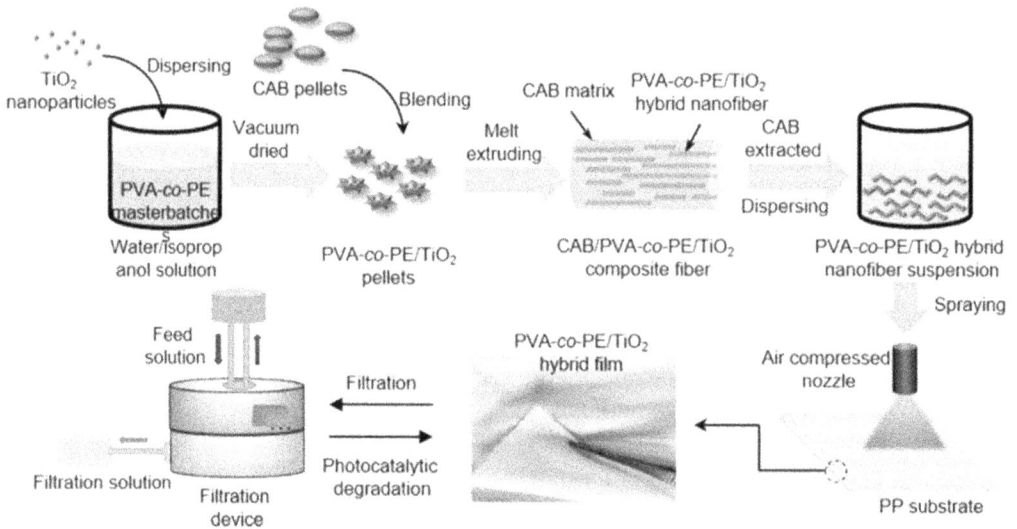

FIGURE 4.9 Schematic diagram of the PVA-co-PE/TiO$_2$ hybrid film production and filtration process. (Reproduced from [62].)

from the solution to the surface of PP-g-DMAEMA/GO. Second, the immobilization of GO on polypropylene (PP) might facilitate the separation and recovery for reuse, maintaining active sites for organic pollutants' capture [60].

Although the membrane filtration technique presents promising performances by itself, some research groups are trying to further improve the water filtration process by combining membrane filtration with other processes for the decontamination of hazardous organics.

Ni et al prepared a hybrid film of poly(vinyl alcohol-co-ethylene) (PVA-co-PE) nanofiber with immobilized TiO$_2$ NPs [62]. Figure 4.9 shows the main steps carried out for the PVA-co-PE/TiO$_2$

hybrid film synthesis and the filtration process. The main idea was to not only filter the polluted solution but also use TiO_2 as a photocatalyst agent. The authors observed that, in addition to presenting satisfactory photocatalytic performance on the degradation of MB under UV lighting, TiO_2 also enabled a self-cleaning of the membrane, which allowed the reuse of the manufactured fiber.

Lee and co-workers proposed a hybrid decontamination method using a nanohybrid material, but in their report, they combined the filtration approach with catalytic ozonation [63]. The authors fabricated a Ce/TiO_x-functionalized catalytic ceramic membrane to be tested in the filtration process. Figure 4.10 shows the proposed mechanism for the reactive oxygen species generation and the schematic mechanisms of the organic pollutant removal in the hybrid catalytic ozonation-membrane filtration process. Results showed that the fabricated membrane presented a great clean water permeability and exhibited a high antifouling property. Besides, the authors emphasized that the nanocatalysts located at the membrane outer surface and inside of it could significantly enhance the accessibility of the reactants to the catalytic sites.

4.6   NANO-ENGINEERED HYBRID MATERIALS FOR PHOTOCATALYSIS

Photocatalysis is another decontamination method that has been widely explored [64,65]. Briefly, the method is described by the chemical interaction between a photocatalytic material and the wastewater, which occurs after the system is irradiated by an illumination source [66,67]. From this chemical interaction, the organic pollutant is then decomposed, and to achieve a great photocatalytic efficiency, some characteristics are expected from the photocatalytic material, such as wettability, low bandgap, i.e., absorption of light in the maximum range as possible, and high surface area [65]. It is difficult to get all these characteristics from just one type of material, so the natural direction followed by the researchers is to combine characteristics from different materials, meaning synthesizing nano-engineered hybrid materials for photocatalysis [68–71].

Wand et al. fabricated graphene foam/TiO_2 nanosheet (G/TiO_2) hybrid structure via hydrothermal method [72]. Figure 4.11 presents the scanning and transmission electron microscopies (SEM and TEM) of the G/TiO_2 hybrid produced. The authors reported that the main reason to combine these two materials was to remove both organic and inorganic pollutants from water. In this way, graphene foam and TiO_2 nanosheets are complementary to each other during the removal of these pollutants [72]. A remarkable result was that after the photocatalytic process, the material could be separated from the aqueous solution and exhibited an excellent recycle stability.

Tao et al. also used TiO_2 in the prepared hybrid structure [73]. They produced multi-layered porous hierarchical TiO_2/g-C_3N_4 hybrid coating. TiO_2 is known to be effective only in the UV range, whereas g-C_3N_4 shows great performance in the visible range. However, the photocatalytic performance of g-C_3N_4 is limited due to the fast recombination of photogenerated electron-hole pairs. The goal of the authors was to combine these two materials to achieve a better organic pollutant photocatalytic performance. Indeed, results showed that the TiO_2/g-C_3N_4 hybrid coating presented a better performance on organic decontamination in the visible range than pure TiO_2 and g-C_3N_4. Besides, the hybrid structure demonstrated stable physicochemical properties, which is suitable for its reuse [73].

Balagji et al. tried to decompose organic pollutants under near-infrared irradiation (NIR). For that, they prepared a lanthanide-doped $NaYF_4$:Yb/Er@CdS core-shell nanostructures [74]. In brief, that core-shell structure was fabricated using a hydrothermal route, and each step is schematized in Figure 4.12a. As main results, the authors found that even though thermal energy was generated by the NIR light, the degradation of the organic pollutants (MO and carbendazim) predominantly occurred by the oxidation of reactive oxygen species formed during the photocatalytic reactions, as shown in Figure 4.12b.

Qu et al. prepared O-g-$C_3N_4/GO/N$-CNT membranes that demonstrated high degradation efficiency for organic pollutants under visible irradiation [75]. The authors attributed the great performance to the presence of GO and nitrogen-doped carbon nanotubes (N-CNT), which worked as

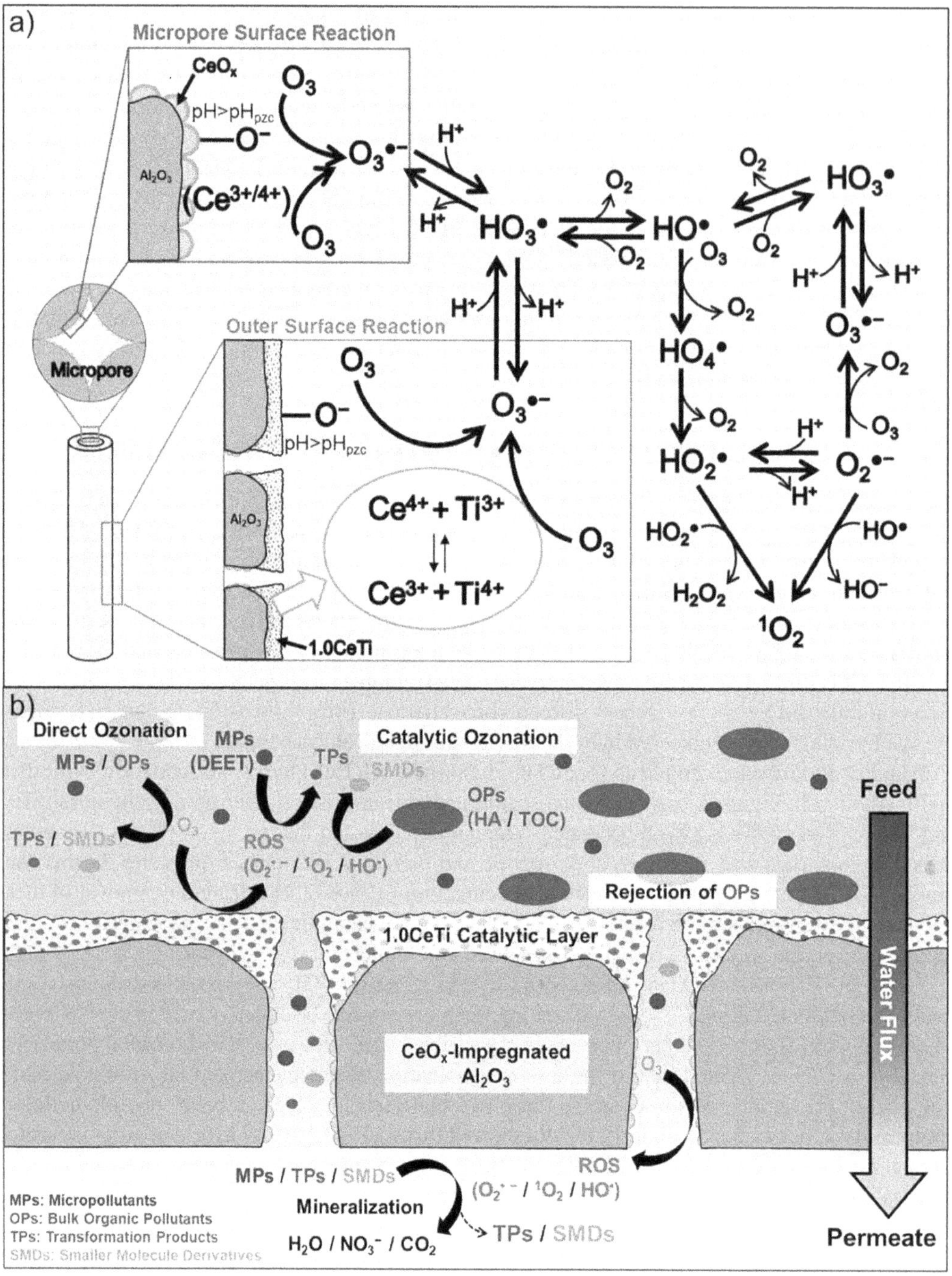

FIGURE 4.10 (a) Proposed mechanism for the reactive oxygen species generation and (b) the schematic mechanisms of the organic pollutant removal in the hybrid catalytic ozonation-membrane filtration process. (Reproduced from [63].)

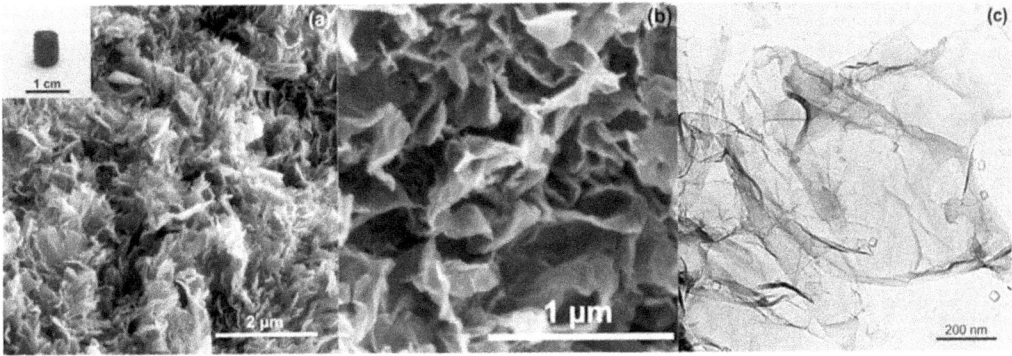

FIGURE 4.11 SEM (a and b) and TEM (c) images, and the inset in (a) is a photography of the G/TiO_2 hybrid. (Reproduced from [72].)

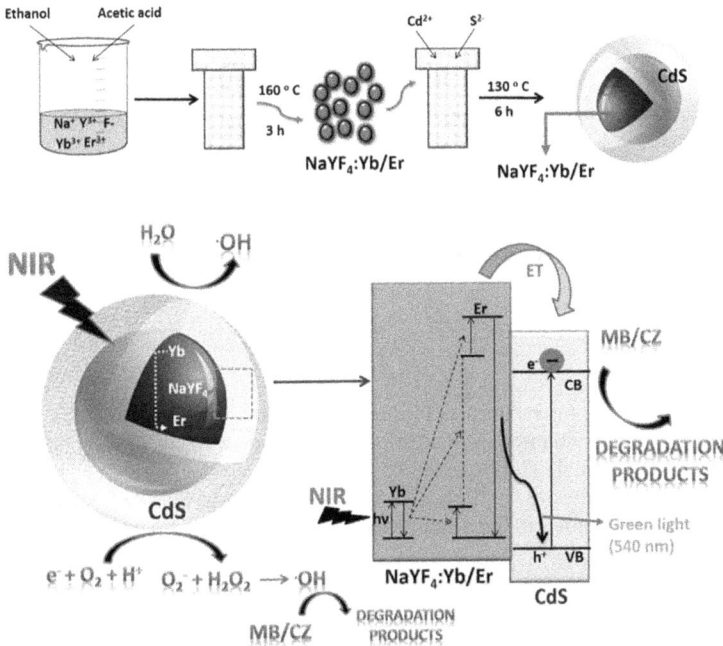

FIGURE 4.12 (a) Schematic illustration for the preparation of the $NaYF_4$:Yb/Er@CdS hybrid core-shell nanocomposites and (b) possible mechanism for the photocatalytic activity of $NaYF_4$:Yb/Er@CdS hybrid nanocomposites in the degradation of pollutants under NIR irradiation. (Reproduced from [74].)

electronic acceptors for monolayer O-g-C_3N_4, and successfully inhibited the recombination rate of photogenerated electron-hole pairs [75].

Mollick and co-workers also performed visible-driven photocatalysis to degrade organic pollutants [76]. They produced a hybrid bromide perovskite combined with metal-organic-framework (HBP@MOF) as their photocatalytic material. HBPs present suitable characteristics for photocatalytic processes, like high absorption coefficient and small exciton binding energy. However, this class of materials also present weak resistance against heat polar solvents, light exposure, and others [76]. In view of this, porous materials as MOFs can act as protecting matrices for HBP-NCs

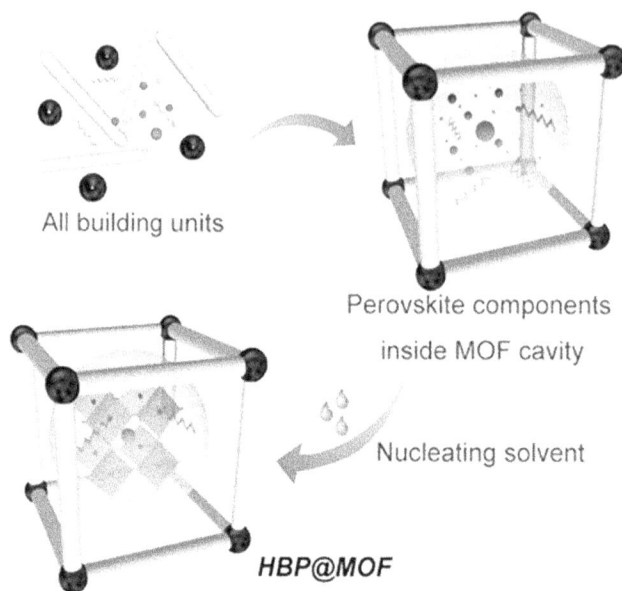

FIGURE 4.13 Schematic illustration highlighting the encapsulation of the HBPs by MOFs. (Reproduced from [76].)

(nanocrystals), and that was the main idea of this report. Mollick et al. choose ZIF-8 as MOF to encapsulate HBP-NCs to protect them from the influence of foreign species, as shown in Figure 4.13. Results confirmed that the prepared hybrid HBP@MOF and the HBP-NCs were successfully protected by the MOFs walls.

REFERENCES

[1] G. Lagaly, Pesticide–clay interactions and formulations, *Appl. Clay Sci.* 18 (2001), pp. 205–209.

[2] A. Stensvand and A. Christiansen, Investigation on fungicide residues in greenhouse-grown strawberries, *J. Agric. Food Chem.* 48 (2000), pp. 917–920.

[3] D. Camacho-Muñoz, J. Martín, J.L. Santos, I. Aparicio and E. Alonso, Distribution and risk assessment of pharmaceutical compounds in river sediments from Doñana Park (Spain), *Water, Air, Soil Pollut.* 224 (2013), p. 1665.

[4] A. Tong, R. Braund, D. Warren and B. Peake, TiO_2-assisted photodegradation of pharmaceuticals – a review, *Open Chem.* 10 (2012), pp. 989–1027.

[5] A.L. Boreen, W.A. Arnold and K. McNeill, Photodegradation of pharmaceuticals in the aquatic environment: A review, *Aquat. Sci.* 65 (2003), pp. 320–341.

[6] S. Kizil and H. Bulbul Sonmez, One-pot fabrication of reusable hybrid sorbents for quick removal of oils from wastewater, *J. Environ. Manage.* 261 (2020), p. 109911.

[7] J. Qi, Y. Hou, J. Hu, W. Ruan, Y. Xiang and X. Wei, Decontamination of methylene Blue from simulated wastewater by the mesoporous rGO/Fe/Co nanohybrids: Artificial intelligence modeling and optimization, *Mater. Today Commun.* 24 (2020), p. 100709.

[8] S.S. Elanchezhiyan, S. Muthu Prabhu, Y. Kim and C.M. Park, Lanthanum-substituted bimetallic magnetic materials assembled carboxylate-rich graphene oxide nanohybrids as highly efficient adsorbent for perfluorooctanoic acid adsorption from aqueous solutions, *Appl. Surf. Sci.* 509 (2020), p. 144716.

[9] S.I. Siddiqui, F. Zohra and S.A. Chaudhry, Nigella sativa seed based nanohybrid composite-Fe_2O_3–SnO_2/BC: A novel material for enhanced adsorptive removal of methylene blue from water, *Environ. Res.* 178 (2019), p. 108667.

[10] M.R.B.M. Rejab, M.H.B.M. Hamdan, M. Quanjin, J.P. Siregar, D. Bachtiar and Y. Muchlis, Historical development of hybrid materials, *Encycl. Renew. Sustain. Mater.* (2020), pp. 445–455.

[11] A. Singh, N. Verma and K. Kumar, *Hybrid Composites: A Revolutionary Trend in Biomedical Engineering*, Elsevier Inc., the Netherlands, 2019.

[12] M. Faustini, L. Nicole, E. Ruiz-Hitzky and C. Sanchez, History of organic–inorganic hybrid materials: Prehistory, art, science, and advanced applications, *Adv. Funct. Mater.* 28 (2018), pp. 1–30.

[13] S.S. Park and C.S. Ha, Hollow mesoporous functional hybrid materials: Fascinating platforms for advanced applications, *Adv. Funct. Mater.* 28 (2018), pp. 1–29.

[14] X. Li, H. Lu, Y. Zhang and F. He, Efficient removal of organic pollutants from aqueous media using newly synthesized polypyrrole/CNTs-CoFe$_2$O$_4$ magnetic nanocomposites, *Chem. Eng. J.* 316 (2017), pp. 893–902.

[15] S. Ganesan, M.P. Bhat, M. Kigga, U.T. Uthappa, H.Y. Jung, T. Kumeria et al., Amine activated diatom xerogel hybrid material for efficient removal of hazardous dye, *Mater. Chem. Phys.* 235 (2019), p. 121738.

[16] M.L. Kalombo, A. Adeniyi, N. Nomadolo, K. Setshedi, M.J. Madito, N. Manyala et al., Preparation and characterisation of polypyrrole nanoparticles for enhancement of granular activated carbon (GAC) as adsorbent, *J. Taiwan Inst. Chem. Eng.* 129 (2021), pp. 264–272.

[17] M.A. Nazir, N.A. Khan, C. Cheng, S.S.A. Shah, T. Najam, M. Arshad et al., Surface induced growth of ZIF-67 at co-layered double hydroxide: Removal of methylene blue and methyl orange from water, *Appl. Clay Sci.* 190 (2020), p. 105564.

[18] L. Seid, D. Lakhdari, M. Berkani, O. Belgherbi, D. Chouder, Y. Vasseghian et al., High-efficiency electrochemical degradation of phenol in aqueous solutions using Ni-PPy and Cu-PPy composite materials, *J. Hazard. Mater.* 423 (2022), p. 126986.

[19] W. Dong, J. Ding, W. Wang, L. Zong, J. Xu and A. Wang, Magnetic nano-hybrids adsorbents formulated from acidic leachates of clay minerals, *J. Clean. Prod.* 256 (2020), p. 120383.

[20] P. Gupta, K. Pandey and N. Verma, Augmented complete mineralization of glyphosate in wastewater via microbial degradation post CWAO over supported Fe-CNF, *Chem. Eng. J.* 428 (2022), p. 132008.

[21] A. Partovinia, A.A. Soorki and M. Koosha, Synergistic adsorption and biodegradation of heavy crude oil by a novel hybrid matrix containing immobilized *Bacillus licheniformis*: Aqueous phase and soil bioremediation, *Ecotoxicol. Environ. Saf.* 222 (2021), p. 112505.

[22] J. Chen, Z. Guo, Y. Xin, Y. Shi, Y. Li, Z. Gu et al., Preparation of efficient, stable, and reusable copper-phosphotriesterase hybrid nanoflowers for biodegradation of organophosphorus pesticides, *Enzyme Microb. Technol.* 146 (2021), p. 109766.

[23] A. Ulu, E. Birhanli, F. Boran, S. Köytepe, O. Yesilada and B. Ateş, Laccase-conjugated thiolated chitosan-Fe$_3$O$_4$ hybrid composite for biocatalytic degradation of organic dyes, *Int. J. Biol. Macromol.* 150 (2020), pp. 871–884.

[24] A. Ahmadi and T. Wu, Towards full cell potential utilization during water purification using Co/Bi/TiO$_2$ nanotube electrodes, *Electrochim. Acta* 364 (2020), p. 137272.

[25] A.J. dos Santos, G. V. Fortunato, M.S. Kronka, L.G. Vernasqui, N.G. Ferreira and M.R.V. Lanza, Electrochemical oxidation of ciprofloxacin in different aqueous matrices using synthesized boron-doped micro and nano-diamond anodes, *Environ. Res.* 204 (2022), p. 112027.

[26] M. Coha, G. Farinelli, A. Tiraferri, M. Minella and D. Vione, Advanced oxidation processes in the removal of organic substances from produced water: Potential, configurations, and research needs, *Chem. Eng. J.* 414 (2021), p. 128668.

[27] X. Du, M.A. Oturan, M. Zhou, N. Belkessa, P. Su, J. Cai et al., Nanostructured electrodes for electrocatalytic advanced oxidation processes: From materials preparation to mechanisms understanding and wastewater treatment applications, *Appl. Catal. B Environ.* 296 (2021), p. 120332.

[28] C.A. Martínez-Huitle and M. Panizza, Electrochemical oxidation of organic pollutants for wastewater treatment, *Curr. Opin. Electrochem.* 11 (2018), pp. 62–71.

[29] M. Shestakova and M. Sillanpää, Electrode materials used for electrochemical oxidation of organic compounds in wastewater, *Rev. Environ. Sci. Biotechnol.* 16 (2017), pp. 223–238.

[30] G.R. Salazar-Banda, G. de O.S. Santos, I.M. Duarte Gonzaga, A.R. Dória and K.I. Barrios Eguiluz, Developments in electrode materials for wastewater treatment, *Curr. Opin. Electrochem.* 26 (2021), p. 100663.

[31] X. Yu, M. Zhou, Y. Hu, K. Groenen Serrano and F. Yu, Recent updates on electrochemical degradation of bio-refractory organic pollutants using BDD anode: A mini review, *Environ. Sci. Pollut. Res.* 21 (2014), pp. 8417–8431.

[32] Y. Hao, H. Ma, F. Proietto, A. Galia and O. Scialdone, Electrochemical treatment of wastewater contaminated by organics and containing chlorides: Effect of operative parameters on the abatement of organics and the generation of chlorinated by-products, *Electrochim. Acta* 402 (2022), p. 139480.

[33] G. Coria, I. Sirés, E. Brillas and J.L. Nava, Influence of the anode material on the degradation of naproxen by Fenton-based electrochemical processes, *Chem. Eng. J.* 304 (2016), pp. 817–825.

[34] G. de O.S. Santos, K.I.B. Eguiluz, G.R. Salazar-Banda, C. Saez and M.A. Rodrigo, Testing the role of electrode materials on the electro-Fenton and photoelectro-Fenton degradation of clopyralid, *J. Electroanal. Chem.* 871 (2020), p. 114291.

[35] S.T. McBeath, D.P. Wilkinson and N.J.D. Graham, Application of boron-doped diamond electrodes for the anodic oxidation of pesticide micropollutants in a water treatment process: A critical review, *Environ. Sci. Water Res. Technol.* 5 (2019), pp. 2090–2107.

[36] F.L. Souza, C. Saéz, M.R.V. Lanza, P. Cañizares and M.A. Rodrigo, The effect of the sp^3/sp^2 carbon ratio on the electrochemical oxidation of 2,4-D with p-Si BDD anodes, *Electrochim. Acta* 187 (2016), pp. 119–124.

[37] F.L. Migliorini, N.A. Braga, S.A. Alves, M.R.V. Lanza, M.R. Baldan and N.G. Ferreira, Anodic oxidation of wastewater containing the Reactive Orange 16 Dye using heavily boron-doped diamond electrodes, *J. Hazard. Mater.* 192 (2011), pp. 1683–1689.

[38] E. Guinea, F. Centellas, E. Brillas, P. Cañizares, C. Sáez and M.A. Rodrigo, Electrocatalytic properties of diamond in the oxidation of a persistant pollutant, *Appl. Catal. B Environ.* 89 (2009), pp. 645–650.

[39] J. Cai, M. Zhou, Y. Pan and X. Lu, Degradation of 2,4-dichlorophenoxyacetic acid by anodic oxidation and electro-Fenton using BDD anode: Influencing factors and mechanism, *Sep. Purif. Technol.* 230 (2020), p. 115867.

[40] A. Arts, M.T. de Groot and J. van der Schaaf, Current efficiency and mass transfer effects in electrochemical oxidation of C1 and C2 carboxylic acids on boron doped diamond electrodes, *Chem. Eng. J. Adv.* 6 (2021), p. 100093.

[41] A. Ahmadi and T. Wu, Electrocatalytic reduction of nitrobenzene using TiO_2 nanotube electrodes with different morphologies: Kinetics, mechanism, and degradation pathways, *Chem. Eng. J.* 374 (2019), pp. 1241–1252.

[42] F. Liu, M. Li, H. Wang, X. Lei, L. Wang and X. Liu, Fabrication and characterization of a Cu-Zn-TiO_2 nanotube array polymetallic nanoelectrode for electrochemically removing nitrate from groundwater, *J. Electrochem. Soc.* 163 (2016), pp. E421–E427.

[43] W. Xie, S. Yuan, X. Mao, W. Hu, P. Liao, M. Tong et al., Electrocatalytic activity of Pd-loaded Ti/TiO_2 nanotubes cathode for TCE reduction in groundwater, *Water Res.* 47 (2013), pp. 3573–3582.

[44] G.G. Bessegato, J.C. Cardoso and M.V.B. Zanoni, Enhanced photoelectrocatalytic degradation of an acid dye with boron-doped TiO_2 nanotube anodes, *Catal. Today* 240 (2015), pp. 100–106.

[45] X. Chang, S.S. Thind and A. Chen, Electrocatalytic enhancement of salicylic acid oxidation at electrochemically reduced TiO2 nanotubes, *ACS Catal.* 4 (2014), pp. 2616–2622.

[46] Y. Yang and M.R. Hoffmann, Synthesis and stabilization of blue-black TiO_2 nanotube arrays for electrochemical oxidant generation and wastewater treatment, *Environ. Sci. Technol.* 50 (2016), pp. 11888–11894.

[47] A. Ahmadi, B. Vogler, Y. Deng and T. Wu, Removal of meropenem from environmental matrices by electrochemical oxidation using Co/Bi/TiO2nanotube electrodes, *Environ. Sci. Water Res. Technol.* 6 (2020), pp. 2197–2208.

[48] X. Yin, Q. Liu, Y. Chen, A. Xu, Y. Wang, Y. Tu et al., Preparation, characterization and environmental application of the composite electrode TiO2-NTs/SnO2—Sb with carbon aerogels, *J. Chem. Technol. Biotechnol.* 94 (2019), pp. 3124–3133.

[49] A.K. Tawade, D. Mohan Kumar, P. Talele, K.K.K. Sharma and S.N. Tayade, Flower-Like ZnO-decorated polyaniline–graphene oxide nanocomposite for electrochemical oxidation of imidacloprid: A hybrid nanocomposite sensor, *J. Electron. Mater.* 48 (2019), pp. 7747–7755.

[50] J. Zhang, W. Zhou, L. Yang, Y. Chen and Y. Hu, Co-N-doped MoO_2 modified carbon felt cathode for removal of EDTA-Ni in electro-Fenton process, *Environ. Sci. Pollut. Res.* 25 (2018), pp. 22754–22765.

[51] W. Yang, M. Zhou, N. Oturan, Y. Li and M.A. Oturan, Electrocatalytic destruction of pharmaceutical imatinib by electro-Fenton process with graphene-based cathode, *Electrochim. Acta* 305 (2019), pp. 285–294.

[52] X. Mi, J. Han, Y. Sun, Y. Li, W. Hu and S. Zhan, Enhanced catalytic degradation by using RGO-Ce/WO_3 nanosheets modified CF as electro-Fenton cathode: Influence factors, reaction mechanism and pathways, *J. Hazard. Mater.* 367 (2019), pp. 365–374.

[53] W. Lai, G. Xie, R. Dai, C. Kuang, Y. Xu, Z. Pan et al., Kinetics and mechanisms of oxytetracycline degradation in an electro-Fenton system with a modified graphite felt cathode, *J. Environ. Manage.* 257 (2020), p. 109968.

[54] X. Wang, Y. Xie, G. Yang, J. Hao, J. Ma and P. Ning, Enhancement of the electrocatalytic oxidation of antibiotic wastewater over the conductive black carbon-PbO_2 electrode prepared using novel green approach, *Front. Environ. Sci. Eng.* 14 (2020), pp. 1–16.

[55] D. Shao, W. Lyu, J. Cui, X. Zhang, Y. Zhang, G. Tan et al., Polyaniline nanoparticles magnetically coated Ti/Sb–SnO$_2$ electrode as a flexible and efficient electrocatalyst for boosted electrooxidation of biorefractory wastewater, *Chemosphere* 241 (2020), pp. 1–11.

[56] N. Yousefi, M. Jones, A. Bismarck and A. Mautner, Fungal chitin-glucan nanopapers with heavy metal adsorption properties for ultrafiltration of organic solvents and water, *Carbohydr. Polym.* 253 (2021), p. 117273.

[57] T.D. Kusworo, A.C. Kumoro, N. Aryanti and D.P. Utomo, Removal of organic pollutants from rubber wastewater using hydrophilic nanocomposite rGO-ZnO/PES hybrid membranes, *J. Environ. Chem. Eng.* 9 (2021), p. 106421.

[58] Y. Ji, Y. Ma, Y. Ma, J. Asenbauer, S. Passerini and C. Streb, Water decontamination by polyoxometalate-functionalized 3D-printed hierarchical porous devices, *Chem. Commun.* 54 (2018), pp. 3018–3021.

[59] I. Gardi and Y.G. Mishael, Designing a regenerable stimuli-responsive grafted polymer-clay sorbent for filtration of water pollutants, *Sci. Technol. Adv. Mater.* 19 (2018), pp. 588–598.

[60] J. Tian, J. Wei, H. Zhang, Z. Kong, Y. Zhu and Z. Qin, Graphene oxide-functionalized dual-scale channels architecture for high-throughput removal of organic pollutants from water, *Chem. Eng. J.* 359 (2019), pp. 852–862.

[61] T.D. Kusworo, D. Soetrisnanto, N. Aryanti, D.P. Utomo, V.D. Tambunan, N.R. Simanjuntak, Evaluation of Integrated modified nanohybrid polyethersulfone-ZnO membrane with single stage and double stage system for produced water treatment into clean water, *J. Water Process Eng.* 23 (2018), pp. 239–249.

[62] Y. Ni, K. Yan, F. Xu, W. Zhong, Q. Zhao, K. Liu et al., Synergistic effect on TiO$_2$ doped poly (vinyl alcohol-co-ethylene) nanofibrous film for filtration and photocatalytic degradation of methylene blue, *Compos. Commun.* 12 (2019), pp. 112–116.

[63] W.J. Lee, Y. Bao, C. Guan, X. Hu and T.T. Lim, Ce/TiOx-functionalized catalytic ceramic membrane for hybrid catalytic ozonation-membrane filtration process: Fabrication, characterization and performance evaluation, *Chem. Eng. J.* 410 (2021), p. 128307.

[64] Z. Wei, J. Liu and W. Shangguan, A review on photocatalysis in antibiotic wastewater: Pollutant degradation and hydrogen production, *Chin. J. Catal.* 41 (2020), pp. 1440–1450.

[65] X. Yang and D. Wang, Photocatalysis: From fundamental principles to materials and applications, *ACS Appl. Energy Mater.* 1 (2018), pp. 6657–6693.

[66] Z. Xiu, M. Guo, T. Zhao, K. Pan, Z. Xing, Z. Li et al., Recent advances in Ti^{3+} self-doped nanostructured TiO$_2$ visible light photocatalysts for environmental and energy applications, *Chem. Eng. J.* 382 (2020), p. 123011.

[67] A. Godoy Junior, A. Pereira, M. Gomes, M. Fraga, R. Pessoa, D. Leite et al., Black TiO$_2$ thin films production using hollow cathode hydrogen plasma treatment: Synthesis, material characteristics and photocatalytic activity, *Catalysts* 10 (2020), p. 282.

[68] Y. Yu, C. Li, S. Huang, Z. Hu, Z. Chen and H. Gao, BiOBr hybrids for organic pollutant removal by the combined treatments of adsorption and photocatalysis, *RSC Adv.* 8 (2018), pp. 32368–32376.

[69] T.T. da Cunha, T.E. de Souza, W.D. do Pim, L.D. de Almeida, G.M. do Nascimento, E. García-España et al., A hybrid catalyst for decontamination of organic pollutants based on a bifunctional dicopper(II) complex anchored over niobium oxyhydroxide, *Appl. Catal. B Environ.* 209 (2017), pp. 339–345.

[70] F. Ji, J. Li, X. Cui, J. Liu, X. Bing and P. Song, Hierarchical C-doped BiPO$_4$/ZnCoAl-LDO hybrid with enhanced photocatalytic activity for organic pollutants degradation, *Appl. Clay Sci.* 162 (2018), pp. 182–191.

[71] M.A. Iqbal, A. Tariq, A. Zaheer, S. Gul, S.I. Ali, M.Z. Iqbal et al., Ti$_3$C$_2$-MXene/bismuth ferrite nanohybrids for efficient degradation of organic dyes and colorless pollutants, *ACS Omega* 4 (2019), pp. 20530–20539.

[72] W. Wang, Z. Wang, J. Liu, Z. Zhang and L. Sun, Single-step one-pot synthesis of graphene foam/TiO$_2$ nanosheet hybrids for effective water treatment, *Sci. Rep.* 7 (2017), pp. 1–8.

[73] W. Tao, M. Wang, R. Ali, S. Nie, Q. Zeng, R. Yang et al., Multi-layered porous hierarchical TiO$_2$/g-C$_3$N$_4$ hybrid coating for enhanced visible light photocatalysis, *Appl. Surf. Sci.* 495 (2019), p. 143435.

[74] R. Balaji, S. Kumar, K.L. Reddy, V. Sharma, K. Bhattacharyya and V. Krishnan, Near-infrared driven photocatalytic performance of lanthanide-doped NaYF$_4$@CdS core-shell nanostructures with enhanced upconversion properties, *J. Alloys Compd.* 724 (2017), pp. 481–491.

[75] L. Qu, G. Zhu, J. Ji, T.P. Yadav, Y. Chen, G. Yang et al., Recyclable visible light-driven OgC$_3$N$_4$/graphene oxide/N-carbon nanotube membrane for efficient removal of organic pollutants, *ACS Appl. Mater. Interfaces* 10 (2018), pp. 42427–42435.

[76] S. Mollick, T.N. Mandal, A. Jana, S. Fajal, A. V. Desai and S.K. Ghosh, Ultrastable luminescent hybrid bromide perovskite@MOF nanocomposites for the degradation of organic pollutants in water, *ACS Appl. Nano Mater.* 2 (2019), pp. 1333–1340.

5 Polyaniline-Based Adsorbents and Photocatalysts for the Elimination of Toxic Heavy Metals

Akbar Samadi and Shuaifei Zhao
Deakin University

CONTENTS

5.1 INTRODUCTION

Heavy metals are very harmful for humans, animals, and other organisms and listed as the priority pollutants by the US Environmental Protection Agency (EPA) [1]. They are originated from the earth crust and penetrate to the environment via various industrial activities and wastewaters [2]. A trace amount of heavy metals can cause serious harms to living organisms due to their biologically non-degradable nature, and can enter into the metabolism systems through food chains and/ or drinking water [3]. The most reported toxic heavy metals include cadmium (Cd), chromium (Cr), copper (Cu), mercury (Hg), nickel (Ni), lead (Pb), zinc (Zn), and metalloids, such as arsenic (As). A variety of technologies, such as adsorption, ion exchange, membrane filtration, photocatalysis, chemical precipitation, coagulation-flocculation, and electrochemical treatment, have been used to treat wastewaters containing heavy metals.

Among these technologies, adsorption by diverse sorbents such as activated carbon and alumina is a promising low-cost technology with a simple and flexible design and process [4]. To overcome the low adsorption capacity and other existing challenges in the cost and regeneration of conventional adsorbents, polyaniline (PANI) and other conjugated polymeric materials have displayed promising effectiveness for pollutants removal [5]. PANI and PANI-based materials have been widely investigated to remove toxic metals mainly owing to their unique morphological features,

DOI: 10.1201/9781003206385-5

exceptional structural characteristics, and the diverse functional groups. The conductive functional groups are rich of electrons and promote adsorption though chelation of pollutants [6,7].

Photocatalytic reduction is a green technology for the removal of toxic metals with high valency (e.g., Cr(VI)) from industrial wastewaters [8,9]. When photocatalysts receive photon with energy higher than the band gaps of the photocatalysts, the photocatalysts become excited and the electrons in the valence bands (VBs) move to the conduction bands (CBs) [10]. The generated electrons cause the reduction of the high valence to its lower valence in aqueous solutions. Transition metals have been widely used as photocatalysts due to their remarkable electronic and physicochemical properties [10–13]. However, they often suffer from fast charge-carrier recombination and low absorption of light [10,13,14]. Their wide band gaps limit absorption only within the UV region [13].

PANI, a novel semiconducting material, acts as a photocatalyst with light (UV or visible), due to its doping/dedoping state [15]. The band gap of the pernigraniline (PB) PANI is 1.4-2.2 eV [16,17]. The band gap of emeraldine salt (ES) PANI is 2.7 eV [18], while the other PANI states (emeraldine base (EB) and leucoemeraldine (LB)) have higher band gaps (3–4 eV) [16,17]. However, visible light active photocatalysts (band gap<3eV) such as PANI cannot perform both reduction of O_2 and oxidation of H_2O at the same time [19]. In order to generate a reductive and oxidative photocatalyst, one way is to combine two photocatalysts in a heterojunction system [19,20]. PANI has a high capability of O_2 reduction due to its high position of CB [15]. Therefore, it can be combined with oxidative photocatalytic materials (e.g., transition metal oxides) to prepare heterojunctionphotocatalysts. Moreover, PANI can enhance the catalytic efficiency of the heterojunction photocatalysts due to the high electron-hole carrying efficiency [13,21]. Other features of PANI, such as facile synthesis, excellent environmental stabilities, controllably reversible properties, and tunable properties, also enhance the heterojunction photocatalysts [22,23].

A key factor for adsorbents or adsorptive photocatalysts is their effective sites for pollutants, which are determined by their specific surface areas and pore size distributions [10,11,24,25]. To increase available sites for adsorption and to prevent sorbents' aggregation, functional and high aspect-ratio nanofillers, such as silica, sawdust, carbon nanotubes (CNTs), graphene oxide (GO), cellulose acetate (CA), and organic molecules, have been introduced into PANI via *in situ* polymerization to form engineered PANI nanostructures [26]. PANI has also been introduced to fabricate nanocomposites with transition metal semiconductors due to its remarkable charge transporting properties, stable photo-sensitizer function, and high thermal and photochemical stabilities. It also provides adsorption-assisted photocatalytic degradation through the high porous system with a high surface area [11,13,27,28].

Herein, we discuss the ways to remove toxic heavy metals by PANI-based adsorbents and photocatalysts. First, the main types of PANI-based materials and synthesis methods are discussed. Then, the relevant mechanisms of the adsorption and photocatalytic reduction by PANI-based adsorbents and photocatalysts are evaluated. Finally, the concluding remarks and future research trends are presented.

5.2 PREPARATION METHODS

PANI is prepared by electrochemical oxidation [29] or chemical [30] polymerization of aniline. Ferric chloride, ammonium peroxydisulfate, and potassium persulfate are often used as the free radical oxidants to initiate the chemical polymerization of the aniline [31–34]. Chemical polymerization of aniline via oxidation is illustrated in Figure 5.1a. Initiation and chain propagation, followed by the termination step, are the main phases in the chemical polymerization [6,35]. Chemical variations of PANI and its doping/dedoping states are illustrated in Figure 5.1b. Electrochemical polymerization can be controlled by the galvanostatic method via applying a constant current (galvanostatic). Moreover, the aniline solution can also be adjusted by other methods, including constant potential orpotential cycling [4,23]. In another kind of electrochemical polymerization, radical polymerization is performed in aqueous electrolyte solutions with high acidity, followed by a mechanism which oxidizes aniline on the electrode and creates anilinium radical cations [6].

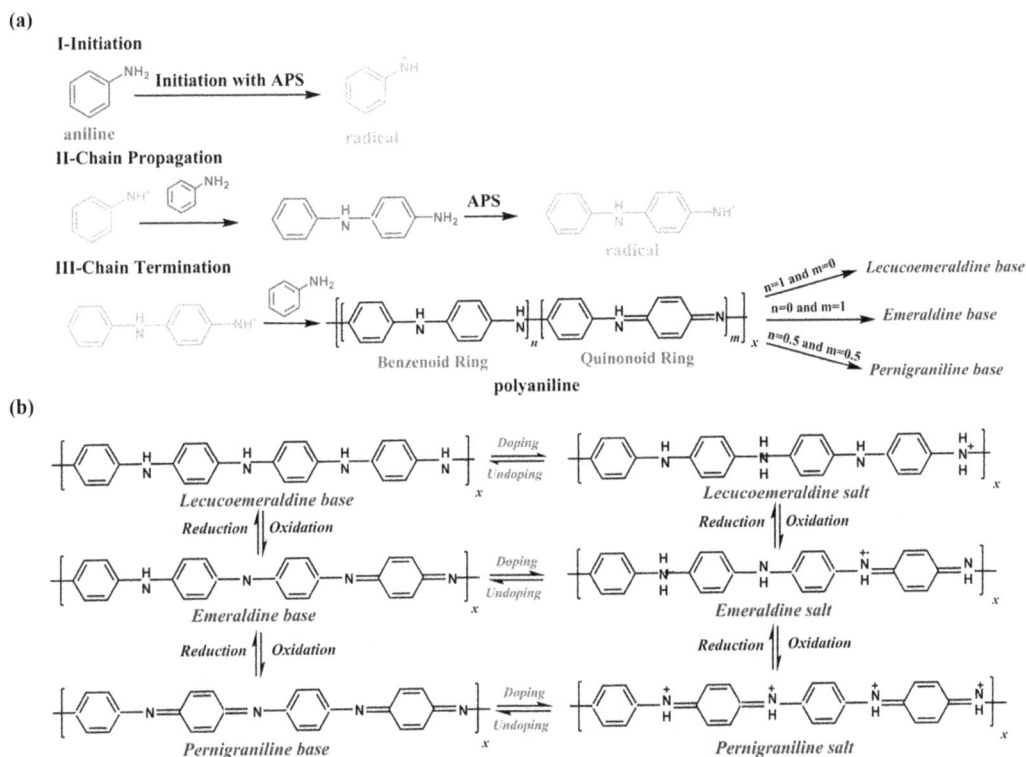

FIGURE 5.1 (a) Main steps in the chemical polymerization of aniline: $x=0$, 0.5, and 1 is for PANI in the form of the leucoemeraldine, emeraldine, and pernigraniline, respectively. (b) PANI structure in various doping states [5].

For PANI-based nanocomposites, *in situ* polymerization of aniline with various organic/inorganic fillers aids low agglomeration of the polymer particles as well as the fillers [10,11,28,36]. However, there are still some challenges to prepare PANI-based nanocomposites with desirable morphologies and features. PANI can be easily synthesized in the presence of diverse 1D, 2D, or 3D materials, and the morphology of the final product depends on many operating parameters, such as the oxidizer type, dopant, solution temperature, oxidation rate, and aniline/filler ratio [5,37,38]. The desirable morphology should have high surface area, to enhance the interaction between the functional groups of the PANI or the filler with the target materials. Other features of PANI, such as thermal and chemical resistance, mechanical strength, electric charge, and electrical conductivity change with the synthesis conditions, should also be carefully considered.

High aspect-ratio fillers (e.g., clay, graphene, and CNTs) have been used in the polymerization process to enhance the surface area of the reactants for adsorption [39–42]. In chemical oxidative *in situ* polymerization, low-cost bio-adsorbents, such as chitosan [43–46] and agriculture wastes [47–49], have been incorporated with PANI and other conjugated polymers to generate functional groups and enhance the surface area of the PANI-based nanocomposites. Recently, ultrasonic irradiation has been introduced in the oxidative polymerization of aniline to promote the production of chitosan-graft-substituted PANI [50].

PANI has an interesting morphology due to its various adjustable shapes (Figure 5.2), with varying pore sizes and pore size distributions as well as its free volume [37]. Three possible intrinsic morphologies for PANI are nanofibers, nanosheets, and nanoparticles, which can be produced by adjusting the synthesis conditions [51]. For example, the solution pH slightly below 7, during the

FIGURE 5.2 Various morphologies of PANI [5]: (a) nanofiber [60], (b) nanotube [58], (c) nanoflake [51], (d) nanoparticle [52], (e) flower-petal [38], (f) nanowire [53], (g) nano-cable [54], (h) nano-needle [55], (i) nano-belt [56], (j) nano-rod [57], (k) nano-stick [58].

polymerization, creates crosslinked ortho-linked chain structures and branched units. Therefore, the decrease in solubility and precipitation creates flat nanoflakes as a result of π–π stacking of individual layers. On the other hand, spherical nanoparticles of PANI are observed when the polymerization is conducted without any acid [52].

PANI-based materials have diverse morphologies, such as PANI nanowire [53], gold/PANI noncable [54], electromagnetic functionalized nano-needle [55], PANI nano-belt [56], PANI nano-rod-coated MWNTs nanocomposites [57], and one-dimensional nano-stick [58], as shown in Figure 5.2. Compared with the pristine PANI, its composites often exhibit better adsorption rates. The enhanced adsorption performance is generally related to the more adsorption sites because of increased surface area and porosity of the composite adsorbents [59]. Adsorption and photocatalysis require sufficient surface area and porosity to bring the reactants together. PANI-based adsorbents and photocatalysts with enhanced surface morphologies are listed in Table 5.1. The surface area of PANI nanocomposites (25–$180\,m^2/g$) is significantly higher than the surface area of pristine PANI ($15\,m^2/g$). Moreover, the PANI-based composite as a photocatalyst has been widely used for chromium reduction due to the high reductive photocatalytic capability of the PANI.

TABLE 5.1

Surface Properties of PANI-Based Adsorbents and Photocatalysts Based on BET Analysis

Material	Application	Surface Area (m²/g)	Pore Volume/Diameter (cm³/g)/(nm)	Pollutants	References
CuFe$_2$O$_4$/PANI	Adsorbent	30.8	0.06/17.8	Hg(II)	[39]
GO/SiO$_2$@PANI	Adsorbent	150.36	0.07/9.32	Cu(II)	[41]
	Adsorbent			Cr(VI)	
PNHM/Fe$_3$O$_4$	Adsorbent	64	0.16/10.3	As(III)	[61]
PANI/Clay	Adsorbent	182.11	1.6/3.8	Cu(II)	[42]
MoS$_2$	Photocatalyst	6.12	-	Cr(VI)	[28]
PANI	Photocatalyst	15.66		Cr(VI)	
PANI@MoS$_2$	Photocatalyst	25.09		Cr(VI)	
SnO$_2$,	Photocatalyst	108.7	-	Cr(VI)	[62]
SnO$_2$/PANI-1%	Photocatalyst	109.1		Cr(VI)	
SnO$_2$/PANI-2%	Photocatalyst	110.5		Cr(VI)	
SnO$_2$/PANI-3%	Photocatalyst	128.8		Cr(VI)	
SnO$_2$/PANI-4%	photocatalyst	121.3		Cr(VI)	

5.3 MATERIALS TYPES

5.3.1 POLYANILINE (PANI)

PANI is a well-known conjugated polymer with highly ordered structures and excellent environmental stability [63]. The repeating units of PANI are reduced benzenoid diamines and oxidized quinoid diamines. Depending on the ratio of these repeating units in the PANI structure, different oxidation states of PANI with varying electric conductivities can be prepared. When the PANI structure only contains benzenoid rings, the emeraldine form is created. The LB form of PANI only contains quinoid rings. The PB form of PANI contains both units (Figure 5.1a) [22,23,64,65]. Predominately, treatment of PANI in acidic solution protonates the amine groups, which provides cationic defects responsible for the reversible conductivity [64]. The base and salt forms of PANI are generated by deprotonation and protonation of PANI, respectively (Figure 5.1b). Emeraldine has the same number of oxidized and reduced units and exhibits the highest stability among different PANI forms [23,31,59,65].

In wastewater treatment, there are diverse pollutants with positive or negative charges. Hence, PANI can be used for the effective removal of these pollutants due to its adjustable electric charges of the active amine and imine groups [66]. PANI in its neutral state (EB) can be protonated by doping with an acidic solution. Thus, it is widely used for adsorption. On the other hand, chemical decomposition is a main drawback of PB and its salts because quinonediimine does not have a sufficient stability in the presence of nucleophiles [23]. Different states of PANI can be easily converted to each other by the addition of chemicals or electrochemical processes for oxidation or reduction [64].

PANI derivatives have pendant functional groups different from PANI. Their solubility in common organic solvents is better than that of PANI. However, they have lower conductivity compared with PANI [7].

5.3.2 PANI-BASED NANOCOMPOSITES

To improve mechanical strength, surface area, and functionality, and harness the synergistic features of nanocomposites, composites of PANI and various materials have been prepared by *in situ*

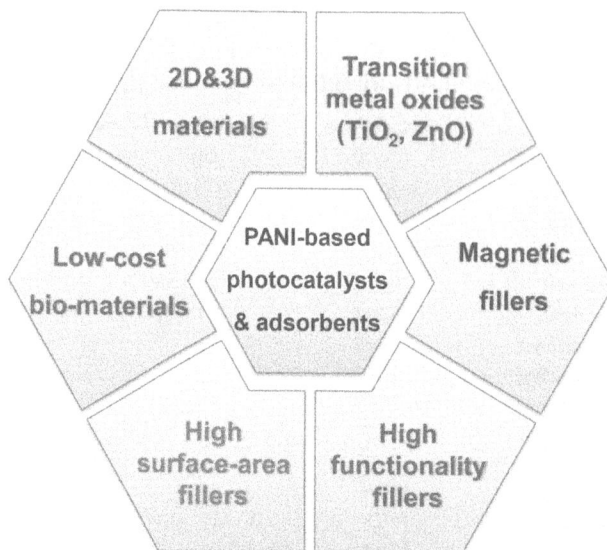

FIGURE 5.3   PANI-based adsorbents and photocatalysts.

polymerization of aniline. Because of the special properties of each filler, the prepared nanocomposites have different features, such as low cost, easy recovery, facile synthesis, enhanced surface area, and high functionality. For PANI/transition metal oxide photocatalysts, PANI offers high surface area and enhanced charged transport properties. PANI-based photocatalysts act as p-n junctions via the combination of PANI as a P-type polymer with metal oxides (e.g., ZnO, TiO$_2$, SnO$_2$, and WO$_3$) as N-type semiconductors. Such photocatalysts are designed to overcome the challenges of the conventional semiconductors, such as leaching, thermal decomposition, low efficiency with visible light, and fast charge recombination [13]. The PANI-metal oxide nanocomposites offer several advantages, such as prohibiting charge recombination, broad light absorption; and low photocorrosion [10]. Various PANI-based nanocomposites are summarized in Figure 5.3.

5.4   REMOVAL OF HEAVY METALS

5.4.1   PHOTOCATALYTIC REMOVAL

Various heavy metals, such as arsenic, mercury, copper, and zinc, are the causes of many environmental problems and health issues. Heavy metals come from diverse industrial activities, such as mining, metal plating, and tanneries. Photocatalysis technology has been employed to remove heavy metals due to its cost-effectiveness, the use of solar energy, capability of oxidation and reduction, and high surface area [3]. However, conventional inorganic photocatalysts have some drawbacks, such as fast charge-carrier recombination, weak light absorption, and wide band gaps [13]. PANI nanocomposites with transition metals have gained interest for photocatalytic treatment of heavy metals owing to the charge transport properties, photochemical stabilities, and adsorption-assisted photocatalytic degradation as listed in Table 5.2 [11,13,27,28].

Chromium ion is a common heavy metal pollutant produced in many industries, such as leather tanning, electroplating, and pigments [69,70]. Cr(VI) is more harmful than Cr(III) and has carcinogenic and mutagenic effects on organisms [71,72]. Adsorption and photocatalytic reduction are common methods for chromium removal [73]. Adsorption-assisted photocatalysis has been used for the reduction of Cr(VI) [10–12,62,68]. The efficiency of this process is dependent on adsorption and diffusion of hexavalent chromium ions on the photocatalyst surface. Transition metals (e.g., TiO$_2$,

TABLE 5.2

Photocatalytic Reduction of Heavy Metals by PANI-Based Photocatalysts

Material	T (°C)	pH	Pollutants	Removal (%)	References
PANI/ZnO			Cr(VI)	81	[67]
			Ni(II)	78	
PANI/ZnO	30	4–7	Cr(VI)	97.2	[10]
PANI/TiO$_2$		3	Cr(VI)	100	[11]
PANI-rGO-SnS$_2$	25	3	Cr(VI)	100	[68]
PANI-SnO$_2$	-	1	Cr(VI)	100	[62]
PANI/MnO$_2$/TiO$_2$	45	2	Cr(VI)	99.9	[12]
PANI/ZnO	-	10	Hg(II)	91.8	[27]
	-	1	Cr(VI)	92.7	

ZnO, CdO, and CdSe) with remarkable electronic properties generally act as good photo-sensitizers owing to their filled VB and empty CB in the ground state [13].

PANIs with high electron-hole carrying efficiency have been employed as stable photo-sensitizers for inorganic semiconductors (e.g., TiO$_2$ and ZnO) due to their remarkable electroactive properties [74,75]. When a photocatalyst is excited by a light-emitting energy greater than its band gap, photo-induced molecular transformation occurs on the catalyst surface and the produced electron-hole pairs generate diverse radicals, such ashydroxyl radicals (-OH), superoxide radical anions (O$_2^-$), and hydroperoxyl radicals (•OOH) [76]. The photo-induced charges have a very short lifetime and can recombine with each other. So, decreasing the recombination rate of charge carriers by tailoring the microstructure and blending with other photocatalytic materials of different band gaps is desirable in the fabrication of semiconductors.

Deng et al. fabricated PANI-TiO$_2$ photocatalysts for the reduction of Cr(VI) and reported an enhanced reduction efficiency due to the adsorption of Cr(VI) ions to the photocatalyst surface with the help of PANI functional groups [11]. They reported 100% reduction efficiency even after ten cycles. As shown in Figure 5.4, by the preparation of heterojunction PANI/TiO$_2$ photocatalyst, the photogenerated electrons remain on the TiO$_2$ surface, while the holes are transferred onto the PANI, resulting in the separation of them. Then, the PANI surface with positively charged amino groups efficiently adsorbs Cr(VI) and forces the product Cr(III) to leave the reaction interface quickly, thereby ensuring the photocatalyst performance.

Poly(aniline-co-anthranilic acid)/ZnO nanocomposites have been applied for photocatalytic reduction of Cr(VI) and Ni(II) [67]. *In situ* chemical oxidative polymerization resulted in a nano-sized distribution of ZnO nanoparticles (15–20 nm) encapsulated in the polymer matrix. Zhang et al. reported an exceptional synergistic effect of PANI/SnS$_2$/N-doped reduced GO nanocomposites on the enhanced photocatalytic activity for the reduction of Cr(VI) under visible light [68]. Bao et al. synthesized PANI from aniline monomers mixed with ZnO nanosheets and prepared ZnO/PANI nanocomposites [10]. The synergistic effect of ZnO/PANI nanocomposites on the photocatalytic performance was verified in their work. As presented in Figure 5.5, the drawbacks of transition metal oxide photocatalysts, such as electron-hole recombination and limited light absorption, can be solved by heterojunction photocatalysts. PANI is a promising candidate for this purpose, due to its capability to electron-hole separation and the other advantages listed in Figure 5.5.

5.4.2 Adsorption

Polymer nanocomposites are promising adsorbents due to their specific tailorable morphologies, and electrical and chemical features. Particularly, the amine and imine functional groups are highly

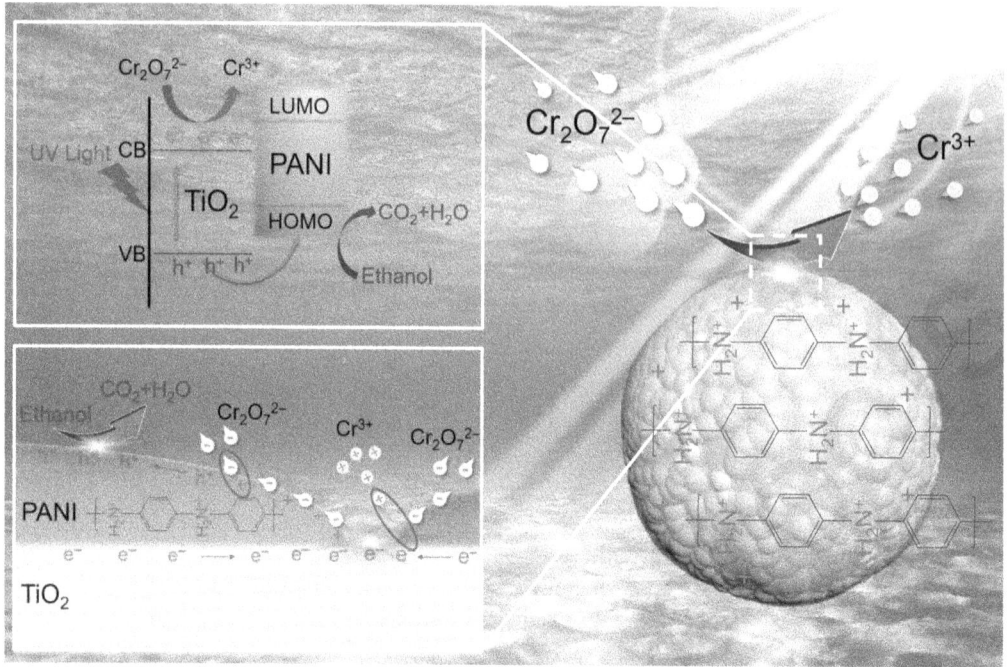

FIGURE 5.4   A proposed adsorption-photoreduction-desorption mechanism for photocatalytic reduction of Cr(VI) assisted by PANI/TiO2 composite [11].

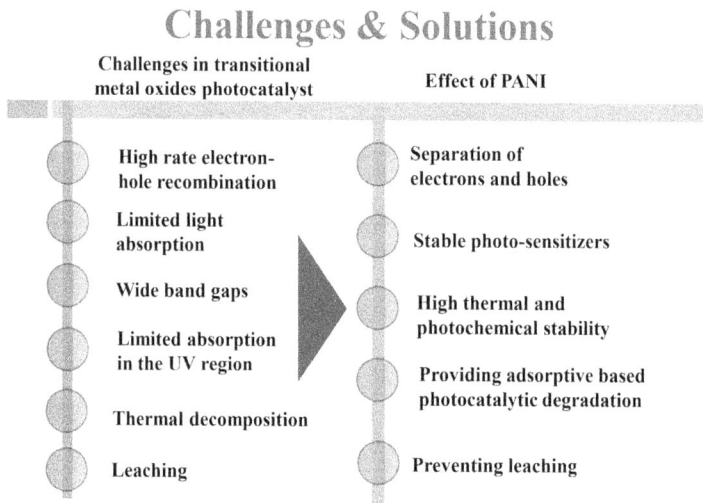

FIGURE 5.5   Challenges and solutions in the application of PANI-based photocatalysts.

reactive adsorption sites. Furthermore, amphoteric carbonaceous surface of PANI can be easily modified by protonation or deprotonation. It also comprises chelating groups (e.g., sulfur, oxygen, and phosphorus) that can adsorb metal ions [77].

Table 5.3 summarizes the heavy metal adsorption results of PANI-based adsorbents with diverse fillers. The pH values in the range of 5–7 lead to optimum adsorption to metal cations because the adsorbent surface is a Schiff base containing negatively charged basic functional groups of imine

TABLE 5.3

Adsorption of Heavy Metals and Metalloids by PANI-Based Adsorbents and Photocatalysts

Material	T (°C)	pH	Pollutants	Q (mg/g)	References
PANI Polymer					
PANI particles	25	7	Cr(VI)	-	[79]
PANI nanosheets	25	2	Cr(VI)	263.2	[33]
PANI derivatives					
PoPDA	30	6	Pb(II)	103.2	[80]
Cu-PmPDA	35	5.2	As(V)	27.4	[81]
PmPDA-Fe$_3$O$_4$	15–45	2–9	Cr(VI)	246.1	[82]
PmPDA-polypyrrole	25	2	Cr(VI)	526	[83]
MnO$_2$@PmPDA	25	5.4	Pb(II)	367	[84]
PpPDA	25	-	Pb(II)	>1,800	[85]
PmPDA/Fe$_3$O$_4$	25	1-13	Cr(VI)	2,750	[86]
PANI Copolymers					
1,8-diaminonaphthalene	25	4	Cr(VI)	150	[87]
pyrrole	25	2	Cr(VI)	227	[88]
PANI grafted chitosan	25	6	Pb(II)	3.2	[89]
m-phenylenediamine	25	6	Cd(II)	-	[90]
Synthetic Polymers					
Polystyrene	30	4.1	Hg(II)	-	[91]
Polystyrene	30	4	Cr(VI)	19	[92]
polyethylene glycol	25	5	Cr(VI)	109.9	[93]
PPy	25	2	Cr(VI)	227	[69]
Polyvinyl alcohol	50	4	Cr(VI)	111.2	[94]
Polyethylene terephthalate	25	2	Cr(VI)	3.0	[95]
Polysaccharide Biopolymers					
PANI/chitosan	25	4.2	Cr(VI)	229.8	[96]
Chitosan-grafted PANI	30	4.2	Cr(VI)	165.6	
PANI/CA	25	3-7	Cu(II)	68.0	[97]
	25	3-7	Pb(II)	251.3	
Chitosan-grafted PANI	30	4.2	Cr(VI)	179.2	[98]
PANI/dextrin	25	3	Cu(II)	-	[99]
PpPDA/chitosan	30	6	Cu(II)	650	[100]
PANI-coated chitin	30	2	Cr(VI)	23	[101]
Sodium alginate/PANI	30	2	Cr(VI)	78.63	[102]
Agriculture Wastes					
PANI/sawdust	25	6	Pb(II)	-	[103]
PANI/jute fiber	20	3	Cr(VI)	62.9	[104]
PANI/sawdust	20	5	Cu(II)	208.8	[105]
PANI/sawdust	25	5	Cd(II)	430	[106]
PANI/rice husk ash	25	9	Hg(II)	-	[107]
PANI/rice husk ash	25	3	Zn(II)	24.3	[108]
PANI-kapok fiber	30	4–5	Cr(VI)	66.2	[109]
PANI/sawdust/polyethylene glycol	25	<2	Cr(VI)	3.2	[110]
Carbon-Based Materials					
Multiwalled CNT	25	4.5	Cr(VI)	31.8	[36]
GO	25	5.8	Cu(II)	38	[111]

(Continued)

TABLE 5.3 (*Continued*)

Adsorption of Heavy Metals and Metalloids by PANI-Based Adsorbents and Photocatalysts

Material	T (°C)	pH	Pollutants	Q (mg/g)	References
GO	30	6.5	Cr(VI)	192	[112]
PANI/GO-CNT	25	2	Cr(VI)	142.9	[113]
PANI/GO	25	7	Zn (II)	210	[114]
Metal Compounds					
Fe_3O_4	25	3	Cr(VI)	6	[115]
Magnetic mesoporous silica	25	8	Cr(VI)	193.9	[116]
Magnetic mesoporous carbon	20	2	Cr(VI)	150	[117]
Clay Minerals					
PANI/sepiolite	25	2	Cr(VI)	206.6	[118]
PANI/palygorskite	15	5.8	Cr(VI)	11.5	[119]
PANI/clay	25	5	Pb(II)	7.4	[120]
PANI/halloysite	25	5.1	Cr(VI)	69.3	[121]
PANI/starch/montmorillonite	25	2	Cr(VI)	208.3	[122]

and amine in that pH range. On the other hand, amine and imine groups are protonated at lower pH values (e.g., acidic solution), which are suitable for adsorbing negatively charged metals such as Cr(VI) [78]. The adsorption capacity is affected by the porosity, surface area, and functional groups of the adsorbent. The adsorption capacity of Pb(II) on the PANI derivatives was in the order of poly(para-phenylenediamine) (PpPDA)>poly(meta-phenylenediamine) (PmPDA)> poly(ortho-phenylenediamine) (PoPDA) [7].

Metal ions can adsorb pollutants by the chelation mechanism and electrostatic attraction of appositively charged pollutants [77]. Diverse factors, such as pH, natural organic matters, ionic strength, and number of functional groups, can affect the chelating mechanism. Electrolytes and metal ions compete during adsorption. Ionic strength can change the interaction between metal ions and the chelating groups, thereby affecting adsorption. In addition, precipitation of a number of metal ions, such as lead, cadmium, zinc, and chromium, takes place at high pH. Therefore, they are more readily to be adsorbed at low pH. Moreover, Cr(VI) and other metal anions can be removed efficiently at lower pH because positive imine groups attract negative oxyanions [73].

As depicted in Figure 5.6, electrostatic interactions, ion exchange, reduction, and chelation are the main adsorption mechanisms [5]. Chromium (VI) is mainly reduced into Cr(III) and then adsorbed via electrostatic interactions and chelation [73]. Zinc is an example of positively charged metal ion. Chelating groups, such as oxygen-containing, nitrogen-containing, and sulfur-containing groups, adsorb these metal cations, leading to the release of negative residual charges on the surfaces of the adsorbents [39,77,123].

In spite of the advantages of PANI-based adsorbents mentioned above, they suffer from some challenging issues such as agglomeration, limited surface area, and low mechanical strength as depicted in Figure 5.7. Therefore, researchers have attempted to overcome these restrictions by incorporating diverse fillers to enhance the surface area, functionality, and adsorption capacity. In recent years, novel 2D nanomaterials, including graphene-based nanomaterials [124], layered double hydroxides [125,126], MoS_2 [127], boron nitride [128–130], and MXenes [131,132], have been widely used to prepare PANI-based composites for adsorption due to their high surface areas, easy exfoliation, and high functionality. Furthermore, titanium dioxide and zinc compounds are of more concern in the modification of PANI adsorbent to introduce antibacterial effects and alter the morphology of the pristine polymer. Finally, PANI-derived carbon [133], 3D MOFs [134,135], and clay minerals [40,42] are the other materials that can be employed.

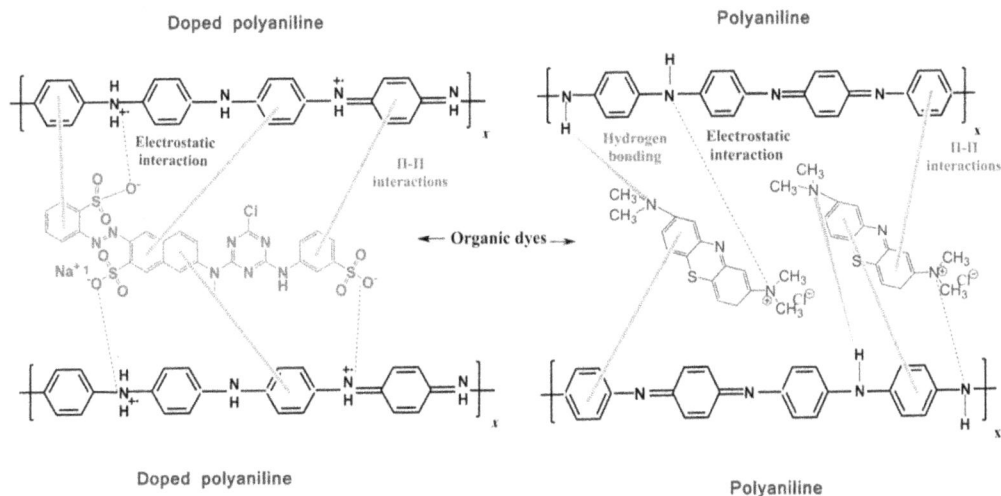

FIGURE 5.6 Proposed heavy metal removal mechanisms by PANI-based adsorbents. In this example, Cr(VI) adsorption is depicted to take place in different ways: (1) attraction of negatively charged Cr(VI) oxyanions on protonated imine groups by means of electrostatic forces, (2) substitution of anions (e.g., Cl⁻) with Cr(VI) oxyanions through ion exchange, and (3) reduction of Cr(VI) oxyanions into Cr(III) cations through the oxidation of amines into quinoid amines, and (4) bounding the Cr(III) cations with oxidized quinoid amines [5].

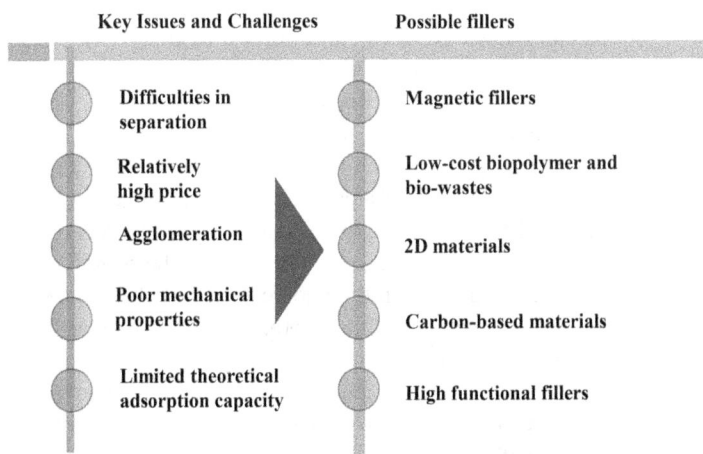

FIGURE 5.7 Challenges and solutions in the application of PANI-based adsorbents.

5.5 HEAVY METAL REMOVAL MECHANISMS

5.5.1 MECHANISM OF PHOTOCATALYSIS

As shown in Figure 5.8a, the band energy (E_g) or band gap of a photocatalyst is determined by the difference between the highest energy level of VB (EV) and the lowest energy level of CB (EC). During photocatalysis, exciting these materials by energy greater than their band gaps causes the transfer of electron from VB to CB, forming excitons, i.e., electrons at the CB and holes at the VB

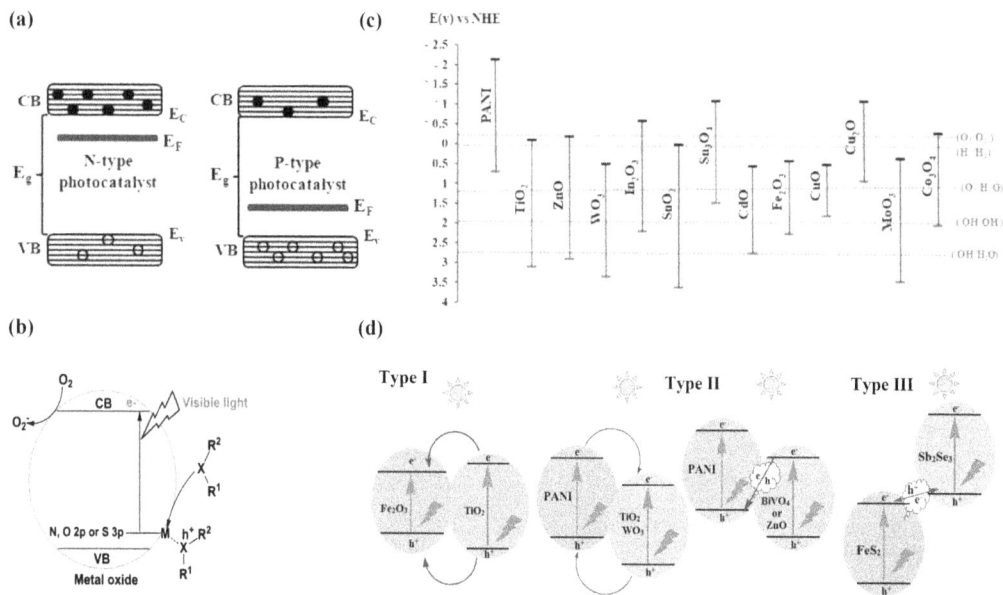

FIGURE 5.8 (a) Band structures of N-type and P-type photocatalysts [15], (b) band positions of VB and CB for metal oxide photocatalysts and PANI [13], (c) surface complexation heteroatoms (X=O, N, or S) on metal (Ti, Nb, or Zn) oxide for visible-light-induced reactions [15], and (d) band alignment for three different types of heterojunctions [140].

(Figure 5.8b). CB electrons can reduce O_2, while holes in VB oxidize H_2O, creating a reductive-oxidative system (ROS) [15]. However, their photocatalytic activity suffers from two main problems: (i) low quantum yield because of fast recombination of electron-hole pairs and (ii) limited absorption only under UV due to the wide band gaps [136,137].

The Fermi level (E_F), the energy level at which the probability of finding an electron is 0.5, lies within the E_g of a photocatalyst. E_F position of a photocatalyst determines its type. For an N-type photocatalysts, E_F is near CB, while for a P-type photocatalyst, E_F is near VB (Figure 5.8a). N-type photocatalysts are more likely to carry electrons, while the P-type photocatalysts have holes as the majority charge carrier. The main challenge in the photocatalysts is the transfer of electron from electrons in the close contact of N-type and P-type photocatalysts. Visible light active photocatalysts cannot reduce O_2 and oxidize H_2O at the same time [19]. Hence, for the generation of ROS, a reductive and an oxidative photocatalysts are required. One solution to generate ROS from visible light active photocatalysts is the combination of a reductive and an oxidative photocatalyst, leading to the creation of a heterojunction photocatalyst [19,20].

The band positions of VB and CB for metal oxide photocatalysts are shown in Figure 5.8c. The main reason for the photocatalytic activity of metal oxides, such as TiO_2, ZnO, and WO_3, is oxygen defects in their crystal structures [138,139]. As seen in Figure 5.8c, PANI has a lower unoccupied molecular orbital (LUMO) in the range of −1 to −2 V vs. normal hydrogen electrode (NHE) and a higher occupied molecular orbital (HOMO) at 0.4–0.6 V vs. NHE [18]. When comparing the LUMO and HOMO values of the transition metal oxide (Figure 5.8c), it is concluded that PANI can be incorporated with most transition metal oxides to prepare type-II heterojunction photocatalysts (Figure 5.8d).

An example of creation of type-II heterojunction photocatalyst was reported by Bao et al. [10]. Figure 5.9 illustrates that the transfer of light-generated electrons from PANI to ZnO surface and transfer of holes in the reverse direction results in the separation of appositely charged excitons.

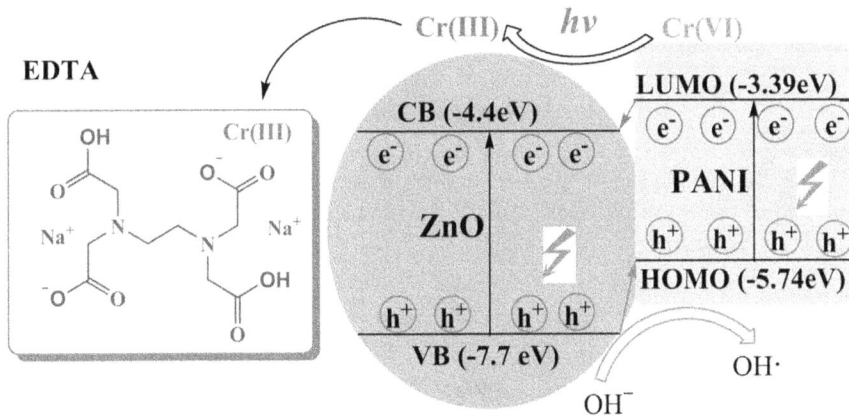

FIGURE 5.9 Proposed mechanism for Cr(VI) photoreduction over ZnO/PANI composites under UV irradiation [10].

Hence, in an acidic medium, the negatively charged HCrO⁻ ions are attracted onto the protonated amine and imine groups of PANI and their oxidizability leads to a partial reduction of Cr(VI) ions into Cr(III) by PANI. Then, the reduced Cr(III) releases into the solution to reach the ion transfer equilibrium. Photocatalysis breaks the equilibrium, and a CB (e^-) and a VB (h^+) are generated by UV light. Thus, the mechanism of Cr(VI) photocatalytic reduction is direct reduction by photogenerated electrons in Cr(VI) photoreduction as shown in the following equations:

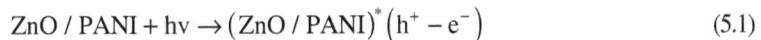

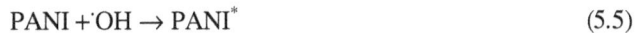

$$ZnO / PANI + hv \rightarrow (ZnO / PANI)^* \left(h^+ - e^-\right) \tag{5.1}$$

$$HCrO^- + 7H^+ + 3e^- \rightarrow Cr^{3+} + 4H_2O \tag{5.2}$$

$$H_2O + h^+ \rightarrow {}^{\cdot}OH + H^+ \tag{5.3}$$

$$OH^- + h^+ \rightarrow {}^{\cdot}OH \tag{5.4}$$

$$PANI + {}^{\cdot}OH \rightarrow PANI^* \tag{5.5}$$

5.5.2 MECHANISM OF ADSORPTION

The adsorption kinetics is a key factor to understand the adsorption mechanisms. The main steps of the diffusion models are film diffusion, intraparticle diffusion, and adsorption-desorption on the active sites [141]. Conversely, reaction models do not consider mass transfer issues and are directly derived from chemical reaction kinetics. The reaction models have also been widely used for the removal of aqueous pollutants by PANI-based adsorbents [42,142]. However, determination of the reaction mechanisms is a complicated process, and thus, the mass transport results from diffusion models and a variety of experimental conditions are used to find clues on the prevailing reactions in the system. In reality, the experimental parameters, such as initial concentrations, adsorbent/adsorbate mass ratios, and time intervals, can change the reaction nature from pseudo-second-order (PSO) to pseudo-first-order (PFO), and vice versa. The PFO model (Eq. 5.6), also called the

Lagergren kinetic model, is the first equation for the study of the adsorption mechanisms and determination of the rate-controlling steps [143].

$$\ln(q_e - q_t) = \ln q_e - Kt \tag{5.6}$$

Where K is the rate constant of PFO adsorption (min^{-1}); q_e and q_t are the adsorption capacities at equilibrium and at time t, respectively (mg/g). However, the PSO model (Eq. 5.7) has been more widely used for adsorption:

$$\frac{t}{q_t} = \frac{1}{K_2 q_e^2} + \frac{t}{q_e} \tag{5.7}$$

where K_2 (g/mg/min) is the rate constant of PSO adsorption.

The PFO model does not deal with the sorbent concentration or the active sites and their effects on the adsorption process. Most adsorption studies have favored the PSO model than the PFO model as the former has better predictions near the equilibrium time [141].

The intraparticle diffusion model was proposed to determine the rate-controlling step and explore the adsorption mechanism [144]:

$$q_t = K_{id} t^{0.5} + C \tag{5.8}$$

where K_{id} is the rate constant and C is a constant. The slope of the plot of q_t vs. $t^{0.5}$ gives the K_{id} value. In some cases, multi-linearity may be observed, suggesting more than one rate-controlling steps. The term C is directly proportional to the boundary-layer thickness [145]. A sharp slope may be seen at the first time interval due to the external diffusion [146,147]. The intraparticle diffusion model has been widely used to investigate the adsorption kinetics of PANI-based sorbents for the removal of pollutants [44,148]. However, in the application of this equation, the adsorption data should be carefully divided into two to three time intervals, and then each step fits individually [142,147]. When the PANI-based adsorbents are highly heterogeneous, the Elovich model can be used to determine the chemisorption by [149]:

$$q_t = \frac{1}{\beta} \ln(\alpha\beta) + \frac{1}{\beta} \ln(t) \tag{5.9}$$

where α is the initial adsorption rate constant (g/mg/min) and β is the desorption rate constant (g/mg). The Elovich model does not provide a particular mechanism insight, but it evaluates the surface heterogeneity of the adsorbents [48]. The Elovich model has been applied for PANI-based adsorption, and the documented regression coefficients (0.85–0.95) were lower than those from the PSO and PFO models [48,150,151]. It can be concluded that morphology of the adsorbent is determinant in the heterogeneous or homogeneous nature of the adsorption and monolayer chemisorption on nitrogen-rich functional amine and imine groups that are dominant [44,142,147,148]. Based on the porous structure of the sorbent, physisorption can contribute to the adsorption [147] and diffusion may control the rate of the adsorption when the system is not well initiated or the adsorbate concentration reduces to very low levels at the end of the adsorption [48,150].

Adsorption-desorption studies determine the regenerability of the exhausted adsorbent after adsorption [152,153]. Moreover, the regenerability of the adsorbent loaded with pollutants should be considered, since it partially determines the overall efficiency of the treatment process [22,25]. Chemical and thermal regeneration processes are widely used for the desorption of a diversity of adsorbents [154]. Chemical regeneration can be effectively achieved for PANI-based adsorbents by changing the solution pH via the doping/dedoping feature of PANI-based sorbents [152,153]. In an effective adsorption-desorption cycle, the polluted solution is treated by PANI-based adsorbents,

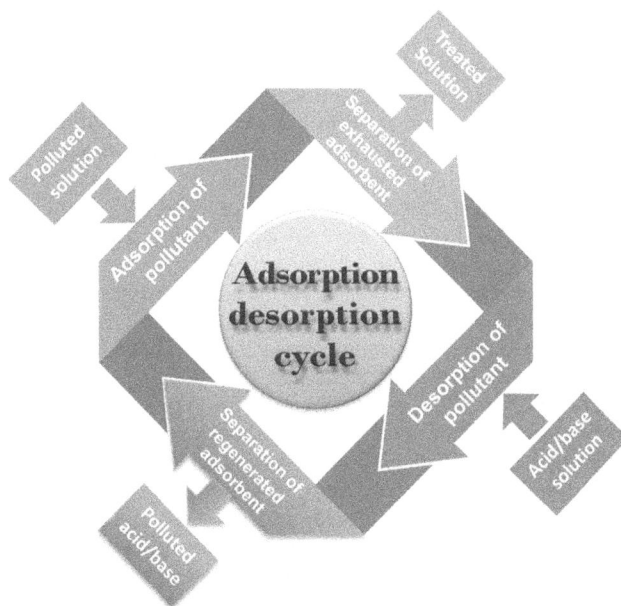

FIGURE 5.10 Schematic representation of adsorption-desorption cycle [5].

and then, the exhausted adsorbent is separated from the treated solution and dissolved in a small amount of acidic or basic solution to desorb the pollutant. Then, the regenerated adsorbent is used to further treat the polluted solution in a cyclic process. The adsorption-desorption cycle is depicted in Figure 5.10. This process consumes a low amount of adsorbents, but could produce secondary pollution due to the use of acid or base that should be further treated or recycled.

5.6 CONCLUDING REMARKS AND PERSPECTIVES

This chapter discusses the removal of toxic heavy metals by PANI-based adsorbents and photocatalysts. PANI, a low-cost conducting polymer, has remarkable electrical characteristics, facile synthesis, diverse morphological features, high functional groups, and environmental stability. Thus, it is a promising material for many applications, especially in adsorption and photocatalytic reduction of pollutants. *In situ* polymerization produces nano-engineered PANI-based adsorbents. Doping/ dedoping offers many variations on its functional groups, surface charge, and porosity. PANI can also enhance the photocatalytic activity of transition metals by the electron-hole separation effect in the form of type-II heterojunction photocatalysts.

PANI nanocomposites with various fillers can remove toxic heavy metals, mainly via electrostatic interactions, ion exchange, reduction, and chelation. Easy regeneration is another advantage of PANI-based adsorbents. The chemical regeneration can be easily performed via adding chemicals to change the solution pH. Nanocomposites of PANI and transition metals exhibit an adsorption-assisted photocatalysis activity.

Incorporation of novel ultrathin fillers, such as 2D graphene-based nanomaterials, layered double hydroxides, and MXenes, into PANI has recently gained attention due to their high surface area, easy exfoliation, and high functionality. Fillers with antibacterial effects, PANI-derived carbon, are also attractive in the fabrication of PANI nanocomposites. The future research efforts should be devoted to the development of green and controllable synthesis methods, the effects of other pollutants on heavy metal removal performance by PANI-based materials, technology scale-up, secondary pollution during regeneration of the adsorbent, cost analysis, and lifecycle assessment of the whole treatment process.

REFERENCES

1. R. K. Gautam, S. K. Sharma, and M. C. Chattopadhyaya, Functionalized Magnetic Nanoparticles for Heavy Metals Removal from Aqueous Solutions, *Royal Society of Chemistry, London* (2014), p. 57.
2. M. Rizwan, S. Ali, M. F. Qayyum, et al., Effect of metal and metal oxide nanoparticles on growth and physiology of globally important food crops—a critical review, *Journal of hazardous materials*, 322(2017), pp. 2–16.
3. M. B. Tahir, H. Kiran, and T. Iqbal, The detoxification of heavy metals from aqueous environment using nano-photocatalysis approach—a review, *Environmental Science and Pollution Research*, 26(2019), pp. 10515–10528.
4. A. Nasar, Polyaniline (PANI) based composites for the adsorptive treatment of polluted water, *Smart Polymers and Composites*, 21(2018), p. 40.
5. A. Samadi, M. Xie, J. Li, et al., Polyaniline-based adsorbents for aqueous pollutants removal—a review, *Chemical Engineering Journal*, 418(2021), p. 129425.
6. T. C. Maponya, M. J. Hato, T. R. Somo, et al., Polyaniline-based nanocomposites for environmental remediation, in *Heavy Metal Ions Removal*, edited by Mario Alfonso Murillo-Tovar, 2019, IntechOpen, London.
7. E. N. Zare, A. Motahari, and M. Sillanpää, Nanoadsorbents based on conducting polymer nanocomposites with main focus on polyaniline and its derivatives for removal of heavy metal ions/dyes—a review, *Environmental Research*, 162(2018), pp. 173–195.
8. R. Vinu and G. Madras, Kinetics of simultaneous photocatalytic degradation of phenolic compounds and reduction of metal ions with nano-TiO_2, *Environmental Science & Technology*, 42(2008), pp. 913–919.
9. M. I. Litter, Treatment of chromium, mercury, lead, uranium, and arsenic in water by heterogeneous photocatalysis, *Advances in Chemical Engineering*, 36(2009), pp. 37–67.
10. C. Bao, M. Chen, X. Jin, et al., Efficient and stable photocatalytic reduction of aqueous hexavalent chromium ions by polyaniline surface-hybridized ZnO nanosheets, *Journal of Molecular Liquids*, 279(2019), pp. 133–145.
11. X. Deng, Y. Chen, J. Wen, et al., Polyaniline-TiO_2 composite photocatalysts for light-driven hexavalent chromium ions reduction, *Science Bulletin*, 65(2020), pp. 105–112.
12. B. Vellaichamy, P. Periakaruppan, and B. Nagulan, Reduction of Cr^{6+} from wastewater using a novel in situ-synthesized $PANI/MnO_2/TiO_2$ nanocomposite—renewable, selective, stable, and synergistic catalysis, *ACS Sustainable Chemistry & Engineering*, 5(2017), pp. 9313–9324.
13. U. Riaz, S. Ashraf, and J. Kashyap, Enhancement of photocatalytic properties of transitional metal oxides using conducting polymers—a mini review, *Materials Research Bulletin*, 71(2015), pp. 75–90.
14. J. Feng, Y. Hou, X. Wang, et al., In-depth study on adsorption and photocatalytic performance of novel reduced graphene oxide-$ZnFe_2O_4$-polyaniline composites, *Journal of Alloys and Compounds*, 681(2016), pp. 157–166.
15. O. S. Ekande and M. Kumar, Review on polyaniline as reductive photocatalyst for the construction of the visible light active heterojunction for the generation of reactive oxygen species, *Journal of Environmental Chemical Engineering*, 9(2021), p. 105725.
16. W. Huang and A. MacDiarmid, Optical properties of polyaniline, Polymer, 34(1993), pp. 1833–1845.
17. H. Huang, Z. Zheng, J. Luo, et al., Internal photoemission in polyaniline revealed by photoelectrochemistry, *Synthetic Metals*, 123(2001), pp. 321–325.
18. S. Chen, D. Huang, G. Zeng, et al., Modifying delafossite silver ferrite with polyaniline—visible-light-response Z-scheme heterojunction with charge transfer driven by internal electric field, *Chemical Engineering Journal*, 370(2019), pp. 1087–1100.
19. F. Imtiaz, J. Rashid, and M. Xu, Semiconductor Nanocomposites for Visible Light Photocatalysis of Water Pollutants, IntechOpen, London, (2019).
20. Q. Xu, L. Zhang, J. Yu, et al., Direct Z-scheme photocatalysts—principles, synthesis, and applications, *Materials Today*, 21(2018), pp. 1042–1063.
21. F. Zhang, T. Ding, Y. Zhang, et al., Polyaniline modified SnS_2 as a novel efficient visible-light-driven photocatalyst, *Materials Letters*, 192(2017), pp. 149–152.
22. S. Mondal, U. Rana, P. Das, et al., Network of polyaniline nanotubes for wastewater treatment and oil/water separation, *ACS Applied Polymer Materials*, 1(2019), pp. 1624–1633.
23. A. Nasar and F. Mashkoor, Application of polyaniline-based adsorbents for dye removal from water and wastewater—a review, *Environmental Science and Pollution Research*, 26(2019), pp. 5333–5356.
24. L. Rizzo, S. Malato, D. Antakyali, et al., Consolidated vs new advanced treatment methods for the removal of contaminants of emerging concern from urban wastewater, *Science of the Total Environment*, 655(2019), pp. 986–1008.

25. D. Q. Zhang, W. Zhang, and Y. Liang, Adsorption of perfluoroalkyl and polyfluoroalkyl substances (PFASs) from aqueous solution—a review, *Science of The Total Environment*, 694(2019), p. 133606.

26. Y. Huang, J. Li, X. Chen, et al., Applications of conjugated polymer based composites in wastewater purification, *Royal Society of Chemistry (RSC)*, 4(2014), pp. 62160–62178.

27. T. Zou, C. Wang, R. Tan, et al., Preparation of pompon-like ZnO-PANI heterostructure and its applications for the treatment of typical water pollutants under visible light, *Journal of Hazardous Materials*, 338(2017), pp. 276–286.

28. Y. Gao, C. Chen, X. Tan, et al., Polyaniline-modified 3D-flower-like molybdenum disulfide composite for efficient adsorption/photocatalytic reduction of Cr (VI), *Journal of Colloid and Interface Science*, 476(2016), pp. 62–70.

29. P. Terangpi and S. Chakraborty, Adsorption kinetics and equilibrium studies for removal of acid azo dyes by aniline formaldehyde condensate, *Applied Water Science*, 7(2016), pp. 3661–3671.

30. P. Chakraborty, A. Kothari, and R. Nagarajan, Highly ordered polyaniline as an efficient dye remover, *Adsorption Science & Technology*, 36(2017), pp. 429–440.

31. C. Xu, H. Chen, and F. Jiang, Adsorption of perflourooctane sulfonate (PFOS) and perfluorooctanoate (PFOA) on polyaniline nanotubes, *Colloids and Surfaces A: Physicochemical and Engineering Aspects*, 479(2015), pp. 60–67.

32. J. Wang, L. Bi, Y. Ji, et al., Removal of humic acid from aqueous solution by magnetically separable polyaniline—adsorption behavior and mechanism, *Journal of Colloid and Interface Science*, 430(2014), pp. 140–146.

33. Y. Tian, H. Li, Y. Liu, et al., Morphology-dependent enhancement of template-guided tunable polyaniline nanostructures for the removal of Cr(VI), *RSC Advances*, 6(2016), pp. 10478–10486.

34. J. Li, Q. Wang, Y. Bai, et al., Preparation of a novel acid doped polyaniline adsorbent for removal of anionic pollutant from wastewater, *Journal of Wuhan University of Technology-Mater. Sci. Ed.*, 30(2015), pp. 1085–1091.

35. P. E. d. G. Alonso, Alternative synthesis methods of electrically conductive bacterial cellulose-polyaniline composites for potential drug delivery application. Doctoral dissertation, 2017.

36. J. Wang, X. Yin, W. Tang, et al., Combined adsorption and reduction of Cr (VI) from aqueous solution on polyaniline/multiwalled carbon nanotubes composite, *Korean Journal of Chemical Engineering*, 32(2015), pp. 1889–1895.

37. A. Samadi, and A. H. Navarchian, Matrimid–polyaniline/clay mixed-matrix membranes with plasticization resistance for separation of CO_2 from natural gas, *Polymers for Advanced Technologies*, 27(2016), pp. 1228–1236.

38. A. Samadi and A. H. Navarchian, Separation of carbon dioxide from natural gas by matrimid-based mixed matrix membranes, *Gas Processing Journal*, 4(2016), pp. 1–18.

39. S. S. Hassan, A. H. Kamel, A. A. Hassan, et al., $CuFe_2O_4$/polyaniline (PANI) nanocomposite for the hazard mercuric ion removal—synthesis, characterization, and adsorption properties study, *Molecules*, 25(2020), p. 2721.

40. S. Kalotra and R. Mehta, Synthesis of polyaniline/clay nanocomposites by in situ polymerization and its application for the removal of Acid Green 25 dye from wastewater, *Polymer Bulletin*, 78(2020), pp. 2439–2463.

41. R. Kumar, M. A. Barakat, M. A. Taleb, et al., A recyclable multifunctional graphene oxide/SiO_2@ polyaniline microspheres composite for Cu(II) and Cr(VI) decontamination from wastewater, *Journal of Cleaner Production*, 268(2020), p. 122290.

42. H. Soltani, A. Belmokhtar, F. Z. Zeggai, et al., Copper(II) removal from aqueous solutions by PANI-clay hybrid material—fabrication, characterization, adsorption and kinetics study, *Journal of Inorganic and Organometallic Polymers and Materials*, 29(2019), pp. 841–850.

43. I. M. Minisy, N. Salahuddin, and M. Ayad, Chitosan/polyaniline hybrid for the removal of cationic and anionic dyes from aqueous solutions, *Journal of Applied Polymer Science*, 136(2019), p. 47056.

44. S. Sahnoun and M. Boutahala, Adsorption removal of tartrazine by chitosan/polyaniline composite—kinetics and equilibrium studies, *International Journal of Biological Macromolecules*, 114(2018), pp. 1345–1353.

45. J. Liu, C. Zhao, G. Yuan, et al., Adsorption of U (VI) on a chitosan/polyaniline composite in the presence of Ca/Mg-U (VI)-CO_3 complexes, *Hydrometallurgy*, 175(2018), pp. 300–311.

46. R. Ahmad, I. Hasan, and A. Mittal, Adsorption of Cr (VI) and Cd (II) on chitosan grafted polyaniline-OMMT nanocomposite—isotherms, kinetics and thermodynamics studies, *Desalin Water Treat*, 58(2017), pp. 144–153.

47. M. Zirpe, H. Bagla, and J. Thakur, Rapid removal of [152+154]Eu(III) using polyaniline/ceria nanocomposite from low level waste, *Journal of Inorganic and Organometallic Polymers and Materials*, 30(2020), pp. 5053–5062.

48. A. Khadir, M. Negarestani, and H. Ghiasinejad, Low-cost sisal fibers/polypyrrole/polyaniline biosorbent for sequestration of reactive orange 5 from aqueous solutions, *Journal of Environmental Chemical Engineering*, 8(2020), p. 103956.

49. X. Liu, B. Wang, G. Jing, et al., Kinetics, isotherms, and mechanism of Cr (VI) adsorption by polyaniline/sunflower stem pith composite adsorbent, *Desalination and Water Treatment*, 141(2019), pp. 197–207.

50. M. S. Alshammari, A. A. Essawy, A. El-Nggar, et al., Ultrasonic-assisted synthesis and characterization of chitosan-graft-substituted polyanilines—promise bio-based nanoparticles for dye removal and bacterial disinfection, *Journal of Chemistry*, 2020(2020), pp. 3297184.

51. H. Navarchian, Z. Hasanzadeh, and M. Joulazadeh, Effect of polymerization conditions on reaction yield, conductivity, and ammonia sensing of polyaniline, *Advances in Polymer Technology*, 32(2013).

52. Q. Shen, M. Mezgebe, F. Li, et al. , Liquids adsorption behavior and surface properties of polyanilines doped by lignosulfonate-modified carbon nanotubes, *Colloids and Surfaces A: Physicochemical and Engineering Aspects*, 390(2011), pp. 212–215.

53. P. Baruah and D. Mahanta, Adsorption and reduction—combined effect of polyaniline emeraldine salt for removal of Cr(VI) from aqueous medium, *Bulletin of Materials Science*, 39(2016), pp. 875–882.

54. K. Huang, Y. Zhang, Y. Long, et al., Preparation of highly conductive, self-assembled gold/polyaniline nanocables and polyaniline nanotubes, *Chemistry–A European Journal*, 12(2006), pp. 5314–5319.

55. Z. Zhang, M. Wan, and Y. Wei, Electromagnetic functionalized polyaniline nanostructures, *Nanotechnology*, 16(2005), p. 2827.

56. G. Li, H. Peng, Y. Wang, et al., Synthesis of polyaniline nanobelts, *Macromolecular Rapid Communications*, 25(2004), pp. 1611–1614.

57. C. Su, G. Wang, and F. Huang, Preparation and characterization of composites of polyaniline nanorods and multiwalled carbon nanotubes coated with polyaniline, *Journal of Applied Polymer Science*, 106(2007), pp. 4241–4247.

58. Q. M. Jia, S. Y. Shan, L. H. Jiang, et al., A effects of polyaniline morphologies on performance for removal rhodamine B, *Advanced Materials Research*, 213(2011), pp. 553–556.

59. A. Muhammad, A.-U.-H. A. Shah, S. Bilal, et al., Basic blue dye adsorption from water using polyaniline/magnetite (Fe_3O_4) composites—kinetic and thermodynamic aspects, *Materials*, 12(2019), p. 1764.

60. A. Salem, The role of polyaniline salts in the removal of direct blue 78 from aqueous solution—a kinetic study, *Reactive and Functional Polymers*, 70(2010), pp. 707–714.

61. S. Dutta, K. Manna, S. K. Srivastava, et al., Hollow polyaniline microsphere/Fe_3O_4 nanocomposite as an effective adsorbent for removal of arsenic from water, *Scientific Reports*, 10(2020), pp. 1–14.

62. J. Li, T. Peng, Y. Zhang, et al., Polyaniline modified SnO_2 nanoparticles for efficient photocatalytic reduction of aqueous Cr (VI) under visible light, *Separation and Purification Technology*, 201(2018), pp. 120–129.

63. Z. A. Boeva and V. G. Sergeyev, Polyaniline—synthesis, properties, and application, *Polymer Science Series C*, 56(2014), pp. 144–153.

64. C. Dhand, N. Dwivedi, S. Mishra, et al., Polyaniline-based biosensors, *Nanobiosensors in Disease Diagnosis*, 4(2015), pp. 25–46.

65. R. Kaur and M. Duhan, Polyaniline as an inceptive dye adsorbent from effluent, in *Advanced Materials for Wastewater Treatment*, 1st edn, edited by Shahid-ul-Salam, Wiley-Scrivener Publishing, Hoboken, (2017), pp. 51–99.

66. J. Li, Y. Huang, and D. Shao, Conjugated polymer-based composites for water purification, in *Fundamentals of Conjugated Polymer Blends, Copolymers and Composites*, edited by Parveen Saini, Wiler-scrivener, 2015.

67. Y. Haldorai, K. Sivakumar, and J. J. Shim, ZnO nanoparticles dispersed poly (aniline-co-o-anthranilic acid) composites—photocatalytic reduction of Cr (VI) and Ni (II), *Polymer Composites*, 35(2014), pp. 839–846.

68. F. Zhang, Y. Zhang, G. Zhang, et al., Exceptional synergistic enhancement of the photocatalytic activity of SnS_2 by coupling with polyaniline and N-doped reduced graphene oxide, *Applied Catalysis B: Environmental*, 236(2018), pp. 53–63.

69. M. Bhaumik, A. Maity, V. V. Srinivasu, et al., Removal of hexavalent chromium from aqueous solution using polypyrrole-polyaniline nanofibers, *Chemical Engineering Journal*, 181–182(2012), pp. 323–333.

70. H. Eisazadeh, Removal of chromium from waste water using polyaniline, *Journal of Applied Polymer Science*, 104(2007), pp. 1964–1967.

71. A. Linos, A. Petralias, C. A. Christophi, et al., Oral ingestion of hexavalent chromium through drinking water and cancer mortality in an industrial area of Greece-an ecological study, *Environmental Health*, 10(2011), pp. 1–8.

72. L. Guo, Y. Xiao, and Y. Wang, Hexavalent chromium-induced alteration of proteomic landscape in human skin fibroblast cells, *Journal of Proteome Research*, 12(2013), pp. 3511–3518.

73. Y. Jiang, Z. Liu, G. Zeng, et al., Polyaniline-based adsorbents for removal of hexavalent chromium from aqueous solution—a mini review, *Environmental Science and Pollution Research*, 25(2018), pp. 6158–6174.

74. G. Yu, J. Gao, J. C. Hummelen, et al., Polymer photovoltaic cells—enhanced efficiencies via a network of internal donor-acceptor heterojunctions, *Science*, 270(1995), pp. 1789–1791.

75. S. Chen, Z. Wei, X. Qi, et al., Nanostructured polyaniline-decorated Pt/C@ PANI core–shell catalyst with enhanced durability and activity, *Journal of the American Chemical Society*, 134(2012), pp. 13252–13255.

76. S. Sarmah and A. Kumar, Photocatalytic activity of polyaniline-TiO_2 nanocomposites, *Indian Journal of Physics*, 85(2011), pp. 713–726.

77. A. Ahmad, H. Ahmad, D. Lokhat, et al., Recent advances in polyaniline-based nanocomposites as potential adsorbents for trace metal ions, in *Polymer-based Nanocomposites for Energy and Environmental Applications*.2018, Elsevier, Cambridge, pp. 597–615.

78. L. Hlekelele, N. E. Nomadolo, K. Z. Setshedi, et al., Synthesis and characterization of polyaniline, polypyrrole and zero-valent iron-based materials for the adsorptive and oxidative removal of bisphenol-A from aqueous solution, *RSC Advances*, 9(2019), pp. 14531–14543.

79. S. M. Riahi, S. M. Borghei, A. Olad, et al., Influence of polyaniline synthesis conditions on its capability for removal and recovery of chromium from aqueous solution, *IJCCE*, 30(2011), pp. 97–100.

80. J. Han, J. Dai, and R. Guo, Highly efficient adsorbents of poly(o-phenylenediamine) solid and hollow sub-microspheres towards lead ions—a comparative study, *Journal of Colloid and Interface Science*, 356(2011), pp. 749–756.

81. S. Dai, B. Peng, L. Zhang, et al., Sustainable synthesis of hollow Cu-loaded poly(m-phenylenediamine) particles and their application for arsenic removal, *RSC Advances*, 5(2015), pp. 29965–29974.

82. T. Wang, L. Zhang, C. Li, et al., Synthesis of core–shell magnetic Fe_3O_4@poly(m-phenylenediamine) particles for chromium reduction and adsorption, *Environmental Science & Technology*, 49(2015), pp. 5654–5662.

83. H. Kera, M. Bhaumik, K. Pillay, et al., m-Phenylenediamine-modified polypyrrole as an efficient adsorbent for removal of highly toxic hexavalent chromium in water, *Materials Today Communications* 15(2018), pp. 153–164.

84. T. Xiong, X. Yuan, H. Wang, et al., Integrating the (311) facet of MnO_2 and the fuctional groups of poly (m-phenylenediamine) in core–shell MnO_2@ poly (m-phenylenediamine) adsorbent to remove Pb ions from water, *Journal of Hazardous Materials*, 389(2020), p. 122154.

85. Y.-L. Min, T. Wang, Y.-G. Zhang, et al., The synthesis of poly(p-phenylenediamine) microstructures without oxidant and their effective adsorption of lead ions, *Journal of Materials Chemistry*, 21(2011), p. 6683.

86. S. Yang, D. Liu, F. Liao, et al., Synthesis, characterization, morphology control of poly (p-phenylenediamine)-Fe_3O_4 magnetic micro-composite and their application for the removal of $Cr_2O_7^{2-}$ from water, *Synthetic Metals*, 162(2012), pp. 2329–2336.

87. Q. Li, Y. Qian, H. Cui, et al., Preparation of poly(aniline-1, 8-diaminonaphthalene) and its application as adsorbent for selective removal of Cr(VI) ions, *Chemical Engineering Journal*, 173(2011), pp. 715–721.

88. M. Bhaumik, A. Maity, V. Srinivasu, et al., Removal of hexavalent chromium from aqueous solution using polypyrrole-polyaniline nanofibers, *Chemical Engineering Journal*, 181(2012), pp. 323–333.

89. R. Karthik and S. Meenakshi, Removal of Pb (II) and Cd (II) ions from aqueous solution using polyaniline grafted chitosan, *Chemical Engineering Journal*, 263(2015), pp. 168–177.

90. N. Zare, M. M. Lakouraj, and A. Ramezani, Effective adsorption of heavy metal cations by superparamagnetic poly(aniline-co-m-phenylenediamine)@Fe_3O_4 nanocomposite, *Advances in Polymer Technology*, 34(2015), pp. 1–11.

91. R. Gupta, R. Singh, and S. Dubey, Removal of mercury ions from aqueous solutions by composite of polyaniline with polystyrene, *Separation and Purification Technology*, 38(2004), pp. 225–232.

92. M. S. Lashkenari, B. Davodi, M. Ghorbani, et al., Use of core-shell polyaniline/polystyrene nanocomposite for removal of Cr (VI), *High Performance Polymers*, 24(2012), pp. 345–355.

93. M. R. Samani, S. M. Borghei, A. Olad, et al., Removal of chromium from aqueous solution using polyaniline – poly ethylene glycol composite, *Journal of Hazardous Materials*, 184(2010), pp. 248–254.

94. R. Karthik and S. Meenakshi, Adsorption study on removal of Cr(VI) ions by polyaniline composite, *Desalination and Water Treatment*, 54(2014), pp. 3083–3093.

95. T. S. Najim and A. J. Salim, Polyaniline nanofibers and nanocomposites—Preparation, characterization, and application for Cr(VI) and phosphate ions removal from aqueous solution, *Arabian Journal of Chemistry*, 10(2017), pp. S3459–S3467.

96. G. Yavuz, E. Dincturk-Atalay, A. Uygun, et al., A comparison study of adsorption of Cr (VI) from aqueous solutions onto alkyl-substituted polyaniline/chitosan composites, *Desalination*, 279(2011), pp. 325–331.

97. N. Jiang, Y. Xu, Y. Dai, et al., Polyaniline nanofibers assembled on alginate microsphere for Cu^{2+} and Pb^{2+} uptake, *Journal of Hazardous Materials*, 215(2012), pp. 17–24.

98. R. Karthik and S. Meenakshi, Facile synthesis of cross linked-chitosan–grafted-polyaniline composite and its Cr (VI) uptake studies, *International Journal of Biological Macromolecules*, 67(2014), pp. 210–219.

99. N. Zare and M. M. Lakouraj, Biodegradable polyaniline/dextrin conductive nanocomposites—synthesis, characterization, and study of antioxidant activity and sorption of heavy metal ions, *Iranian Polymer Journal*, 23(2014), pp. 257–266.

100. N. A. Abdelwahab, E. A. Al-Ashkar, and M. A. A. El-Ghaffar, Preparation and characterization of eco-friendly poly(p-phenylenediamine) and its composite with chitosan for removal of copper ions from aqueous solutions, *Transactions of Nonferrous Metals Society of China*, 25(2015), pp. 3808–3819.

101. R. Karthik and S. Meenakshi, Synthesis, characterization and Cr(VI) uptake study of polyaniline coated chitin, *International Journal of Biological Macromolecules*, 72(2015), pp. 235–242.

102. R. Karthik and S. Meenakshi, Removal of Cr(VI) ions by adsorption onto sodium alginate-polyaniline nanofibers, *International Journal of Biological Macromolecules*, 72(2015), pp. 711–717.

103. R. Ansari and F. Raofie, Removal of mercuric ion from aqueous solutions using sawdust coated by polyaniline, *E-Journal of Chemistry*, 3(2006), pp. 35–43.

104. A. Kumar, S. Chakraborty, and M. Ray, Removal and recovery of chromium from wastewater using short chain polyaniline synthesized on jute fiber, *Chemical Engineering Journal*, 141(2008), pp. 130–140.

105. D. Liu, D. Sun, and Y. Li, Removal of Cu(II) and Cd(II) from aqueous solutions by polyaniline on sawdust, *Separation Science and Technology*, 46(2010), pp. 321–329.

106. M. S. Mansour, M. E. Ossman, and H. A. Farag, Removal of Cd (II) ion from waste water by adsorption onto polyaniline coated on sawdust, *Desalination*, 272(2011), pp. 301–305.

107. M. Ghorbani, M. S. Lashkenari, and H. Eisazadeh, Application of polyaniline nanocomposite coated on rice husk ash for removal of Hg (II) from aqueous media, *Synthetic Metals*, 161(2011), pp. 1430–1433.

108. M. Ghorbani, Removal of zinc ions from aqueous solution using polyaniline nanocomposite coated on rice husk, *Iranica Journal of Energy & Environment*, 3(2012), pp. 66–71.

109. Y. Zheng, W. Wang, D. Huang, et al., Kapok fiber oriented-polyaniline nanofibers for efficient Cr (VI) removal, *Chemical Engineering Journal*, 191(2012), pp. 154–161.

110. M. R. Samani and D. Toghraie, Removal of hexavalent chromium from water using polyaniline/wood sawdust/poly ethylene glycol composite—an experimental study, *Journal of Environmental Health Science and Engineering*, 17(2019), p. 53.

111. Y. Liu, L. Chen, Y. Li, et al., Synthesis of magnetic polyaniline/graphene oxide composites and their application in the efficient removal of Cu (II) from aqueous solutions, *Journal of Environmental Chemical Engineering*, 4(2016), pp. 825–834.

112. D. K. L. Harijan and V. Chandra, Polyaniline functionalized graphene sheets for treatment of toxic hexavalent chromium, *Journal of Environmental Chemical Engineering*, 4(2016), pp. 3006–3012.

113. M. O. Ansari, R. Kumar, S. A. Ansari, et al., Anion selective pTSA doped polyaniline@ graphene oxide-multiwalled carbon nanotube composite for Cr (VI) and Congo red adsorption, *Journal of Colloid and Interface Science*, 496(2017), pp. 407–415.

114. M. Ramezanzadeh, M. Asghari, B. Ramezanzadeh, et al., Fabrication of an efficient system for Zn ions removal from industrial wastewater based on graphene oxide nanosheets decorated with highly crystalline polyaniline nanofibers (GO-PANI)—Experimental and ab initio quantum mechanics approaches, *Chemical Engineering Journal*, 337(2018), pp. 385–397.

115. H. Gu, S. B. Rapole, J. Sharma, et al., Magnetic polyaniline nanocomposites toward toxic hexavalent chromium removal, *RSC Advances*, 2(2012), p. 11007.

116. L. Tang, Y. Fang, Y. Pang, et al., Synergistic adsorption and reduction of hexavalent chromium using highly uniform polyaniline–magnetic mesoporous silica composite, *Chemical Engineering Journal*, 254(2014), pp. 302–312.

117. G. Yang, L. Tang, Y. Cai, et al., Effective removal of Cr(vi) through adsorption and reduction by magnetic mesoporous carbon incorporated with polyaniline, *RSC Advances*, 4(2014), pp. 58362–58371.

118. J. Chen, X. Hong, Q. Xie, et al., Highly efficient removal of chromium(VI) from aqueous solution using polyaniline/sepiolite nanofibers, *Water Science and Technology*, 70(2014), pp. 1236–1243.

119. J. Wang, X. Han, Y. Ji, et al., Adsorption of Cr(VI) from aqueous solution onto short-chain polyaniline/palygorskite composites, *Desalination and Water Treatment*, 56(2014), pp. 356–365.

120. S. Piri, Z. A. Zanjani, F. Piri, et al., Potential of polyaniline modified clay nanocomposite as a selective decontamination adsorbent for Pb(II) ions from contaminated waters; kinetics and thermodynamic study, *Journal of Environmental Health Science and Engineering*, 14(2016), p. 20.

121. T. Zhou, C. Li, H. Jin, et al., Effective adsorption/reduction of Cr (VI) oxyanion by halloysite@ polyaniline hybrid nanotubes, *ACS applied materials & interfaces*, 9(2017), pp. 6030–6043.

122. A. Olad, M. Bastanian, and H. Bakht Khosh Hagh, Thermodynamic and kinetic studies of removal process of hexavalent chromium ions from water by using bio-conducting starch–montmorillonite/polyaniline nanocomposite, *Journal of Inorganic and Organometallic Polymers and Materials*, 29(2019), pp. 1916–1926.

123. B. Gao, Y. Li, and Z. Chen, Adsorption behaviour of functional grafting particles based on polyethyleneimine for chromate anions, *Chemical Engineering Journal*, 150(2009), pp. 337–343.

124. K. Araki, T. Kato, U. Kumar, et al., Dielectric properties of a hydrogen-bonded liquid-crystalline sidechain polymer, *Macromolecular Rapid Communications*, 16(1995), pp. 733–739.

125. K. Maeda, Y. Takeyama, K. Sakajiri, et al., Nonracemic dopant-mediated hierarchical amplification of macromolecular helicity in a charged polyacetylene leading to a cholesteric liquid crystal in water, *Journal of the American Chemical Society*, 126(2004), pp. 16284–16285.

126. J. W. Goodby, P. J. Collings, T. Kato, et al., *Handbook of Liquid Crystals, 8 Volume Set*, John Wiley & Sons, Wiley-VCH, Weinheim, 2014.

127. J. Tiddy, D. L. Mateer, A. P. Ormerod, et al., Highly ordered aggregates in dilute dye-water systems, *Langmuir*, 11(1995), pp. 390–393.

128. P. Karthikeyan, S. S. Elanchezhiyan, J. Preethi, et al., Mechanistic performance of polyaniline-substituted hexagonal boron nitride composite as a highly efficient adsorbent for the removal of phosphate, nitrate, and hexavalent chromium ions from an aqueous environment, *Applied Surface Science*, 511(2020), p. 145543.

129. S. Ahmad, A. Sultan, W. Raza, et al., Boron nitride based polyaniline nanocomposite—preparation, property, and application, *Journal of Applied Polymer Science*, 133(2016), pp. 62160–62178.

130. S. Shahabuddin, R. Khanam, M. Khalid, et al., Synthesis of 2D boron nitride doped p-olyaniline hybrid nanocomposites for photocatalytic degradation of carcinogenic dyes from aqueous solution, *Arabian journal of chemistry*, 11(2018), pp. 1000–1016.

131. A. C. Stuart, W. T. Huck, J. Genzer, et al., Emerging applications of stimuli-responsive polymer materials, *Nature Materials*, 9(2010), pp. 101–113.

132. W. Zhang, W. Jin, T. Fukushima, et al., Supramolecular linear heterojunction composed of graphite-like semiconducting nanotubular segments, *Science*, 334(2011), pp. 340–343.

133. J. M. Park and S. H. Jhung, Polyaniline-derived carbons—remarkable adsorbents to remove atrazine and diuron herbicides from water, *Journal of Hazardous Materials*, 396(2020), p. 122624.

134. M. Zhou, T. J. Kidd, R. D. Noble, et al., Supported lyotropic liquid-crystal polymer membranes—promising materials for molecular-size-selective aqueous nanofiltration, *Advanced Materials*, 17(2005), pp. 1850–1853.

135. D. K. Yoo, N. Abedin Khan, and S. H. Jhung, Polyaniline-loaded metal-organic framework MIL-101(Cr)—promising adsorbent for CO_2 capture with increased capacity and selectivity by polyaniline introduction, *Journal of CO_2 Utilization*, 28(2018), pp. 319–325.

136. V. Subramanian, E. E. Wolf, and P. V. Kamat, Catalysis with TiO_2/gold nanocomposites. Effect of metal particle size on the Fermi level equilibration, *Journal of the American Chemical Society*, 126(2004), pp. 4943–4950.

137. M.-R. Huang, H.-J. Lu, and X.-G. Li, Efficient multicyclic sorption and desorption of lead ions on facilely prepared poly (m-phenylenediamine) particles with extremely strong chemoresistance, *Journal of Colloid and Interface Science*, 313(2007), pp. 72–79.

138. Z.-F. Huang, J. Song, X. Wang, et al., Switching charge transfer of $C_3N_4/W_{18}O_{49}$ from type-II to Z-scheme by interfacial band bending for highly efficient photocatalytic hydrogen evolution, *Nano Energy*, 40(2017), pp. 308–316.

139. S. Wang, Z. Teng, Y. Xu, et al., Defect as the essential factor in engineering carbon-nitride-based visible-light-driven Z-scheme photocatalyst, *Applied Catalysis B: Environmental*, 260(2020), p. 118145.

140. S. N. F. M. Nasir, H. Ullah, M. Ebadi, et al., New insights into Se/$BiVO_4$ heterostructure for photo-electrochemical water splitting—a combined experimental and DFT study, *The Journal of Physical Chemistry C*, 121(2017), pp. 6218–6228.

141. H.-A. Tayebi, Z. Dalirandeh, A. Shokuhi Rad, et al., Synthesis of polyaniline/Fe_3O_4 magnetic nanoparticles for removal of reactive red 198 from textile waste water—kinetic, isotherm, and thermodynamic studies, *Desalination and Water Treatment*, 57(2016), pp. 22551–22563.

142. Y.-D. Liang, Y.-J. He, Y.-H. Zhang, et al., Adsorption property of alizarin red S by $NiFe_2O_4$/polyaniline magnetic composite, *Journal of Environmental Chemical Engineering*, 6(2018), pp. 416–425.

143. A. Dhanagopal, L. Dheenathayalan, and V. Thiyagarajan, Kinetic, thermodynamic and isotherm studies for the removal of acid red 88 from aqueous medium by polyaniline-$CuCl_2$ composite, *Oriental Journal of Chemistry*, 35(2019), pp. 1774–1781.

144. W. J. Weber and J. C. Morris, Kinetics of adsorption on carbon from solution, *Journal of the Sanitary Engineering Division*, 89(1963), pp. 31–60.

145. M. Laabd, A. Hallaoui, N. Aarb, et al., Removal of polycarboxylic benzoic acids using polyaniline-polypyrrole copolymer—experimental and DFT studies, *Fibers and Polymers*, 20(2019), pp. 896–905.

146. T. Lingeswari and T. Vimala, Isotherm, kinetics and thermodynamic study of adsorption of phthalocyanine and azo dyes by $CoCl_2$ doped polyaniline, *Asian Journal of Chemistry*, 32(2020), pp. 746–752.

147. H. Javadian, Application of kinetic, isotherm and thermodynamic models for the adsorption of Co(II) ions on polyaniline/polypyrrole copolymer nanofibers from aqueous solution, *Journal of Industrial and Engineering Chemistry*, 20(2014), pp. 4233–4241.

148. Z. Ayazi, Z. M. Khoshhesab, F. F. Azhar, et al., Modeling and optimization of adsorption removal of reactive orange 13 on the alginate–montmorillonite–polyaniline nanocomposite via response surface methodology, *Journal of the Chinese Chemical Society*, 64(2017), pp. 627–639.

149. D. Sparks, *Kinetics of Reactions in Pure and in Mixed Systems*, edited by Sparks, D. L., (1986), CRC Press Inc., Boca Raton, Florida, pp. 83–145.

150. W. Lyu, M. Yu, J. Feng, et al., Highly crystalline polyaniline nanofibers coating with low-cost biomass for easy separation and high efficient removal of anionic dye ARG from aqueous solution, *Applied Surface Science*, 458(2018), pp. 413–424.

151. M. Tanzifi, S. H. Hosseini, A. D. Kiadehi, et al., Artificial neural network optimization for methyl orange adsorption onto polyaniline nano-adsorbent—kinetic, isotherm and thermodynamic studies, *Journal of Molecular Liquids*, 244(2017), pp. 189–200.

152. F. Ishtiaq, H. N. Bhatti, A. Khan, et al., Polypyrole, polyaniline and sodium alginate biocomposites and adsorption-desorption efficiency for imidacloprid insecticide, *International Journal of Biological Macromolecules*, 147(2020), pp. 217–232.

153. E. Igberase, P. Osifo, and A. Ofomaja, The adsorption of copper (II) ions by polyaniline graft chitosan beads from aqueous solution—equilibrium, kinetic and desorption studies, *Journal of Environmental Chemical Engineering*, 2(2014), pp. 362–369.

154. E. Gagliano, M. Sgroi, P. P. Falciglia, et al., Removal of poly-and perfluoroalkyl substances (PFAS) from water by adsorption—role of PFAS chain length, effect of organic matter and challenges in adsorbent regeneration, *Water Research*, 171(2020), p. 115381.

6 Emerging MXene-Based Materials for the Removal of Environmental Pollutants

Dana Susan Abraham and Margandan Bhagiyalakshmi
Central University of Kerala

Mari Vinoba
Kuwait Institute for Scientific Research

CONTENTS

6.1 INTRODUCTION

Over the past several years, with the increasing population, industrialization and globalization have much contaminated the environment and disturbed the ecology to a greater extent. At this instance, it is realized that all the chemicals used to uplift human lifestyles have polluted the environment. Hence, environmental regulations are enforced to control and monitor the impact of pollutants on the environment. In this regard, sustainable researchers have a responsibility to provide the required solution to streamline the quality of the domain. One of the most severe environmental problems is water pollution, which is generally vulnerable due to industrial and domestic effluent. Water pollution could dangerously affect the ecosystem of the earth. Highly toxic and carcinogenic heavy metal effluents and organic pollutants such as pesticides, fertilizers, hydrocarbons, polyphenols, detergents, oils, and pharmaceuticals are also the causes of pollution [1–3]. Especially, highly complex dyes and radioactive effluents are the major pollutants identified in the water, which can cause waste-borne diseases and affect the livelihood of the human community. Either physical or chemical adsorption process can remove all the pollutants. Thus, an efficient sorbent material is an ultimate quest by researchers for solving problems related to water pollution.

Recently developed two-dimensional (2D) materials possess unique physicochemical properties than conventional bulk materials and are reported to be efficient absorbents to remove pollutants from water. The discovery of graphene in 2004 [4] is well investigated as adsorbent; however, because of the van der Waals force of attraction, graphene is readily restacked into graphite, limiting its potential. Over the past few years, an entirely new class of 2D-transition metal carbide/nitride materials known as MXene has garnered experimentalists, theoretical chemists, and physicists. Professor Yuri Gogotsi and his colleagues at Drexel University deserve all of the credit for

DOI: 10.1201/9781003206385-6

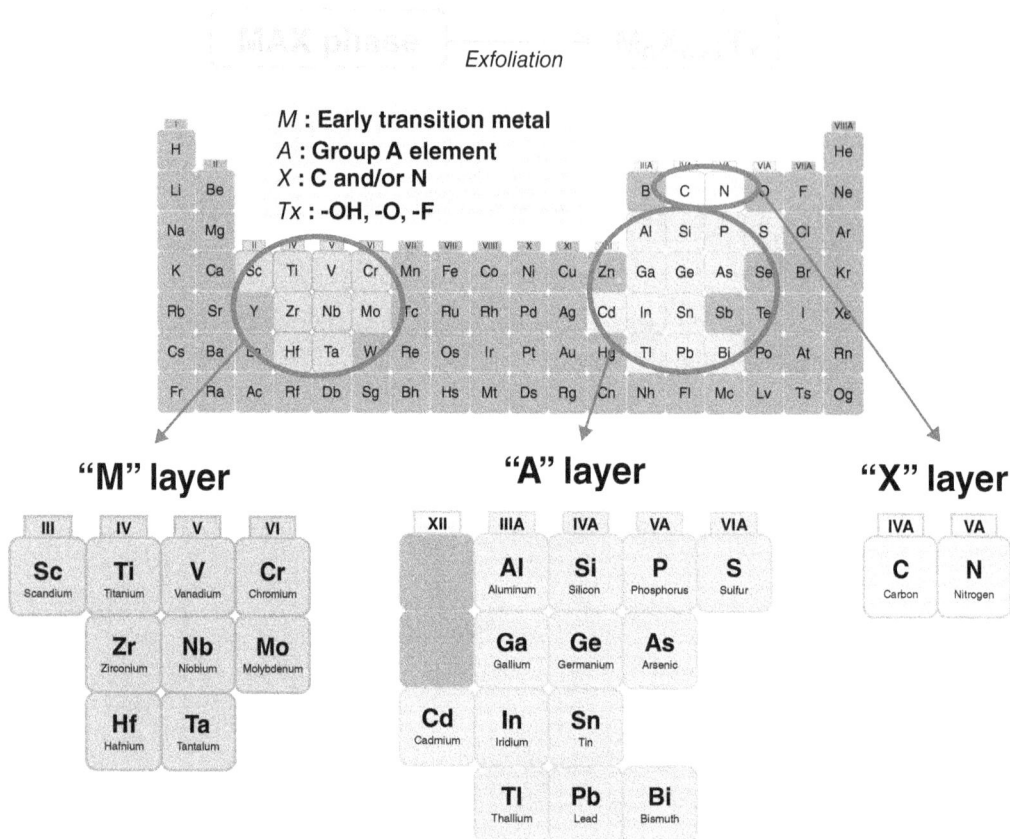

FIGURE 6.1   The elements of the periodic table that are used to synthesize MXenes.

discovering MXenes [5]. Researchers and academicians have been fascinated with MXenes, and their mechanical, magnetic, electrical, and chemical properties have made them a hot topic since their discovery.

MXenes were derived from MAX phase precursors by etching the A atom layers with appropriate reagents, mostly fluorine-containing acids such as HF, a combination of LiF and HCl, and ammonium bifluoride. MAX phases are layered, hexagonal carbides and nitrides, which have the general formula: $M_{n+1}AX_n$, (MAX), where $n=1$ to 4, M in the MAX phase refers to transition metal elements (such as Sc, Ti, Zr, Hf, V, Nb, Ta, and Cr), A refers to the main-group element (predominantly IIIA or IVA) 'A,' and X is carbon or nitrogen. Figure 6.1 demonstrates the plausible elements of the periodic table used to synthesize MAX phase and MXenes [6]. The general formula of MXene is $M_{n+1}X_nT_x$, where 'M' and 'X' have the same meaning as in the MAX phase, and 'T' represents the surface functional groups (–F, –OH, and =O), which arise during the etching process. Removal of 'A' from the MAX phase implants the broken charge neutrality in MXene. Thus, MXenes possess a maximum conductivity due to surface functional groups and stoichiometry [7]. Also, the functional groups allow MXenes to have tuned properties, a layered structure, and customizable space, which will enable them to adapt to different application scenarios. MXene has superior mechanical properties to other 2D materials due to its M-X solid bond. Additionally, they noted that the delocalization effect of the electrons in the 'd' orbital of the 'M' atoms combined with the surface stacking geometry resulted in excellent mechanical properties.

MXene also exhibits good magnetic properties because of the magnetic instability caused by the high Fermi energy level [8]. Due to its partially naked surface, made up of transition metal elements, MXene is chemically stable at average temperatures and pressure. It also has strong redox properties when exposed to water and oxygen. MXenes also exhibit high thermal conductivity, high adsorption, and strong hydrophilicity [9]. The layered MXene has water molecules or metal ions that gets easily embedded in its layers. The surface functional groups and highly electropositive edges of MXene materials make hydrophilic nanosheets highly ion conductive. Thus, surface functional groups make MXenes highly solvent stable and allow tailorable adsorption properties, facilitating highly robust composite materials [10]. As a result of the hydrophilic nature of the functional groups in MXene, interlayered spaces provide ultrafast water potential within the structural galleries, which aids in the rapid kinetic mechanism of environmental contaminant adsorption and sequestration.

This chapter provides an overview of the current state of research concerning MXene and MXene-based materials for adsorption-based environmental remediation. Herein, a detailed literature review on the use of MXene-based material as adsorbent for heavy metal ion removal, organic dye degradation, and elimination of radionuclide from wastewater is listed.

6.2 MXENE AND MXENE-BASED MATERIALS FOR ADSORPTION-BASED ENVIRONMENTAL REMEDIATION

6.2.1 Heavy Metal Removal from Wastewater

Heavy metal has a density greater than 5 g/cc, and the atomic weights of these elements range from 63.5 to 200.6 [11]. In the periodic table, elements of the fourth period are called to be heavy metals, primarily chromium (Cr), cobalt (Co), nickel (Ni), copper (Cu), zinc (Zn), arsenic (As); occasionally, lead (Pb), cadmium (Cd), and mercury (Hg) also come in this category [12]. In nature, heavy metals are primarily found as carbonates, oxides, silicates, and sulfides. In most cases, these compounds are not soluble in water and break down gradually by weathering and being exposed to precipitation [13]. Currently, their concentrations are elevated due to the overabundance of industrial waste [14]. Particularly, mines, metallurgy, electroplating, industrial processes, agricultural practices, and household waste contribute to the release of heavy metals into the water. Microorganisms are incapable of degrading heavy metals once released into the environment. It leads to an accumulation in the food chain, resulting in adverse consequences for humans and the ecosystem [15]. Globally, heavy metal pollution is considered a major environmental issue. It is necessary to remove heavy metals because they are toxic and carcinogenic. Heavy metal ions are removed from wastewater using various techniques such as coagulation [16], chemical oxidation [17], photocatalytic reduction [18], electrochemical processes [19], ion exchange [20], and ultrafiltration [21]. These methods are still efficient and effective in high concentration conditions but suffer from drawbacks such as high energy consumption and high-cost process with second pollutants [22].

Apart from these methods, the adsorption process has been widely employed to remove heavy metal ions from wastewater. In recent years, numerous research and development activities have been focused on developing both theoretically and practically adsorption systems. Adsorption offers multitudinous advantages, including ease of operation, remarkable efficiency, regenerative properties, a broad spectrum of available adsorbents, low environmental impact, and cost-effectiveness. Adsorbents should have a high specific surface area, and proper functional groups are usually adequate. As of today, a variety of porous materials are used as adsorbents for removing environmental pollutants, including activated carbon [23], carbon nanomaterials [24], ordered mesoporous silica [25], kaolin [26], zeolites [27], and metal-organic frameworks [28]. Although these adsorbents have high efficiencies, they have some downsides. For instance, activated carbon has relatively poor adsorption capacity because of low mass transfer [29], and unmodified zeolites also exhibit poor adsorption selectivity [30].

Since MXenes and MXene-based materials have superior structural and chemical stabilities and are hydrophilic on surfaces, they can potentially be employed as adsorbents for removing environmental pollutants. Metal ions can be effectively captured by the surface-active sites Ti-OH and Ti-F present on MXene [31,32]. There have been numerous demonstrations of the ability of MXenes to adsorb heavy metals, including Cu^{2+}, Cd^{2+}, Hg^{2+}, Ni^{2+}, Pb^{2+}, and Cr^{6+} from water. MXene nanosheets with <2Å interlayer spacing offer an opportunity to trap heavy metal ions <4.5Å. Moreover, MXene surface functionalization dramatically enhances its ability to absorb heavy metals. Titanium-based MXene ($Ti_3C_2T_x$) is widely used to adsorb heavy metals [33]. The unique layered structure of MXene nanoplatelets combined with active functional groups and available binding sites makes it an excellent adsorbent for removing heavy metals from water [34]. Thus, based on the excellent adsorption properties and reusability, MXenes are found to be a suitable adsorbent for removing heavy metals from wastewater.

Feng et al. [34] created a new atomic-scale alk-MXene/LDH adsorbent via *in situ* mechanical self-assembly and intercalation using alkali-level compounds that effectively removed Ni^{2+} from aqueous solutions. Ni^{2+} was adsorbed chemically on alk-MXene/LDH with an adsorption capacity of 222.717 mg/g, and its removal efficiency is more than 97.35%. Their study found that the alk-MXene/LDH nanocomposite showed a significant regeneration capacity after eight successive adsorption-desorption cycles (>85.32%).

In another report, Yang et al. [35] incorporated nanoparticles of Fe_3O_4 into MXene ($Ti_3C_2T_x$) to improve their structural properties. To fabricate Fe_3O_4@MXene membranes, Fe_3O_4 nanoparticles and MXene nanosheets were self-assembled on cellulose acetate (CA) membranes. It was found that Fe_3O_4 nanoparticles were evenly distributed on MXene sheets, thereby expanding the nanochannel of MXene. Compared with the pure MXene membrane, this nanocomposite membrane substantially showed an improvement in removing heavy metal ions. The composite NF membrane removed Cu^{2+}, Cd^{2+}, and Cr^{6+} from wastewater at maximum efficiencies of 63.2%, 64.1%, and 70.2%, respectively, based on the combined effect of adsorption and layer sieving. In addition, the Fe_3O_4@MXene membrane demonstrated good recyclability after being washed with HCl.

MXenes exhibit a limited capacity to adsorb metal ions, primarily due to their restricted adsorption sites. Dong et al. [36] synthesized MXene/alginate composites to address this issue. Alginate contains many amino and carboxyl groups with substantial chelating properties for heavy metal ions, increasing their adsorption capacity. Further, the 2D lamellar structure of the composites increases the efficiency of heavy metal ion transport. It drastically reduces its time to reach equilibrium in the adsorption process. MXene/alginate composite displayed a significant adsorption capacity of 382.7 mg/g for Pb^{2+} and 87.6 mg/g for Cu^{2+}; also, it took only 15 minutes to reach equilibrium. In addition, the composite could be regenerated by adding an acidic solution and can be reutilized without any apparent degradation in performance. MXene/alginate composites are thus an excellent adsorbent for heavy metal ions thanks to their substantial capacity, high efficiency, and ease of regeneration.

Shahzad et al. [37] synthesized a novel adsorbent, MXene core ($Ti_3C_2T_x$) shell aerogel spheres (MX-SA) for removing Hg^{2+} from aqueous solution. Several different compositions of MXene and sodium alginate (SA) powders (% w/w) were synthesized. In 0.5–1.0 M HNO_3, MX-SA$_{4:20}$ displayed excellent performances with affinity and binding capacities for Hg^{2+} that matched or outperformed the most effective mercury sorbents available. They observed that MX-SA$_{4:20}$ had a remarkable adsorption capacity of 932.84 mg/g for Hg^{2+}. Adsorbent removal efficiencies were excellent, with 100% removal efficiency for Hg^{2+}, and greater than 90% for Cd^{2+}, Cr^{3+}, Pb^{2+}, and Cu^{2+}. The unique structure within the microspheres, the extensive porosity, their high surface areas, the presence of oxygenated functional groups present in MXene nanosheets, and their available active sites make them ideal for removing heavy metals from wastewater.

Adsorption enhancement with rotating magnetic fields (RMF) enhances transport and alters hydrogen bonds and surface functional groups. In combination with the intrinsic magnetic properties of the adsorbent, the magnetic field (MF) can tailor the adsorption process for selective adsorption of pollutants. Ren et al. [38] developed SA/MXene/$CoFe_2O_4$ (SA/MX/CFO) through cross-linking SA and

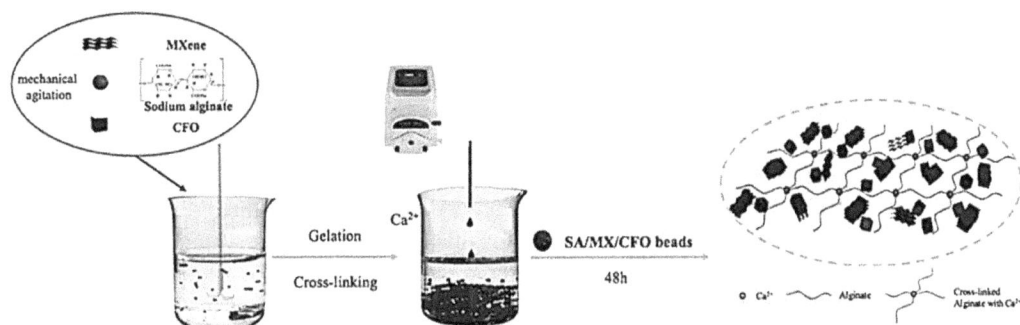

FIGURE 6.2 Schematic representation of the synthesis of sodium alginate/MXene/CoFe$_2$O$_4$ (SA/MX/CFO).

used them as adsorbents to remove contaminants using external MFs. Figure 6.2 shows the schematic representation of SA/MX/CFO synthesis. MF increased the adsorption capacity of ciprofloxacin by 24.19% but only marginally increased Cu^{2+} adsorption. Despite the lack of magnetic effect for the adsorption capacity of Cu^{2+}, it revealed that its adsorption rates were boosted by 359.76% for ciprofloxacin and 371% for Cu^{2+}. The rate of adsorption and selectivity were greatly enhanced by applying an MF in conjunction with the properties inherent to the adsorbent.

MXene/polymer composites (Ti$_3$C$_2$T$_x$ – PDOPA) were used to remove heavy metal ions from an aqueous solution [22]. Ti$_3$C$_2$T$_x$ was blended with levodopa (DOPA), which undergoes self-polymerization from poly (DOPA). Compared to pure Ti$_3$C$_2$T$_x$, the resultant Ti$_3$C$_2$T$_x$ – PDOPA composites were more effective in removing Cu^{2+}. Adsorption of Cu^{2+} on Ti$_3$C$_2$T$_x$ – PDOPA was highly correlated with the pH of the solution. At pH 11, the solution was found to have the highest adsorption capacity, and Cu^{2+} removal efficiency was 46.6 mg/g and 93.2%, respectively.

Thermally induced cross-linking approach was accompanied by hydroxylation of MXene with alkali to form alk-MXene membranes used to remove heavy metals (Pb^{2+}, Cu^{2+}, Cd^{2+}), as well as anions (Cl$^-$ and NO$_3^-$) in wastewater [39]. Hydroxylated MXene membranes performed better than pristine MXene membranes because of their negatively charged surfaces. The enhanced negative charge allowed heavy metal cations such as Pb^{2+}, Cu^{2+}, and Cd^{2+} to be removed by 99.5%. Surface charges further allowed NO$_3^-$, Cl$^-$ to be released as well at ~97%. A high level of operational stability has been noted for almost 70 minutes during wastewater purification by utilizing the thermally induced cross-linking technique.

The comparison of multilayer (ML) and delaminated (DL) MXene nanosheets for Cu adsorption was reported by Shahzad et al. [40]. The DL-Ti$_3$C$_2$T$_x$ displayed a higher Cu uptake rate than the multilayer (ML)-Ti$_3$C$_2$T$_x$. The increased surface area, discursiveness, hydrophilicity, and surface functionalities of DL-Ti$_3$C$_2$T$_x$ provided excellent Cu removal performance. MXene, with its oxygenated moieties, enabled Cu^{2+} to be reductively adsorbed to Cu$_2$O and CuO. It was found that the adsorption capacity was maximum at 78.45 mg/g, and 80% of the total metal content was absorbed within a minute. DL-Ti$_3$C$_2$T$_x$ had a 2.7-fold higher adsorption capacity than commercial activated carbon. Using the DL-Ti$_3$C$_2$T$_x$ as a heavy-metal adsorbent has also been demonstrated to be time-efficient, given that equilibrium was reached within 3 minutes. When regenerated after three adsorption cycles, it also displayed good adsorption performance.

The microstructure and surface properties of Ti$_3$C$_2$T$_x$ MXene-based films for the removal of multiple negatively and positively charged heavy metal ions from water without pressure were studied by Xie et al. [41]. The insertion of reduced graphene oxide is responsible for the optimization of the microstructure. This caused a progressive hydroxylation of the film surface, increased its wettability, and facilitated heavy metal ion adsorption and transfer. This combined effect of improving the microstructure and controlling the surface properties of $Ti_3C_2T_x$-based films enhanced its ability to remove various heavy metals from water, and the removal abilities were 84, 890, 1,241, and 1,172 mg/g corresponding to Cr(VI), Pd(II), Au(III), and Ag(I)

The multilayered oxygen-functionalized Ti_3C_2 nanosheets ($Ti_3C_2O_x$) (designated as M-Ti_3C_2) were reported by Hu et al for the effective removal of Hg (II) from water [42]. It is demonstrated through the experiments and DFT calculations that crystalline Hg_2Cl_2 has formed instead of the volatile Hg as a result of adsorption and catalytic reduction. They found that the Ti atoms on the [001] faces of M-Ti_3C_2 are primarily responsible for this removal process. Among the features of M-Ti_3C_2 that ensure efficient Hg (II) removal are rapid removal kinetics, high capacity, high selectivity, and a broad pH range. In this study, a simple thermal treatment system was developed to recover solid mercury from M-Ti_3C_2 surfaces. Mercury is recovered as crystallized Hg_2Cl_2 with an efficiency of up to 95%. The distinct interaction of M-Ti_3C_2 and OH was transformed into TiO_2/C nanocomposites. Furthermore, TiO_2/C nanocomposites showed a better ability to degrade organic pollutants than conventional P25 (Degussa, Germany). With these outstanding properties combined with mercuric recycling properties, M-Ti_3C_2 is a formidable candidate for rapidly removing and recovering Hg (II).

Specifically, Jun et al. [43] tested MXenes' ability to adsorb Pb (II) from wastewater. As a control for MXene, they used powder-activated carbon (PAC) [43]. Although MXenes have a lower surface area than PAC, their surface charge made them more effective for adsorption. Pb (II) adsorption by MXene was systematically studied under kinetic and isothermal conditions. The pseudo-second-order kinetic and Freundlich isotherm models were found to explain adsorption well. Moreover, MXene had a rapid adsorption rate; equilibrium was reached in just 30 minutes. An experimental study using various pH levels, different concentrations of humic acid, and ionic strengths endorsed that electrostatic attraction is the primary adsorption mechanism for Pb (II). Heavy metals, such as Pb (II), Cu (II), Zn (II), and Cd (II), were efficiently adsorbed by MXene from a single-electrolyte solution. A further benefit of MXene is that it exhibited good reusability when subjected to four adsorption/desorption cycles.

Wang et al. [44] functionalized Ti_2CT_x MXene nanosheets with different bio-surfactants, namely, a cationic surfactant chitosan (CS), an anionic surfactant lignosulfonate (LS), and a non-ionic surfactant enzymatic hydrolysis lignin (EHL), and studied Pb^{2+} adsorption capacity [44]. Ti_2CT_x nanosheets were not only protected from being oxidized and restacked by bio-surfactants, but their hydrophilic properties were also enhanced. Compared to pure Ti_2CT_x, CS and lignosulfonate-functionalized Ti_2CT_x nanosheets, Ti_2CT_x-EHL nanosheets exhibited improved morphology and enhanced adsorption performance. A maximum adsorption capacity of 232.9 mg/g for Pb^{2+} ions was achieved with EHL-functionalized Ti_2CT_x. Ti_2CT_x nanosheets functionalized by EHL are protected from restacking and can adsorb even more effectively thanks to their active functional groups. EHL's non-ionic properties, Ti_2CT_x nanosheets, are more easily prepared, resulting in improved adsorption efficiency and ion exchange. Furthermore, EHL promotes a synergistic interaction between functional groups and chelation to Pb^{2+}, resulting in enhanced chemisorption.

Using an ultrasonic technique, Hu et al. [45] peeled the etched surface of the MAX phase, which generated a great deal of oxygen-containing functional groups on the surface. The adsorbent was found to have a high selectivity for Hg^{2+}. Adsorption studies at various pH and temperatures were carried out in mercuric chloride and mercuric nitrate solutions. The maximum Langmuir adsorption capacity for mercuric nitrate is 1,057.3 mg/g, and that of mercuric chloride is 773.29 mg/g at pH 5.0 and 30°C. With Pb^{2+}, Cd^{2+}, Ni^{2+}, and Cu^{2+} as competing ions, 100% Hg was removed, while Pb at 47.79%, Cd at 18.41%, Ni at 9.8%, and Cu at 7.59% removal rates. The adsorption capacity was also high even at a low pH (pH = 2). It was observed that electrostatic interaction and interactions between hydroxyl groups bound to Ti and Hg^{2+} control the adsorption of Hg on the adsorbent.

To explore the synergistic role of the oxygenated terminals of $Ti_3C_2T_x$ and MoS_2 in mercury adsorption, Shahzad et al. [46] functionalized $Ti_3C_2T_x$ nanosheets with nanolayered MoS_2 (MoS_2/ MX-II). Figure 6.3 exhibits a schematic illustration of MoS_2/MX-II nanocomposite fabrication.

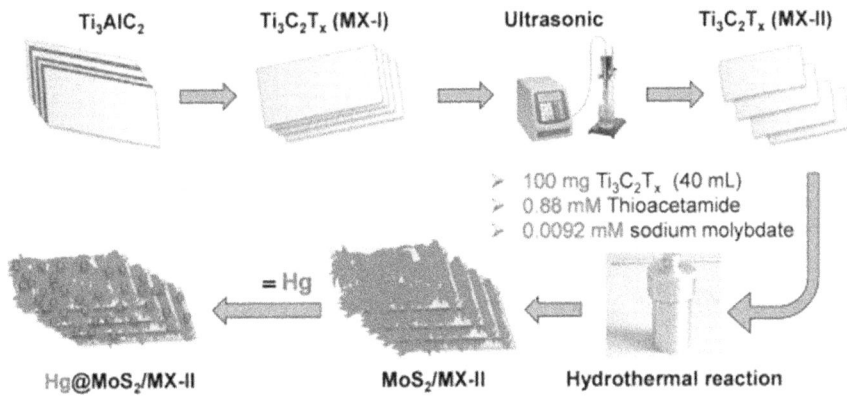

FIGURE 6.3   Schematic illustration of the fabrication of MoS$_2$/MX-II nanocomposite.

The Ti$_3$C$_2$T$_x$ nanosheets acquired higher surface areas and interlayer distances through ultrasonication, which helped in the removal process. In both the aqueous and vapor phases, MoS$_2$/MX-II had outstanding adsorption characteristics for mercury due to its synergistic properties. MoS$_2$/MX-II nanocomposite adsorbed Hg^{2+} at 7.16 mmol/g and Hg^0 at 1.65 mmol/g. Using MoS$_2$/MX-II, Hg^{2+} was efficiently removed from the aqueous solution in a matter of 120 seconds, and the system performed exceptionally well even under highly acidic (pH = 2) and alkaline conditions (pH > 10). Additionally, MoS$_2$-MX-II displayed excellent recycling properties after five cycles.

Khan et al. [47] used hydrothermal and solution-treatment procedures to fabricate three-dimensional flower-like appearing δ-MnO$_2$, MXene, and δ-MnO$_2$/MXene *in situ* hybrid (IH) composites. These adsorbents fabricated were quantified to capture Cr (VI). Optimum conditions were determined based on the effects of pH, contact time, and Cr (VI) ion concentration. Among these adsorbents, IH captured an enormous amount of Cr (VI), followed by MXene, and then δ-MnO$_2$. Comparatively to pristine δ-MnO$_2$ and MXene, the IH composite exhibited greater adsorption capacities by over 50% and 29%, respectively. Adsorption is enhanced by polar groups (–OH, –F, –O) present in the adsorbent and its larger surface area. It was found that the IH adsorbent effectively removed Cr (VI) at pH 3 with its highest capacity (353.87 mg/g), considerably superior to MXene and δ-MnO$_2$ owing to the interpenetrating flower-like δ-MnO$_2$ in the layer of MXene. Cr (VI) removal kinetics studies revealed a pseudo-second-order model, whereas data from isothermal experiments demonstrated a Freundlich model. Table 6.1 shows a comparative study of MXene and MXene-based materials in removing heavy metal ions.

6.2.2   DYE DEGRADATION BY MXENES

Dye is a class of organic compounds that impart solid and bright colors to other substances at the molecular level or dispersion [48]. Dyestuffs, textiles, papers, and plastics are some industries that color their products with dyes, and they also use a great deal of water. They consequently release significant quantities of colored wastewater [49]. Contact with dye effluent can irritate the skin. Dye effluents can cause permanent eye damage or even burns if they contact the eye, whether an animal or a human [50]. If dyes are ingested, they can cause extreme sweat, dizziness, methemoglobinemia, burns on the mouth, and nausea [51]. Many dyes are inherently toxic and sometimes carcinogenic, resulting in grave health risks for aquatic organisms [52]. Thus before discharge, the wastewater must be treated to remove dyes.

Water containing synthetic dyes can be treated in various ways: by photocatalysis [53], Fenton photolysis [54], ozonolysis [55], sonocatalytic degradations [56], reductions with zero-valent metals [57], membrane processes [58], and biodegradation [59]. Nevertheless, these techniques have

TABLE 6.1
Comparative Study of MXene and MXene-Based Materials in Removing Heavy Metal Ions

MXene	Heavy Metal	Adsorption Conditions			Adsorption Capacity (mg/g)	Isotherms/Kinetics	Removal Efficiency	Regeneration	Ref.
		pH	Time (min)	Temp (°C)					
Ti$_3$C$_2$T$_x$-PDOPA nanocomposites	Cu^{2+}	11	60	25	46.6	Freundlich isotherm/pseudo-first-order	93.2%		[22]
Alk-MXene/LDH	Ni^{2+}	7	120	25	222.7171	Redlich-Peterson adsorption isotherm/pseudo-second-order kinetics	>97.35%	>85.32%	[34]
Fe$_3$O$_4$@MXene composite NF membrane	a. Cu^{2+} b. Cd^{2+} c. Cr^{6+}							d. ~76%	[35]
MXene/alginate composites		≥5	15	60	g. 382.7 h. 87.6	Langmuir isotherm/pseudo-second-order-kinetics		Adsorption loss rates i. 8.9% j. 5.4%	[36]
MXene-sodium alginate	Hg^{2+}	4.5	120	25	932.84	Redlich-Peterson adsorption isotherm/pseudo-second-order kinetics	100%	99.69%	[37]
sodium alginate/MXene/CoFe$_2$O$_4$	Cu^{2+}	5.5	24 hours	25	371%	Langmuir isotherm/pseudo-second-order kinetics		80% after five cycles	[38]
Alk-MXene membrane	a. Pb^{2+},Cu^{2+}, Cd^{2+}, b. Cl$^-$, NO$_3^-$								[39]
(DL)-Ti$_3$C$_2$T$_x$	Cu^{2+}	5	3	25	78.45	Freundlich isotherm/pseudo-second-order kinetics	98.29%	30% after third cycle	[40]
Ti$_3$C$_2$T$_x$-based films	m. Cr(VI) n. Pd(II) o. Au(III) p. Ag(I)				q. 84 r. 890 s. 1,241 t. 1,172				[41]
M-Ti$_3$C$_2$ nanosheets	Hg^{2+}	>11	12 hours	25	4,806	Langmuir model	>99.6%		[42]

(Continued)

TABLE 6.1 (Continued)
Comparative Study of MXene and MXene-Based Materials in Removing Heavy Metal Ions

MXene	Heavy Metal	Adsorption Conditions			Adsorption Capacity (mg/g)	Isotherms/Kinetics	Removal Efficiency	Regeneration	Ref.
		pH	Time (min)	Temp (°C)					
Ti_2CT_x-EHL nanosheets	Pb^{2+}	3.2	24 hours	30	232.9	Langmuir isotherm/pseudo-second-order kinetics	97.9%		[44]
MXene	Hg^{2+}	5		30	u. Mercuric nitrate: 1,057.3 v. Mercuric chloride: 7739	Langmuir model/quasi-second-order kinetic	100%	99.8%–96.3% from one to nine cycles.	[45]
MoS_2/MX-II	Hg^{2+}	6.5	120 seconds		7.16 mmol/g	Langmuir isotherm/pseudo-second-order kinetics	99.75%		[46]
δ-MnO_2/MXene in situ hybrid (IH) composites	Cr(VI)	3	30		353.87	Freundlich isotherm/pseudo-second-order kinetics		Deteriorate to <50% after the fifth cycle.	[47]

several shortcomings, including high costs, inconvenient by-products, and a low removal rate. Of the numerous dye removal techniques that have been tested, adsorption has proven to be the most effective at removing almost any dye. The advantages of adsorption over other treatment methods for synthetic dyes in an aqueous solution include greater removal efficiency, ease of operation, being generally reusable, and being relatively inexpensive.

Based on the above, it is obvious that dye adsorption in wastewaters is a crucial issue in water treatment. However, low efficiency and difficult recycling still pose obstacles to its practical application. MXene is suitable for treating dye-polluted wastewater, where conventional adsorbents are associated with economic shortcomings. MXene has physicochemical properties that make it an excellent adsorbent for removing organic pollutants: tunable surface functional groups, substantial surface area, fine structure, hydrophilicity, chemical stability, and metallic conductivity [60]. Dye molecules adhere to MXenes primarily through their interactions with the functional groups on their surfaces. We report on studies involving the use of MXene to remove toxic dyes.

Wei et al. presented the method of treating multilayered MXene using alkaline solution [61] to modify the adsorption performance of $Ti_3C_2T_x$. It was found that this treatment had the positive effect of expanding the interlayer spacing of MXenes and altering the surface functional group, which increased its adsorption abilities for methylene blue (MB) and ultimately accelerated their removal. The adsorptive properties of pristine and alkali-treated $Ti_3C_2T_x$ (with LiOH, NaOH, and KOH) were evaluated for MB removal. The alk-$Ti_3C_2T_x$ removed more than the pristine $Ti_3C_2T_x$, while NaOH-$Ti_3C_2T_x$ exhibited the most remarkable adsorption capacity of 189 mg/g among the alk-$Ti_3C_2T_x$. The combination of surface adsorption and intercalation adsorption is believed to be responsible for the high MB removal rate of NaOH-$Ti_3C_2T_x$.

Cai et al. [59] successfully synthesized phytic acid (PA)-doped MXene composites with a single-step hydrothermal method. In light of the distinctive stacked nanorod structure, the obtained PA-MXene composites showed high adsorption capacities for MB and Rhodamine B (RhB). Due to the large surface area and abundant hydroxyl functional groups in PA-MXene-12, its adsorption capacity is much greater than MXene itself. PA-MXene-12 has an adsorption capacity of 42 and 22 mg/g for MB and RhB, respectively. Synthesized PA-MXene composites displayed high stability and reproducibility. This material may be helpful as an adsorbent for wastewater treatment.

Accordion-like V_2CT_x was demonstrated as an adsorbent for removing MB from wastewater [62]. MB could spontaneously intercalate into the lamellar space of the accordion-like V_2CT_x due to its large surface area, electrostatic interactions, and hydrogen bonds with V_2CT_x. Figure 6.4 illustrates the mechanism of MB adsorption on accordion-like MXene V_2CT_x. At 25°C, V_2CT_x exhibited an adsorption capacity of 111.11 mg g^{-1}. V_2CT_x and MB adhere to each other electrostatically and fortify the hydrogen bond to enhance adsorption and selectivity between cationic and anionic dyes.

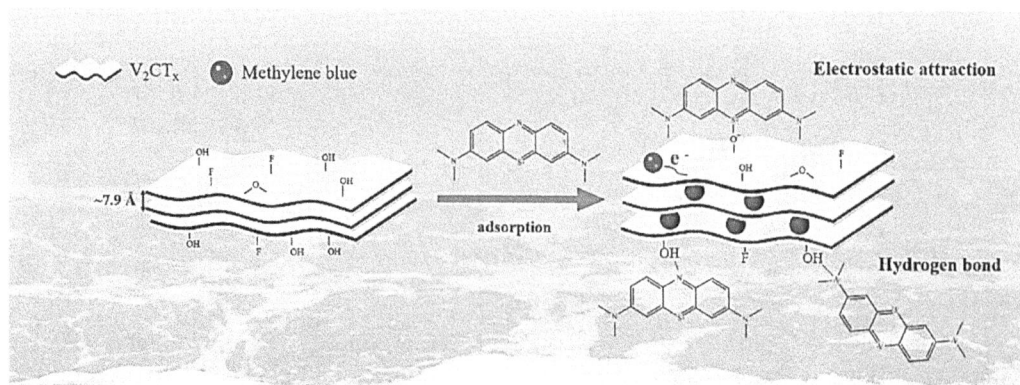

FIGURE 6.4   Diagram illustrating the mechanism of MB adsorption on Accordion-like MXene V_2CT_x.

Considering the potential reusability and the ability to remove cationic dyes from textile dye-bearing wastewater, V_2CT_x is believed to be a promising candidate for this application.

Jun et al. [60] employed $Ti_3C_2T_x$ as a selective adsorbent to treat synthetic dyes in wastewater. MXene was dispersed by ultrasonication (US) at two different frequencies of 28 and 580 kHz. MXene mediated at 28 kHz was significantly more oxygenated and dispersed than MXene moderated at 580 kHz and pristine MXene; due to collapsed cavitation bubbles induced by shock waves and shear force. The adsorption capacity of US-assisted MXene surpassed that of stirring-assisted MXene. MB and methyl orange (MO) were removed using US-modified MXene as an adsorbent. In contrast to stirrer-assisted MXene, US-assisted MXene performed better for only MB. MXene with US assistance features the following advantages: rapid onset at low doses, excellent adsorption performance, specificity for positively charged targets, and high regeneration characteristics.

Electrosorption is a low-cost, high-efficiency adsorption technique that can be controlled quickly and is electrochemically accelerated. Yao et al. [63] are credited with developing the first electrosorption method for organic dye on MXene-based films. Porous MXene/single-walled carbon nanotube (p-MXene/SWCNTs) films were explored as a freestanding electrode for organic dye electrosorption in wastewater. The specific surface area, porous structure, and abundance of functional groups contribute to the dynamic electrostatic interactions and hydrogen bonding forces in the p-MXene/SWCNTs (0.6) film. The film exhibits excellent selectivity and efficiency in removing organic dyes in the as-prepared form. MB can spontaneously adhere to film surfaces and interlaminate with them as the voltage is applied. p-MXene/SWCNT films exhibit remarkable adsorption capacities for MB with a total of 1,068.8 mg/g when exposed to −1.2 V, which is significantly larger than the open circuit capacity of 55.8 mg/g. A notable characteristic is its high reusability without secondary contamination, and its adsorption capacity can reach up to 28,403.7 mg/g.

The *in situ* growth method was used by Zhu et al. [64] to successfully synthesize a new 2D MXene embellished with Fe_3O_4 to remove the dye MB. A study of the adsorption properties of 2D-MX@ Fe_3O_4 at different temperatures was conducted. Enhanced interface interactions between 2D-MX@ Fe_3O_4 and MB were observed because 2D-MX@Fe_3O_4 exhibited a negatively charged surface on a neutral medium and achieved 91.93% of the decolorization at 55°C. At higher temperatures (40°C and 55°C), the Freundlich isotherm model gives good predictions about the removal process of MB, while the Langmuir isotherm is accurate at lower temperatures (25°C). Observation of thermodynamic studies has shown that 2D-MX@Fe_3O_4 can dramatically increase its ability to remove MB via exothermic and chemisorption reactions. As a result of hydrogen bonding (Ti-OH-N) and electrostatic attraction on the 2D-MX@Fe_3O_4 surface, the abundant Ti–OH groups accelerated MB decolorization at higher temperatures, whereas at 25°C, surface adsorption was responsible for MB removal.

MXene-COOH@$(PEI/PAA)_n$ composites based on chemically modified MXene were successfully fabricated by Li et al. [65] via layer-by-layer self-assembly. The adsorption capacities of the synthesized nanocomposite for three dyes, Safranine T (ST), MB, and Neutral red (NR), were studied. As evidenced by pseudo-second-order calculations and thermodynamic fitting, well-defined removal rates were recorded for three dye models used. The composite was additionally stable and reusable. Combined with the MXene-functionalized and self-assembled composite material, this experiment provided a novel idea for studying core-shell composite materials.

Luo et al. [66] fabricated MXene composite loaded with Co_3O_4 nanocrystals by *in situ* solvothermal synthesis. The uniform anchoring of the Co_3O_4 nanocomposites on the surfaces of Ti_3C_2 sheets enhanced their catalytic activity. In addition, the Ti_3C_2 sheets were prevented from being stacked again by this. The synthesized MXene-Co_3O_4 nanocomposites demonstrated a superior catalytic activity for RhB and MB degradation. The catalytic performance of this sample was excellent after eight successive catalytic cycles, indicating that this nanocomposite is stable and repeatable. The results show that MXene-Co_3O_4 nanocomposites can be employed in a wide range of applications.

A high-efficiency nanohybrid structure comprising bismuth ferrite nanoparticles $BiFeO_3$ and $Ti_3C_2T_x$ sheets were constructed by Iqbal et al. [67] to degrade aqueous organic pollutants such as Congo Red (CR) and acetophenone. Double-solvent solvothermal synthesis was adopted to synthesize $BiFeO_3$ (BFO)/Ti_3C_2 (MXene) nanohybrids. Under visible light illumination, the hybrid photocatalytically degraded 100% of organic dye (CR) in 42 minutes and a colorless aqueous pollutant (acetophenone) in 150 minutes. At the same time, the same pollutant underwent only 60% degradation under dark. This hybrid system offered a large surface area of $147 \, m^2/g$ and a short recombination time. High surface area and low recombination rates lead to a quick and efficient degradation of organic molecules, making the BFO/MXene nanohybrid an ideal photocatalyst for many applications in the future.

A large amount of $Ti_3C_2T_x$ MXene, predominantly F-terminated, may be an effective adsorbent for MB in wastewater [68]. The F-terminated MXene adsorbed an incredibly high 92% MB within 5 minutes. According to experiments, the strong MB adsorption capability of MXenes with mostly F-terminated ends was due to electrostatic interactions between negatively charged surfaces of MXenes and cationic molecules of MB. Additionally, this MXene featured excellent recyclability, allowing it to be used repeatedly. Table 6.2 shows a comparative study of MXene and MXene-based materials for dye adsorption.

6.2.3 RADIONUCLIDE ELIMINATION BY MXENES

Several nuclear power plants have been built due to the advent of nuclear technology. Energy security, energy independence, and the reduction of fossil fuel imports all depend on the development of atomic energy [69]. The use of nuclear technology to meet global energy demands also has a downside: it produces radioactive waste [70]. Significant environmental concerns have been raised regarding the adverse effects of nuclear energy utilization. These concerns include radioactive pollution caused by the nuclear fuel cycle and catastrophic releases caused by atomic accidents [71]. Nuclear waste is a huge problem that is causing concern in many countries and needs to be resolved urgently. Radioactive elements radiation and toxic chemical effects may threaten human health and ecosystems. A nuclear waste treatment process effectively removes half-life nuclear radioactive materials from the natural environment, such as uranium, barium, cesium, thorium, palladium, and strontium.

Solvent extraction [72], co-precipitation [73], ion exchange [74], membrane separation [75], and biological treatments [76] have been traditionally used for separating and enriching the high concentration of radionuclides. These methods are not appropriate for the remediation of soils and waters contaminated by low and medium levels of radioactive wastes. The adsorption method stands out as one of the most effective methods for the purification of radioactive metals, owing primarily to its simplicity, ease of operation, cost-effectiveness, high efficiency, and can be implemented on a large scale [77]. Several organic-inorganic adsorbents have been identified to sequester radioactive waste due to the rapid emergence of nuclear waste governance [18].

MXenes and their derivatives possess several exciting properties, including multiple sorption sites, unique layer configurations, superior hydrophilicity, polarity control, redox activity, and radiation tolerance. These properties make them promising adsorbent alternatives for removing radionuclides from the environment [60]. MXenes also exhibit prominent d-spacing, which helps in the adsorption mechanism, roughly tuned by encapsulating radionuclides with selective intercalants. Additionally, it is suggested that the de-lamination of more stacked MXenes transforms them into nanosheets with thinner walls, which would significantly increase their ability to coordinate radionuclides. In addition to metallic surfaces, etchants can aid in capturing radioactive substances. As a result of the higher radioactivity scalability of MXenes and chemical compatibility with molten salt environments, they are an excellent candidate to be selected as a material for the management of nuclear waste material [78].

TABLE 6.2

Comparative Study of MXene and MXene-Based Materials for Dye Adsorption

MXene	Adsorption Conditions				Adsorption Capacity (mg/g)	Isotherms/Kinetics	Removal Efficiency	Regeneration	Ref.
	Dye	pH	Time (min)	Temp (°C)					
NaOH-$Ti_3C_2T_x$	MB	8.8–9	24 hours	25	189	Langmuir isotherm			[61]
PA-MXene composite	w. MB x. RhB	7		25	y. 42 z. 22	Langmuir isotherm/pseudo-second-order kinetics		After 12 cycles a. 85% b. 84%	[85]
Accordion-like V_2CT_x	MB	11	24	25	111.11	Langmuir isotherm/pseudo-second-order kinetics	97.1%	80.0% after second cycle	[62]
US-assisted MXenes	MB	7	30	20		Freundlich isotherm/pseudo-second-order kinetics		Four cycles	[86]
p-MXene/SWCNTs	MB				28,403.7			95.2% after five cycles	[63]
2D-MX@Fe_3O_4	MB		12 hours	55	11.68	Freundlich isotherm	91.93%		[64]
MXene-COOH@(PEI/PAA)n	c. ST d. MB e. NR		f. 180 g. 200 h. 250	298	35.599 81.9672 46.1255	Langmuir isotherm/pseudo-second-order kinetics		64.4% after eight consecutive cycles	[65]
Ti_3C_2–SO_3H	MB	7	70	298	111.11	Langmuir isotherm/pseudo-first-order			[87]
MXene-Co_3O_4 nanocomposites	i. MB j. RhB		k. 240 l. 80	25				a. 92.37% b. after eight cycles	[66]
Multilayer $Ti_3C_2T_x$	MB	2	5		64.3		91.9%		[68]

The studies were carried out using dry and hydrated $Ti_3C_2T_x$ to remove U(VI) from aqueous solutions [79]. DMSO was used to intercalate MXene, which exhibited five times greater adsorption capacity than dry $Ti_3C_2T_x$. As a result of the more flexible nature and the much larger interlayer space of hydrated $Ti_3C_2T_x$, U(VI) removal was significantly better than that of dry $Ti_3C_2T_x$. The data demonstrate that $Ti_3C_2T_x$ is an excellent material candidate for capturing and encapsulating U(VI) and may develop high-performance nuclear energy systems with highly efficient waste treatment and disposal.

MXenes were also investigated to determine the ability to remove Th(IV) under different storage conditions, such as dry and hydrated forms [80]. Hydrated Ti_2CT_x displayed an enhanced sorption capability than its dry counterpart due to more extensive interlayer space, and hence, ions could quickly enter and diffuse past. Ti_2CT_x-hydrated sorbents can have an adsorption capacity of 213.2 mg/g. Further, Ti_2CT_x-hydrated is highly selective toward Th(IV) adsorption even when competing metals are present. According to these results, Ti_2CT_x-hydrated may be a good candidate for the purification and concentration of thorium.

$Ti_3C_2T_x$ was explored as a scavenger for Cs^+ ions in contaminated water [81]. In this study, the superior characteristics of the layered structure and the surface functional groups of MXene contributed to the rapid steady-state adsorption of Cs^+ to 25.4 mg/g at room temperature within a minute. Pseudo-second-order kinetics were observed for the adsorption of Cs^+ ions. It was found that the Freundlich adsorption isotherm was in agreement with the adsorption data, indicating heterogeneity at the surface of $Ti_3C_2T_x$. Despite five adsorption-desorption cycles, $Ti_3C_2T_x$ was well able to regenerate. In addition to removing Cs^+ ions from contaminated water, the high regeneration and adsorption capacities of MXene suggest that this material can also remove other pollutants from aqueous solutions.

Mu et al. [82] produced an alk-$Ti_3C_2T_x$ material that performed well in removing barium ions from an aqueous environment because of its wide layer spacing and abundance of active adsorption sites. Alk-$Ti_3C_2T_x$ binds Ba^{2+} at a maximum rate of 46.46 mg/g, almost three times higher than the adsorption by unmodified $Ti_3C_2T_x$. Ba^{2+} removal from simulated mixed nuclear wastewater exhibits an extremely high level of selectivity that makes it highly suitable for environmental wastewater treatment.

Mu and his co-workers used various temperature treatments to study palladium adsorption on MXene materials [83]. Relatively high specific surface area and substantial interlayer spacing of these materials allowed them to successfully remove Pd^{2+} from the HNO_3 aqueous solution, especially those prepared at higher temperatures (450°C). Based on the Langmuir model, Pd^{2+} adsorption capacity is 184.56 mg/g. It is also noteworthy that the prepared MXene demonstrated an excellent regeneration capacity and reusability, with no noticeable degradation in the Pd^{2+} adsorption capacity after five cycles. Results indicate that the MXene material can effectively remove palladium from wastewater.

MXene-derived hierarchical titanate nanostructures (HTNs) were fabricated by Zhang et al. [84] by using Ti_2CT_x MXene as a precursor. Figure 6.5 displays the radionuclide sequestration with the as-prepared MXene-derived HTNs. Thanks to their well-maintained layered structures and abundantly exchangeable guest cations, these HTNs were able to remove Eu(III) with high sorption capacities efficiently. A large amount of free water and abundant hydroxyl groups are present on the surface of the titanate nanostructures, which poses an opportunity for Eu(III) uptake and the formation of active ion-exchange sites. In environmental clean-up of radioactive pollution, these HTNs have promising properties based on their excellent sorption capacity and facile synthesis under mild conditions.

These 2D materials have tremendous potential as radionuclide adsorbents. MXenes are observed to be good at removing radionuclides, and they seem to outperform other pollutant adsorbents and 2D materials.

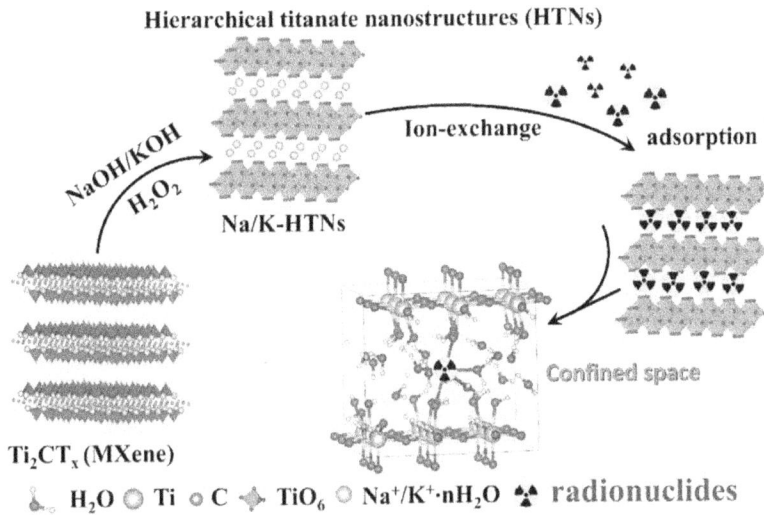

FIGURE 6.5 Radionuclide sequestration with MXene derived HNT.

6.3 CONCLUSION

MXenes' multifaceted characteristics, such as excellent mechanical and electrical properties, hydrophilicity, high adsorption capacity, high surface area, and fast adsorption equilibrium, are crucial for being a potential adsorbent for wastewater treatment. It can be concluded that MXenes and MXene-based composites possess good stability with high adsorption capacities from effluent water containing heavy metals, dyes, and radionuclides after adsorption. Among the MXene families, MXene ($Ti_3C_2T_x$) is extensively investigated for environmental applications, including heavy metal sequestration, organic dye adsorption, and radionuclide adsorption. There is a high probability that many other MXenes other than $Ti_3C_2T_x$ can also be used as adsorbents for environmental applications. Thus, MXenes with appropriate functionalization, desirable physiochemical properties, superior absorptivity, and excellent stability should be taken up to attain highly efficient remediation of environmental pollutants.

REFERENCES

[1] I. Ali, and H.Y. Aboul-Enein, *Chiral Pollutants: Distribution, Toxicity and Analysis by Chromatography and Capillary Electrophoresis*, John Wiley & Sons, Hoboken, New Jersey, 2004.

[2] R.A. Meyers, *Encyclopedia of Environmental Pollution and Cleanup*, Wiley, Hoboken, New Jersey, 1999.

[3] D. Barceló, *Emerging Organic Pollutants in Waste Waters and Sludge*, Vol. 5, Springer Science & Business Media, Berlin, Germany, 2005.

[4] K.S. Novoselov, A.K. Geim, S.V. Morozov, D.-E. Jiang, Y. Zhang, S.V. Dubonos, I.V. Grigorieva, and A.A. Firsov, Electric field effect in atomically thin carbon films, *Science* 306 (2004), pp. 666–669.

[5] M. Naguib, M. Kurtoglu, V. Presser, J. Lu, J. Niu, M. Heon, L. Hultman, Y. Gogotsi, and M.W. Barsoum, Two-dimensional nanocrystals produced by exfoliation of Ti_3AlC_2, *Advanced Materials* 23 (2011), pp. 4248–4253.

[6] A. Zamhuri, G.P. Lim, N.L. Ma, K.S. Tee, and C.F. Soon, MXene in the lens of biomedical engineering: synthesis, applications and future outlook, *Biomedical Engineering Online* 20 (2021), pp. 1–24.

[7] L. Gao, C. Li, W. Huang, S. Mei, H. Lin, Q. Ou, Y. Zhang, J. Guo, F. Zhang, and S. Xu, MXene/polymer membranes: synthesis, properties, and emerging applications, *Chemistry of Materials* 32 (2020), pp. 1703–1747.

[8] M. Khazaei, M. Arai, T. Sasaki, C.Y. Chung, N.S. Venkataramanan, M. Estili, Y. Sakka, and Y. Kawazoe, Novel electronic and magnetic properties of two-dimensional transition metal carbides and nitrides, *Advanced Functional Materials* 23 (2013), pp. 2185–2192.

[9] R. Niu, R. Han, Y. Wang, L. Zhang, Q. Qiao, L. Jiang, Y. Sun, S. Tang, and J. Zhu, MXene-based porous and robust 2D/2D hybrid architectures with dispersed $Li_3Ti_2(PO_4)_3$ as superior anodes for lithium-ion battery, *Chemical Engineering Journal* 405 (2021), p. 127049.

[10] H. Riazi, S.K. Nemani, M.C. Grady, B. Anasori, and M. Soroush, Ti_3C_2 MXene–polymer nanocomposites and their applications, *Journal of Materials Chemistry A* 9 (2021), pp. 8051–8098.

[11] M. Barakat, New trends in removing heavy metals from industrial wastewater, *Arabian Journal of Chemistry* 4 (2011), pp. 361–377.

[12] A. Azimi, A. Azari, M. Rezakazemi, and M. Ansarpour, Removal of heavy metals from industrial wastewaters: a review, *ChemBioEng Reviews* 4 (2017), pp. 37–59.

[13] J.G. Dean, F.L. Bosqui, and K.H. Lanouette, Removing heavy metals from waste water, *Environmental Science & Technology* 6 (1972), pp. 518–522.

[14] V.K. Gupta, and I. Ali, *Environmental Water: Advances in Treatment, Remediation and Recycling*, Newnes, Amsterdam, Netherlands, 2013.

[15] J. Yang, B. Hou, J. Wang, B. Tian, J. Bi, N. Wang, X. Li, and X. Huang, Nanomaterials for the removal of heavy metals from wastewater, *Nanomaterials* 9 (2019), p. 424.

[16] L. Charerntanyarak, Heavy metals removal by chemical coagulation and precipitation, *Water Science and Technology* 39 (1999), pp. 135–138.

[17] D. Mantzavinos, and E. Psillakis, Enhancement of biodegradability of industrial wastewaters by chemical oxidation pre-treatment, *Journal of Chemical Technology & Biotechnology: International Research in Process, Environmental & Clean Technology* 79 (2004), pp. 431–454.

[18] H. Deng, Z.-J. Li, L. Wang, L.-Y. Yuan, J.-H. Lan, Z.-Y. Chang, Z.-F. Chai, and W.-Q. Shi, Nanolayered Ti_3C_2 and $SrTiO_3$ composites for photocatalytic reduction and removal of uranium (VI), *ACS Applied Nano Materials* 2 (2019), pp. 2283–2294.

[19] S. Vasudevan, J. Lakshmi, and G. Sozhan, Studies on the Al–Zn–In-alloy as anode material for the removal of chromium from drinking water in electrocoagulation process, *Desalination* 275 (2011), pp. 260–268.

[20] A. Bashir, L.A. Malik, S. Ahad, T. Manzoor, M.A. Bhat, G. Dar, and A.H. Pandith, Removal of heavy metal ions from aqueous system by ion-exchange and biosorption methods, *Environmental Chemistry Letters* 17 (2019), pp. 729–754.

[21] E. Katsou, S. Malamis, and K.J. Haralambous, Industrial wastewater pre-treatment for heavy metal reduction by employing a sorbent-assisted ultrafiltration system, *Chemosphere* 82 (2011), pp. 557–564.

[22] D. Gan, Q. Huang, J. Dou, H. Huang, J. Chen, M. Liu, Y. Wen, Z. Yang, X. Zhang, and Y. Wei, Bioinspired functionalization of MXenes (Ti_3C_2TX) with amino acids for efficient removal of heavy metal ions, *Applied Surface Science* 504 (2020), p. 144603.

[23] S. Wong, N. Ngadi, I.M. Inuwa, and O. Hassan, Recent advances in applications of activated carbon from biowaste for wastewater treatment: a short review, *Journal of Cleaner Production* 175 (2018), pp. 361–375.

[24] R. Thines, N. Mubarak, S. Nizamuddin, J. Sahu, E. Abdullah, and P. Ganesan, Application potential of carbon nanomaterials in water and wastewater treatment: a review, *Journal of the Taiwan Institute of Chemical Engineers* 72 (2017), pp. 116–133.

[25] D.A. Giannakoudakis, I. Anastopoulos, M. Barczak, E. Antoniou, K. Terpiłowski, E. Mohammadi, M. Shams, E. Coy, A. Bakandritsos, and I.A. Katsoyiannis, Enhanced uranium removal from acidic wastewater by phosphonate-functionalized ordered mesoporous silica: surface chemistry matters the most, *Journal of Hazardous Materials* 413 (2021), p. 125279.

[26] S. Mustapha, J. Tijani, M. Ndamitso, S. Abdulkareem, D. Shuaib, A. Mohammed, and A. Sumaila, The role of kaolin and kaolin/ZnO nanoadsorbents in adsorption studies for tannery wastewater treatment, *Scientific Reports* 10 (2020), pp. 1–22.

[27] J. Shi, Z. Yang, H. Dai, X. Lu, L. Peng, X. Tan, L. Shi, and R. Fahim, Preparation and application of modified zeolites as adsorbents in wastewater treatment, *Water Science and Technology* 2017 (2018), pp. 621–635.

[28] L. Joseph, B.-M. Jun, M. Jang, C.M. Park, J.C. Muñoz-Senmache, A.J. Hernández-Maldonado, A. Heyden, M. Yu, and Y. Yoon, Removal of contaminants of emerging concern by metal-organic framework nanoadsorbents: a review, *Chemical Engineering Journal* 369 (2019), pp. 928–946.

[29] M.A. Atieh, Removal of chromium (VI) from polluted water using carbon nanotubes supported with activated carbon, *Procedia Environmental Sciences* 4 (2011), pp. 281–293.

[30] B. Szala, T. Bajda, and A. Jeleń, Removal of chromium (VI) from aqueous solutions using zeolites modified with HDTMA and ODTMA surfactants, *Clay Minerals* 50 (2015), pp. 103–115.

[31] Q. Peng, J. Guo, Q. Zhang, J. Xiang, B. Liu, A. Zhou, R. Liu, and Y. Tian, Unique lead adsorption behavior of activated hydroxyl group in two-dimensional titanium carbide, *Journal of the American Chemical Society* 136 (2014), pp. 4113–4116.

[32] A. Yaqub, Q. Shafiq, A.R. Khan, H.M.S. Muhammad, and F. Shahzad, Recent advances in the adsorptive remediation of wastewater using two-dimensional transition metal carbides (MXenes): a review, *New Journal of Chemistry* (2021).

[33] I. Ihsanullah, MXenes (two-dimensional metal carbides) as emerging nanomaterials for water purification: progress, challenges and prospects, *Chemical Engineering Journal* 388 (2020), p. 124340.

[34] X. Feng, Z. Yu, R. Long, X. Li, L. Shao, H. Zeng, G. Zeng, and Y. Zuo, Self-assembling 2D/2D (MXene/LDH) materials achieve ultra-high adsorption of heavy metals Ni2+ through terminal group modification, *Separation and Purification Technology* 253 (2020), p. 117525.

[35] X. Yang, Y. Liu, S. Hu, F. Yu, Z. He, G. Zeng, Z. Feng, and A. Sengupta, Construction of Fe_3O_4@ MXene composite nanofiltration membrane for heavy metal ions removal from wastewater, *Polymers for Advanced Technologies* 32 (2021), pp. 1000–1010.

[36] Y. Dong, D. Sang, C. He, X. Sheng, and L. Lei, Mxene/alginate composites for lead and copper ion removal from aqueous solutions, *RSC Advances* 9 (2019), pp. 29015–29022.

[37] A. Shahzad, M. Nawaz, M. Moztahida, J. Jang, K. Tahir, J. Kim, Y. Lim, V.S. Vassiliadis, S.H. Woo, and D.S. Lee, Ti_3C_2Tx MXene core-shell spheres for ultrahigh removal of mercuric ions, *Chemical Engineering Journal* 368 (2019), pp. 400–408.

[38] J. Ren, Z. Zhu, Y. Qiu, F. Yu, T. Zhou, J. Ma, and J. Zhao, Enhanced adsorption performance of alginate/MXene/$CoFe_2O_4$ for antibiotic and heavy metal under rotating magnetic field, *Chemosphere* 284 (2021), p. 131284.

[39] S. Wang, F. Wang, Y. Jin, X. Meng, B. Meng, N. Yang, J. Sunarso, and S. Liu, Removal of heavy metal cations and co-existing anions in simulated wastewater by two separated hydroxylated MXene membranes under an external voltage, *Journal of Membrane Science* 638 (2021), p. 119697.

[40] A. Shahzad, K. Rasool, W. Miran, M. Nawaz, J. Jang, K.A. Mahmoud, and D.S. Lee, Two-dimensional $Ti_3C_2T_x$ MXene nanosheets for efficient copper removal from water, *ACS Sustainable Chemistry & Engineering* 5 (2017), pp. 11481–11488.

[41] X. Xie, C. Chen, N. Zhang, Z.-R. Tang, J. Jiang, and Y.-J. Xu, Microstructure and surface control of MXene films for water purification, *Nature Sustainability* 2 (2019), pp. 856–862.

[42] K. Fu, X. Liu, D. Yu, J. Luo, Z. Wang, and J.C. Crittenden, Highly efficient and selective Hg (II) removal from water using multilayered $Ti_3C_2O_x$ MXene via adsorption coupled with catalytic reduction mechanism, *Environmental Science & Technology* 54 (2020), pp. 16212–16220.

[43] B.-M. Jun, N. Her, C.M. Park, and Y. Yoon, Effective removal of Pb (II) from synthetic wastewater using $Ti_3C_2T_x$ MXene, *Environmental Science: Water Research & Technology* 6 (2020), pp. 173–180.

[44] S. Wang, Y. Liu, Q.-F. Lü, and H. Zhuang, Facile preparation of biosurfactant-functionalized Ti_2CT_x MXene nanosheets with an enhanced adsorption performance for Pb (II) ions, *Journal of Molecular Liquids* 297 (2020), p. 111810.

[45] X. Hu, C. Chen, D. Zhang, and Y. Xue, Kinetics, isotherm and chemical speciation analysis of Hg (II) adsorption over oxygen-containing MXene adsorbent, *Chemosphere* 278 (2021), p. 130206.

[46] A. Shahzad, J. Jang, S.-R. Lim, and D.S. Lee, Unique selectivity and rapid uptake of molybdenumdisulfide-functionalized MXene nanocomposite for mercury adsorption, *Environmental Research* 182 (2020), p. 109005.

[47] A.R. Khan, S.K. Awan, S.M. Husnain, N. Abbas, D.H. Anjum, N. Abbas, M. Benaissa, C.R. Mirza, S. Mujtaba-ul-Hassan, and F. Shahzad, *3D Flower like δ-MnO₂/MXene Nano-Hybrids for the Removal of Hexavalent Cr from Wastewater*, Ceramics International (2021), United Kingdom.

[48] N. Zhang, A. Ishag, Y. Li, H. Wang, H. Guo, P. Mei, Q. Meng, and Y. Sun, Recent investigations and progress in environmental remediation by using covalent organic framework-based adsorption method: a review, *Journal of Cleaner Production* (2020), p. 123360.

[49] G. Crini, Non-conventional low-cost adsorbents for dye removal: a review, *Bioresource Technology* 97 (2006), pp. 1061–1085.

[50] T. Robinson, G. McMullan, R. Marchant, and P. Nigam, Remediation of dyes in textile effluent: a critical review on current treatment technologies with a proposed alternative, *Bioresource Technology* 77 (2001), pp. 247–255.

[51] M.A.M. Salleh, D.K. Mahmoud, W.A.W.A. Karim, and A. Idris, Cationic and anionic dye adsorption by agricultural solid wastes: a comprehensive review, *Desalination* 280 (2011), pp. 1–13.

[52] C. O'Neill, F.R. Hawkes, D.L. Hawkes, N.D. Lourenço, H.M. Pinheiro, and W. Delée, Colour in textile effluents–sources, measurement, discharge consents and simulation: a review, *Journal of Chemical Technology & Biotechnology: International Research in Process, Environmental & Clean Technology* 74 (1999), pp. 1009–1018.

[53] S. Natarajan, H.C. Bajaj, and R.J. Tayade, Recent advances based on the synergetic effect of adsorption for removal of dyes from waste water using photocatalytic process, *Journal of Environmental Sciences* 65 (2018), pp. 201–222.

[54] H. Kusic, N. Koprivanac, and L. Srsan, Azo dye degradation using Fenton type processes assisted by UV irradiation: a kinetic study, *Journal of Photochemistry and photobiology A: Chemistry* 181 (2006), pp. 195–202.

[55] A. Tehrani-Bagha, N.M. Mahmoodi, and F. Menger, Degradation of a persistent organic dye from colored textile wastewater by ozonation, *Desalination* 260 (2010), pp. 34–38.

[56] A. Khataee, M. Sheydaei, A. Hassani, M. Taseidifar, and S. Karaca, Sonocatalytic removal of an organic dye using TiO_2/montmorillonite nanocomposite, *Ultrasonics Sonochemistry* 22 (2015), pp. 404–411.

[57] L. Zhang, Q. Shao, and C. Xu, Enhanced azo dye removal from wastewater by coupling sulfidated zerovalent iron with a chelator, *Journal of Cleaner Production* 213 (2019), pp. 753–761.

[58] L. Yao, L. Zhang, R. Wang, S. Chou, and Z. Dong, A new integrated approach for dye removal from wastewater by polyoxometalates functionalized membranes, *Journal of Hazardous Materials* 301 (2016), pp. 462–470.

[59] R. Pereira, M. Pereira, M. Alves, and L. Pereira, Carbon based materials as novel redox mediators for dye wastewater biodegradation, *Applied Catalysis B: Environmental* 144 (2014), pp. 713–720.

[60] B.-M. Jun, S. Kim, J. Heo, C.M. Park, N. Her, M. Jang, Y. Huang, J. Han, and Y. Yoon, Review of MXenes as new nanomaterials for energy storage/delivery and selected environmental applications, *Nano Research* 12 (2019), pp. 471–487.

[61] Z. Wei, Z. Peigen, T. Wubian, Q. Xia, Z. Yamei, and S. ZhengMing, Alkali treated $Ti_3C_2T_x$ MXenes and their dye adsorption performance, *Materials Chemistry and Physics* 206 (2018), pp. 270–276.

[62] H. Lei, Z. Hao, K. Chen, Y. Chen, J. Zhang, Z. Hu, Y. Song, P. Rao, and Q. Huang, Insight into adsorption performance and mechanism on efficient removal of methylene blue by accordion-like V_2CT_x MXene, *The Journal of Physical Chemistry Letters* 11 (2020), pp. 4253–4260.

[63] C. Yao, W. Zhang, L. Xu, M. Cheng, Y. Su, J. Xue, J. Liu, and S. Hou, A facile synthesis of porous MXene-based freestanding film and its spectacular electrosorption performance for organic dyes, *Separation and Purification Technology* 263 (2021), p. 118365.

[64] Z. Zhu, M. Xiang, L. Shan, T. He, and P. Zhang, Effect of temperature on methylene blue removal with novel 2D-magnetism titanium carbide, *Journal of Solid State Chemistry* 280 (2019), p. 120989.

[65] K. Li, G. Zou, T. Jiao, R. Xing, L. Zhang, J. Zhou, Q. Zhang, and Q. Peng, Self-assembled MXene-based nanocomposites via layer-by-layer strategy for elevated adsorption capacities, *Colloids and Surfaces A: Physicochemical and Engineering Aspects* 553 (2018), pp. 105–113.

[66] S. Luo, R. Wang, J. Yin, T. Jiao, K. Chen, G. Zou, L. Zhang, J. Zhou, L. Zhang, and Q. Peng, Preparation and dye degradation performances of self-assembled MXene-Co_3O_4 nanocomposites synthesized via solvothermal approach, *ACS Omega* 4 (2019), pp. 3946–3953.

[67] M.A. Iqbal, A. Tariq, A. Zaheer, S. Gul, S.I. Ali, M.Z. Iqbal, D. Akinwande, and S. Rizwan, Ti_3C_2-MXene/bismuth ferrite nanohybrids for efficient degradation of organic dyes and colorless pollutants, *ACS Omega* 4 (2019), pp. 20530–20539.

[68] N.M. Tran, Q.T.H. Ta, A. Sreedhar, and J.-S. Noh, Ti_3C_2Tx MXene playing as a strong methylene blue adsorbent in wastewater, *Applied Surface Science* 537 (2021), p. 148006.

[69] Y. Chen, F. Chao, and T. Jiejuan, Necessity of developing nuclear energy in China, *Nuclear Power Engineering* (2014), p. S1.

[70] H. Chen, D. Shao, J. Li, and X. Wang, The uptake of radionuclides from aqueous solution by poly (amidoxime) modified reduced graphene oxide, *Chemical Engineering Journal* 254 (2014), pp. 623–634.

[71] Q.-H. Hu, J.-Q. Weng, and J.-S. Wang, Sources of anthropogenic radionuclides in the environment: a review, *Journal of Environmental Radioactivity* 101 (2010), pp. 426–437.

[72] K. Zhang, Y. Zeng, Z. Liu, H. Cao, W. Zhu, and Q. Guo, Removal of radionuclide uranium from South China's ion-adsorption rare earth leach liquor using solvent extraction with naphthenic acid, *Solvent Extraction Research and Development, Japan* 27 (2020), pp. 39–47.

[73] J. Zhu, Q. Liu, J. Liu, R. Chen, H. Zhang, R. Li, and J. Wang, Ni–Mn LDH-decorated 3D Fe-inserted and N-doped carbon framework composites for efficient uranium (VI) removal, *Environmental Science: Nano* 5 (2018), pp. 467–475.

[74] Y. Sun, S. Yang, G. Sheng, Z. Guo, X. Tan, J. Xu, and X. Wang, Comparison of U (VI) removal from contaminated groundwater by nanoporous alumina and non-nanoporous alumina, *Separation and Purification Technology* 83 (2011), pp. 196–203.

[75] J. Gao, S.-P. Sun, W.-P. Zhu, and T.-S. Chung, Chelating polymer modified P84 nanofiltration (NF) hollow fiber membranes for high efficient heavy metal removal, *Water Research* 63 (2014), pp. 252–261.

[76] M. Narayani, and K.V. Shetty, Chromium-resistant bacteria and their environmental condition for hexavalent chromium removal: a review, *Critical Reviews in Environmental Science and Technology* 43 (2013), pp. 955–1009.

[77] J. Li, X. Wang, G. Zhao, C. Chen, Z. Chai, A. Alsaedi, T. Hayat, and X. Wang, Metal–organic framework-based materials: superior adsorbents for the capture of toxic and radioactive metal ions, *Chemical Society Reviews* 47 (2018), pp. 2322–2356.

[78] J.A. Kumar, P. Prakash, T. Krithiga, D.J. Amarnath, J. Premkumar, N. Rajamohan, Y. Vasseghian, P. Saravanan, and M. Rajasimman, Methods of synthesis, characteristics, and environmental applications of Mxene: a comprehensive review, *Chemosphere* 286 (2022), p. 131607.

[79] L. Wang, W. Tao, L. Yuan, Z. Liu, Q. Huang, Z. Chai, J.K. Gibson, and W. Shi, Rational control of the interlayer space inside two-dimensional titanium carbides for highly efficient uranium removal and imprisonment, *Chemical Communications* 53 (2017), pp. 12084–12087.

[80] S. Li, L. Wang, J. Peng, M. Zhai, and W. Shi, Efficient thorium (IV) removal by two-dimensional Ti_2CT_x MXene from aqueous solution, *Chemical Engineering Journal* 366 (2019), pp. 192–199.

[81] A.R. Khan, S.M. Husnain, F. Shahzad, S. Mujtaba-ul-Hassan, M. Mehmood, J. Ahmad, M.T. Mehran, and S. Rahman, Two-dimensional transition metal carbide ($Ti_3C_2T_x$) as an efficient adsorbent to remove cesium (Cs^+), *Dalton Transactions* 48 (2019), pp. 11803–11812.

[82] W. Mu, S. Du, Q. Yu, X. Li, H. Wei, and Y. Yang, Improving barium ion adsorption on two-dimensional titanium carbide by surface modification, *Dalton Transactions* 47 (2018), pp. 8375–8381.

[83] W. Mu, S. Du, X. Li, Q. Yu, H. Wei, Y. Yang, and S. Peng, Removal of radioactive palladium based on novel 2D titanium carbides, *Chemical Engineering Journal* 358 (2019), pp. 283–290.

[84] P. Zhang, L. Wang, L.-Y. Yuan, J.-H. Lan, Z.-F. Chai, and W.-Q. Shi, Sorption of Eu (III) on MXene-derived titanate structures: the effect of nano-confined space, *Chemical Engineering Journal* 370 (2019), pp. 1200–1209.

[85] C. Cai, R. Wang, S. Liu, X. Yan, L. Zhang, M. Wang, Q. Tong, and T. Jiao, Synthesis of self-assembled phytic acid-MXene nanocomposites via a facile hydrothermal approach with elevated dye adsorption capacities, *Colloids and Surfaces A: Physicochemical and Engineering Aspects* 589 (2020), p. 124468.

[86] B.-M. Jun, S. Kim, H. Rho, C.M. Park, and Y. Yoon, Ultrasound-assisted $Ti_3C_2T_x$ MXene adsorption of dyes: removal performance and mechanism analyses via dynamic light scattering, *Chemosphere* 254 (2020), p. 126827.

[87] Y. Lei, Y. Cui, Q. Huang, J. Dou, D. Gan, F. Deng, M. Liu, X. Li, X. Zhang, and Y. Wei, Facile preparation of sulfonic groups functionalized Mxenes for efficient removal of methylene blue, *Ceramics International* 45 (2019), pp. 17653–17661.

7 Metal Oxide-Biochar Nanocomposites for the Effective Removal of Environmental Contaminants

Jnyanashree Darabdhara and Md. Ahmaruzzaman
National Institute of Technology Silchar

CONTENTS

7.1 INTRODUCTION

Over the last two decades, the rapid growth in industrialization and various other technologies has caused the exponential increase in various contaminants found in our natural resources [1]. Our household and industries release about 80% of waste effluents to our water systems by direct or indirect means without any necessary pretreatment, which declines the water quality [2]. The presence of various organic and inorganic contaminants, including dyes, pharmaceuticals, agricultural wastes, and heavy metals, in our natural resources has become a topic of great concern globally and has received significant research importance [3,4]. The non-biodegradable nature of heavy metals results in their accumulation in living species, while the high persistent, toxic, problematic removal, and easily transferable nature of organic wastes make their presence a serious concern for human

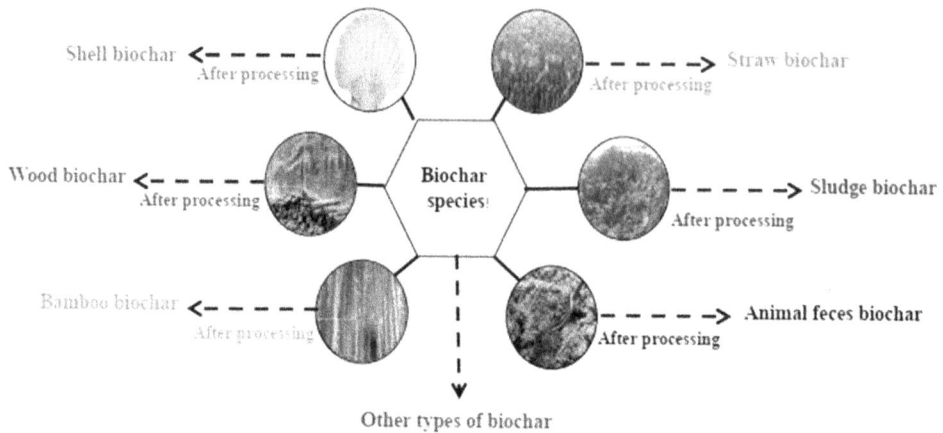

FIGURE 7.1 Biochar classification. (From Ref. [11]: Y. Dai, N. Zhang, C. Xing, Q. Cui and Q. Sun, The adsorption, regeneration and engineering applications of biochar for removal organic pollutants: a review, *Chemosphere* 223 (2019), pp. 12–27.)

health [5]. Thus, in order to minimize this issue, it has become very necessary to find out cost-effective solutions with very less negative impact on the environment.

Biochar is a carbonaceous material obtained by pyrolysis of biomass feedstock either in the absence or in the presence of very limited supply of oxygen. Biochar has gained much attention from researchers around the world due to its ability to tackle various environmental issues, including climate change and removal of pollutants, which improves the growth of plants and soil fertility [6–8]. The generation of biochar is considered to be an environment-friendly, inexpensive, and efficient approach as it involves reuse of the waste materials under limited supply of energy at temperatures below 700°C and also as the feedstock for biochar production such as various solid and agricultural waste biomasses is readily available [9,10]. Figure 7.1 shows the different sources of feedstock from which biochar can be obtained [11].

The surface properties of biochar, including the high specific surface area, presence of sufficient amount of oxygen-containing functional groups, good ion-exchange capability, and high mineral content, make them efficient for applicable as an adsorbent in wastewater treatment process [12,13]. Application of biochar is considered as a sustainable approach for the removal of both the organic and inorganic contaminants present in wastewater, particularly the toxic heavy metals and metalloids [14,15]. Owing to the advantageous properties attributed by biochar, they have earned a wide recognition for mitigating various environmental issues such as reducing the emission of greenhouse gases [16], catalyst [17], in remediation of soil [18,19], as an adsorbent [20], in wastewater treatment process [21], etc. Despite several beneficiary aspects of biochar, they also possess certain disadvantages, including their very small particle size, difficulty in separating them from water, and poor density. Hence, in order to minimize these issues and improve the adsorption performance for the elimination of toxic chemicals, both physical and chemical methods could be applied to bring improvement in the properties of biochar by increasing the surface area and the number of functional groups [22,23]. In order to improve the adsorption performance of biochar, several surface modifications of biochar have been reported, which include the use of metal oxide nanoparticles such as Fe_2O_3, ZnO, Al_2O_3, and many more to bring modifications in the surface charge, application of activating agents such as H_2SO_4, $ZnCl_2$ and NaOH, etc., to enhance the specific surface area and metal oxide support, including Fe_3O_4, MgO and Fe^0, etc., which can potentially react with the contaminants [24–26].

Nanoparticles involve the materials that fall within the size range of 1–100 nm; they include metal nanoparticles, metal oxides, graphene, carbon nanotubes, zero-valent metal atoms, etc. [27,28]. However, this chapter will mainly focus on metal oxides and other metal nanoparticles.

Metal nanoparticles show high activity for several pollutants particularly towards heavy metals and organic molecules owing to their exceptional mechanical properties, high specific surface area, and presence of large number of adsorption sites [29]. Hence, they are used as an adsorbent, catalyst, reductant, and oxidant for removing various organic and inorganic contaminants, including the heavy metal found in aqueous medium [30,31]. However, one of the disadvantages of metal nanoparticles is their ability to agglomerate, thus resulting in the formation of large-scale particles, which in turn decreases the activity and exceptional properties of metal nanoparticles [32,33]. The agglomeration of nanoparticles could be reduced with the help of porous supports such as activated carbon, biochar, silica, etc. However, among all these supports, biochar having large surface area and excellent ion-exchange capabilities is most often used as support for the metal nanoparticles [34,35]. The formation of novel biochar-supported metal oxide nanoparticle (MNPs@BC) combines the beneficial aspects of both biochar and metal oxide nanoparticles and reduces the disadvantages of both the materials [36]. The formation of MNPs@BC composite minimizes the drawbacks of metal oxide nanoparticles by preventing their aggregation and leaching through proper distribution and stabilization of the loaded nanoparticles [37]. Besides this, MNPs@BC composite changes the properties of both the materials and escalates the number of surface active sites and oxygen functionalities [32]. As compared to pristine biochar, the metal oxide biochar nanocomposites display small crystal size, better dispersibility, and better electron transfer capacity [38]. Nanometal oxide-biochar composites are very effective for the adsorptive removal of toxic pollutants such as antibiotics, dyes, and other such organic pollutants as they consist of sufficiently high numbers of adsorption sites and surface functionalities [39]. Besides this, biochar-based photocatalyst support shows high adsorption of a wide variety of organic compounds, and as compared with unmodified nanoparticles, biochar-supported photocatalyst shows higher efficient photodegradation of toxic organic contaminants owing to its higher adsorption potential [40]. This chapter focuses on the application of metal oxide-biochar nanocomposites for the removal of both organic and inorganic wastes from our environment.

7.2 FORMATION OF METAL OXIDE-BIOCHAR COMPOSITES

7.2.1 IMPREGNATION

It is one of the most widely used techniques for the preparation of metal oxide-biochar composites. This method produces biochar composites with large specific surface area, high porosity, good cation-exchange capacity, and presence of large quantity of functional groups [41]. In this method, solid biochar is dipped into solutions consisting of active particles, which latter cling with the solid material [42]. Some of the most commonly used solvents for the impregnation of biochar are $KMnO_4$ [43], $MgCl_2$ [44], and $Fe_2SO_4 7H_2O$ [43]. One such example is the formation of hydrous manganese oxide biochar composite obtained by the impregnation of biochar from peeled pine wood with Mn solutions. The presence of hydrous manganese oxide greatly influenced the adsorption uptake of the biochar. As compared to the pristine biochar, the manganese oxide-biochar composite containing 3.65% Mn showed high adsorption capacity and accelerated the adsorption rate for the uptake of Pb(II) ions, thereby improving the removal efficiency of Pb(II) ions to 98.9%. The manganese oxide-biochar composite presented five times higher monolayer adsorptions as compared to the original biochar [45].

7.2.2 CO-PRECIPITATION METHOD

This is an economic way of producing highly pure nanoparticles having precise stoichiometry and homogeneity. Proper control over the nucleation growth of the nanoparticles and their size or shape could be achieved with this method [46,47]. This method involves precipitation of metal salts of the solution followed by clinging to the biochar surface through pH adjustments, reduction, and other approaches [48]. An example of this method is the formation of nano-MnO_2-biochar composite containing self-assembled spherical MnO_2 nanoparticles with a diameter of 200–500 nm. The MnO_2-biochar composite was obtained by the dispersal of corn stalks biochar in deionized water with $MnSO_4.7H_2O$. The suspension was then allowed for ultrasonic mixing for 1 hour followed by a dropwise addition of $KMnO_4$ solution. The solution was then allowed for aging at an ambient environment for 24 hours, followed by filtering, washing, and drying the obtained product. The MnO_2-biochar composite showed better sorption of di-n-butyl phthalate and oxytetracycline compared to the original biochar [49]. Similarly, Saravanakumar and his co-workers prepared magnetic NiO/biochar composite by dispersing the plant-based biochar PJ8C (activated by $CaCO_3$) with crystals of $NiNO_3$ for 30 minutes in water followed by the dropwise addition of NaOH solution. This led to the precipitation of green color $Ni(OH)_2$ on the carbon mixture, and the surplus filtrate was allowed to decant, which was followed by sonication of the rest mixture to achieve a homogeneous distribution of the composite $Ni(OH)_2$-carbon. The resultant was washed and further allowed for heating in a muffle furnace to get the desired product NiO-PJ8C nanocomposite. The NiO-PJ8C nanocomposite showed reduced specific surface area and pore volume than the original biochar; however, the existence of oxygen resulted in an increase in the pore diameter. The NiO-PJ8C nanocomposite also showed higher adsorption capacity for Pb(II) ions (43 mg/g) than the original biochar (28 mg/g) [50].

7.2.3 Pyrolysis

This method requires directly pyrolyzing the biomass enriched with the desired metal components under anaerobic conditions [51]. Li and his co-workers designed ZnO/ZnS-modified biochar by direct pyrolysis of Zn-polluted corn stover biomass under anaerobic conditions in a reactor at temperatures of 500–800°C. The Zn-modified biochar had a porous structure with high Brunauer-Emmett-Teller (BET) surface area and total pore volume than the common biochar. Moreover, the ZnO/ZnS-modified biochar also displayed a greater effective removal of heavy metals such as Pb(II), Cu(II), and Cr(VI) with high sorption capacities than the common biochar [52]. Yao and his co-workers designed Mg-integrated biochar by direct pyrolysis of Mg-enriched tomato plants in the presence of N_2 at 600°C. The characterization results revealed the presence of both MgO and $Mg(OH)_2$ particles within the biochar matrix. The Mg-biochar composite showed good phosphate sorption from aqueous solutions [51].

7.2.4 Ball Milling Method

It is an economic and eco-friendly approach, which could be adopted by researchers for the modification of biochars. Zheng and his co-workers prepared MgO-biochar composite by ball milling of MgO particles with Hickory wood biochar at 500 rpm by alternating the direction of rotation for 12 hours. The resultant MgO-biochar composite consisted of uniformly distributed MgO nanoparticles on the biochar surface. The ball milling of MgO nanoparticles with the biochar reduced the crystalline size of the nanoparticles from 80 to 20 nm. The MgO-biochar composite was effective for the removal of methylene blue (MB) and phosphate from aqueous medium [53].

In addition to the above-mentioned method, the sol-gel method has also been adopted by researches for the formation of metal oxide-biochar nanocomposites. Khataee and his co-workers reported the formation of ZrO_2-biochar nanocomposite by the sol-gel method with $Zr(OC_3H_7)_4$ as the precursor for Zr. The biochar of wheat husk was ultrasonically dispersed in ethanol solution for 15 minutes followed by the addition of Zr precursor to the above mixture. The suspension was allowed to stir magnetically for an hour before the dropwise addition of the ethanol-HCL mixture. The suspension was then filtered and dried at 100°C after an hour of agitation. The resultant ZrO_2-biochar composite contained spherical-shaped nanoparticles of ZrO_2 on the matrix composite with diameters ranging from 5 to 40 nm [54].

The different techniques discussed so far for the preparation of metal-oxide biochar composite are shown in Figure 7.2 [39].

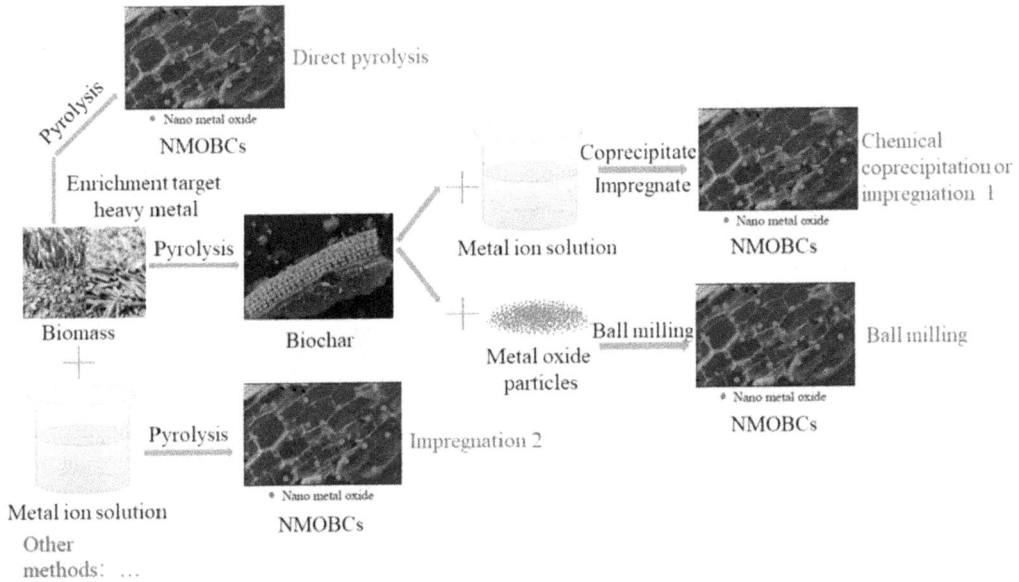

FIGURE 7.2 Graphical representation of various processes adopted for the formation of metal oxide-biochar nanocomposites. (From Ref. [39]: C. Zhao, B. Wang, B.K.G. Theng, P. Wu, F. Liu, S. Wang, X. Lee, M. Chen, L. Li, X. Zhang, Formation and mechanisms of nano-metal oxide-biochar composites for pollutants removal: a review, Sci. Total Environ. 767 (2021), p. 145305.)

7.2.5 APPLICATION OF ULTRASOUND

It is one of the recent methods that could be used for the formation of biochar photocatalyst support. Lisowski and his co-workers used ultrasound for the synthesis of two TiO_2 biochar composites obtained from soft wood pellets and miscanthus straw pellet through a multiple-step process. They were also regarded as mesoporous structures owing to the presence of cracks in the surface [55].

7.3 MORPHOLOGICAL CHANGES IN THE METAL-OXIDE COMPOSITE

The formation of metal-oxide biochar composites produces morphological changes in the biochar. According to a study conducted by Sun and his co-workers, Fe_3O_4-biochar composite showed a morphological change from the smooth surface of the acid-treated biochar to a rugged surface owing to the aggregation of particles, which confirmed the presence of granulated Fe_3O_4 particles on the biochar surface. Besides this, from their transmission electron microscopy (TEM) analysis, they confirmed the presence of only one phase in the acid-treated biochar, while the metal oxide biochar composites revealed the existence of two phases, viz., the biochar and Fe_3O_4 particles [56].

7.4 APPLICATION OF METAL OXIDE-BIOCHAR COMPOSITES AS AN ADSORBENT FOR THE REMOVAL OF EMERGING CONTAMINANTS

7.4.1 REMOVAL OF ORGANIC POLLUTANTS

Organic pollutants are most commonly found in the aquatic systems; they are mostly responsible for water coloring and degrade the water quality, thereby making it unfavorable for drinking and other useful purposes. The most common types of organic contaminants that are found in our water systems include dyes, pharmaceutical compounds, aromatic organic compounds, and agricultural wastes. Textile industries is one of the most common sources, which releases nearly tones of dyes in a year to our water systems. The wastewaters released from the textile industries are rich in synthetic dyes that are harmful for our environment. Moreover, the dyes present in industrial effluents are toxic and carcinogenic in nature. Hence, their exposure can bring huge casualty to human beings and animals; hence, it is very important to find out ways to remove them from our natural resources. Till today, a number of attempts have been made for their removal from industrial wastewater by applying the various techniques, including adsorption, membrane separation, and advanced oxidation processes. However, among the various methods adopted so far, the adsorption process due to the easy operation and good efficiency is considered to be one of the most efficient ways for the removal of both organic and inorganic contaminants from wastewater [57]. Chaukura and his co-workers were successful in removing methyl orange (MO) by using nanocomposite of Fe_2O_3-biochar as an adsorbent. They prepared the Fe_2O_3-biochar nanocomposite by pyrolyzing the pulp and paper biochar impregnated with $FeCl_3$. Calculation of the BET surface area showed that Fe_2O_3-biochar nanocomposite had very low specific surface area ($15.32\,m^2/g$) than the original biochar ($174.29\,m^2/g$). Such low specific surface area of Fe_2O_3-biochar nanocomposite suggested the presence of Fe_2O_3 nanoparticles occupying the interstitial pores of the biochar. In spite of such low specific surface area, the Fe_2O_3-biochar nanocomposite was very effective for the removal of MO and showed 52.79% high removal capacity compared to the simple biochar. The adsorption data of MO on Fe_2O_3-biochar nanocomposite were best fitted to the Freundlich adsorption isotherm with an adsorption capacity of 20.53 mg/g after 30 minutes of reaction time in solutions of pH 8 [58]. A Fe_xO_y/biochar composite was prepared by a facile green synthesis method ultrasonically from the extract of banana peel, and $FeSO_4$, was found to be very effective for the removal of MB. Fourier transform infrared and energy dispersive X-ray spectroscopic (FTIR) and EDS studies revealed the formation of Fe_xO_y/biochar composite through the appearance of a new peak at $586\,cm^{-1}$ originating from the Fe-O bond and by the presence of uniformly distributed Fe and O on the surface of the biochar matrix. The Fe_xO_y/biochar composite showed the successful removal of MB, and the adsorption of MB on Fe_xO_y/biochar could be best explained by the Langmuir adsorption isotherm and through the pseudo-second-order kinetics. The Fe_xO_y/biochar composite showed higher adsorption of MB than the original biochar over a wide pH range from 2.05 to 9.21, with a maximum adsorption capacity of 862 mg/g for solutions with pH 6.1 corresponding to the temperature of 313 K. As per the X-ray photoelectron spectroscopy (XPS) analysis, such high adsorptions of MB resulted from the electrostatic interactions between the MB and the Fe_xO_y/biochar composite. Moreover, the Fe_xO_y/biochar composites also presented good reusability where they maintained high adsorption efficiency even after five adsorption/desorption cycles [59]. Sufficiently high removal of the anionic dye reactive red (RR120) was observed with CuO-biochar nanocomposite obtained by the balling of CuO powder with hickory wood biochar. TEM images suggested the presence of uniformly distributed CuO particles on the biochar surface; the occurrence of a new peak at $485\,cm^{-1}$ due to the vibration of Cu-O bond in the FTIR spectra of the CuO-biochar composite confirmed the formation of the composite. The presence of CuO particles on the biochar surface facilitated the electrostatic attraction between the composite and RR120, with sufficiently high adsorption capacity of 1,399 mg/g observed for the 10%-CuO-biochar composite. Moreover, the CuO-biochar composite also showed a relatively faster removal of RR120 whereby the equilibrium was reached after 3 hours of contact time. The adsorption rate of RR120 to the biochar composite could be best explained by the pseudo-second-order kinetic model. Such excellent removal of RR120 was attributed to the electrostatic attraction existing between the CuO nanoparticles embedded on the biochar and the pollutant RR120 [60].

Antibiotics are a class of medicinal compounds widely used in humans and animals for the treatment of various infections and also in cattle husbandry to prevent diseases and promote growth. However, a major portion of administered antibiotics get discharged as unmetabolized active products, which potentially contaminate our natural water system. Some common antibiotics are found to possess enduring half-life and persistent in nature. The presence of vestigial antibiotics can result in the development of resistant properties of some microorganisms, which are very vicious as they can lead to the inefficiency of other antibiotics for curing new diseases [61]. Li and his co-workers designed a magnetic manganese oxide-biochar composite (MMB) by using the raw biochar obtained from potato leaves through pyrolysis. MMB showed a satisfactory application as an adsorbent for the removal of three fluoroquinolone antibiotics, viz., norfloxacin (NOR), ciprofloxacin (CIP), and enrofloxacin (ENR), present in aqueous medium. EDS analysis confirmed the uniform distribution of Mn and Fe on the biochar composite. MMB showed an increase in BET surface area and pore volume and a decrease in pore diameter compared with the simple biochar. Langmuir adsorption isotherm and pseudo-second-order kinetics could best describe the adsorption of the three antibiotics on the MMB surface with the maximum adsorption capacity value of 6.94 mg/g for NOR, 8.37 mg/g for CIP, and 7.19 mg/g for ENR; moreover, the adsorption of CIP, ENR, and NOR on MMB was enhanced by 1.87, 1.92, and 1.80 times in comparison with the original biochar. The removal of fluoroquinolone drugs by the MMB was greatly influenced by the change in pH and ionic strength of the solution, and there was a decrease in the adsorption capacity of MMB with the rise of the solution pH and ionic strength between 3–10 and 0.001–0.1 [62]. The removal of two antibiotics levofloxacin (LEV) and tetracycline (TC) present in wastewater was studied with two distinct types of iron oxide-loaded biochar composite obtained from corn husk. The two different types of iron oxide-loaded biochar composite were synthesized by the combination of pyrolysis and impregnation methods. Depending on the method sequences used in the preparation, pyrolysis-impregnation (PI) or impregnation-pyrolysis (IP) resulted in the formation of two different types of iron oxide on the biochar composite. The XRD analysis revealed the presence of γ-Fe_2O_3 particles in the biochar composite developed by IP, whereas the biochar composite developed by PI showed a broad band corresponding to the presence of hydrous amorphous iron oxide. The biochar composite developed by PI showed a superior adsorption uptake for both the antibiotics TC and LEV than the biochar composite developed by IP; such superior adsorption behavior was attributed to the presence of more active OH groups in the biochar composite designed by PI method; however, such OH groups are more prone for interacting with LEV in preference to TC. According to the XPS and FTIR analysis, the adsorption of LEV onto the composite surface was mainly governed by hydrogen bonding, electrostatic forces, F replacement, and bridging bidentate complexation, while hydrogen bonding, electrostatic interaction, and carbonyl-Fe interaction mainly directed the mechanism of interaction between the adsorbent and the TC [63]. Heo and his co-workers studied the adsorption of endocrine-disrupting agent bisphenol A (BPA) and antibiotic sulfamethoxazole (SMX) by using a novel $CuZnFe_2O_4$-biochar composite obtained by hydrothermal process from the biochar obtained from bamboo feedstock and Fe_2O_3, ZnO, and CuO powders. The $CuZnFe_2O_4$-biochar composite presented superior textural properties, including high surface area (61.5 m^2/g), pore volume (0.157 cm^3/g), and pore diameter (10.2 nm) as compared to the simple biochar (S_{BET} = 24.56 m^2/g, pore volume = 0.048 cm^3/g and pore diameter = 7.74 nm). The $CuZnFe_2O_4$-biochar composite showed a relatively faster adsorption of both BPA and SMX by reaching the equilibrium in 30 minutes owing to the presence of sufficiently high number of active sites and hence presented more activity towards the two pollutants in comparison with the original biochar. The experimental adsorption results showed best fitting to the pseudo-second-order kinetics and Freundlich adsorption isotherm. The maximum adsorption efficiency for the adsorption of BPA by $CuZnFe_2O_4$ was 263.2 mg/g, which is much higher in terms of the adsorption capacity of the biochar alone (185.2 mg/g). Calculations of the thermodynamics parameters revealed that the adsorption of both BPA and SMX was spontaneous and endothermic. The mechanisms of interaction for holding the BPA and SMX molecules on the $CuZnFe_2O_4$-biochar composite include hydrogen bonding, π-π electron donor-acceptor mechanism, and hydrophobic forces [64].

7.4.2 Removal of Inorganic Pollutants

The rise in industrial growth and activities of human beings has considerably increased the concentration of toxic heavy metals in waste effluents. Discharge of such waste effluents contaminated with toxic heavy metals like Pb, Hg, As, Cd, etc., to our natural resources is considered as a serious threat for both human health and our environment. The carcinogenic and non-biodegradable nature of heavy metals can cause serious health problems in living beings. Many recent experiments have focused on the removal of heavy metal ions by implying the methods, including adsorption, advanced oxidation processes, membrane separation, and magnetic field application [65]. Yang and his co-workers designed three biochar composite, viz., the NaOH-biochar, MnO_x-biochar, and the FeO_x-biochar, for the removal of Cd(II) ions present in aqueous solutions. The MnO_x-biochar composite presented the maximum adsorption capacity (81.10 mg/g) and rate of adsorption (14.46 g/(mg h)) for Cd(II) removal as compared to the other two biochar composite and the pristine biochar. Impregnation of MnO_x particles on the surface of the biochar matrix improved the microporous structure by enhancing specific surface area, pore volume, and oxygen functionalities on the surface. Immobilization of MnO_x particles developed adsorption sites on the biochar surface, which were more salient than the increased surface area for the high removal of Cd(II) ions [66]. The removal of heavy metal (As) from aqueous solution was studied with Fe-Mn oxide-immobilized biochar composite developed by the impregnation of corn stem biochar with $FeSO_4.7H_2O$ and $KMnO_4$ solutions. The removal of As on the Fe-Mn oxide-biochar surface occurred through adsorption-oxidation process with the initial adsorption of a small amount of As(III) on the Fe and Mn oxide particles followed by oxidation of the remaining As(III) to As(V) with the composite showing more preferential adsorption to As(V). The Fe-Mn oxide-biochar composite showed high affinity for As removal by displaying a maximum adsorption capacity of 8.80 mg/g, which is much higher than the adsorption capacity displayed by the original biochar (2.89 mg/g). Moreover, the adsorption data for As removal showed best fitting to the Freundlich isotherm model [43]. A rice straw biochar coated with MnO_x was prepared from the aqueous-phase reaction between $KMnO_4$ and the rice straw biochar under a neutral environment. The scanning electron microscopy (SEM) and EDS analyses confirmed the presence of MnO_x on the biochar surface, while an increase in O-H functionalities on the biochar surface was confirmed by FTIR analysis. The MnO_x-layered biochar composite showed a pH-dependent removal of Pb(II) ions with an increase in Pb(II) sorption upon an increase in solution pH from 2.62 to 4.78. The adsorption data followed the pseudo-second-order kinetics and the Langmuir adsorption isotherm showing a maximum sorption capacity of 305.25 mmol/g. The MnO_x-coated biochar was very effective for Pb(II) removal, showing very less influenced adsorptions even in the presence of highly concentrated Ca(II) ions along with good reusability by maintaining the removal efficiency of Pb(II) ions above 90% even after four cycles [67].

As per the previous analysis that has been reported so far, unmodified biochar shows very poor removal of various inorganic ions such as phosphate, fluoride, ammonia, and many more owing to the presence of negative charge on the biochar surface. However, incorporation of metal oxide nanoparticles such as MgO, MnO_2, Fe_xO_y, etc., promoted the ability of the modified biochar to bind successfully to ligands by enhancing the positive charge for removing inorganic ions including phosphate, fluoride, ammonia, etc., from waste effluents [68]. CaO-biochar composite was prepared from egg shell and rice straw feedstocks by ball milling and thermal decomposition at 800°C. In this biochar preparation, the egg shell acted as the source of Ca, while the carbon source was the rice straw. XRD diffraction patterns corresponding to the CaO-biochar composite confirmed the presence of CaO and CaOH on the biochar surface, and the CaO-biochar composite showed improved adsorptive removal of phosphate ions in aqueous solutions than the raw biochar. The removal of phosphate ions by CaO-biochar composite could be best explained by the pseudo-second-order model, which indicated chemisorption to be the dominant mechanism for phosphate removal. Excellent phosphate adsorption capacity (231 mg/g) calculated from the Langmuir isotherm model was observed with the CaO-biochar composite, which could be attributable to the enhanced surface area and the presence of Ca nanoparticles on the biochar surface. The removal of the phosphate ions was also influenced by the pH of the solution and the temperature, the adsorption capacity was increased with the increase in solution pH from 3 to 11, and the similar trend was also followed corresponding to the rise of temperature from 298 to 318 K [69]. The removal of nitrate and fluoride ions was studied with α-Fe_2O_3- and Fe_3O_4-immobilized Douglas fir biochar composite; the biochar composite was obtained by the impregnation of the pristine biochar with $FeCl_3$ solution followed by pyrolysis at 600°C. As depicted by the SEM images, the α-Fe_2O_3/Fe_3O_4-biochar composite contained both octahedral- and semi-spherical-shaped Fe_3O_4 particles along with spindle-shaped α-Fe_2O_3 particles. The nitrate and fluoride removal capacities were maintained high over a wide pH range (pH 2–10), and the composite showed a faster removal rate of both the ions and Langmuir maximum adsorption capacity of 15.5 mg/g for nitrate and 9.04 mg/g for fluoride [70].

7.5 CATALYTIC REMOVAL OF EMERGING CONTAMINANTS

7.5.1 GENERAL MECHANISM OF PHOTOCATALYTIC DEGRADATION

7.5.1.1 Adsorption

Catalytic nanoparticles' incorporation on biochar improves the adsorption of organic contaminants and finally enhances the organic compounds' photodegradation than the degradation by nanoparticles alone. The extent of biochar present in photocatalysts performs a significant character for adsorption and photodegradation. The surface adsorption property deposits the organic contaminants on the surface of the catalyst rather than transferring them to the center of decomposition. Hence, the reactive oxygen species generated under light illumination does not migrate from the activation position. However, in some cases, a portion of the organic pollutants get absorbed without contact with the photocatalyst. Such particles are attacked by the reactive oxygen species. It can be assumed that diffusion of reactive oxygen species occurs from the activation site to the reaction site present on the composite surface. Hence, the photodegradation process is enhanced by the adsorption of organic contaminants of the composite surface [40].

7.5.1.2 Photodegradation

Absorption of photons with energy less than or equal to the energy difference of the band gap of the photocatalyst results in excitation of an electron to the conduction band from the valence band. This produces an electron-hole in the valence band upper edge, while the electron occupies the conduction band lower edge. However, diffusion or recombination of the h^+ and e^- might occur. These h^+s and e^-s generated may scavenge OH, O_2, and H_2O molecules to generate the reactive species $OH^{\cdot}$ and $O_2^{\cdot-}$ that can mineralize the organic pollutants [71,72].

7.5.1.3 Ozonization

Ozonation plays a crucial role by controlling the speed and reaction rate during the photodegradation by the biochar-supported photocatalysts. Hence, the combined effect of adsorptions, photodegradation, and ozonization improves the degradation of organic pollutants to a higher extent [54]. $^{\cdot}OH$ and H_2O_2 are generated upon exposure to the photon energy. So instead of recombination of the electron, it generates $O_2^{\cdot-}$ and $^{\cdot}OH$ upon reaction with H_2O_2 and oxygen [73]. Therefore, the reaction rate gets accelerated owing to the increase in the reactive oxygen species, thereby resulting in enhanced pollutants' reduction [74].

7.5.2 PHOTOCATALYTIC APPLICATIONS OF BIOCHAR

Besides the application of biochars as adsorbents for the removal of toxic chemicals from our natural resources, they also find their application as favorable supports/catalysts that could be used in several fields such as environment remediation, energy generation, agricultural aspects, and many more. The high specific surface area of biochar imparts a large number of active sites that could facilitate reactions; moreover, the presence of large number of functional groups on the biochar surface enhances the reactivity of biochar with a wide number of contaminants. These physicochemical properties of biochar make them very promising as a precursor for catalyst for various

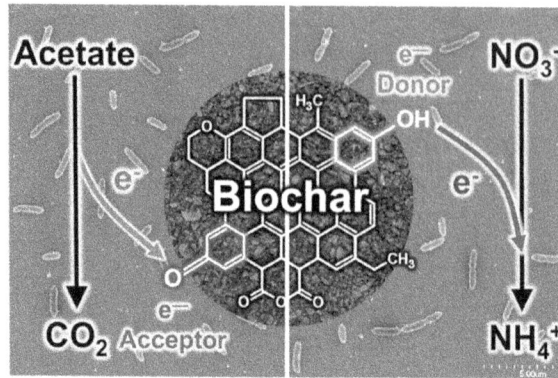

FIGURE 7.3 Biochar as electron reservoir and electron shuttle through its aromatic framework. (From Ref. [76]: J.M. Saquing, Y.H. Yu, and P.C. Chiu, Wood-derived black carbon (biochar) as a microbial electron donor and acceptor, Environ. Sci. Technol. Lett. 3 (2016), pp. 62–66).

environmental purposes [75]. Biochar acts as an appropriate support for a large number of nanoparticles; photocatalytic nanomaterials generally have low surface area. However, the formation of hybrids of these nanomaterials with biochar results in improving the surface area and the number of active sites. Dispersion of nanoparticles on the biochar surface improves the degradation of organic contaminants upon enhancing the number of active sites and the scattering of light. Biochar helps in electron transmission during photocatalysis; the aromatic framework of biochar causes electron delocalization from the donor site to the contaminants. Figure 7.3 shows the electron transmission in biochar [76]. This electron transport property of biochar reduces the recombination of electrons-holes, thus improving the photodegradation efficiency of a wide number of biochar-supported photocatalysts. Biochar also acts as electron reservoir and facilitates the separation between the electrons and the holes; most of the semiconductor photocatalysts get deactivated due to fast recombination between the charges. However, the combination of the photocatalyst with biochar increases the time required for recombination [40]. Zhang and his co-workers performed the degradation of Reactive Brilliant Blue KN-R with TiO_2/biochar composite photocatalyst. TEM and XRD analyses confirmed the formation of TiO_2 film belonging to the anatase phase with a particle size ranging between 15 and 20 nm. The TiO_2/biochar composite showed a sufficient decolorization of KN-R with highest decolorization of 99.71% and 96.99% taking place at high acidic and basic conditions (pH 1 and at pH 11) in 60 minutes, respectively. The decolorization of KN-R molecule was primarily dependent on the adsorptions by the biochar surface and the catalytic photodegradation by TiO_2. The degradation of the KN-R molecules occurred through the following steps: at first, the porous structure of the biochar and adsorption of dye molecules on its surface at the time of close vicinity of the KN-R molecules with the TiO_2 film enhances the interaction between them. Adsorption of photons by TiO_2 film with energy greater than or equal to its band gap generates holes (h^+) and electrons (e^-) whose recombination could be prevented by the biochar. Meanwhile, hydroxyl ($\cdot$OH) radicals and superoxide ion radicals are produced by the reaction between the holes (h^+) and the electrons (e^-) with water/OH^- ions and O_2 on the TiO_2 film surface. The two radicals thus produced cause the degradation of the KN-R dye molecule and its intermediates upon oxidation. The whole process is schematically represented in Figure 7.4 [77].

Ball milling of ZnO nanoparticles with bamboo stakes biochar for the formation of ZnO/biochar nanocomposites was found to be highly effective for the removal of MB from aqueous solution through the process of adsorption and photocatalytic degradation. The ZnO(25)/biochar (25 is the ZnO weight ratio) nanocomposite showed 95.19% removal of MB under the illumination of 250-W LED lamp with MB initial concentration of 160 mg/L. Adsorption dominated by electrostatic interactions played a key role for the removal of MB, while photocatalysis was also influencing the MB

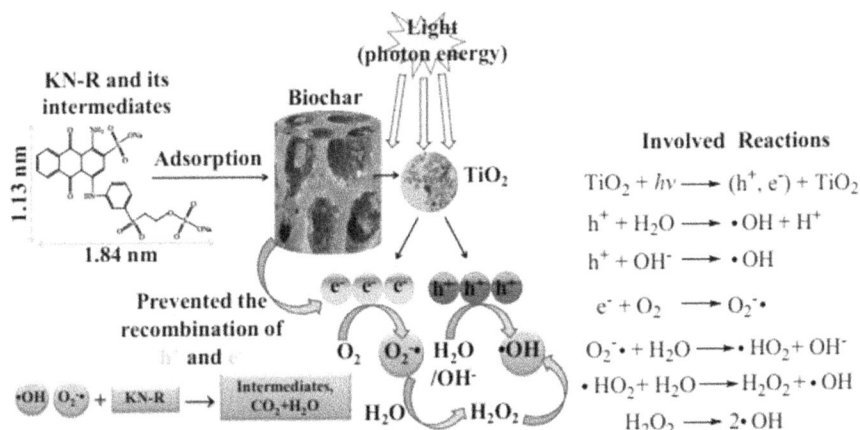

FIGURE 7.4 KN-R decolorization mechanism. (From Ref. [77]: S. Zhang, and X. Lu, Treatment of wastewater containing Reactive Brilliant Blue KN-R using TiO_2/BC composite as heterogeneous photocatalyst and adsorbent, *Chemosphere* 206 (2018), pp. 777–783.)

removal due to the degradation of the MB molecules, which were adsorbed or present in free state by the photogenerated free radical. Besides stabilizing the ZnO nanoparticles, the biochar plays an important role during photodegradation by improving the transport of photogenerated electrons [78]. Zhang and his co-workers used TiO_2-supported biochar for the degradation of antibiotic SMX; the TiO_2-supported biochar was obtained by the sol-gel method using acid-pretreated reed straw biochar and calcination at distinct temperatures mainly at 300°C, 400°C, and 500°C. SEM images described the vessel-shaped structure of the pre-acid-treated biochar (p-BC) with smooth walls, while the TiO_2-supported biochar (calcined at 300°C) reflected the same vessel structure but with rough surface sidewalls comprising TiO_2 nanoparticles attached to the external and internal surface of the tube. However, calcination at higher temperature (500°C) resulted in agglomeration of the TiO_2 nanoparticles with deformation of the original morphology of the biochar. The removal efficiency of the TiO_2-supported biochar was decreased upon increasing the calcination temperature; the TiO_2-supported biochar calcined at 300°C showed approximately 1.52 times higher removal efficiency than the TiO_2-supported biochar calcined at 500°C. For a solution of pH 4 with a catalyst dosage of 0.2 g under the presence of UV light, the TiO_2-supported biochar (300°C) showed a superior degradation of SMX (91.27%) than pure TiO_2 (58.47%). The removal efficiency of TiO_2-supported biochar (300°C) support was still maintained at 65.70% in the presence of other anions, including SO_4^{2-}, Cl^-, and NO_3^- in real water sample, although the SO_4^{2-}, Cl^-, and NO_3^- ions are believed to inhibit the SMX degradation on reaction with the hydroxyl radicals and the holes. Also, the photocatalyst showed an excellent regenerability by maintaining 86% SMX removal even after the fifth successive cycle. The combined aspects of adsorption and photodegradation made the TiO_2-supported biochar more effective towards the removal of SMX than the pure TiO_2 [79].

7.6 ENVIRONMENTAL ASPECTS OF METAL OXIDE/BIOCHAR COMPOSITE

Despite the use of biochar for removing the emerging contaminants from our environmental resources, we must also give emphasis to the potential risks of biochar towards our environment, and the application of biochar gets limited owing to the risk caused by the biochar during its use. Biochar contains contaminants such as polyaromatic hydrocarbons, heavy metals, VOCs (volatile organic compounds), and dioxins. Release and transport of these biochar contaminants is detrimental for living organisms, including both plants and microorganisms [80]. Thus, in order to avoid the negative risk of biochar during environmental applications, it is very essential to take necessary measurements during manufacturing of biochar in such a way that the product biochar contains very less compositions of these contaminants. Excessive application of biochar in soil can have an adverse effect for microorganism survival and influencing the lives of animals and plants. Therefore, it is very necessary to test the effects of biochar before its application in the environment to minimize its negative impacts [81]. Biochar application leads to the enhancement of soil quality; however, it reduces the potency of herbicides, thereby arising the need to use more herbicides – which in turn enhances the herbicides concentration in soil – thus causing more pollution [82].

Furthermore, we also could not ignore the potent risks of the modifying materials added to improve the applications of biochar. In addition to the several beneficial aspects depicted by the composites formed between biochar and nanoparticles, we also could not ignore the potent risk of metal oxide nanoparticles due to their toxicity towards the environment and organisms [83]. Some of the metal oxides used for the modification of biochars such as TiO_2, ZnO, and SiO_2 are considered to be toxic towards microorganisms [84]. Thus, the risk of leaching out of these nanoparticles after the application of biochar in water/soil is very much harmful to the environment and microorganisms. Improving the stability of the biochar nanocomposites and minimizing the toxic nanoparticle release are very much essential prior to the application of the biochar-based nanocomposites in order to improve the biochar application and minimize the unforeseen adverse effects towards environment.

7.7 CONCLUSION

This chapter presents the various synthetic approaches that could be adopted for the preparation of metal oxide-biochar composites. Combining the beneficial aspects of both the biochar and the nanoparticles, we can assume that metal oxide-biochar composites show tremendous applicability for the removal of both organic and inorganic wastes/contaminants present in our environmental resources. The biochar composites have emerged as a potent sorbent for the removal of contaminants due to their large surface area, pore volume, and the presence of large functionalities on its surface. However, the removal of the pollutants is influenced by different factors such as pH, temperature, and preparation technique, etc. π-π interaction, electrostatic attractions, pore filling, and ion exchange are some of the mechanisms responsible for the removal of hazardous pollutants by the biochar composites. The metal oxide-biochar composite could efficiently remove both the organic and inorganic contaminants by both adsorption and photocatalytic degradation. However, we also could not ignore the disadvantages exhibited by these composites due to their extensive use in our environment; therefore, it is essential for our researchers to conduct more experiments for the economic synthesis of biochar from appropriate feedstocks in a way to reduce the inherent contaminants in the biochar so that the biochar composites could be potentially applied for removing the unwanted hazardous compounds present in water/soil by eliminating their unforeseen negative impacts.

REFERENCES

1. K. Rajeshwar, M.E. Osugi, W. Chanmanee, C.R. Chenthamarakshan, M.V.B. Zanoni, P. Kajitvichyanukul, and R. Krishnan-Ayer, Heterogeneous photocatalytic treatment of organic dyes in air and aqueous media, *J. Photochem. Photobiol. C: Photochem. Rev.* 9 (2008), pp. 171–192.
2. B. Chen, Q. Ma, C. Tan, T.T. Lim, L. Huang, and H. Zhang, Carbon-based sorbents with three-dimensional architectures for water remediation, *Small* 11 (2015), pp. 3319–3336.
3. W. Ben, B. Zhu, X. Yuan, Y. Zhang, M. Yang, and Z. Qiang, Occurrence, removal and risk of organic micropollutants in wastewater treatment plants across China: comparison of wastewater treatment processes, *Water. Res.* 130 (2018), pp. 38–46.
4. L. Zhu, L. Meng, J. Shi, J. Li, X. Zhang, and M. Feng, Metal-organic frameworks/carbon-based materials for environmental remediation: a state-of-the-art mini-review, *J. Environ. Manage.* 232 (2019), pp. 964–977.

5. M. Houde, D.C.G. Muir, K.A. Kidd, S. Guildford, K. Drouillard, M.S. Evans, X. Wang, D.M. Whittle, D. Haffner and H. Kling, Influence of lake characteristics on the biomagnification of persistent organic pollutants in lake trout food webs, *Environ. Toxicol. Chem.* 27 (2008), pp. 2169–2178.

6. J. Lehmann, A handful of carbon, *Nature* 447 (2007), pp. 143–144.

7. S. Tripathi, S.K. Sonkar and S. Sarkar, Growth stimulation of gram (*Cicer arietinum*) plant by water soluble carbon nanotubes. *Nanoscale* 3 (2011), pp. 1176–1181.

8. M.V. Khodakovskaya, B.S. Kim, J.N. Kim, M. Alimohammadi, E. Dervishi, T. Mustafa and C.E. Cerniglia, Carbon nanotubes as plant growth regulators: effects on tomato growth, reproductive system, and soil microbial community. *Small* 9 (2013), pp. 115–123.

9. J. Lehmann, J. Gaunt, and M. Rondon, Biochar sequestration in terrestrial ecosystems-a review, *Mitig. Adapt. Strat. Glob Change* 11 (2006), pp. 403–427.

10. N.L. Panwar, A. Pawar, and B.L. Salvi, Comprehensive review on production and utilization of biochar, *SN Appl. Sci.* 1 (2019), p. 168.

11. Y. Dai, N. Zhang, C. Xing, Q. Cui and Q. Sun, The adsorption, regeneration and engineering applications of biochar for removal organic pollutants: a review, *Chemosphere* 223 (2019), pp. 12–27.

12. X. Zhang, W. Fu, Y. Yin, Z. Chen, R. Qiu, M.-O. Simonnot and X. Wang, Adsorption reduction removal of Cr(VI) by tobacco petiole pyrolytic biochar: batch experiment, kinetic and mechanism studies, *Bioresour. Technol.* 268 (2018), pp. 149–157.

13. Y. Zhang, B. Cao, L. Zhao, L. Sun, Y. Gao, J. Li and Fan Yang, Biochar-supported reduced graphene oxide composite for adsorption and coadsorption of atrazine and lead ions, *Appl. Surf. Sci.* 427 (2018), pp. 147–155.

14. K. Henryk, C. Jarosław and Ż. Witold, Peat and coconut fiber as biofilters for chromium adsorption from contaminated wastewaters, *Environ. Sci. Pollut. Res. Int.* 23 (2016), pp. 527–534.

15. K. Vikrant, K.H. Kim, Y.S. Ok, D.C.W. Tsang, Y.F. Tsang, B.S. Giri, R.S. Singh, Engineered/designer biochar for the removal of phosphate in water and wastewater, *Sci. Total Environ.* 616–617 (2018), pp. 1242–1260.

16. D. Matovic, Biochar as a viable carbon sequestration option: global and Canadian perspective, *Energy* 36 (2011), pp. 2011–2016.

17. J. Lee, K.H. Kim, and E.E. Kwon, Biochar as a catalyst, *Renew. Sust. Energ. Rev.* 77 (2017), pp. 70–79.

18. W.T. Tsai, S.C. Liu, H.R. Chen, Y.M. Chang and Y.L. Tsai, Textural and chemical properties of swine-manure-derived biochar pertinent to its potential use as a soil amendment, *Chemosphere* 89 (2012), pp. 198–203.

19. J. Liu, Y. Ding, L. Ma, G. Gao, and Y. Wang, Combination of biochar and immobilized bacteria in cypermethrin-contaminated soil remediation, *Int. Biodeterior. Biodegr.* 120 (2017), pp. 15–20.

20. H.Y. Fong, C.P. Te, S.C. Hao, L.S. Lien, S.L. Ping, Z. Yuan and Q.C. Sheng, Microwave pyrolysis of rice straw to produce biochar as an adsorbent for CO_2 capture, *Energy* 84 (2015), pp. 75–82.

21. B. Wang, Z. Wang, Y. Jiang, G. Tan, N. Xu, and Y. Xu, Enhanced power generation and wastewater treatment in sustainable biochar electrodes based bioelectrochemical system, *Bioresour. Technol.* 241 (2017), pp. 841–848.

22. G. Lian, B. Wang, X. Lee, L. Li, T. Liu, and W. Lyu, Enhanced removal of hexavalent chromium by engineered biochar composite fabricated from phosphogypsum and distillers grains, *Sci. Total Environ.* 697 (2019), p. 134119.

23. L. Zhang, S. Tang, F. He, Y. Liu, W. Mao, and Y. Guan, Highly efficient and selective capture of heavy metals by poly(acrylic acid) grafted chitosan and biochar composite for wastewater treatment, *Chem. Eng. J.* 378 (2019), p. 122215.

24. M.B. Ahmed, J.L. Zhou, H.H. Ngo, W. Guo, and M. Chen, Progress in the preparation and application of modified biochar for improved contaminant removal from water and wastewater, *Bioresour Technol.* 214 (2016), pp. 836–851

25. T. Sizmur, T. Fresno, G. Akgül, H. Frost and E.M. Jiménez, Biochar modification to enhance sorption of inorganics from water, *Bioresour. Technol.* 246 (2017), pp. 34–47

26. F.R. Oliveira, A.K. Patel, D.P. Jaisi, S. Adhikari, H. Lu, and S.K. Khanal, Environmental application of biochar: current status and perspectives, *Bioresour. Technol.* 246 (2017), pp. 110–122.

27. C.Y. Ning, L.W. Yu, L.Y. Ping, W.Y. Xin, C.Y. Rong, X. Wei, Z. Li, Z.J. Chao, L. Hui, Modification, application and reaction mechanisms of nano-sized iron sulfide particles for pollutant removal from soil and water: a review, *Chem. Eng. J.* 362 (2019), pp. 144–159.

28. C. Lei, Y. Sun, D.C.W. Tsang and D. Lin, Environmental transformations and ecological effects of iron-based nanoparticles, *Environ Pollut.* 232 (2018), pp. 10–30.

29. Y.H. Teow and A.W. Mohammad, New generation nanomaterials for water desalination: a review, *Desalination* 451 (2019), pp. 2–17.

30. Y. Zou, X. Wang, A. Khan, P. Wang, Y. Liu, A. Alsaedi, T. Hayat, and X. Wang, Environmental remediation and application of nanoscale zero-valent iron and its composites for the removal of heavy metal ions: a review, *Environ. Sci. Technol.* 50 (2016), pp. 7290–7304.

31. H. Lu, H. Dong, W. Fan, J. Zuo and X. Li, Aging and behavior of functional TiO_2 nanoparticles in aqueous environment, *J. Hazard. Mater.* 325 (2017), pp. 113–119.

32. O.M.R. -Narvaez, J.M.P. Hernandez, A. Goonetilleke and E.R. Bandala, Biochar-supported nanomaterials for environmental applications, *J. Ind. Eng. Chem.* 78 (2019), pp. 21–33.

33. X.-F. Tan, Y.-G. Liu, Y.-L. Gu, Y. Xu, G.-M. Zeng, X.-J. Hu, S.-B. Liu, X. Wang, S.-M. Liu, J. Li, Biochar-based nano-composites for the decontamination of wastewater: a review, *Bioresour. Technol.* 212 (2016), pp. 318–333.

34. K.S.D. Premarathna, A.U. Rajapaksha, B. Sarkar, E.E. Kwon, A. Bhatnagar, Y.S. Oke, M. Vithanage, Biochar-based engineered composites for sorptive decontamination of water: a review, *Chem. Eng. J.* 372 (2019), pp. 536–550.

35. K.R. Thines, E.C. Abdullah, N.M. Mubarak and M. Ruthiraan, Synthesis of magnetic biochar from agricultural waste biomass to enhancing route for wastewater and polymer application: a review, *Renew. Sust. Energ. Rev.* 67 (2017), pp. 257–76.

36. R. Li, J.J. Wang, L.A. Gaston, B. Zhou, M. Li, R. Xiao, Q. Wang, Z. Zhang, H. Huang, W. Liang, H. Huang, X. Zhang, An overview of carbothermal synthesis of metal–biochar composites for the removal of oxyanion contaminants from aqueous solution, *Carbon* 129 (2018), pp. 674–687.

37. S.H. Ho, S. Zhu and J.S. Chang, Recent advances in nanoscale-metal assisted biochar derived from waste biomass used for heavy metals removal, *Bioresour. Technol.* 246 (2017), pp. 123–134.

38. S. Wang, M. Zhao, M. Zhou, Y.C. Li, J. Wang, B. Gao, S. Sato, K. Feng, W. Yin, A. D. Igalavithana, P. Oleszczuk, X. Wang, Y.S. Ok, Biochar-supported nZVI (nZVI/BC) for contaminant removal from soil and water: a critical review, *J. Hazard. Mater.* 373 (2019), pp. 820–834.

39. C. Zhao, B. Wang, B.K.G. Theng, P. Wu, F. Liu, S. Wang, X. Lee, M. Chen, L. Li, X. Zhang, Formation and mechanisms of nano-metal oxide-biochar composites for pollutants removal: a review, *Sci. Total Environ.* 767 (2021), p. 145305.

40. M. Ahmaruzzaman, Biochar based nanocomposites for photocatalytic degradation of emerging organic pollutants from water and wastewater, *Mater. Res. Bull.* 140 (2021), p. 111262.

41. B. Wang, B. Gao and J. Fang, Recent advances in engineered biochar productions and applications. *Crit. Rev. Environ. Sci. Technol.* 47 (2017), pp. 2158–2207.

42. X. Zhang, P. Sun, K. Wei, X. Huang and X. Zhang, Enhanced H_2O_2 activation and sulfamethoxazole degradation by Fe-impregnated biochar, *Chem. Eng. J.* 385 (2020), p. 123921.

43. L. Lin, Z. Song, Y. Huang, Z.H. Khan, and W. Qiu, Removal and oxidation of arsenic from aqueous solution by biochar impregnated with Fe-Mn oxides, *Water Air Soil Pollut.* 230 (2019), p. 105.

44. R. Xiao, J.J. Wang, R. Li, J. Park, Y. Meng, B. Zhou, S. Pensky, Z. Zhang, Enhanced sorption of hexavalent chromium [Cr(VI)] from aqueous solutions by diluted sulfuric acid assisted MgO-coated biochar composite, *Chemosphere* 208 (2018), pp. 408–416.

45. M.C. Wang, G.D. Sheng, and Y.P. Qiu, A novel manganese-oxide/biochar composite for efficient removal of lead(II) from aqueous solutions, *Int. J. Environ. Sci. Technol.* 12 (2015), pp. 1719–1726.

46. A.-X. Li, Y.-L. Wang, X. Xiong, H.-F. Liu, N.-N Wu, R. Liu, Microstructure and synthesis mechanism of dysprosia-stabilized zirconia nanocrystals via chemical coprecipitation, *Ceram.Int.* 46 (2020), pp. 13331–13341.

47. E. Gharibshahian, The effect of polyvinyl alcohol concentration on the growth kinetics of $KTiOPO_4$ nanoparticles synthesized by the co-precipitation method, *High Tech Innov. J.* 1 (2020), pp. 187–193.

48. H. Luo, Q. Lin, X. Zhang, Z. Huang, H. Fu, R. Xiao, S.-S. Liu, Determining the key factors of nonradical pathway in activation of persulfate by metal-biochar nanocomposites for bisphenol A degradation, *Chem. Eng. J.* 391 (2020), p. 123555.

49. M. Gao, Y. Zhang, X. Gong, Z. Song and Z. Guo, Removal mechanism of di-n-butyl phthalate and oxytetracycline from aqueous solutions by nano-manganese dioxide modified biochar, *Environ Sci Pollut Res Int.* 25 (2018), pp. 7796–7807.

50. R. Saravanakumar, K. Muthukumaran and N. Selvaraju, Enhanced Pb(II) ions removal by using magnetic NiO/biochar composite, *Mater. Res. Express* 6 (2019), p. 105504.

51. Y. Yao, B. Gao, J. Chen, M. Zhang, M. Inyang, Y. Li, A. Alva, and L. Yang, Engineered carbon (biochar) prepared by direct pyrolysis of Mg-accumulated tomato tissues: characterization and phosphate removal potential, *Bioresour. Technol.* 138(2013), pp. 8–13.

52. C. Li, L. Zhang, Y. Gao, and A. Li, Facile synthesis of nano ZnO/ZnS modified biochar by directly pyrolyzing of zinc contaminated corn stover for Pb(II), Cu(II) and Cr(VI) removals, *Waste Manage.* 79(2018), pp. 625–637.

53. Y. Zheng, Y. Wan, J. Chen, H. Chen, and B. Gao, MgO modified biochar produced through ball milling: a dual-functional adsorbent for removal of different contaminants, *Chemosphere*. 243(2020), p. 125344.

54. A. Khataee, B. Kayan, P. Gholami, D. Kalderis, S. Akay, and L. Dinpazhoh, Sonocatalytic degradation of reactive yellow 39 using synthesized ZrO_2 nanoparticles on biochar, *Ultrason. Sonochem*. 39(2017), pp. 540–549.

55. P. Lisowski, J.C. Colmenares, O. Maˇseket, W. Lisowski, D. Lisovytskiy, A. Kaminska, and D. Lomot, Dual Functionality of TiO_2/biochar hybrid materials: photocatalytic phenol degradation in liquid phase and selective oxidation of methanol in gas phase, *ACS Sustainable Chem. Eng*. 5(2017), pp. 6274–6287.

56. P. Sun, C. Hui, R.A. Khan, J. Du, Q. Zhang, and Y.H. Zhao, Efficient removal of crystal violet using Fe_3O_4-coated biochar: the role of the Fe_3O_4 nanoparticles and modeling study their adsorption behaviour, *Sci. Rep*. 5(2015), p. 12638.

57. M.J. Uddin, R.E. Ampiaw, and W. Lee, Adsorptive removal of dyes from wastewater using a metal-organic framework: a review, *Chemosphere*. 284(2021), p. 131314.

58. N. Chaukura, E.C. Murimba, and W. Gwenzi, Synthesis, characterisation and methyl orange adsorption capacity of ferric oxide–biochar nano-composites derived from pulp and paper sludge, *Appl. Water Sci*. 7(2017), pp. 2175–2186.

59. P. Zhang, D. O'Connor, Y. Wang, L. Jiang, T. Xia, L. Wang, D.S.W. Tsang, Y.S. Ok, and D. Hou, A green biochar/iron oxide composite for methylene blue removal. *J. Hazard. Mater*. 384(2020), p. 121286.

60. X.Wei, X. Wang, B. Gao, W. Zou, and L. Dong, Facile ball-milling synthesis of CuO/biochar nanocomposites for efficient removal of reactive red 120. *ACS Omega*. 5(2020), pp. 5748–5755.

61. M.N. Alnajrani, and O.A. Alsager, Removal of antibiotics from water by polymer of intrinsic microporosity: isotherm, kinetics, thermodynamics, and adsorption mechanism, *Sci. Rep*. 10(2020), p. 794.

62. R. Li, Z. Wang, X. Zhao, X. Li, and X. Xie, Magnetic biochar-based manganese oxide composite for enhanced fluoroquinolone antibiotic removal from water, *Environ. Sci. Pollut. Res*. 25(2018), pp. 31136–31148.

63. Y. Chen, J. Shi, Q. Du, H. Zhang, and Y. Cui, Antibiotic removal by agricultural waste biochars with different forms of iron oxide, *RSC Adv*. 9(2019), pp. 14143–14153.

64. J. Heo, Y. Yoon, G. Lee, Y. Kim, J. Han, and C.M. Park, Enhanced adsorption of bisphenol A and sulfamethoxazole by a novel magnetic $CuZnFe_2O_4$–biochar composite, *Bioresour. Technol*. 281(2019), pp. 179–187.

65. N.A.A. Qasem, R.H. Mohammed, and D.U. Lawal, Removal of heavy metal ions from wastewater: a comprehensive and critical review, *NPJ Clean Water*. 4(2021), p. 36.

66. B. Li, L. Yang, C.Q. Wang, Q.P Zhang, Q.C. Liu, Y.D. Li, and R. Xiao, Adsorption of Cd(II) from aqueous solutions by rape straw biochar derived from different modification processes, *Chemosphere*. 175(2017), pp. 332–340.

67. G. Tan, Y. Wu, Y. Liu, and D. Xiao, Removal of Pb (II) ions from aqueous solution by manganese oxide coated rice straw biochar A low-cost and highly effective sorbent, *J. Taiwan Inst. Chem. Eng*. 84(2018), pp. 85–92.

68. J. Liu, J. Jiang, Y. Meng, A. Aihemaiti, Y. Xu, H. Xiang, Y. Gao, and X. Chen, Preparation, environmental application and prospect of biochar-supported metal nanoparticles: a review, *J. Hazard. Mater*. 388(2020), p. 122026.

69. X. Liu, F. Shen, and X. Qi, Adsorption recovery of phosphate from aqueous solution by CaO-biochar composites prepared from eggshell and rice straw. *Sci. Total Environ*. 666(2019), pp. 694–702.

70. N.B. Dewage, A.S. Liyanage, C.U. Pittman Jr, D. Mohan, and T. Mlsna, Fast nitrate and fluoride adsorption and magnetic separation from water on α-Fe_2O_3 and Fe_3O_4 dispersed on Douglas fir biochar, *Bioresour. Technol*. 263(2018), pp. 258–265.

71. A. Molla, M. Sahu, and S. Hussain, Under dark and visible light: fast degradation of methylene blue in the presence of Ag-In-Ni-S nanocomposites, *J. Mater. Chem. A*. 3(2015), pp. 15616–15625.

72. Q. Lu, Y. Zhang, and S. Liu, Graphene quantum dots enhanced photocatalytic activity of zinc porphyrin toward the degradation of methylene blue under visible-light irradiation, *J. Mater. Chem. A*. 3(2015), pp. 8552–8558.

73. X. Yang, W. Chen, J. Huang, Y. Zhou, Y. Zhu, and C. Li, Rapid degradation of methylene blue in a novel heterogeneous Fe_3O_4@rGO@TiO_2-catalysed photo-Fenton system, *Sci. Rep*. 5(2015), p. 10632.

74. J.C. Colmenares, R.S. Varma, and P. Lisowski, Sustainable hybrid photocatalysts: titania immobilized on carbon materials derived from renewable and biodegradable resources, *Green Chem*. 18(2016), pp. 5736–5750.

75. Z. Zhang, Z. Zhu, B. Shen, and L. Liu, Insights into biochar and hydrochar production and applications: a review, *Energy*. 171(2019), pp. 581–598.

76. J.M. Saquing, Y.H. Yu, and P.C. Chiu, Wood-derived black carbon (biochar) as a microbial electron donar and acceptor, *Environ. Sci. Technol. Lett.* 3(2016), pp. 62–66.

77. S. Zhang, and X. Lu, Treatment of wastewater containing Reactive Brilliant Blue KN-R using TiO_2/BC composite as heterogeneous photocatalyst and adsorbent, *Chemosphere.* 206(2018), pp. 777–783.

78. F. Yu, F. Tian, H. Zou, Z. Ye, C. Peng, J. Huang, Y. Zheng, Y. Zhang, Y. Yang, X. Wei, and B. Gao, ZnO/biochar nanocomposites via solvent free ball milling for enhanced adsorption and photocatalytic degradation of methylene blue, *J. Hazard. Mater.* 415(2021), p. 125511.

79. H. Zhang, Z. Wang, R. Li, J. Guo, Y. Li, J. Zhu, and X. Xie, TiO_2 supported on reed straw biochar as an adsorptive and photocatalytic composite for the efficient degradation of sulfamethoxazole in aqueous matrices, *Chemosphere.* 185(2017), pp. 351–360.

80. H. Zheng, C. Zhang, B. Liu, G. Liu, M. Zhao, G. Xu, X. Luo, F. Li and B. Xing, Biochar for water and soil remediation: production, characterization, and application, in *A New Paradigm for Environmental Chemistry and Toxicology*, G. Jiang and X. Li, eds., Springer, Singapore, 2020, pp. 153–196.

81. S.M. Ndirangu, Y. Liu, K. Xu, and S. Song. Risk evaluation of pyrolyzed biochar from multiple wastes, *J. Chem.* 2019(2019), pp. 1–28.

82. X. Yang, S. Zhang, M. Ju, and L. Liu, Preparation and modification of biochar materials and their application in soil remediation, *Appl. Sci.* 9(2019), p. 1365.

83. W. Jiang, H. Mashayekhi, and B. Xing, Bacterial toxicity comparison between nano-and micro-scaled oxide particles, *Environ. Pollut.* 157(2009), pp. 1619–1625.

84. L.K. Adams, D.Y. Lyon, and P.J.J. Alvarez, Comparative eco-toxicity of nanoscale TiO_2, SiO_2, and ZnO water suspensions, *Water Res.* 40(2006), pp. 3527–3532.

8 Metal Organic Framework (MOF)-Based Advanced Materials for Clean Environment

Hossam E. Emam
National Research Centre, Textile Research and Technology Institute

Reda M. Abdelhameed
National Research Centre, Chemical Industries Research Institute

Hanan B. Ahmed
Helwan University

CONTENTS

8.1 INTRODUCTION

Numerous studies were extensively made on porous materials owing to their various uses, such as storage, drug delivery templates, magnetism, separation, and sensing, in addition to catalysis. Owing to their cost-effectiveness and ease of synthesis, the amorphous porous solidified materials (like plastic materials and gels) are highly applicable, in spite of their construction of randomly ordered repeated units [1]. Their main disadvantages are the variations in their molecular topography with non-expectable channels in addition to the lack of regularity in ordering of the repeating units, mainly resulting in lowering the mechanical stabilization. However, the porous crystalline solid materials are characterized with highly applicability owing to their ordered composition with reproducible topography, to give these type of structures the high mechanical and thermal stability [2]. Zeolites and nanoporous silica are special examples of ordered porous solid materials that have expectable characteristics with highly accessible porous structure, dimensionality, and topology [3].

Based on the porosity, the porous solid materials are categorized into three main classes: macroporous (>500 Å), mesoporous (20–500 Å), and microporous (5–20 Å) [4]. It was reported that either natural or synthetic inorganic microporous zeolites are composed of aluminum-based silicates and

DOI: 10.1201/9781003206385-8

are applied in various purposes. However, the pristine zeolites have higher applicability, although they are somehow limited owing to their hardness, thus having an impact on the removal of the impurities. The synthetic zeolites were reported to be applicable in separation, adsorption, and catalysis, because they could be tailored under controllable conditions to be suitable for the market requirements. It was shown that the zeolite $4\,\text{Å}$ ($Na_{12}Al_{12}Si_{12}O_{48}$) is commonly prepared as a drying reagent and zeolite ZSM-5 functionalized as a cracking reagent. Additionally, for the removal of petrol from water, silicalite-1 was applied, which act as a porous sorbent [5]. Although zeolites are used for various purposes, they also have some disadvantages, the difficulty in adjusting the preparation conditions and having very limited number of geometrical features and structural compositions.

In recent years, a novel category of porous solid materials that acted as metal organic framework (MOF) were investigated, whereas they showed high applicability for adsorption and separation [6–8]. These porous solid materials showed special and superior characteristics such as the largest pore size and the widest surface area compared to the others [9]. MOFs are considered as the organic analogues of inorganic zeolites, in which the oxygen atoms are replaced with stiff organic linkers, which act as bridges for cations to be especially composed of one-, two-, and three-dimensional (3D) constructions [10]. MOF is reported as a type o coordination polymer, and it was investigated in the 1950s [11]; however, the main consideration with such reagents was observed later in the 1990s by Omar Yaghi et al., as they studied the compatibility of multi-dentate carboxylate ligands to replace 4,4′-bipyridine moiety to be known as a neutral framework or MOFs [12].

Such newly discovered materials were known with different properties, such as organic zeolite analogues, MOFs, hybrid organic-inorganic materials, coordination polymers, and coordination networks, since the middle of the 1990s [13]. Yaghi described the uses of MOF for this novel category as hybrids with various structures, geometries, and characteristics. Eventually, in accordance with IUPAC system, MOF could be used as a coordination polymer that is continuously expanded in one, two, or three dimensions via coordination bonding, i.e., highly accessible coordination networks or coordinating polymers with opened framework constructs with accessible voids [14]. Coordination entities are neutral molecules or ions that are consisted of cation in center, mainly a metal ion, which can connect with the surrounding cross-linkers. MOFs act as linker of both inorganic and organic units in order to obtain the as-required porous composition.

In the last decade, various types of MOFs were investigated, owing to various advantageous like their wide surface area attributing to their ultra-micro-porosities, and the flexibilities that could be varied to obtain different structures with diversities in their pore size, topography, and optical characteristics [15–17]. These advantageous characteristics are responsible for superior applicability. Mostly, these functionalities were correlated with their porosity (up to $7,000\,\text{m}^2/\text{g}$) , such as their exploitation in storing, catalysis, and separation [18]. The most applicable types that were reported in literature in 1999 were Cu-BTC and MOF- [19,20]. Then, the extensively stable MIL-101 was discovered as a zeolite imidazole framework in 2002 and contains imidazolate material [19]. Recently, the researchers were extensively considered with the modification of different types of MOFs for various applications with high stability, like MIL-100, ZIFs, UiO-66, and MIL-125.

Fuel combustion is known to be harmfully affecting the surrounding medium via the elimination of S-, N-, and O-based compounds as the hazardous compounds. For example, the evolved CO_2, NO_2, and SO_2 within the fuel combustion were recorded to be 164,000, 448, and 1,122 pounds per billion British thermal unit (BTU), respectively [21]. The environmental and economic impacts correlated with the energy cost effects, the storage of hydrogen, oxygen, nitrogen, and methane gases as fuel, in addition to the filtration fuel, were highly considered as attractive niche research areas of interest. Thus, this chapter is focused on concluding the common synthetic techniques of MOFs, their uses, and their composites; and the recent reports in literature describing their adsorption potentiality for gas capturing, gas storing, and ultra-filtration fuel. These techniques could clarify the superiority of exploiting MOFs in getting a clean environment. Removing or separating some of the fuel pollutants via hydrogen bonds for inactive materials could be performed, so MOFs with their extensively high applicability could eliminate the fuel contaminants through the hydrogen bonding for the improvement of the adsorption capacities [22].

8.2 SYNTHETIC APPROACHES

There are different synthetic approaches that were successfully investigated for the synthesis of different types of MOFs, such as low evaporation, solvo-thermal, hydrothermal, microwave-assisted, electrochemical, mechano-chemical, and sono-chemical synthetic approaches (Figure 8.1) [19]. Slow evaporation is ascribed as a conventional methodology and is not required any of external source of energy, making this technique one of the favorable synthetic techniques. However, extending reaction time of this type of synthetic approach compared to the others is considered one of its main drawbacks. At ambient condition, a solution of the synthetic equivalents is concentrated by employing the low evaporation method on the reaction liquor [19].

Solvo-thermal technique is the highly applied methodology for the preparation of MOFs, attributing to its simplicity and controllability. However, there are some disadvantageous as it requires long duration and the prepared MOFs were shown to have large pore sizes. In such methodology, MOFs were prepared under pressure, and above the boiling points of the applied solvents, the synthetic equivalents were shown with some unexpected alterations in their chemical composition. Therefore, high boiling point organic solvents were preferably used in such type of synthetic methodologies, like CH_3OH, $(CH_3)_2CO$, $(C_2H_5)_2CONH_2$, acetonitrile, $(CH_3)_2CONH_2$, and C_2H_5OH. In the hydrothermal technique, H_2O is exploited as the solvent and proceeded under a high pressure in autoclaves. One of the characteristic properties of hydrothermal technique is the generation of highly stabilized MOFs with an efficient crystal growth. The disadvantages of the hydrothermal technique are the expensiveness and its difficulty in the observation of the crystal growth [19].

In microwave-assisted synthetic technique, the characteristics of MOFs prepared are similar to those synthesized via the solvo-thermal techniques, whereas the processing is observably faster. This class of synthetic techniques could be described as one of the energy saving methodologies. This methodology has advantages, such as small particle sizes, high efficient, regular morphological characteristics, and phase selectivity [23]. Electrochemical synthetic methodology does not require the application of metal salts. However, it gives a continuous growth of crystals for MOFs, which makes this process industrially applicable. Electrochemical methodology relies on providing the

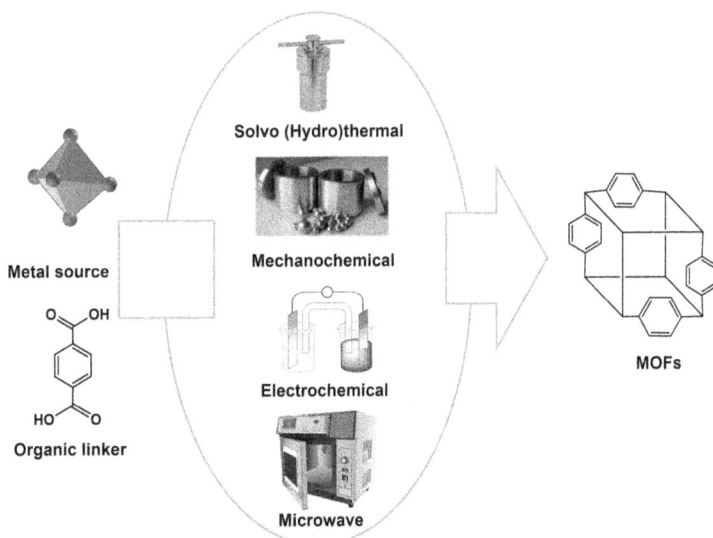

FIGURE 8.1 Polyhedron structures for different MOFs.

metal ions by anodic dissolving process in synthetic liquor that contains organic linking reagents and electrolytes [23].

Mechano-chemical synthetic technique is ascribed as solventless methodology for the production of MOFs, as it can be exploitable for the accelerated production of MOFs via the liquid-assisted grinding, where few amounts of solvent are required for the solid-liquid interaction. It was reported that by altering the type of solvent, one-dimensional (1D), two-dimensional (2D), or three-dimensional (3D) coordination polymeric matrices could be prepared from the same reaction liquors [24]. This technique could be applied for the preparation of some zeolitic imidazolate frameworks (ZIFs). In sono-chemical synthetic methodology, the ultrasound is exploited for the promotion of the synthetic reaction. Sono-chemical synthetic technique of MOFs is exhibited with many beneficial characteristics; the used equipment is lower in cost than microwave units and the consumption of energy is low since no motivation time is required before the stabilization of sonicator, and can also be described as timeless technique.

In case of functional MOFs, the applied cross-linking agent might be exploited as starting equivalent or MOFs could be primarily treated with the functionalized reacting reagents. MIL-101 (Cr)-SO_3H, UiO-66 (Zr)-COOH, MIL-101 (Fe)-SO_3H, UiO-66 (Zr)-NH_2, UiO-66 (Zr)-NH_3Cl, UiO-66 (Zr)-COOH, and MIL-125 (Ti)-NH were all considered as functional MOFs that were produced using various types of functionalized ligands [25–29]. MIL-101 (Cr) and MIL-101 (Cr)-SO_3H acquired their functions by post-treatment with ethanolamine (EA), ethylene diamine, and tris(2-aminoethyl)-amine (TAEA), receptively, to prepare MIL-101 (Cr)-EA, MIL-101 (Cr)-ED, and MIL-101 (Cr)-SO_3H-TAEA [25,27,30].

For MOF-based composite, two main categories were investigated in accordance with their composition: MOF modified with central atoms like metal salts or nanoparticles and MOF modified with polymers. For the first category, the central atoms were supposed to be enclosed in three ways: by impregnation within the pores of MOF, or via physical uploading on MOF surface, or through the coating of MOF with particles in such structure of core-shell [27,31–38]. Ingrowth of MOF for the preparation of MOF@MOF as a core-shell structure was recently performed [31–34]. Abdelhameed et al. and Emam et al., uploaded the nanoparticles of Ag_3PO_4, Ag_3VO_4, and Ag_2WO_4 on MIL-125 (Ti)-NH_2 surface to prepare a photocatalytically active composite [35,36]. Ag_3PO_4/MIL-101 (Cr)/$NiFe_2O_4$ (APO/MOF/NFO) composite was in situ synthesized via uploading hypothesis [39]. The hydrogel of Fe_3O_4@u-BTC that is magnetically responsive composite was obtained via the dry gel conversion synthesis technique [40]. BiOBr@UiO-66 (Zr)-NH_2 composite was prepared via co-precipitation methodology [41]. Graphite oxide (GO) and phosphor-tungstic acid (PWA) were uploaded onto Cu-BTC and/or MIL-101 (Cr) to prepare higher surface area/active composite [GO@Cu-BTC, GO/MIL-101, PWA@Cu-BTC, PWA@MIL-101 (Cr)] [42–45].

Moreover, immediate and ex situ techniques are two basic preparative approaches for the synthesis of MOF modified with polymeric structures. For the synthesis of a composite with higher MOF content, different MOFs [Cu-BTC, Ln (Eu^{3+}, Tb^{3+}), MIL-68 (In)-NH_2, MIL-125 (Ti)-NH_2, MIL-53-NH_2 UiO-66 (Zr)-$(COOH)_2$, ZIF-8 (Zn), ZIF-67 (Co)] were ingrained within the yarn (cotton, wool, viscose, linen, silk, Nylon, PET) for the preparation of MOF@yarn composite by in situ/hydrothermal and in situ/infrared techniques [46–53]. The obtained composites through the in situ techniques were shown to be characterized with significantly higher applicabilities compared to those obtained from the ex situ methodology. Polysulfone@Cu-BTC composites were prepared via the immediate methodology via interaction of the metal salt with organic ligand in DMF with polysulfone [54]. Melamine@UiO-66 (Zr)-COOH and Cu-BTC@cotton were ex situ prepared via modifying MOF within melamine or fabric matrix, respectively [27,55].

8.3 CAPTURING OF TOXIC GASES

In accordance with the recent approaches, MOF was demonstrated to have a characteristic degree of stabilization when become exposed to chemicals or heat; therefore, they can be ascribed as compatible reagents for clean environment attributing to their characterization with structural variations. Recent approaches investigated the superiority of MOF with high absorptivity of toxic gases

(Table 8.1), such as CO_2, carbon monoxide (CO), NO_2, and hydrogen disulfide (H_2S) [56,57], as represented in Figure 8.2.

CO_2 sequestration is extensively required for the reduction of the anthropogenic nature impacting the greenhouse effects. CO_2 separation after combusting from the plant flue gas is important for mitigation of the phenomenon of global warming [58]. CO_2 partial pressure in a typical flue gas is shown to be about 0.15 bar, which is lower than the atmospheric pressure. Recent reports showed

TABLE 8.1
Summary of Gases Adsorption onto MOFs and MOF-Containing Composites Reported in Literature

Gases	MOF/Composite	Gas Uptake	Condition (Temp/Pressure)	Reference	Reference
CO_2	MOF-74 (Ni)	6.0 mM/g	298 K, 0.1 atm	[60]	Liu et al. (2010)
	MOF-74 (Mg)	4.1 mM/g	298 K, 0.1 atm	[60]	Liu et al. (2010)
	Cu-BTC	69.0 mL/g	295 K	[61]	Xie et al. (2012)
	MOF-74 (Mg)	5.3 mM/g	303 K, 150 mbar	[63]	Mason et al. (2011)
	ABCD MOF (In)	3.6 mM/g	196 K	[62]	Choi et al. (2019)
	MIL-101 (Cr)-SO_3H-TAEA	2.3 mM/g	303 K, 150 mbar	[25]	Li et al. (2016)
	MIL-101 (Cr)-SO_3H-TAEA	1.1 mM/g	293 K, 0.4 mbar	[25]	Li et al. (2016)
	MOF (Ni)	64.0 cm³/g	195 K, 1 atm	[64]	Huang et al. (2016)
	Li-exchanged MOF (Ni)	113.0 cm³/g	195 K, 1 atm	[64]	Huang et al. (2016)
NO_2	UiO-66 (Zr)	40.0–73.0 g/kg		[26]	Ebrahim and Bandosz (2014)
	UiO-66 (Zr)-COOH	79.0–118.0 g/kg		[26]	Ebrahim and Bandosz (2014)
	Melamine@UiO-66 (Zr)	3.0–10.0 g/kg		[26]	Ebrahim and Bandosz (2014)
	Melamine@UiO-66 (Zr)-COOH	41.0–93.0 g/kg		[26]	Ebrahim and Bandosz (2014)
	Cu-BTC	92.0 g/kg	Moist conditions	[67]	Petit et al. (2012)
	GO@Cu-BTC	200.0 g/kg	Moist conditions	[67]	Petit et al. (2012)
	Cu-BTC	110.0 g/kg	Dry conditions	[67]	Petit et al. (2012)
	GO@Cu-BTC	135.0 g/kg	Dry conditions	[67]	Petit et al. (2012)
CO	MOF-74 (Mg)	4.6 mM/g	298 K, 1.2 bar	[65]	Bloch et al. (2014)
	MOF-74 (Mn)	3.2 mM/g	298 K, 1.2 bar	[65]	Bloch et al. (2014)
	MOF-74 (Fe)	6.0 mM/g	298 K, 1.2 bar	[65]	Bloch et al. (2014)
	MOF-74 (Co)	6.0 mM/g	298 K, 1.2 bar	[65]	Bloch et al. (2014)
	MOF-74 (Ni)	5.8 mM/g	298 K, 1.2 bar	[65]	Bloch et al. (2014)
	MOF-74 (Zn)	2.0 mM/g	298 K, 1.2 bar	[65]	Bloch et al. (2014)
NH_3	MIL-101 (Fe)-SO_3H	18.0 mM/g	298 K, 1 bar	[27]	Van Humbeck et al. (2014)
	MIL-101 (Fe)-SO_3H	5.0 mM/g	298 K, 5 mbar	[27]	Van Humbeck et al. (2014)
	UiO-66 (Zr)-NH_2	10.5 mM/g	298 K, 1 bar	[27]	Van Humbeck et al. (2014)
	UiO-66 (Zr)-NH_2	2.0 mM/g	298 K, 5 mbar	[27]	Van Humbeck et al. (2014)
	UiO-66 (Zr)-NH_3Cl	12.0 mM/g	298 K, 1 bar	[27]	Van Humbeck et al. (2014)
	UiO-66 (Zr)-NH_3Cl	4.0 mM/g	298 K, 5 mbar	[27]	Van Humbeck et al. (2014)
CH_4	Cu-BTC	227 cm³/cm³	298 K, 35 bar	[87]	Peng et al. (2013)
	Cu-BTC	267 cm³/cm³	298 K, 65 bar	[87]	Peng et al. (2013)
	MOF-74 (Ni)	230 cm³/cm³	298 K, 35 bar	[88]	Mason et al. (2014)
	MOF-74 (Ni)	260 cm³/cm³	298 K, 65 bar	[88]	Mason et al. (2014)
	MOF-74 (Co)	221 cm³/cm³	298 K, 35 bar	[88]	Mason et al. (2014)
	MOF-74 (Co)	249 cm³/cm³	298 K, 65 bar	[88]	Mason et al. (2014)
	MOF-74 (Mg)	200 cm³/cm³	298 K, 35 bar	[88]	Mason et al. (2014)
	MOF-74 (Zn)	188 cm³/cm³	298 K, 35 bar	[89]	Wu et al. (2009)
	MIL-53 (Al)	0.7 mM/g	298 K, 0.1 MPa	[76]	Mollmer et al. (2012)
	Cu(Me-4py-trz-ia)	1.1 mM/g	298 K, 0.1 MPa	[76]	Mollmer et al. (2012)

(Continued)

TABLE 8.1 (*Continued*)

Summary of Gases Adsorption onto MOFs and MOF-Containing Composites Reported in Literature

Gases	MOF/Composite	Gas Uptake	Condition (Temp/Pressure)	Reference	Reference
H_2	MOF-5 (Zn)	10.0% (wt%)	77 K, 100 bar	[77]	Kaye et al. (2007)
	Cu-BTC	3.6% (wt%)	77 K, 50 bar	[78]	Panella et al. (2006)
	MOF-177 (Zn)	7.5% (wt%)	77 K, 70 bar	[79]	Wong-Foy et al. (2006
	MOF-74 (Ni)	3.0% (wt%)	77 K, 10 bar	[80]	Villajos et al. (2015)
	MOF-74 (Co)	3.2% (wt%)	77 K, 10 bar	[80]	Villajos et al. (2015)
	MOF-177 (Zn)	0.6% (wt%)	298k, 10 MPa	[71]	Li and Yang (2007)
	MOF-177 (Zn)/catalyst	1.5%–2.5% (wt%)	298k, 10 MPa	[71]	Li and Yang (2007)
	MOF (Ni)	96.0 cm³/g	77 K, 1 atm	[64]	Huang et al. (2016)
	Li-exchanged MOF (Ni)	115.0 cm³/g	77 K, 1 atm	[64]	Huang et al. (2016)
N_2	MIL-53 (Al)	0.2 mM/g	298 K, 1 MPa	[76]	Mollmer et al. (2012)
	Cu(Me-4py-trz-ia)	0.3 mM/g	298 K, 1 MPa	[76]	Mollmer et al. (2012)
	MOF (Ni)	13.0 cm³/g	77 K, 1 atm	[64]	Huang et al. (2016)
	Li-exchanged MOF (Ni)	32.0 cm³/g	77 K, 1 atm	[64]	Huang et al. (2016)

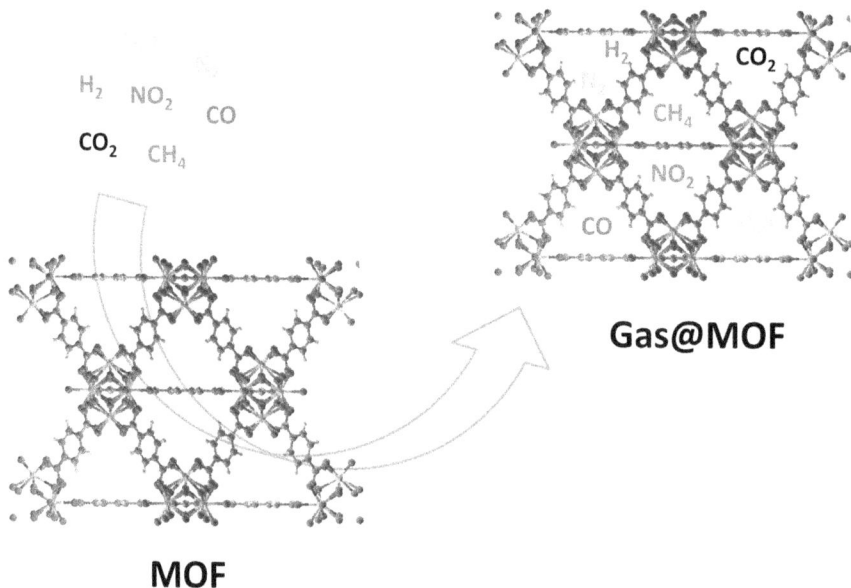

FIGURE 8.2 General schematic summarizes the different synthetic methods for [a] MOFs, [b & c] MOF-containing composites.

the utilization of the experiments and simulation for the examination of MOFs for their highest CO_2 capacity at about 0.1 atm [59]. The researchers confirmed that Ni/DOBDC and Mg/DOBDC (Ni-MOF-74 and Mg-MOF-74 or CPO-27-Ni and CPO-27-Mg) have the highest CO_2 capacity at 0.1 atm and 298 K, which were 5.95 and 4.07 mol/kg. Besides surface area or free volume, the researchers also investigated that MOFs with high density of opening metal site like Ni/DOBDC and Mg/DOBDC are promisingly applicable as candidates for capturing of CO_2 from gas. They reported that Ni/DOBDC exhibited higher CO_2 capacities than 5A zeolites and NaX at 0.1 atm and 25°C. In case of Ni/DOBDC, water does not show any effects on the adsorption of CO_2 as in 5A zeolites and

NaX, and it is shown to be easier for the liberation of water from Ni/DOBDC by regeneration of heat [60]. So, under the wet conditions, Ni/DOBDC could adsorb more CO_2 than zeolites. At 22°C, the absorbed CO_2 on Cu-BTC (surface area BET of 934 m²/g) was estimated to be 69 mL/g [61].

3D ABDC (In) MOFs were prepared via the interaction of di-topic azobenzene-4,4'-dicarboxylic acid (H2ABDC) with 1-ethyl-3-methylimidazolium tetrafluoroborate ([EMIM][BF4]) [62]. 3D ABDC (In) MOFs were applicable in the adsorption of CO_2 gas, and the pre-adsorbed CO_2 molecules improved the sequential adsorption of newly impregnated CO_2 molecules. At 196 K, CO_2 adsorption with ABDC (In) MOFs was 81.3 cm³/g (3.63 mM/g) [62]. Quite higher adsorption of CO_2 (5.3 mM/g) was estimated by using MOF-74 (Mg) at 40°C and 150 mbar [63]. The exploitation of MIL-101 (Cr)-SO₃H-TAEA [tris(2-aminoethyl) amine] resulted in lower adsorptive capacity (2.3 mM/g) at 40°C and 150 mbar that furtherly diminished to 1.1 mM/g via lowering the reaction temperature and pressure to 20°C and 0.4 mbar, respectively [25]. In accordance with Li et al., CO_2 adsorption capability onto the 3D anionic (Ni) MOF was observably enhanced after the cationic exchanging process [64]. At 195 K and 1 atm, the adsorbed CO_2 was estimated to be 64 and 113 cm³/g, before and after the cationic exchange, respectively [64]. MOF-74 with various centered metals (Mg, Mn, Fe, Co, Ni, and Zn) was exploited in the adsorption of CO with a capacity estimated to be 2.0–6.0 mM/g at 298 K and 1.2 bar [65].

After modification with melamine (melamine@UiO-66 (Zr)-COOH and melamine@UiO-67 (Zr)-COOH), respectively, NO_2 adsorption onto UiO-66 (Zr)-COOH and UiO-67 (Zr)-COOH was diminished from 40–73 and 79–118 to 3–10 and 41–93 g/kg [26]. Wang et al. investigated the application of UiO-66 (Zr)-(COOH)₂-containing membrane in the separation of H_2S/CH_4 mixture [66]. By modification with GO, in moisture, the capacity for adsorption using Cu-BTC composite was significantly increased from 92 to 200 g H_2S/kg and from 110 to 135 g NO_2/kg in dry condition [45]. However, NO_2 and H_2S adsorption onto GO@Cu-BTC composite was performed under ambient condition [67], showed increasing capacity of adsorption by 12% and 50%, respectively. MOFs exhibited opened metal sites that might coordinate with the oxygen-containing groups of GO, leading to the organization of porous interfacial spaces, resulting in the improvement of the capacity of adsorption [67]. Recently, Cui et al., affirmed the superiority of MOFs in the purification and separation of light hydrocarbons (LHs), including the alkanes, alkenes, and alkynes [8]. Recently, the adsorption for gas-phase mixtures of toxic volatile organic compounds like aliphatic aldehydes was investigated via various MOFs like UiO-66-NH₂, MOF-5, UiO-66, and MOF-199 [68].

8.4 STORAGE OF GASES

Liberation of harmful gases to the surrounding environment was approved to show global security threats. Effective capturing of gases is importantly required for the human protection. The capability of MOFs for storage of gas is considerably studied to be extensively exploited for storing of H_2 O_2, N_2, and CH_4 (Table 8.1) that could be exploited as alternative fuel, owing to the environmental impacts and economic effects correlated with the cost-effectiveness. In 1997, $M_2(4,4'$-bpy)₃(NO₃)₄ (where M is Co, Ni, or Zn) as 1D MOF was investigated for (O_2, CH_4, and N_2) gas storage at high pressure [69]. The layered zinc terephthalate $Zn_2(BDC)_4(H_2O)_2$ (MOF-2) with its permanent porosity was furtherly prepared via N_2 gas isotherm at low pressure and temperature [70]. Consequently, the progression in the investigation of the superior applicability of the first robust 3D MOF known as MOF-5 ($Zn_4O(BDC)_6$) affirmed its compatibility in gas storage purpose [12]. MOF-5 exhibited high porosity (63% of volume of cell unit) and a surface area (2,320 m²/g), which significantly affected the potency of the zeolites. Interestingly, researchers demonstrated various approaches for controllable management of the porosity in the efficient potency of MOFs [23,71,72].

Recent approaches represented the superiority of six different MOFs, namely, MOF-199, MOF-5, 177MOF-74, MOF-177, IRMOF-62, and IRMOF-3 for the capacity of adsorption dynamically to different hazardous gases, like CO, SO_2, NH_3, tetra-hydro-thiophene, Cl_2, dichloromethane, and ethylene oxide [73]. As a key factor, the porosity showed an observable effect for the measurement of the performance of MOFs for the dynamic adsorption. For example, MOF-199, with attractive exploitation, outperformed BPL carbon besides Cl_2. However, MOF-74 and IRMOF-3 were better in the sorption of NH_3 and SO_2, respectively [73]. In recent reports, Ln (SION105-Eu) was used as a recoverable adsorbing reagent for the capture of NH_3, as it was superiorly exploited for five repetitive cycles with no effects on its crystallinity or luminescence [74].

The capacity of adsorption for H_2 gas was 0.62% (wt/v) using MOF-177 (Zn) and improved to be 1.5%–2.5% (wt/v) at 298 K and 10 MPa, with the addition of a catalyst for H_2 dissociation [75]. N_2 and O_2 gas adsorption processes at equilibrium were estimated at 1 atm pressure to be 0.10 and 0.18 mM/g. Cu(Me-4py-trz-ia) and MIL-53 (Al) adsorbed N_2 in range of 0.22–0.30 mM/g, respectively, at 298 K and 0.1 MPa [76]. After cationic exchange, 3D anionic (Ni) MOFs were prepared to exhibit a better capability for the adsorption of H_2 and N_2 gases, compared to the original Ni-based MOF [64]. Adsorption of N_2 and H_2 gases was increased from 13 and 96 to 32 and 115 cm^3/g, respectively, at 77 K and 1 atm, which was attributable to the cationic exchange [64]. At the normal conditions, NH_3 adsorption onto graphene oxide@Cu-BTC composite increased by 4% compared to Cu-BTC [67]. H_2 gas storage was widely carried out using MOF-74 (Ni and Co), MOF-177 (Zn), and Cu-BTC, MOF-5 (Zn) [77–80].

Environmental and economic impacts for the natural gases were extensively concerned, comprising mainly methane gas as a transportation fuel and especially as a substitution for gasoline (petrol) [81,82]. For the environmental impact, methane roughly delivered the duplicated form of the energy compared to the coal with releasing the same amounts of CO_2 [83]. High capacity for methane storage was demonstrated via the exploitation of MOFs, which intensively attracted the attention of the researchers and industries to exploit such type of materials as on-board tank of CH_4 for vehicles. Additional challenge for the exploitation of CH_4 as a source of energy is its mass- and volume-efficient storing and its delivering at room temperature [84,85]. Nowadays, CH_4 gas is stored at 250 bars in the vehicles. It is suitable for buses as large vehicles, but is not fit well for their application for cars [86]. It was investigated that storage of 263 V/V is needed for the practical applicability. Thus, in recent years, investigation of an alternative storage material is extensively needed. In this point of view, compared to the empty tanks, MOFs have demonstrated superior ability for storing high amount of CH_4 at low pressure. The archetypical Cu-BTC was characterized with the highest volume up taking up to 267 V/V storage at 65 bars to obtain the targeting molecule [87]. MIL-53 (Al) and Cu(Me-4py-trz-ia) were adsorbed 0.71 and 1.12 mM/g from CH_4, at 298 K and 0.1 MPa, respectively [76]. CH_4 adsorption onto MOF-74 (Ni, Co, Mg, and Zn) and the capacity of adsorption were mainly depended on the pressure, and these properties were affirmed by Mason et al. [88] and Wu et al. [89].

8.5 PURIFICATION OF FUEL

In the last decade, owing to the extensive global demanding of energy, non-conventional and new energy resources required innovation. Till now, the main energy resource is the fossil fuel. However, fossil fuel could be identified as the oil/petroleum, coal, and natural gas. Petroleum, as the most applicable type, represents 40% of electricity, which is globally generated from the coal. The exploitation of fossil fuel dangerously affected the environment by the spreading of toxic chemicals with burning [90]. This impendence is shown to be increased via the increment in the energy requesting. Non-conventional fossil fuels that contain massive quantities of contaminants like sulfur-containing contaminants (SCCs), nitrogen-containing contaminants (NCCs), and oxygen-containing contaminants (OCCs). Thus, the increment in fossil fuel utilization could be described as a serious problem.

Upon burning of fossil fuels (like crude oil, gasoline, diesel, jet fuel, furnace oil, etc), CO_2, NOx, and SOx gases were liberated, which could cause different environmental harms like global warming, severe air and water pollution by evolution of gases to the atmospheric air, and other serious environmental problems. Moreover, when such toxic gases were mixed with the rain, it could result in acidified rains, which exhibit harmful effects on the infrastructures and the surrounding environment [91].

However, it was found that the adsorption could be identified as the highly efficient/common methodology for removing the fuel pollutants, and the adsorbents basically were selected in accordance with their capability for interaction with the pollutants as shown in Figure 8.3.

Sulfur-containing contaminants in the fossil fuel are found to be represented as thiophene and its derivatives, e.g., dibenzothiophene, benzothiophene, 4-methyl-benzothiophene, 3,7-dimethyl-dibenzothiophene, 4,6-dimethyl-dibenzothiophene, and 2,8-dimethyl-dibenzothiophene. There are different gaseous sulfur-based contaminants that could generate SO_3, H_2S, and SO_2 with burning or via decomposition. Post-combustion and pre-combustion methodologies are the techniques that could be proceeded in order to diminish the evolution of SO. Pre-combustion is more favorable; owing to that, the post-combustion process is disadvantageous due to expensive and the existence of hot effluent with corrosion [92]. The bio-desulfurization and oxidation were represented with high potentiality for the removal of sulfur-containing contaminants; however, such techniques are proceeded in liquid phase and the oxidation products should be eliminated, and this resulted in limitation of their applications [93–95]. Hydro-desulfurization mainly required catalyst and high temperature and pressure to be proceeded that could reflect in its expensiveness [96–98]. Liquid-phase adsorption is the best technique for removing SCCs as it is characterized with mild processing conditions, without the need of the oxidizing or reducing reagents, and could be potentially industrially applied.

The most applicable adsorbents in the de-sulfurization of liquid fuel are the activated carbon structures [96,99–105], zeolites [106,107], and some of mesoporous materials [97,108]. MOFs containing composites were recently used in desulfurization of liquid fuel. Zhou et al. reported that HPMo (phosphomolybdic acid) could be successfully doped within the building blocks of MOF-199 as a type of MOF in order to prepare HPMo-x@MOF-199 that could be subsequently applicable as a highly stabilized catalyst for desulfurization via oxidation reaction [109]. It was investigated that the abundant intermolecular cavities of MOF-199 can be stably occupied with HPMo, which could

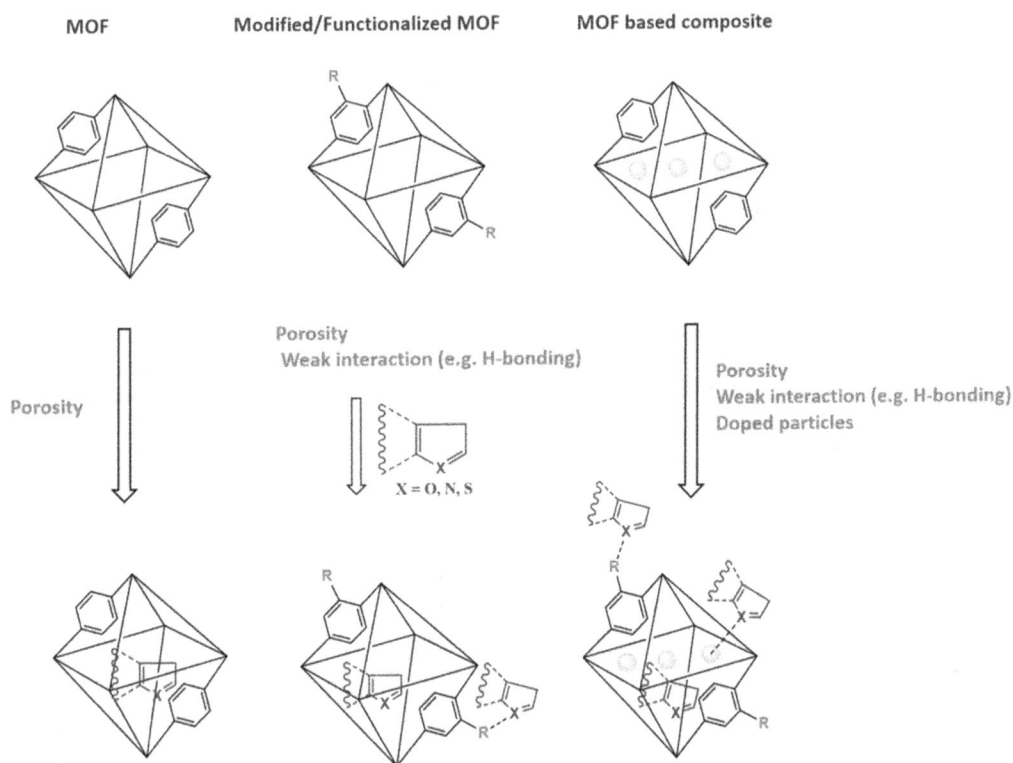

FIGURE 8.3 Adsorption of gases onto MOFs and MOF-containing composites through pore structure.

have a significant effect on increment of the uploaded amounts of HPMo and promote the stabilization of the reactive sites for catalysis. Owing to confining effects, HPMo-x@MOF-199 showed excellent potentiality and stabilization in catalytic desulfurization via oxidation for 4,6-dimethyldibenzothiophene, which could be ascribed as one of the most stubborn sulfur-containing containments in liquid fuel. Additionally, the catalyst could be reused for nine cycles, exhibiting a significant enhancement in the stability compared to unconfining samples. Meanwhile, the oxidative products – high-valuable products – could be superiorly separated. Additionally, in the case of desulfurization of the real diesel, the HPMo-x@MOF-199 catalyst exhibited a satisfied catalytic potency (90.2%), showing its potential applicability [109].

As represented in Table 8.2, different MOFs (e.g., UiO-66, Cu-BTC, Ni-BTC, ZIF-67, ZIF-8, CPO-27, MIL-53, MIL-101, and Y-BTC), functional MOFs (e.g., UiO-66-NH$_2$ and UiO-66-COOH), and MOF-based composites (e.g., MIL-53-NH$_2$/fabric, Cu/MIL-100, Fe$_3$O$_4$/Cu-BTC, GO/Cu-BTC) were superiorly applicable in the adsorption of dibenzothiophene, benzothiophene, and thiophene [40,110–118]. However, the exploitation of MOFs was shown observable enhancement in capacity of

TABLE 8.2
Summary of MOFs and MOF-Containing Composites Used in S-Compounds Removal from Liquid Fuel Mentioned in Literature

SCCs	MOF	Maximum Capacity (g/kg)	Reference	Reference
Thiophene	Y-BTC	31	[115]	Xiang et al. (2014)
	Ni-BTC	95–114	[113	Tian et al. (2015)
	Cu-BTC	29, 72, 210	[110,112,113]	Liu et al. (2012)
				Tian et al. (2015)
				Peralta et al. (2012)
	ZIF-8 (Zn)	25	[110]	Peralta et al. (2012)
	ZIF-67 (Co)	51	[110]	Peralta et al. (2012)
	CPO-27 (Ni)	252	[110]	Peralta et al. (2012)
	MIL-101 (Ni)	25	[116]	Aslam et al. (2017)
	UiO-66 (Zr)	17	[118]	Zhang et al. (2018)
	UiO-66 (Zr)-NH$_2$	30	[118]	Zhang et al. (2018)
	UiO-66 (Zr)-COOH	22	[118]	Zhang et al. (2018)
	Graphite oxide@Cu-BTC	61	[114]	Chen et al. (2018)
	Fe$_3$O$_4$@Cu-BTC	52	[40]	Tan et al. (2017)
	Cu+@MIL-100 (Fe)	154	[117]	Khan and Jhung (2012)
	MIL-53 (Al)-NH$_2$@Fabric	469–517	[49]	Emam et al. (2019)
Benzothiophene	Cu-BTC	18–118	[112,121]	Wu et al. (2014)
				Liu et al. (2012)
	MIL-53 (Al)	8	[121]	Wu et al. (2014)
	MIL-53 (Cr)	24	[121]	Wu et al. (2014)
	UiO-66 (Zr)	20	[118]	Zhang et al. (2018)
	UiO-66 (Zr)-NH$_2$	36	[118]	Zhang et al. (2018)
	UiO-66 (Zr)-COOH	23	[118]	Zhang et al. (2018)
	PWA@CuBTC	149	[123]	Khan and Jhung (2012)
Dibenzothiophene	Cu-BTC	33, 274	[112,122]	Liu et al. (2012)
				Liu et al. (2015)
	MIL-101 (Cr)	33	[132]	Jia et al. (2013)
	PWA@MIL-101 (Cr)	137	[132]	Jia et al. (2013)

the adsorption, owing to their porous crystalline composition. Recently, copper-based MOFs with and without opened metal site like HKUST-1 (Cu-BTC) and Cu-JAST-1, respectively, were prepared and exploited in the removal of dimethyl sulfide traces [119]. In case of HKUST-1, a similar capacity of adsorption for dimethyl sulfide was obtained with prolongation of the duration of adsorption (more than 15 hours) was required compared to Cu-JAST-1, which was reported to require shorter duration of adsorption ($\approx$2 hours) [119].

Cu-BDC, Cr-BTC, Cu-BTC, and Cr-BDC [BTC = 1,3,5-benzene tricarboxylic acid; BDC = 1,4-di-carboxy-benzene] as four examples of MOFs were synthesized via the utilization of two different ligands and metals, while the adsorption ability of the prepared MOFs for removing thiophenes in diesel oil was monitored. Except Cu-BDC, all the studied types were characterized with the adsorption capabilities high toward DBT at ambient temperature. Adsorption capacity for the examined MOFs was found to be ordered as follows: Cu-BTC > Cr-BDC > Cr-BTC $\gg$ Cu-BDC. Q_{max} of Cu-BTC, Cr-BDC, and Cr-BTC was 56.1, 41.0, and 30.7 g/kg, respectively [120]. Adsorption capacity of thiophene and benzothiophene was realized to exhibit nearly the same ordering by using Cu-BTC [112,113,121]. Moreover, significantly high adsorption of sulfur-containing containments (210.4–273.5 g/kg) was estimated when Cu-BTC as highly porous type was exploited [122].

Xu et al. studied the compatibility of 4 types of MOFs, namely, Cu-BTC (HKUST-1), Cu-BDC, Zn-BTC, and Zn-BDC, for desulfurization of DBT via adsorption technique [123]. Adsorption capacities of 8 mg S/g MOF on Cu-BTC were estimated, which were significantly better than those of Cu-BDC (0.32 mg S/g), Zn-BTC (0.14 mg S/g), and Zn-BDC (0.33 mg S/g) attributing to the high surface area. Ma et al. monitored the efficiency of Cu-MOF-199 for desulfurization of mercaptan and approved the adsorptive capacity of 74.7 mg/g [124]. Tian et al. synthesized HKUST-1 with a high surface area of 1,347 m²/g. The adsorption capacity of this compound was 0.141 mmol/g [113].

Magnetic MOF composites that are composed of MOF and magnetic metallic nanostructures have been investigated. This kind of adsorbents could be easily removed from the fuel by using external magnetic field. Chen et al. also represented a study for the incorporation of Fe_3O_4 within the matrix of HKUST-1 (Fe_3O_4/HKUST-1) for desulfurization of thiophene via adsorption [125]. It approved that the adsorption capacity was estimated to be 30.67 mg/g owing to Lewis acid-base interaction and bi-complexation. Jin et al. reported that magnetic MOF composite (Fe3O4@PAA@MOF-199) could be synthesized in order to be applicable for the adsorptive removal of BT and DBT [126]. It showed an adsorption capacity of 35 and 15.9 mg/g for DBT and BT, respectively. Habimana et al. represented a study of immobilization of Europium (Eu) metal within the matrix of MOF for desulfurization via adsorption for a sample containing 1,000 µg/g of thiophene. It exhibited an adsorption capacity of 24.59 mg/g with five cycles [127].

Jafarinasab et al. demonstrated a report for studying the preparation of nanocrystals of Co-ZIF-67 as a cobalt-based MOF, in order to be applicable as adsorbent for the removal of dibenzothiophene (DBT) from liquid fuel [128]. By exploitation of the activated Co-ZIF-67/M-Al (SBET = 2,478.8 m²/g), a maximum adsorption capacity of 550 mg/g was achieved at a temperature of 65°C and $M_{adsorbent}/V_{fuel}$ = 0.8 mg/mL, with opportunity of recycling for four repeated cycles.

Huang et al. ascribed that Ag@CPO-27-Ni composites were synthesized via solvo-thermal methodology, differing in the amounts of AgNPs loaded upon CPO-27-Ni to be characterized with cylinder-like one-dimensional cavities with a diameter of ca. 11 Å composed of 2,5-dihydroxybenzene-1,4-dicarboxylate and nickel ions and the unsaturated metallic sites [129]. Adsorptive desulfurization was shown to exhibit the capacity of Ag@CPO-27-Ni (however, silver was 1.23 wt%), which was observably improved compared to CPO-27-Ni. 35 ppmw$_s$ DBT (or benzothiophene-containing oil (20 g) was de-sulfurized to 2.01 ± 0.27 (or 4.05 ± 0.26) ppmw$_s$ exploiting 0.2 (or 0.5) g of the composite, respectively; the composite exhibited much better selectivity factors of DBT adsorption for naphthalene than CPO-27-Ni.

Jin et al. had uniquely reported a new investigation for the preparation of water-resistant and bimetallic MOFs known as Cu-ZIF-8, via the coordination synthetic methodology. They affirmed that the pore size could be controlled via adjusting the amount of Zn(II) and Cu(II) that could be doped within the centered cubic crystal lattice of ZIF-8 [130]. The results showed that Cu-ZIF-8 was characterized with significant uptake capacities for benzothiophene (BT) compared with those previously investigated adsorbents in the literature [130]. Cu-ZIF-8 showed high stability with water, octane, and benzene, exhibiting more than 95% for the initial uptake capacities after five repetitive cycles.

Adsorption of benzothiophene was raised from 117.6 g/kg for pure Cu-BTC to 149.2 g/kg for composite of phospho-tungstic acid (PWA)/Cu-BTC, whereas uploading of PWA leaded to 26% increment in Q_{max} [131]. Owing to the porosity enhancement, the synthesized (GO)@Cu-BTC composite via solvo-thermal methodology improved the adsorption of thiophene from 38 to 61 g/kg [114]. The direct impregnation of PWA within MIL-101(Cr) leaded to the increment of the DBT adsorption (136.5) by 4.2 times than that of MIL-101(Cr) [132].

MIL-53-NH$_2$ as one of the MOFs with cotton and wool composites was prepared via the infrared-assisted methodology with the direct ingrowth of MOF within the fabric matrices [49]. The prepared MIL-53-NH$_2$/fabric composites exhibited the most efficiency for the removal thiophene from n-heptane and the adsorption capacity maximized to be in the range of 469.4–516.5 mg/g [49]. The capacity of adsorption decreased by 28.4%–43.6% only, after applying for four recycles.

On the other hand, the nitrogen-based contaminants that exist in the liquid fuel can be categorized into two types: basic nitrogenated contaminants as six-membered rings (like pyridine, acridines, quinolone, and tetrahydroquinoline) and non-basic nitrogenated contaminants as five-membered rings (like pyrrole, indole, and carbazole) besides their derivatives (Table 8.3). Coal-obtained fuels intensively contain high amounts of nitrogen-based contaminants [133]. Additionally, nitrogen-based contaminants competes with sulfur-based contaminants via hydro-desulfurization (HDS), which is ascribed as the conventional methodology for the removal of sulfur-based contaminants, as the nitrogen-based contaminants act in the poisoning of HDS catalysts [134]. Therefore, nitrogen-based contaminants must formerly be removed before HDS. Many adsorbing reagents, such as activated carbon [135,136], zeolites [137–139], silicate [140–144], activated alumina [135,141,145], silica-alumina [146,147], modified polymers, and resins [148,149], were used in the removal of nitrogen-based contaminants from liquid fuel. But MOFs were shown with advantageous characteristics in terms of their significantly high capacity of adsorption [150,151].

The application of UiO-66 (Zr) was shown to be effective in achieving high efficiency in the adsorption (an adsorption capacity of 199–213 g/kg) of indole from fuel [28,29]. However, UiO-66 functionalized via adding amino or sulfate groups to the building block of MOF, in order to realize much higher adsorption of nitrogen-based contaminants, owing to the high affinity for adsorption. UiO-66 (Zr)-NH$_2$ (33), UiO-66 (Zr)-NH$_2$ (67), UiO-66 (Zr)-NH$_2$ (100), and UiO-66 (Zr)-SO$_3$H (18) were exfoliated successfully for the removal of indole from fuel with concentrations of 265, 277, 312, and 239 g/kg, respectively [28,29]. UiO-66 (Zr)-COOH exhibited higher efficient adsorption capacity of indole (130 g/kg) and pyrrole (142 g/kg) than UiO-66 (Zr) [152]. Seo et al., reported that the adsorption of nitrogenated contaminants (indole, quinoline, pyrrole, pyridine, and methylpyrrole) in the range of 67–170 g/kg using UiO-66 (Zr) with free COOH groups was achieved, while hydrogen bonding is the key factor for enhancement of the capacity of adsorption [153]. Zhang et al. also reported UiO-66 (Zr)-NH2 as a Zr-containing MOF [118]. They affirmed that the mentioned MOF had an adsorption capacity of 29.5 and 35.7 mg S/g with TH and BT, respectively, owing to hydrogen bond-donating sites. Moreover, functionalization of UiO-66 (Zr) leaded to further enhancement of the capacity of adsorption to 1.6–2.2 times and the exploited UiO-66-NH-SO$_3$H has successfully removed 369 g/kg of indole [152]. Ye et al. demonstrated a green technique for the fabrication of bimetallic UiO-66(Hf-Zr) with high dispersity of Hf/Zr-OH-defect locations under solventless condition [154]. UiO-66(0.13Hf-Zr) showed superior activities for the oxidation of sulfur containments (1,000 ppm S, with sulfur removal efficiencies of 99.8%) in fuel at ambient conditions for only 15 minutes. As the sulfur content increased from 1,000 to 2,000 ppm, only slightly longer duration (within 40 minutes) was required for the complete removal of sulfur containments [154].

PWA/MIL-101 (Cr) was prepared and functionalized within the nitrogenated contaminants' adsorption. PWA (1%)/MIL-101 (Cr) showed an increase in the adsorption capacity by 20% for

TABLE 8.3

Summary of MOFs and MOF-Containing Composites Used in Nitrogenated Compounds Removal from Liquid Fuel Mentioned in Literature

NCCs	MOFs	Maximum Capacity (g/kg)	Reference	Reference
Indole	Cu-BTC	220	[114]	Chen et al. (2018)
	MIL-100 (Cr)	149–416	[44,153,157,160]	Ahmed and Jhung (2014), Seo et al. (2016b), Seo et al. (2016a), Ahmed and Jhung (2016)
	MIL-100 (Fe)	162–357	[26,44,160]	Ahmed and Jhung (2014), Ebrahim and Bandosz (2014)
	UiO-66 (Zr)	199–213	[157]	Seo et al. (2016b)
	MIL-125 (Ti)	264	[152]	Ahmed, Khan et al. (2017)
	MIL-101 (Cr)-ED	336	[25]	Li et al. (2016)
	MIL-125 (Ti)-NH$_2$	502	[152]	Ahmed, Khan et al. (2017)
	P-MIL-125 (Ti)-NH$_2$	583	[152]	Ahmed, Khan et al. (2017)
	UiO-66 (Zr)-NH$_2$	265–312	[29,157]	Ahmed and Jhung (2015) Seo et al. (2016b)
	UiO-66 (Zr)-COOH	130	[157]	Seo et al. (2016b)
	UiO-66 (Zr)-SO$_3$H	239	[29]	Ahmed and Jhung (2015)
	UiO-66 (Zr)-NH-SO$_3$H	369	[152]	Ahmed et al. (2017)
	PWA@MIL-101 (Cr)	152	[42]	Ahmed et al. (2013)
	GO@MIL-101 (Cr)	319	[43]	Ahmed, Khan et al. (2013)
	GnO@MIL-101 (Cr)	542–593	[44]	Ahmed and Jhung (2016)
	P-Ade@MIL-101 (Cr)	532	[158]	Sarker et al. (2018)
	CuCl@MIL-100 (Cr)	171	[160]	Ahmed and Jhung (2014)
	polysulfone@Cu-BTC	676	[54]	Emam et al. (2019)
	MIL-53 (Al)-NH$_2$@Fabric	179–204	[52]	Abdelhameed et al. (2018c)
Quinoline	MIL-100 (Cr)	229–498	[42,134,171]	Maes et al. (2011) Ahmed and Jhung (2013) Almarri et al. (2009),
	MIL-100 (Fe)	357	[171]	Maes et al. (2011)
	MIL-125 (Ti)	103	[152]	Ahmed, Khan et al. (2017)
	UiO-66 (Zr)	190	[153]	Seo, Ahmed et al. (2016)
	MIL-101 (Cr)-ED	301	[25]	Li et al. (2016)
	MIL-101 (Cr)-NH$_2$	301	[157]	Seo et al. (2016b)
	MIL-125 (Ti)-NH$_2$	460	[152]	Ahmed, Khan et al. (2017)
	UiO-66 (Zr)-COOH	170	[157]	Seo et al. (2016b)
	P-MIL-125 (Ti)-NH$_2$	546	[152]	Ahmed, Khan et al. (2017)
	PWA@MIL-101 (Cr)	274	[42]	Ahmed et al. (2013)
	GO@MIL-101 (Cr)	549	[43]	Ahmed, Khan et al. (2013)
	GnO@MIL-101 (Cr)	484–498	[44]	Ahmed and Jhung (2016)
	CuCl@ MIL-100 (Cr)	457	[160]	Ahmed and Jhung (2014)
	P-Ade@MIL-101 (Cr)	511	[158]	Sarker et al. (2018)
	polysulfone@Cu-BTC	619	[54]	Emam et al. (2019)
	MIL-53 (Al)-NH$_2$@Fabric	149–164	[52]	Abdelhameed et al. (2018c)
Pyrrole	UiO-66 (Zr)	124	[152]	Ahmed. Khan et al. (2017)
	UiO-66 (Zr)-COOH	142	[152]	Ahmed. Khan et al. (2017)

quinolone (274 g/kg) compared to MIL-101 (Cr) (229 g/kg). However, indole adsorption was non-observably diminished from 162 to 152 g/kg via the immobilization of PWA within MOF matrix [42]. Recently, modification of NH_2-MIL-101 with oxalyl chloride (OC-ED-A-M101) resulted in an increase in adsorption capacity for indole and quinoline by 11.7 and 9.3 times rather than the active carbon structures [155]. Compared to the other MOFs, OC-ED-A-M101 was characterized by the highly efficient capacity of adsorption (714 mg/g) of indole [155]. MIL-101 (Cr) was prepared in the presence of GO to produce GO/MIL-101 (Cr) composite to be applied for nitrogen-based contaminants' adsorption [43]. Compared to MIL-101, surface area of the prepared GO (0.25%)/MIL-101 composite was enhanced by 22%. Subsequently, the indole and quinoline adsorption was significantly increased from 244 and 481 g/kg for MIL-101 to 319 and 549 g/kg for GO/MIL-101 composite, respectively [43]. Using GO@MIL-101 composite, much higher capacity of adsorption for indole (542–593 g/kg) and observable decrease in adsorption of quinoline (484–498 g/kg) were detected via the comparison with GO/MIL-101 composite [44].

Shi et al. investigated that MOF-505 was synthesized and impregnated with polyethylene glycol (PEG) matrix to produce mixed matrices, which could be used as membranes for desulfurization of fuel [156]. Membrane with a weight percent of 3 for MOF-505 exhibited the optimal desulfurization potency (a permeation flux of 2.66 kg/(m²·h)) and an increment of the sulfur factor of 8.15, which were increased by 158% and 25% versus the native membrane, respectively.

Owing to the possibility of formation of H-bonding, modification of MIL-101 (Cr) with NH_2 groups leaded to the significant enhancement in the capacity of adsorption for indole by a factor of 1.7 and 2.3 more than that of non-modified MIL-101 (Cr) and modified MIL-101 (Cr) with butyl groups, respectively [157]. Uploading of adenine on MIL-101 (Cr) was combined with the improvement in the adsorption of quinoline and indole by a factor of 1.3 and 1.4, respectively, by referring to virgin MIL-101 (Cr) [158]. High capacity of adsorption for P-Ade@MIL-101 (Cr) was estimated owing to H-bonding between MOF- and nitrogen-based contaminants with cation-pi interaction [158]. After modifying MIL-125 (Ti) by amination and then protonation, elimination of nitrogen-based contaminants was enhanced by 1.9–2.2 times for indole and 4.5–5.3 times for quinoline [159]. MIL-125 (Ti)-NH_2 (100) and P-MIL-125 (Ti)-NH_2(100) exhibited an affinity for the adsorption of nitrogen-based contaminants by 460–502 and 546–583 g/kg, respectively [159].

Copper and aluminum chlorides were separately impregnated with MIL-100 (Cr) and MIL-100 (Fe), respectively, for the adsorptive removal of nitrogen-based contaminants [160]. Impregnation of copper chloride within the MOF to prepare CuCl@MIL-100 (Cr) composites resulted in an increment of quinoline and indole adsorption by 9% and 15%, respectively. However, the immobilization of aluminum chloride resulted in an enhancement of MIL-100 (Fe) affinity for the quinoline adsorption by 17%. The quinoline amounts removed by $AlCl_3$@MIL-100 (Fe), MIL-100 (Cr), MIL-100 (Fe), and CuCl@ MIL-100 (Cr) were estimated to be 417, 420, 357, and 457 g/kg, respectively [160].

Polysulfone@Cu-BTC films with different contents of Cu-BTC were exploited in the removal of indole and quinoline [54]. Impregnation of Cu-BTC resulted in the significant enhancement in the adsorption capacity of indole and quinoline from 676 and 619 to 220 and 188 g/kg, respectively [54]. MIL-53-NH_2 increased more than that of cotton and wool as natural textiles, and the resulted composite was functionalized in the removal of nitrogen-based contaminants [52]. The capacity of adsorption for the removal of indole and quinoline as nitrogen-based contaminants was evaluated to be 149–204 g/kg. After recycling for four cycles, the adsorption capacity for MIL-53-NH_2/fabric composite was diminished by 19.2%–40.9% [52].

Oxygen-based contaminants are mostly the unfavorable contaminants, and large amounts of them could be transported to diesel fuel within refining, leading to serious problems, like the containers' corrosion (Table 8.4) [161]. Phenol is identified as one of the oxygenated compounds in fuel, which is known as one of the highly toxic compounds to living organisms [162]. Within the combustion, phenols are transformed to gummy compounds within engine, resulting in blocking of the filters [163]. For that serious problem, the removal of phenols from fuel is essentially required for refining of crude oil.

TABLE 8.4

Summary of MOFs and MOF-Containing Composites Used in Adsorption of Phenols According to Literature

Phenols	MOF	Maximum Capacity (g/kg)	Medium	Reference	Reference
Phenol	MOF-5 (Zn)	25–93	Water	[169]	Xie et al. (2015)
	MIL-100 (Fe)	28	Water	[170]	Baojian Liu et al. (2014)
	MIL-100 (Cr)	27	Water	[170]	Baojian Liu et al. (2014)
	MIL-101 (Cr)	250	n-octane	[30]	Bhadra et al. (2016)
	HO-MIL-101 (Cr)	330	n-octane	[30]	Bhadra et al. (2016)
	MIL-101 (Al)-NH$_2$	26	Water	[170]	Baojian Liu et al. (2014)
	Cu-BTC@viscose	333	n-octane	[53]	Abdelhameed et al. (2017)
	Cu-BTC@wool	303	n-heptane	[53]	Abdelhameed et al. (2017)
p-nitrophenol	MIL-100 (Fe)	32	Water	[170]	Baojian Liu et al. (2014)
	MIL-100 (Cr)	26	Water	[170]	Baojian Liu et al. (2014)
	MIL-101 (Al)-NH$_2$	193	Water	[170]	Baojian Liu et al. (2014)
Bisphenol A	MIL-53 (Cr)	421	Water	[172]	Park et al. (2013)
	MIL-53 (Al)	325	Water	[173]	Zhou et al. (2013)
	MIL-53 (Al)-F127	465	Water	[173]	Zhou et al. (2013)

Complexion, precipitation, and liquid-liquid extraction are the most conventional techniques for the removal of phenols from petroleum [164–168]. These techniques are disadvantageous in terms of high consumption of chemical reagents, leaching out the phenols in the wastewater, requiring hazardous solvents, and requiring many steps. Most of the mentioned drawbacks of the adsorption technique for the removal of phenol is its wide-scale applicability.

MOF-5 (Zn) was functionalized in the removal of nearly 97% of phenolic contaminants from the neutral media at 40°C [169]. MIL-100 (Fe), MIL-100 (Cr), and MIL-101 (Al)-NH$_2$ were exploited for phenol and p-nitrophenol (PNP) removal from water via adsorption technique [170]. Similarly, low capacity for the removal of phenolic compounds via the adsorption technique was investigated for three types of the MOFs. However, a quite high adsorption of PNP was detected onto MIL-101 (Al)-NH$_2$ owing to higher possibility for the formation of H-bonding [170]. MIL-53 (Cr) as water-soluble MOF was exploited for the adsorptive removal of phenol, p-cresol, and bisphenol-A from wastewater [171,172]. By comparing with the activated carbon, MIL-53 (Cr) exhibited highly accelerated adsorption reaction of bisphenol-A with higher capacity of adsorption for MIL-53 [172]. Additionally, exploiting MIL-53 (Cr) was much preferable for the removal of bisphenol even at very low concentration [172]. It was found that MIL-53 (Al) exhibited higher efficiency for the adsorption of bisphenol A (329 g/kg) than active carbon (130–263 g/kg) from aqueous medium [173]. Further increase in MOF capacity of adsorption (473 g/kg) was estimated for the treated MIL-53 (Al) [MIL-53 (Al)-F127]. Time of equilibrium for the removal of bisphenol A diminished from 90 minutes for MIL-53 (Al) to 30 minutes for MIL-53(Al)-F127 [173].

In accordance with the literature, the previous approaches that were employed in the elimination of phenols from liquid fuel were shown to be very limited. MIL-101 (Cr) represented a significant capacity of adsorption for phenol with 2.7 times higher than that of the activated carbon structures [30]. It was also investigated that more hydroxyl groups were demonstrated within the construction of the MOF via modification with EA to obtain EA-MIL-101 (Cr) [HO-MIL-101 (Cr)] [30]. Moreover, after the addition of hydroxyl groups to the building blocks of the MOF, the capacity of adsorption was raised up to ~3.7 times of activated carbon, owing to hydrogen bonding. Hydrogen bonds can exist between the oxygen atoms of HO-MIL-101 (Cr) and hydrogen atoms of the phenol groups, which was affirmed via the comparative adsorption studies of phenols on HO-MIL-101 (Cr) and MIL-101 (Cr). Uptake of phenols using activated carbon, MIL-101 (Cr), and HO-MIL-101 (Cr) was estimated to be 91, 250 ,and 330 g/kg, respectively [30]. Wool and viscose were modified with Cu-BTC via both immediate and ex situ methodologies, and characterized by the uptake of phenol to be 333 and 303 mg/g for Cu-BTC@viscose and Cu-BTC@wool, respectively [53]. The phenols' adsorption was performed via the physical deposition within the pores, H-bonding between cellulose functional groups, and chelating of phenols with Cu (II) through the coordinating bonds [53].

8.6   WATER TREATMENT

In order [174] to obtain a pure water system, we must also develop effective approaches for concurrent removal of existing pollutants, like organic dyes and heavy metals, from contaminated water. In wastewater, both organic and heavy dyes usually exist, causing several problems with their removal at the same time. The use of MOFs as the absorbent to remove the contaminant attracts considerable attention, due to their wide surface and adjustable pore size, in recent years. It is thus desirable to remove dyes and heavy metals from the wastewater. These materials are also widely used in gas separation, photo-electrocatalysis, the delivery of medication, sensing, adsorbing and gas storage, as inorganic polymers with unique chemical and physical properties. MOFs may also be used to remove organic substances, such as dyes, pesticides, heavy metals, phenolic compounds, and aromatic rings.

Heterogeneous catalysis is a catalytic process in which catalysts and reactants are in different phases, mainly including liquid-solid reaction and gas-solid reaction. Taking the liquid-solid reaction as an example, solid catalysts have great advantages on the recycle of catalysts. Usage of simple physical approaches such as filtration and centrifugation will isolate the used catalysts from the substrates. However, heterogeneous catalysts are typically less catalytic than homogenous catalysts because of a decrease in the active sites and the interface between catalysts and reactants. MOFs have been highly attracted by the pore structures, which promotes adsorption and concentration of the substrate at catalytic sites. As a kind of MOFs, MIL-125 is a promising catalytic candidate (Figure 8.4). They have a plenty of Lewis acids from rare earth ions. Owing to the complex organized actions of rare earth ions, ligand selection is more complex in order to achieve different functions such as acid-base sites, light adsorption, and sophisticated channel morphology. In particular, the MIL-125 networks are stable at a high temperature or even in harsh acidic and alkaline environment. Additionally, MIL-125 is also suitable for carrying metal species as a container. Ideally, for the control of the catalytic operation, the electronic interaction and/or synergistic effect of metal loads and MOF is used. Within the following reaction section, we will discuss the catalytic and photocatalytic application of MOFs in detail. In recent years, MOFs are being used as heterogeneous catalysts for many organic reactions due to the fact that these materials offer a considerable level of structural flexibility in terms of composition with regard to the nature of the transition metal and the substituents present in the organic ligands [175].

Sun et al. [176] studied the synthesis for a series of M/MIL-125(Ti)-NH2 materials (M=Pt and Au), which were employed in the reduction of CO_2 under irradiation of visible light. The insertion of Pt improved the photocatalytic performance of MIL-125(Ti)-NH_2. Mixed transition metal nanoparticles (MTMNPs) are then found to be more efficient, in which metal/vanadate (MeVOs where I=Ag, Bi, Fe, Ca, etc) are commonly known to be very interesting due to their various applications in the catalytic field, the antibacterial strength, the multi-ferrous behavior, and the nanobiology.

Nasalevich et al. [177] investigated a synthetic strategy for the efficient encapsulation in MOF (NH_2-MIL-125(Ti)) of a derivative with a well-defined cobaloxime proton reduction catalyst. Co@ MOF is an efficient, fully recyclable noble metal-free catalytic system for the development of light-driven hydrogen in water under visible light. The resulting hybrid Co@MOF system is a solid heterogeneous composite material. The facile synthesis of a multifunctional catalyst@MOF composite (Co@MOF) through a 'ship-in-a-bottle' approach was approved. Effective photocatalytic processing of H_2 from water under visible light illumination was achieved via employing Co@MOF composite as a robust material (Figure 8.4). Co@MOF is explored to be a recyclable material compared with the pristine NH_2-MIL-125(Ti); the introduction of a molecular CCAS results in a 20-fold increase in H_2 evolution, which demonstrates the synergy between a stable photo-active MOF structure and a degenerated CCCAP catalyst. The composite has extensively characterized by the formation of cobaloxime within the MOF pores. The precise cobalt species structure remains unaccounted. The noble metal-free catalyst Co@MOF exceeds any photocatalytic MOF system as a simple example of multifunctional MOF-based systems. With a modular design behind a co-operative photoactive matrix and a catalytically active guest, similar composites can be applied to other photocatalytic composite applications.

Han et al. [178] investigated that Au nanoparticles supported on amino-functionalized Ti-benzene dicarboxylate MOFs (Au/NH_2-MIL-125(Ti)) could be successfully prepared by a facile ultrasonic method. The obtained nanocomposites demonstrated an excellent hydrazine oxidation electrostatic behavior, as a result of their large specific surface area and good conductivity. We also found that pH solution specifically influences Au/NH_2-MIL-125(Ti) electro-catalytical behavior in the direction of hydrazine oxidation. Therefore, this technique of electrochemical method for the detection of hydrazine could be expressed as simple, sensitive, selective, and cheap. The linear dynamic

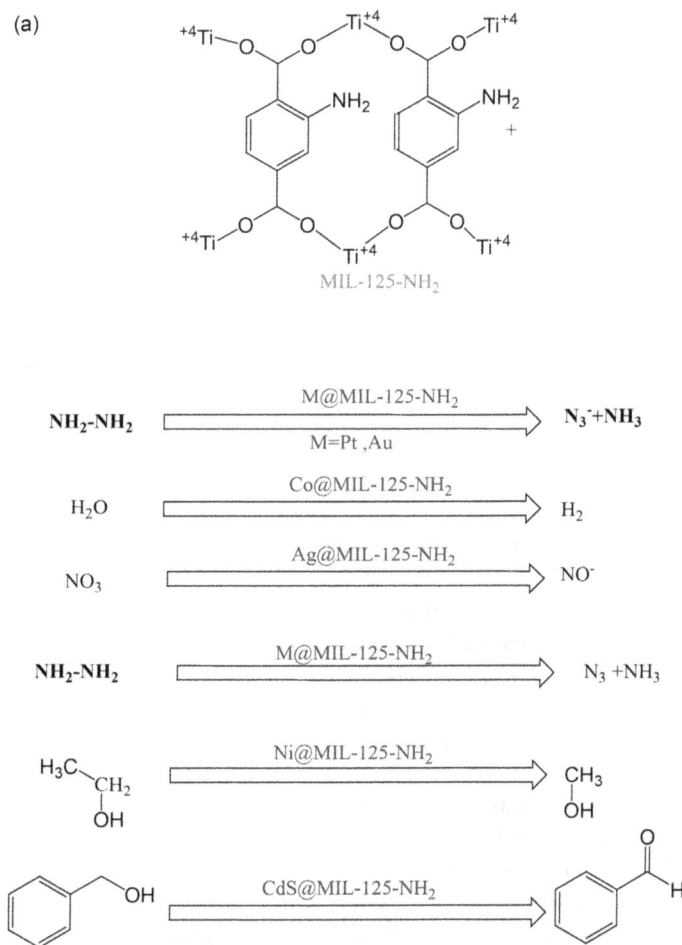

FIGURE 8.4 Adsorption of gases onto MOFs and MOF-containing composites through pore structure.

(*Continued*)

(b)

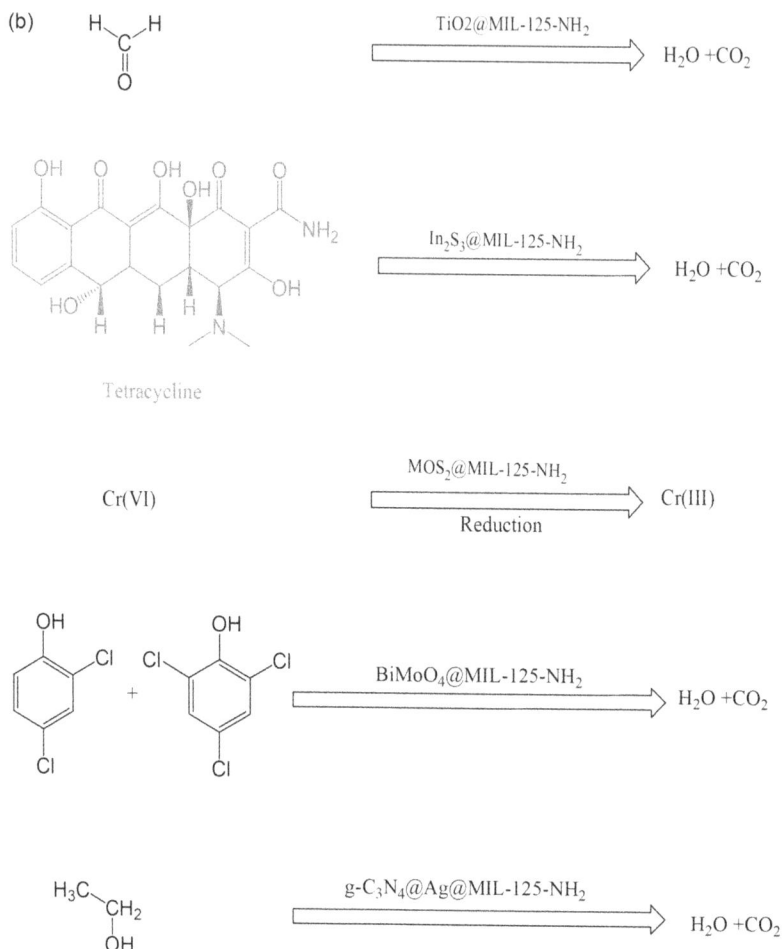

FIGURE 8.4 (*CONTINUED*)   Adsorption of gases onto MOFs and MOF-containing composites through pore structure.

range was obtained between 10 and 100 mM with a maximum range of 0.5 nM. Manufacturing of AuNPs on amino-functional Ti-based MOFs was shown to be promising for hydrazine sensing. It was decided that the electroanalytical chemistry of metal nanoparticles/MOF nanocomposites in the future could be widely applied. The synthesized Au/NH_2-MIL-125(Ti) nanocomposites by a simple and fast ultrasonic method, and their electrochemical characteristics were investigated. It was found that the composites displayed excellent electro-catalytic tendency for hydrazine oxidation, which has a high surface area and excellent electrochemical behavior. In addition, it is also confirmed that hydrazine oxidization was a proton-participating process. Based on this point, it could be constructed as a facile electrochemical sensor for hydrazine detection. The constructed sensor possessed a wide variety of linear 10 nM–100 mM and a low detection limit of 0.5 nM, which is superior compared to the other reported studies in literature.

In 2015, Zhu et al. [179] reported that new MOF-like organotitanium polymer Ag-doped (Ag@ NH_2-MOF(Ti)) was synthesized with 2-aminoterephthalic acid (H_2ATA) and tetra-n-butyl titanate ($Ti(OC_4H_9)_4$) by the microwave-supported coordination and polymerization reaction. The composite of Ag@NH_2-MOF(Ti) exhibited an excellent activity in visible light-driven photocatalytic NO oxidization, which is much higher than N-doped TiO_2 by two times. The NH_2-MOF(Ti) is active in visible light and absorbed in conjunction with HO• and O_2 to produce photoelectrons and holes.

AgNPs permitted the assembly of NH_2-MOF(Ti) polymers to favor multiple reflected light absorption, and also enabled the transfer of photoelectrons to reduce the recombination of photoelectron-holes, leading to an increased photocatalytic activity (PCA) for NO oxidation and inactivating bacteria. Therefore, the high stability may be used repeatedly.

Rui Zhang et al. [180] showed that CdS-MIL-125(Ti) could be synthesized from MIL-125(Ti) and decoration with CdS nanoparticles. The CdS nanoparticles have been heavily dispersed on the MIL-125(Ti) surface, and the CdS-MIL-125(Ti) composite will facilitate the separation and increase the CdS dispersion. In selective oxidation of benzyl alcohol into benzaldehyde with the molecular oxygen as oxidant, the nanocomposites were subsequently used as photocatalysts. In the photocatalytic reaction under visible light irradiation, the nanoparticles showed excellent efficiency. The conversion of benzyl alcohol to benzaldehyde was about 20.1% with no by-products observed.

In 2016, Yanghe et al. [181] showed that the selective photocatalytic oxidation of several alcohols to their corresponding aldehydes using O_2 as oxidant can be achieved over Ni-doped NH2-MIL-125(Ti) upon visible light irradiation. Therefore, research in progress provides guidance for the production of photocatalysts for an environmental, sustainable, and thus green alcoholic process, in which the chemical industry is of the greatest importance. The selective oxidation and photocatalytic activity of NH2-MIL-125(Ti) could be possibly enhanced to better visible light reaping and successful electron transfer by doping NiNPs into NH2-MIL-125(Ti), which is the most efficient. However, the efficiency in the selective oxidation of alcohol of the non-doped NH2-MIL-125(Ti) should be significantly improved for industrial use. In 2016, Yang et al. [182] also reported that photocatalytic light synthesis, under sunlight, is a beneficial chemical method. As an effective bifunctional visible light reaction catalyst for the photo-reduction of nitro-compounds and the oxidation from alcohols, a heteroengineered MIL-125/Ag/g-C_3N_4 nanocomposite has been presented. A picture was put on the surface of the G-C_3N_4 and MIL-125, to increase visible light absorption by the surface plasmon resonance. The reactive effectiveness of the photocatalyst depends on two major factors: the light adsorption of catalysts and AgNPs.

In 2015, Wang et al. [183] explored that solvo-thermal method was successfully applied for the preparation of In_2S_3@MIL-1125(Ti) (MLS) photocatalytic adsorbent. Hybrids are suggested to be the center and 3D In2S3 network MIL-125(Ti), as the shell has a high surface area, mesoporous structure, increased electronegativity, and a visible light absorption. In the MLS, tetracycline (TC) was extracted from water with excellent adsorption efficiency. The process of adsorption is dependent on the solution pH, ionic tolerance, and initial TC concentration. It could explain the adsorption and adsorption kinetics in the Langmuir isotherm and pseudo-second-order mode. The surface complexation, π-π interactions, hydrogen bonding, and adsorption mechanism are primarily responsible for the dynamics of electrostatic. Furthermore, in TC degradation experiments under visible light exposure in the presence of core-shell MLS, the optimal additive content of MIL-125(Ti) in synthesis process was 0.1 g, and the corresponding photodegradation efficiency for TC was 63.3%, which was higher than that of pure In_2S_3 and pure MIL-125(Ti). The advanced photocatalytic efficiency has largely been attributed to the opened pore structure, the efficient transfer of photo-produced carriers, the transmission of Ti^{3+}-Ti^{4+} electron interval, and the synergistic effect of MIL-125(Ti) and In_2S_3.

According to Zhang et al. [184], mesoporous heterojunctions with surface defects have been manufactured using a simple solvo-thermal method for preparing mesoporous NH_2-MIL-125(Ti)@ Bi_2Mo_6. With a relaxation area of 87.7 m^2/gL and a narrow pore size of 8.2 nm, the mesoporous core-shell structure extended the photo-response to visible light because of the wide strip gap of 1.89 eV. The highly toxic dichlorophenol and trichlorophenol, respectively, had a visible light-driven photocatalytic degradation efficiency of 93.28% and 92.19%, and the corresponding rates were 8 and 17 times higher than the levels reached by NH_2-MIL-125(Ti) pure. The photocatalytic oxygen production rate increased to 171.3 mmol/g. Recycling implies high durability over many cycles, which is useful for practical use. The excellent photocatalytic efficiency can be due to core-shell heterojunctions and surface defects supporting load separation and visible light absorption; the mesopore structure provided a sufficient number of active areas of surface and mass transfer.

The new mesoporous photocatalyst gave a new insight into the development of other high-performance photocatalyst core structures in the area.

Qianqian Huang et al. [185] successfully synthesized the photocatalytic composite TiO_2@NH_2-MIL-125 by *in situ* solvo-thermal method via near interface interaction between NH_2-MIL-125 and TiO_2, to be applicable for the removal of formaldehyde (HCHO) under UV-radiation. The results showed that TiO_2 particles were strongly scattered on the surface of NH_2-MIL-125, and there was a strong electronic interaction between elements of the composite. The photocatalytic results for HCHO removal in TiO_2@NH_2-MIL-125 were substantially improved compared to pure TiO_2 and NH_2-MIL-125, due to the combined NH_2-MIL-125 high adsorption power, lower TiO_2 dispersion, and efficient interface charge transmission between TiO_2 and NH_2-MIL-125. In addition, after 12 hours of continuous photocatalytic reaction, the HCHO removal rate over the composite was 90%, suggesting a high catalyst stability. TiO_2@NH_2-MIL-125 nanocomposite was synthesized with a simple solvo-thermal in situ method and was first used as a titanium source of MOFs by $TiCl_4$. The composite samples have a wide surface area and pored volume and a microporous structure (including mesoporous structure) that facilitate the dispersion of TiO_2. In general, the photoactivity performance of TiO_2@NH_2-MIL-125 heterostructure was higher than the pure NH_2-MIL-125 and TiO_2, respectively because of their interface, which enhanced the promotion of efficient interfacial transfer of charges carriers. The high HCHO adsorption power will shorten the travel distance of the photocatalytic action on the adsorption sites and catalytic active sites.

Wang et al. [179] showed that metal sulfides-sensitized MIL-125(Ti) heterostructured hybrids were fabricated via a facile photo-deposition strategy, by which Ag_2S, CdS, and CuS QDs and graphene-like MoS_2 sheets were uniformly deposited on the MIL-125(Ti). It was found that the heterostructure hybrids can be utilized as an efficient photocatalyst for Cr (VI) reduction under irradiation of visible light. The photocatalytic mechanisms over the as-obtained heterostructures were explored, and the improved visible light absorption and electron-hole pair separation were responsible for the remarkably enhanced visible photocatalytic performance. It is anticipated that our work can provide a new insight in the fabrication of semiconductor/MOFs heterogeneous catalyst for solar-chemical energy conversion applications.

The other is the picture-generated load carrier's separating capacity. AgNPs could allow the direct migration of photoelectrons from g-C_3N_4 to MIL-125 and could inhibit the recombination of electron-hole systems as an electron-leading bridge at the interface between MIL125 and g-C_3N_4. That is why, relative to MIL-125, g-C3N4, MIL-125/Ag, and MIL-125/g-C3N4, MIL-125/Ag displayed the highest degree of PA. There was a detailed description of a photocatalyst of such reactions. In addition, nitro-compounds and alcoholic oxidation have been photo-reduced and catalyst recycled four times with excellent conversion and selectivity. The assumption is that in the field of selective organic transformations, MIL-125/Ag/g-C_3N_4 will be a promising visible light catalyst. The normal MIL-125 photocatalytic operation is enhanced effectively. In addition, under visible light radiation, the MIL-125/Ag/g-C_3N_4 showed good recyclability. As a useful visible light photocatalyst for organic transformations, MIL-125/Ag/g-C_3N_4 nanocomposites are expected to be efficient and economical sustainable catalysts for green synthesis of amines and aldehydes derivatives.

In 2014 [186], $FeCl_3$-NH_2-MIL-125(Ti) with different $FeCl_3$ was investigated for its humidity-sensing properties. $FeCl_3$-NH_2-MIL-125(Ti) was more effective in sensing humidity compared to NH_2-MIL-125(Ti). $FeCl_3$-NH_2-MIL-125(Ti) (8 wt%) provided the highest linearity for the optimum mixing ratio. $FeCl_3$-NH_2-MIL-125(Ti) impedance (8% wt) is reduced by approximately three orders of magnitude in the range of 11%–95% RH. The FeCl3-NH2-MIL-125(Ti) response time and recovery time (8 wt%) for the humidity sensor are 11 and 86 seconds, respectively. These findings indicate that $FeCl_3$-NH_2-MIL-125(Ti) could be included in moisture sensors.

Hongxu et al. [187] declared that, in particular, MIL-125(Ti) could be synthesized by the hydrothermal method, and the composites displayed an excellent visible absorption of light with decoration of composite with sliver nanoparticles, which appears scattering on the surface of MIL-125(Ti) as AgNPs were uploaded. It is estimated that 3wt% loading amount of Ag as co-catalyst on MIL-125(Ti) enhanced the catalytic activity slightly by promoting efficient photodegradation of Rhodamine B (Figure 8.5). However, experiments using radical scavenger have shown that O^{-2} and electron are the main reactants, with no noticeable loss of catalytic activity. A system is also being provided on Ag@MIL-125(Ti) for photocatalytic organic degradation. The photocatalyst activity was improved by raising the loaded level of Ag as a co-catalyst, which more easily promoted RhB degradation, measured as 3 wt%. The key reactive species were also described with $\bullet O_2$ and $\bullet OH$ using radical scavengers. Without a substantial loss of catalytic activity, the as-prepared catalyst demonstrated excellent stability and could be re-used five times. In addition, a possible photocatalytic mechanism was also proposed for biodegradation by Ag@MIL-125(Ti). It could be also shown to provide a new insights into MOF-125(Ti)-assisted photocatalytic content in environmental pollution control by modified AgNPs using a visible light photocatalytic action.

Hou et al. [188] reported that MOFs and graphic carbon hybrids (g-C_3N_4) have successfully been prepared using a simple solvo-thermal method and the photocatalyst for the heterostructures of g-C_3N_4/Ti-benzene dicarboxylate (MIL-125(Ti)). The hybrids have large surfaces, mesoporous structures, and thermal stability and enhance the absorption of visible light. The composites showed a more powerful photocatalytic efficiency of Rhodamine B degradation from aqueous solution in visible light irradiation in comparison with pure MIL-125(Ti) and g-C_3N_4. The optimum g-C_3N_4 content was estimated to be 7.0 wt% for g-C_3N_4/MIL-125(Ti), while the corresponding RhB degradation rate was approximately 0.0624 min^{-1}, approximately 2.1 and 24 times higher than that for pure g-C_3N_4 and MIL-125(Ti). The indirect dye photosensitization, the Ti^{3+}, Ti^{4+} intervalence electron transfer, and the synergistic effect between MIL-125(Ti) and g-C_3N_4 were the three reasons for improved photo-degradation performance. Therefore, it can clearly be assumed to have tremendous potential for environmental remediation in metal-free semiconductor/MOF photocatalysts. The reusability and stability of photocatalysts for dye degradation were also demonstrated by cyclical experiments. However, the efficiency of MOFs can easily be highly enhanced by differentiating the metal ions of the ingredients and bridging organic linkers, while more effective photocatalysts for environmental remediation applications would be required to be metal-free semiconductors.

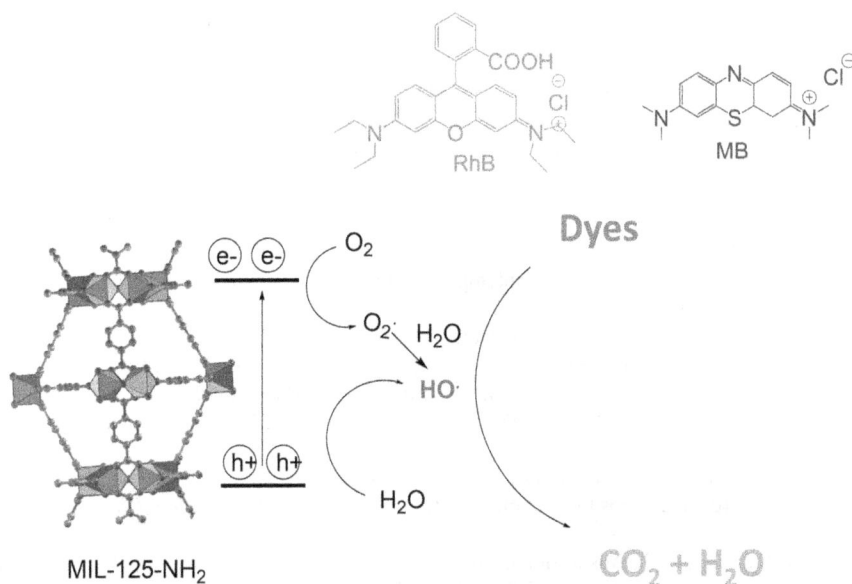

FIGURE 8.5 Removal of fuel contaminants (SCCs, NCCs, and OCCs) by MOFs, functionalized MOF, and MOF-containing composites by adsorption process, clarifying the factors affected the adsorption capacity.

Yuan et al.'s [189] study showed that for the construction of Ag/rGO/MIL-125(Ti), an efficient one-pot/self-assembly and photo-reducing strategy was developed, integrating negative GO with positive MIL-125(Ti) on the basis of electrostatic attractive interactions and GO photoreduction and Ag NP deposition, thus leading to ternary hetero-heterogeneous Ag/rGO/MIL-125(Ti). The photocatalyst heterostructure showed an increased pH to degrade RhB by visible light irradiation. In Ag/rGO/MIL-125(Ti), the photo-degradation rate of RhB in RhB is 0.0644 min^{-1}, which is 1.62 times higher than in MIl-125(Ti) (Figure 8.5). In addition, indirect dye photosensitization, locally located surface plasmon resonance of Ag NP, Ti^{3+}-Ti^{+4} plus electron interval transfer, and the synergistic effects between MIL-125(Ti), AgNPs, and rGO were attributed to improved photocatalytic performance. Cyclic studies have also shown the reusable RhB degradation photocatalyst.

In 2018, Abdelhameed et al. [190] studied the use of modified NH2-MIL-125-core and Ag$_3$PO$_4$ NPs as shell, to produce Ag$_3$PO$_4$@NH$_2$-MIL-125 nanocomposites, which were employed in the degradation of MB and RhB under visible light (Figure 8.5). According to the literature, the most used photocatalyst is TiO$_2$ because of its chemical stability, non-toxicity, and its strong oxidizing ability for photo-induced holes. Ag$_3$PO$_4$ NP is one of the most known examples of (2.45 eV) semiconductor narrow band gaps, with good PCA in the visible region, but with an inherent fast loading recombination; its efficiency is relatively low. The big downside of Ag$_3$PO$_4$ NPs as a photo-catalyst was that during the photocatalytic reaction, they are likely to be self-reduced to silver. Ag$_3$PO$_4$ was higher than P25 and NH$_2$-MIL-125, with a higher degradation rate of MB and RhB (80%, or 74%), as it has a significant optical absorption under visible light. Nanocomposite Ag$_3$PO$_4$@NH$_2$-MIL-125 was much more active than pure Ag$_3$PO$_4$ NPs with MB and RhB, which might be due to the growth of heterojunction between Ag$_3$PO$_4$ and NH$_2$-MIL-125.

8.7   CONCLUSION

This chapter concluded the adsorption potentiality of MOFs, functionalized MOFs, and MOF-based materials for clean environment. More than one hundred and sixty recent researches were reported on gas capturing (CO$_2$, NO$_2$, SO$_2$), gas storage (H$_2$, N$_2$, O$_2$, NH$_4$), fuel ultrafiltration from pollutants (sulfur, nitrogen, oxygen-based contaminants), and removal of toxic pollutants from water. In accordance with the reported approaches, the adsorption capacity basically depended on the pore size and the accessibility to form H-bonding with functional groups and with lower relevance on π-π interaction. The crucial outcome of this chapter is the superiority of composites of MOFs/MOFs in cleaning the environments through the use of clean resources of energy that directly resulted in rescuing living organisms from the toxic-eliminated contaminants.

REFERENCES

1. Gangu, K.K., et al., A review on contemporary metal–organic framework materials. *Inorganica Chimica Acta*, 2016. 446: 61–74.
2. Motkuri, R.K., et al., Metal organic frameworks-synthesis and applications. In *Industrial Catalysis and Separations*, Apple Academic Press, New York, pp. 61–103, 2014. ISBN: 9780429170287.
3. Noshadi, I., *Development of Functionalized Nanoporous Materials for Biomass Transformation to Chemicals and Fuels*. 2015. Doctoral Dissertations. p. 653. https://opencommons.uconn.edu/dissertations/653
4. Sing, K.S., Reporting physisorption data for gas/solid systems with special reference to the determination of surface area and porosity (recommendations 1984). *Pure and Applied Chemistry*, 1985. 57(4): 603–619.
5. Erdem-Şenatalar, A., et al., Adsorption of methyl tertiary butyl ether on hydrophobic molecular sieves. *Environmental Engineering Science*, 2004. 21(6): 722–729.
6. Ban, Y., N. Cao, and W. Yang, Metal-organic framework membranes and membrane reactors: Versatile separations and intensified processes. *Research*. 2020. 2020: 1583451.

7. Dou, Y., W. Zhang, and A. Kaiser, Electrospinning of metal–organic frameworks for energy and environmental applications. *Advanced Science*, 2020. 7(3): 1902590.

8. Cui, W.G., T.L. Hu, and X.H. Bu, Metal–organic framework materials for the separation and purification of light hydrocarbons. *Advanced Materials*, 2020. 32(3): 1806445.

9. Férey, G., Hybrid porous solids: Past, present, future. *Chemical Society Reviews*, 2008. 37(1): 191–214.

10. Tranchemontagne, D.J., et al., Secondary building units, nets and bonding in the chemistry of metal–organic frameworks. *Chemical Society Reviews*, 2009. 38(5): 1257–1283.

11. Knobloch, F.W. and W.H. Rauscher, Coordination polymers of copper (II) prepared at liquid-liquid interfaces. *Journal of Polymer Science*, 1959. 38(133): 261–262.

12. Li, H., et al., Design and synthesis of an exceptionally stable and highly porous metal-organic framework. *Nature*, 1999. 402(6759): 276.

13. James, S.L., Metal-organic frameworks. *Chemical Society Reviews*, 2003. 32(5): 276–288.

14. Batten, S.R., et al., Terminology of metal–organic frameworks and coordination polymers (IUPAC recommendations 2013). *Pure and Applied Chemistry*, 2013. 85(8): 1715–1724.

15. Abdelhameed, R.M., et al., Building light-emitting metal-organic frameworks by post-synthetic modification. *ChemistrySelect*, 2017. 2(1): 136–139.

16. Abdelhameed, R.M., et al., Designing near-infrared and visible light emitters by postsynthetic modification of Ln^{+3}–IRMOF-3. *European Journal of Inorganic Chemistry*. 2014. 2014(31): 5285–5295.

17. Abdelhameed, R.M. and I. El Radaf, Self-cleaning lanthanum doped cadmium sulfide thin films and linear/nonlinear optical properties. *Materials Research Express*, 2018. 5(6): 066402.

18. Furukawa, H., et al., The chemistry and applications of metal-organic frameworks. *Science*, 2013. 341(6149): 1230444.

19. Stock, N. and S. Biswas, Synthesis of metal-organic frameworks (MOFs): Routes to various MOF topologies, morphologies, and composites. *Chemical Reviews*, 2011. 112(2): 933–969.

20. Abdelhameed, R.M., et al., Separation of bioactive chamazulene from chamomile extract using metal-organic framework. *Journal of Pharmaceutical and Biomedical Analysis*, 2017. 146: 126–134.

21. Viswanathan, B., *Energy Sources: Fundamentals of Chemical Conversion Processes and Applications.* 1st Edition, Elsevier, Amsterdam, 2016, ISBN: 9780444563606.

22. Seo, P.W., et al., Adsorptive removal of pharmaceuticals and personal care products from water with functionalized metal-organic frameworks: Remarkable adsorbents with hydrogen-bonding abilities. *Scientific Reports*, 2016. 6: 34462.

23. Sun, Y. and H.-C. Zhou, Recent progress in the synthesis of metal–organic frameworks. *Science and Technology of Advanced Materials*, 2015. 16(5): 054202.

24. Beldon, P.J., et al., Rapid room-temperature synthesis of zeolitic imidazolate frameworks by using mechanochemistry. *Angewandte Chemie International Edition*, 2010. 49(50): 9640–9643.

25. Li, H., et al., Incorporation of alkylamine into metal–organic frameworks through a brønsted acid–base reaction for CO_2 capture. *ChemSusChem*, 2016. 9(19): 2832–2840.

26. Ebrahim, A.M. and T.J. Bandosz, Effect of amine modification on the properties of zirconium–carboxylic acid based materials and their applications as NO_2 adsorbents at ambient conditions. *Microporous and Mesoporous Materials*, 2014. 188: 149–162.

27. Van Humbeck, J.F., et al., Ammonia capture in porous organic polymers densely functionalized with brønsted acid groups. *Journal of the American Chemical Society*, 2014. 136(6): 2432–2440.

28. Ahmed, I., et al., Adsorption of nitrogen-containing compounds from model fuel over sulfonated metal–organic framework: Contribution of hydrogen-bonding and acid–base interactions in adsorption. *The Journal of Physical Chemistry C*, 2015. 120(1): 407–415.

29. Ahmed, I. and S.H. Jhung, Effective adsorptive removal of indole from model fuel using a metal-organic framework functionalized with amino groups. *Journal of Hazardous Materials*, 2015. 283: 544–550.

30. Bhadra, B.N., I. Ahmed, and S.H. Jhung, Remarkable adsorbent for phenol removal from fuel: Functionalized metal–organic framework. *Fuel*, 2016. 174: 43–48.

31. Wu, S., et al., Shape control of core–shell MOF@ MOF and derived MOF nanocages via ion modulation in a one-pot strategy. *Journal of Materials Chemistry A*, 2018. 6(37): 18234–18241.

32. McDonald, K.A., et al., Polymer@ MOF@ MOF: "grafting from" atom transfer radical polymerization for the synthesis of hybrid porous solids. *Chemical Communications*, 2015. 51(60): 11994–11996.

33. Zhang, L., et al., Internally extended growth of core–shell NH_2-MIL-101(Al)@ ZIF-8 nanoflowers for the simultaneous detection and removal of Cu (II). *Journal of Materials Chemistry A*, 2018. 6(42): 21029–21038.

34. Gu, Y., et al., Controllable modular growth of hierarchical MOF-on-MOF architectures. *Angewandte Chemie International Edition*, 2017. 56(49): 15658–15662.

35. Abdelhameed, R.M., et al., Enhanced photocatalytic activity of MIL-125 by post-synthetic modification with CrIII and Ag nanoparticles. *Chemistry–A European Journal*, 2015. 21(31): 11072–11081.

36. Emam, H.E., et al., Doping of silver vanadate and silver tungstate nanoparticles for enhancement the photocatalytic activity of MIL-125-NH_2 in dye degradation. *Journal of Photochemistry and Photobiology A: Chemistry*, 2019. 383: 111986.

37. Zhu, S.-R., et al., Enhanced photocatalytic performance of BiOBr/NH_2-MIL-125 (Ti) composite for dye degradation under visible light. *Dalton Transactions*, 2016. 45(43): 17521–17529.

38. Du, Q., et al., Selective photodegradation of tetracycline by molecularly imprinted ZnO@ NH_2-UiO-66 composites. *Chemical Engineering Journal*, 2020: 124614.

39. Zhou, T., et al., Highly efficient visible-light-driven photocatalytic degradation of rhodamine B by a novel Z-scheme Ag_3PO_4/MIL-101/$NiFe_2O_4$ composite. *Catalysis Science & Technology*, 2018. 8(9): 2402–2416.

40. Tan, P., et al., Fabrication of magnetically responsive HKUST-1/Fe_3O_4 composites by dry gel conversion for deep desulfurization and denitrogenation. *Journal of Hazardous Materials*, 2017. 321: 344–352.

41. Bibi, R., et al., Hybrid BiOBr/UiO-66-NH_2 composite with enhanced visible-light driven photocatalytic activity toward RhB dye degradation. *RSC Advances*, 2018. 8(4): 2048–2058.

42. Ahmed, I., et al., Adsorptive denitrogenation of model fuels with porous metal-organic frameworks (MOFs): Effect of acidity and basicity of MOFs. *Applied Catalysis B: Environmental*, 2013. 129: 123–129.

43. Ahmed, I., N.A. Khan, and S.H. Jhung, Graphite oxide/metal–organic framework (MIL-101): Remarkable performance in the adsorptive denitrogenation of model fuels. *Inorganic Chemistry*, 2013. 52(24): 14155–14161.

44. Ahmed, I. and S.H. Jhung, Remarkable adsorptive removal of nitrogen-containing compounds from a model fuel by a graphene oxide/MIL-101 composite through a combined effect of improved porosity and hydrogen bonding. *Journal of Hazardous Materials*, 2016. 314: 318–325.

45. Petit, C., et al., Reactive adsorption of acidic gases on MOF/graphite oxide composites. *Microporous and Mesoporous Materials*, 2012. 154: 107–112.

46. Abdelhameed, R.M., M. Rehan, and H.E. Emam, Figuration of Zr-based MOF@ cotton fabric composite for potential kidney application. *Carbohydrate Polymers*, 2018. 195: 460–467.

47. Emam, H.E., O.M. Darwesh, and R.M. Abdelhameed, In-growth metal organic framework/synthetic hybrids as antimicrobial fabrics and its toxicity. *Colloids and Surfaces B: Biointerfaces*, 2018. 165: 219–228.

48. Emam, H.E. and R.M. Abdelhameed, In-situ modification of natural fabrics by Cu-BTC MOF for effective release of insect repellent (N, N-diethyl-3-methylbenzamide). *Journal of Porous Materials*, 2017. 24(5): 1175–1185.

49. Emam, H.E., et al., Non-invasive route for desulfurization of fuel using infrared-assisted MIL-53 (Al)-NH_2 containing fabric. *Journal of Colloid and Interface Science*, 2019. 556: 193–205.

50. Emam, H.E., H.N. Abdelhamid, and R.M. Abdelhameed, Self-cleaned photoluminescent viscose fabric incorporated lanthanide-organic framework (Ln-MOF). *Dyes and Pigments*, 2018. 159: 491–498.

51. Emam, H.E. and R.M. Abdelhameed, Anti-UV radiation textiles designed by embracing with nano-MIL (Ti, In)–metal organic framework. *ACS Applied Materials & Interfaces*, 2017. 9(33): 28034–28045.

52. Abdelhameed, R.M., et al., Applicable strategy for removing liquid fuel nitrogenated contaminants using MIL-53-NH_2@ natural fabric composites. *Industrial & Engineering Chemistry Research*, 2018. 57(44): 15054–15065.

53. Abdelhameed, R.M., et al., Cu-BTC metal-organic framework natural fabric composites for fuel purification. *Fuel Processing Technology*, 2017. 159: 306–312.

54. Emam, H.E., A.E. Abdelhamid, and R.M. Abdelhameed, Refining of liquid fuel from N-containing compounds via using designed polysulfone@ metal organic framework composite film. *Journal of Cleaner Production*, 2019. 218: 347–356.

55. Abdelhameed, R.M., et al., Cu–BTC@ cotton composite: Design and removal of ethion insecticide from water. *RSC Advances*, 2016. 6(48): 42324–42333.

56. Martínez-Ahumada, E., et al., MOF materials for the capture of highly toxic H_2S and SO_2. *Organometallics*, 2020. 39(7): 883–915.

57. Evans, A., R. Luebke, and C. Petit, The use of metal–organic frameworks for CO purification. *Journal of Materials Chemistry A*, 2018. 6(23): 10570–10594.

58. Zou, R., et al., Storage and separation applications of nanoporous metal–organic frameworks. *CrystEngComm*, 2010. 12(5): 1337–1353.

59. Yazaydın, A.O., et al., Screening of metal– organic frameworks for carbon dioxide capture from flue gas using a combined experimental and modeling approach. *Journal of the American Chemical Society*, 2009. 131(51): 18198–18199.

60. Liu, J., et al., CO_2/H_2O adsorption equilibrium and rates on metal– organic frameworks: HKUST-1 and Ni/DOBDC. *Langmuir*, 2010. 26(17): 14301–14307.

61. Xie, J., et al., Synthesis, characterization and experimental investigation of Cu-BTC as CO_2 adsorbent from flue gas. *Journal of Environmental Sciences*, 2012. 24(4): 640–644.

62. Choi, I.-H., et al., Gas sorption properties of a new three-dimensional In-ABDC MOF with a diamond net. *Frontiers in Materials*, 2019. 6: 218.

63. Mason, J.A., et al., Evaluating metal–organic frameworks for post-combustion carbon dioxide capture via temperature swing adsorption. *Energy & Environmental Science*, 2011. 4(8): 3030–3040.

64. Huang, Y.L., L. Jiang, and T.B. Lu, Modulation of gas sorption properties through cation exchange within an anionic metal–organic framework. *ChemPlusChem*, 2016. 81(8): 780–785.

65. Bloch, E.D., et al., Reversible CO binding enables tunable CO/H_2 and CO/N_2 separations in metal-organic frameworks with exposed divalent metal cations. *Journal of the American Chemical Society*, 2014. 136(30): 10752–10761.

66. Wang, S., et al., Computational exploration of H_2S/CH_4 mixture separation using acid-functionalized UiO-66 (Zr) membrane and composites. *Chinese Journal of Chemical Engineering*, 2015. 23(8): 1291–1299.

67. Petit, C. and T.J. Bandosz, Exploring the coordination chemistry of MOF–graphite oxide composites and their applications as adsorbents. *Dalton Transactions*, 2012. 41(14): 4027–4035.

68. Vikrant, K., et al., Utilization of metal–organic frameworks for the adsorptive removal of an aliphatic aldehyde mixture in the gas phase. *Nanoscale*, 2020. 12(15): 8330–8343.

69. Kondo, M., et al., Three-dimensional framework with channeling cavities for small molecules:{[M_2 (4, 4'-bpy)$_3$ (NO_3)$_4$] xH_2O} n (M = Co, Ni, Zn). *Angewandte Chemie International Edition in English*, 1997. 36(16): 1725–1727.

70. Li, H., et al., Establishing microporosity in open metal– organic frameworks: Gas sorption isotherms for Zn (BDC)(BDC= 1, 4-benzenedicarboxylate). *Journal of the American Chemical Society*, 1998. 120(33): 8571–8572.

71. Li, H., et al., Recent advances in gas storage and separation using metal–organic frameworks. *Materials Today*, 2018. 21(2): 108–121.

72. Tsivadze, A.Y., et al., Metal-organic framework structures: Adsorbents for natural gas storage. *Russian Chemical Reviews*, 2019. 88(9): 925.

73. Britt, D., D. Tranchemontagne, and O.M. Yaghi, Metal-organic frameworks with high capacity and selectivity for harmful gases. *Proceedings of the National Academy of Sciences of the United States of America*, 2008. 105(33): 11623–11627.

74. Nguyen, T.N., et al., A recyclable metal-organic framework for ammonia vapour adsorption. *Chemical Communications*, 2020.

75. Li, Y. and R.T. Yang, Gas adsorption and storage in metal– organic framework MOF-177. *Langmuir*, 2007. 23(26): 12937–12944.

76. Möllmer, J., et al., Pure and mixed gas adsorption of CH_4 and N_2 on the metal–organic framework Basolite® A100 and a novel copper-based 1, 2, 4-triazolyl isophthalate MOF. *Journal of Materials Chemistry*, 2012. 22(20): 10274–10286.

77. Kaye, S.S., et al., Impact of preparation and handling on the hydrogen storage properties of Zn4O (1, 4-benzenedicarboxylate) 3 (MOF-5). *Journal of the American Chemical Society*, 2007. 129(46): 14176–14177.

78. Panella, B., et al., Hydrogen adsorption in metal–organic frameworks: Cu-MOFs and Zn-MOFs compared. *Advanced Functional Materials*, 2006. 16(4): 520–524.

79. Wong-Foy, A.G., A.J. Matzger, and O.M. Yaghi, Exceptional H_2 saturation uptake in microporous metal– organic frameworks. *Journal of the American Chemical Society*, 2006. 128(11): 3494–3495.

80. Villajos, J.A., et al., Co/Ni mixed-metal sited MOF-74 material as hydrogen adsorbent. *International Journal of Hydrogen Energy*, 2015. 40(15): 5346–5352.

81. Bromberg, L. and W.K. Cheng, *Methanol as an Alternative Transportation fuel in the US: Options for Sustainable and/or Energy-Secure Transportation*. Cambridge, MA: Sloan Automotive Laboratory, Massachusetts Institute of Technology, 2010.

82. Khan, M.I., T. Yasmin, and A. Shakoor, Technical overview of compressed natural gas (CNG) as a transportation fuel. *Renewable and Sustainable Energy Reviews*, 2015. 51: 785–797.

83. Keller, M., W.A. Kaplan, and S.C. Wofsy, Emissions of N_2O, CH_4 and CO_2 from tropical forest soils. *Journal of Geophysical Research: Atmospheres*, 1986. 91(D11): 11791–11802.

84. Liu, R., *Synthesis, Characterisation and Decomposition Properties of Manganese-Based Borohydrides for Hydrogen Storage*. 2012, PhD thesis, University of Birmingham, UK.

85. Carné Sánchez, A., *A New Synthetic Method for Nanoscale Metal-Organic Frameworks and Their Application as Contrast Agents for Magnetic Resonance Imaging*, PhD thesis, Autonomous University of Barcelona, Spain.

86. Chandra, V., *Fundamentals of Natural Gas: An International Perspective*. 2017. PennWell Corporation, University of Michigan, ISBN: 9781593700881.

87. Peng, Y., et al., Methane storage in metal–organic frameworks: Current records, surprise findings, and challenges. *Journal of the American Chemical Society*, 2013. 135(32): 11887–11894.

88. Mason, J.A., M. Veenstra, and J.R. Long, Evaluating metal–organic frameworks for natural gas storage. *Chemical Science*, 2014. 5(1): 32–51.

89. Wu, H., W. Zhou, and T. Yildirim, High-capacity methane storage in metal– organic frameworks M_2 (dhtp): The important role of open metal sites. *Journal of the American Chemical Society*, 2009. 131(13): 4995–5000.

90. Colvile, R., et al., The transport sector as a source of air pollution. *Atmospheric Environment*, 2001. 35(9): 1537–1565.

91. Pawelec, B., et al., Retracted article: Towards near zero-sulfur liquid fuels: A perspective review. *Catalysis Science & Technology*, 2011. 1(1): 23–42.

92. Blanco-Brieva, G., et al., Removal of refractory organic sulfur compounds in fossil fuels using MOF sorbents. *Global NEST Journal*, 2010. 12(12): 296–304.

93. Campos-Martin, J.M., et al., Oxidative processes of desulfurization of liquid fuels. *Journal of Chemical Technology & Biotechnology*, 2010. 85(7): 879–890.

94. Capel-Sanchez, M.C., J.M. Campos-Martin, and J.L. Fierro, Removal of refractory organosulfur compounds via oxidation with hydrogen peroxide on amorphous Ti/SiO_2 catalysts. *Energy & Environmental Science*, 2010. 3(3): 328–333.

95. Monticello, D.J., Riding the fossil fuel biodesulfurization wave. *Chemtech*. 1998. 28(7): 38–45.

96. AL-Hammadi, S.A., A.M. Al-Amer, and T.A. Saleh, Alumina-carbon nanofiber composite as a support for MoCo catalysts in hydrodesulfurization reactions. *Chemical Engineering Journal*, 2018. 345: 242–251.

97. Saleh, T.A., S.A. AL-Hammadi, and A.M. Al-Amer, Effect of boron on the efficiency of MoCo catalysts supported on alumina for the hydrodesulfurization of liquid fuels. *Process Safety and Environmental Protection*, 2019. 121: 165–174.

98. Saleh, T.A., et al., Synthesis of molybdenum cobalt nanocatalysts supported on carbon for hydrodesulfurization of liquid fuels. *Journal of Molecular Liquids*, 2018. 272: 715–721.

99. Deliyanni, E., M. Seredych, and T.J. Bandosz, Interactions of 4, 6-dimethyldibenzothiophene with the surface of activated carbons. *Langmuir*, 2009. 25(16): 9302–9312.

100. Jeon, H.-J., et al., Removal of refractory sulfur compounds in diesel using activated carbon with controlled porosity. *Energy & Fuels*, 2009. 23(5): 2537–2543.

101. Zhou, A., X. Ma, and C. Song, Liquid-phase adsorption of multi-ring thiophenic sulfur compounds on carbon materials with different surface properties. *The Journal of Physical Chemistry B*, 2006. 110(10): 4699–4707.

102. Saleh, T.A. and G.I. Danmaliki, Adsorptive desulfurization of dibenzothiophene from fuels by rubber tyres-derived carbons: Kinetics and isotherms evaluation. *Process Safety and Environmental Protection*, 2016. 102: 9–19.

103. Danmaliki, G.I. and T.A. Saleh, Effects of bimetallic Ce/Fe nanoparticles on the desulfurization of thiophenes using activated carbon. *Chemical Engineering Journal*, 2017. 307: 914–927.

104. Danmaliki, G.I., T.A. Saleh, and A.A. Shamsuddeen, Response surface methodology optimization of adsorptive desulfurization on nickel/activated carbon. *Chemical Engineering Journal*, 2017. 313: 993–1003.

105. Saleh, T.A., Simultaneous adsorptive desulfurization of diesel fuel over bimetallic nanoparticles loaded on activated carbon. *Journal of Cleaner Production*, 2018. 172: 2123–2132.

106. Hernández-Maldonado, A.J. and R.T. Yang, Desulfurization of transportation fuels by adsorption. *Catalysis Reviews*, 2004. 46(2): 111–150.

107. Yang, R.T., A.J. Hernández-Maldonado, and F.H. Yang, Desulfurization of transportation fuels with zeolites under ambient conditions. *Science*, 2003. 301(5629): 79–81.

108. Ko, C.H., et al., Surface status and size influences of nickel nanoparticles on sulfur compound adsorption. *Applied Surface Science*, 2007. 253(13): 5864–5867.

109. Zhou, S., et al., Metal-organic framework encapsulated high-loaded phosphomolybdic acid: A highly stable catalyst for oxidative desulfurization of 4, 6-dimethyldibenzothiophene. *Fuel*, 2022. 309: 122143.

110. Peralta, D., et al., Metal–organic framework materials for desulfurization by adsorption. *Energy & Fuels*, 2012. 26(8): 4953–4960.

111. Rui, J., et al., Adsorptive desulfurization of model gasoline by using different Zn sources exchanged NaY zeolites. *Molecules*, 2017. 22(2): 305.

112. Liu, B., et al., Adsorption equilibrium of thiophenic sulfur compounds on the Cu-BTC metal–organic framework. *Journal of Chemical & Engineering Data*, 2012. 57(4): 1326–1330.

113. Tian, F., et al., Thiophene adsorption onto metal–organic framework HKUST-1 in the presence of toluene and cyclohexene. *Fuel*, 2015. 158: 200–206.

114. Chen, M., et al., Adsorptive desulfurization of thiophene from the model fuels onto graphite oxide/metal-organic framework composites. *Petroleum Science and Technology*, 2018. 36(2): 141–147.

115. Xiang, L., et al., Synthesis of rare earth metal-organic frameworks (Ln-MOFs) and their properties of adsorption desulfurization. *Journal of Rare Earths*, 2014. 32(2): 189–194.

116. Aslam, S., et al., Dispersion of nickel nanoparticles in the cages of metal-organic framework: An efficient sorbent for adsorptive removal of thiophene. *Chemical Engineering Journal*, 2017. 315: 469–480.

117. Khan, N.A. and S.H. Jhung, Low-temperature loading of Cu^+ species over porous metal-organic frameworks (MOFs) and adsorptive desulfurization with Cu^+-loaded MOFs. *Journal of Hazardous Materials*, 2012. 237: 180–185.

118. Zhang, X.-F., et al., Adsorptive desulfurization from the model fuels by functionalized UiO-66 (Zr). *Fuel*, 2018. 234: 256–262.

119. Morita, M., et al., Direct observation of dimethyl sulfide trapped by MOF proving efficient removal of sulfur impurities. *RSC Advances*, 2020. 10(8): 4710–4714.

120. Zhang, H.-X., et al., Adsorption behavior of metal–organic frameworks for thiophenic sulfur from diesel oil. *Industrial & Engineering Chemistry Research*, 2012. 51(38): 12449–12455.

121. Wu, L., et al., A combined experimental/computational study on the adsorption of organosulfur compounds over metal–organic frameworks from fuels. *Langmuir*, 2014. 30(4): 1080–1088.

122. Liu, Y., et al., Screening of desulfurization adsorbent in metal–organic frameworks: A classical density functional approach. *Chemical Engineering Science*, 2015. 137: 170–177.

123. Xu, W., et al., Facile room temperature synthesis of metal–organic frameworks from newly synthesized copper/zinc hydroxide and their application in adsorptive desulfurization. *RSC Advances*, 2016. 6(44): 37530–37534.

124. Ma, X., et al., Reactive adsorption of low concentration methyl mercaptan on a Cu-based MOF with controllable size and shape. *RSC Advances*, 2016. 6(99): 96997–97003.

125. Chen, M., et al., Magnetic hybridized Fe_3O_4/HKUST-1 composite modified with graphite oxide to remove thiophene from model fuels. *Petroleum Science and Technology*, 2019. 37(22): 2260–2268.

126. Jin, T., et al., Promoting desulfurization capacity and separation efficiency simultaneously by the novel magnetic Fe_3O_4@ PAA@ MOF-199. *RSC Advances*, 2014. 4(79): 41902–41909.

127. Habimana, F., et al., Synthesis of europium metal–organic framework (Eu-MOF) and its performance in adsorptive desulfurization. *Adsorption*, 2016. 22(8): 1147–1155.

128. Jafarinasab, M., et al., An efficient Co-based metal–organic framework nanocrystal (Co-ZIF-67) for adsorptive desulfurization of dibenzothiophene: Impact of the preparation approach on structure tuning. *Energy & Fuels*, 2020. 34(10): 12779–12791.

129. Huang, C., et al., Combination of coordinatively unsaturated metal sites and silver nano-particles in a Ni-based metal-organic framework for adsorptive desulfurization. *Microporous and Mesoporous Materials*, 2021. 323: 111241.

130. Jin, Y., et al., Highly efficient capture of benzothiophene with a novel water-resistant-bimetallic Cu-ZIF-8 material. *Inorganica Chimica Acta*, 2020. 503: 119412.

131. Khan, N.A. and S.H. Jhung, Adsorptive removal of benzothiophene using porous copper-benzenetricarboxylate loaded with phosphotungstic acid. *Fuel Processing Technology*, 2012. 100: 49–54.

132. Jia, S.-Y., et al., Adsorptive removal of dibenzothiophene from model fuels over one-pot synthesized PTA@ MIL-101 (Cr) hybrid material. *Journal of Hazardous Materials*, 2013. 262: 589–597.

133. Li, N., et al., Analysis and comparison of nitrogen compounds in different liquid hydrocarbon streams derived from petroleum and coal. *Energy & Fuels*, 2010. 24(10): 5539–5547.

134. Almarri, M., X. Ma, and C. Song, Role of surface oxygen-containing functional groups in liquid-phase adsorption of nitrogen compounds on carbon-based adsorbents. *Energy & Fuels*, 2009. 23(8): 3940–3947.

135. Almarri, M., X. Ma, and C. Song, Selective adsorption for removal of nitrogen compounds from liquid hydrocarbon streams over carbon-and alumina-based adsorbents. *Industrial & Engineering Chemistry Research*, 2008. 48(2): 951–960.

136. Anisuzzaman, S., et al., Adsorptive denitrogenation of fuel by oil palm shells as low cost adsorbents. *Journal of Applied Sciences*, 2014. 14(23): 3156–3161.

137. Hong, X. and K. Tang, Absorptive denitrogenation of diesel oil using a modified NaY molecular sieve. *Petroleum Science and Technology*, 2015. 33(15–16): 1471–1478.

138. Hernández-Maldonado, A.J. and R.T. Yang, Denitrogenation of transportation fuels by zeolites at ambient temperature and pressure. *Angewandte Chemie*, 2004. 116(8): 1022–1024.

139. Li, C., B.-X. Shen, and J.-C. Liu, The removal of organic nitrogen compounds in naphtha by adsorption. *Energy Sources, Part A: Recovery, Utilization, and Environmental Effects*, 2013. 35(24): 2348–2355.

140. Yang, H., et al., Inhibition of nitrogen compounds on the hydrodesulfurization of substituted dibenzothiophenes in light cycle oil. *Fuel Processing Technology*, 2004. 85(12): 1415–1429.

141. Feng, Y., A study on the process conditions of removing basic nitrogen compounds from gasoline. *Petroleum Science and Technology*, 2004. 22(11–12): 1517–1525.

142. Mushrush, G.W., et al., Post-refining removal of organic nitrogen compounds from diesel fuels to improve environmental quality. *Journal of Environmental Science and Health, Part A*, 2011. 46(2): 176–180.

143. Lee, S.-W., J.W. Ryu, and W. Min, SK hydrodesulfurization (HDS) pretreatment technology for ultralow sulfur diesel (ULSD) production. *Catalysis Surveys from Asia*, 2003. 7(4): 271–279.

144. Min, W.-S., et al., Method for manufacturing cleaner fuels. 2001, Google Patents.

145. Kim, J.H., et al., Ultra-deep desulfurization and denitrogenation of diesel fuel by selective adsorption over three different adsorbents: A study on adsorptive selectivity and mechanism. *Catalysis Today*, 2006. 111(1–2): 74–83.

146. Shiraishi, Y., A. Yamada, and T. Hirai, Desulfurization and denitrogenation of light oils by methyl viologen-modified aluminosilicate adsorbent. *Energy & Fuels*, 2004. 18(5): 1400–1404.

147. Wang, Y. and R. Li, Denitrogenation of lubricating base oils by solid acid. *Petroleum Science and Technology*, 2000. 18(7–8): 965–973.

148. Cronauer, D.C., et al., Shale oil denitrogenation with ion exchange. 3. Characterization of hydrotreated and ion-exchange isolated products. *Industrial & Engineering Chemistry Process Design and Development*, 1986. 25(3): 756–762.

149. Misra, P., et al., Selective removal of nitrogen compounds from gas oil using functionalized polymeric adsorbents: Efficient approach towards improving denitrogenation of petroleum feedstock. *Chemical Engineering Journal*, 2016. 295: 109–118.

150. Furukawa, H., et al., Ultrahigh porosity in metal-organic frameworks. *Science*, 2010. 329(5990): 424–428.

151. Li, J.-R., R.J. Kuppler, and H.-C. Zhou, Selective gas adsorption and separation in metal–organic frameworks. *Chemical Society Reviews*, 2009. 38(5): 1477–1504.

152. Ahmed, I., N.A. Khan, and S.H. Jhung, Adsorptive denitrogenation of model fuel by functionalized UiO-66 with acidic and basic moieties. *Chemical Engineering Journal*, 2017. 321: 40–47.

153. Seo, P.W., I. Ahmed, and S.H. Jhung, Adsorptive removal of nitrogen-containing compounds from a model fuel using a metal–organic framework having a free carboxylic acid group. *Chemical Engineering Journal*, 2016. 299: 236–243.

154. Ye, G., et al., Defect-rich bimetallic UiO-66 (Hf-Zr): Solvent-free rapid synthesis and robust ambient-temperature oxidative desulfurization performance. *Applied Catalysis B: Environmental*, 2021. 299: 120659.

155. Mondol, M.M.H., et al., A remarkable adsorbent for removal of nitrogenous compounds from fuel: A metal-organic framework functionalized both on metal and ligand. *Chemical Engineering Journal*, 2020, 404: 126491.

156. Shi, W., et al., Enhanced desulfurization performance of polyethylene glycol membrane by incorporating metal organic framework MOF-505. *Separation and Purification Technology*, 2021. 272: 118924.

157. Seo, P.W., I. Ahmed, and S.H. Jhung, Adsorption of indole and quinoline from a model fuel on functionalized MIL-101: Effects of H-bonding and coordination. *Physical Chemistry Chemical Physics*, 2016. 18(22): 14787–14794.

158. Sarker, M., et al., Adsorptive removal of indole and quinoline from model fuel using adenine-grafted metal-organic frameworks. *Journal of Hazardous Materials*, 2018. 344: 593–601.

159. Ahmed, I., et al., Protonated MIL-125-NH$_2$: Remarkable adsorbent for the removal of quinoline and indole from liquid fuel. *ACS Applied Materials & Interfaces*, 2017. 9(24): 20938–20946.

160. Ahmed, I. and S.H. Jhung, Adsorptive denitrogenation of model fuel with CuCl-loaded metal–organic frameworks (MOFs). *Chemical Engineering Journal*, 2014. 251: 35–42.

161. Balster, L.M., et al., Analysis of polar species in jet fuel and determination of their role in autoxidative deposit formation. *Energy & Fuels*, 2006. 20(6): 2564–2571.

162. Park, J.-S., M.T. Brown, and T. Han, Phenol toxicity to the aquatic macrophyte *Lemna paucicostata*. *Aquatic Toxicology*, 2012. 106: 182–188.

163. Zabarnick, S. and M. Mick, Inhibition of jet fuel oxidation by addition of hydroperoxide-decomposing species. *Industrial & Engineering Chemistry Research*, 1999. 38(9): 3557–3563.

164. Jiao, T., et al., The new liquid–liquid extraction method for separation of phenolic compounds from coal tar. *Chemical Engineering Journal*, 2015. 266: 148–155.

165. Jiao, T., et al., Separation of phenolic compounds from coal tar via liquid–liquid extraction using amide compounds. *Industrial & Engineering Chemistry Research*, 2015. 54(9): 2573–2579.

166. Guo, W., et al., Separation of phenol from model oils with quaternary ammonium salts via forming deep eutectic solvents. *Green Chemistry*, 2013. 15(1): 226–229.

167. Hou, Y., et al., Separation of phenols from oil using imidazolium-based ionic liquids. *Industrial & Engineering Chemistry Research*, 2013. 52(50): 18071–18075.

168. Gao, J., et al., Efficient separation of phenol from oil by acid–base complexing adsorption. *Chemical Engineering Journal*, 2015. 281: 749–758.

169. Xie, K., et al., Study of adsorptive removal of phenol by MOF-5. *Desalination and Water Treatment*, 2015. 54(3): 654–659.

170. Liu, B., et al., Adsorption of phenol and p-nitrophenol from aqueous solutions on metal–organic frameworks: Effect of hydrogen bonding. *Journal of Chemical & Engineering Data*, 2014. 59(5): 1476–1482.

171. Maes, M., et al., Extracting organic contaminants from water using the metal–organic framework Cr^{III} (OH)·$\{O_2C-C_6H_4-CO_2\}$. *Physical Chemistry Chemical Physics*, 2011. 13(13): 5587–5589.

172. Park, E.Y., et al., Adsorptive removal of bisphenol-A from water with a metal-organic framework, a porous chromium-benzenedicarboxylate. *Journal of Nanoscience and Nanotechnology*, 2013. 13(4): 2789–2794.

173. Zhou, M., et al., The removal of bisphenol A from aqueous solutions by MIL-53 (Al) and mesostructured MIL-53 (Al). *Journal of Colloid and Interface Science*, 2013. 405: 157–163.

174. Shi, Z., et al., Magnetic metal organic frameworks (MOFs) composite for removal of lead and malachite green in wastewater. *Colloids and Surfaces A: Physicochemical and Engineering Aspects*, 2018. 539: 382–390.

175. Dhakshinamoorthy, A., A.M. Asiri, and H. Garcia, Catalysis by metal–organic frameworks in water. *Chemical Communications*, 2014. 50(85): 12800–12814.

176. Sun, D., et al., Noble metals can have different effects on photocatalysis over metal–organic frameworks (MOFs): A case study on M/NH_2-MIL-125 (Ti)(M= Pt and Au). *Chemistry–A European Journal*, 2014. 20(16): 4780–4788.

177. Nasalevich, M.A., et al., Co@ NH_2-MIL-125 (Ti): Cobaloxime-derived metal–organic framework-based composite for light-driven H_2 production. *Energy & Environmental Science*, 2015. 8(1): 364–375.

178. Han, Y., et al., Ultrasonic synthesis of highly dispersed Au nanoparticles supported on Ti-based metal–organic frameworks for electrocatalytic oxidation of hydrazine. *Journal of Materials Chemistry A*, 2015. 3(28): 14669–14674.

179. Zhu, W., et al., Visible-light-induced aerobic photocatalytic oxidation of aromatic alcohols to aldehydes over Ni-doped NH_2-MIL-125 (Ti). *Applied Catalysis B: Environmental*, 2015. 172: 46–51.

180. Zhang, R., G. Li, and Y. Zhang, Photochemical synthesis of CdS-MIL-125 (Ti) with enhanced visible light photocatalytic performance for the selective oxidation of benzyl alcohol to benzaldehyde. *Photochemical & Photobiological Sciences*, 2017. 16(6): 996–1002.

181. Fu, Y., et al., Visible-light-induced aerobic photocatalytic oxidation of aromatic alcohols to aldehydes over Ni-doped NH_2-MIL-125 (Ti). *Applied Catalysis B: Environmental*, 2016. 187: 212–217.

182. Yang, Z., et al., Construction of heterostructured MIL-125/Ag/g-C_3N_4 nanocomposite as an efficient bifunctional visible light photocatalyst for the organic oxidation and reduction reactions. *Applied catalysis B: Environmental*, 2017. 205: 42–54.

183. Wang, H., et al., In situ synthesis of In_2S_3@ MIL-125 (Ti) core–shell microparticle for the removal of tetracycline from wastewater by integrated adsorption and visible-light-driven photocatalysis. *Applied Catalysis B: Environmental*, 2016. 186: 19–29.

184. Zhang, S., et al., Surface-defect-rich mesoporous NH_2-MIL-125 (Ti)@ Bi_2MoO_6 core-shell heterojunction with improved charge separation and enhanced visible-light-driven photocatalytic performance. *Journal of Colloid and Interface Science*, 2019. 554: 324–334.

185. Abazari, R., A.R. Mahjoub, and J. Shariati, Synthesis of a nanostructured pillar MOF with high adsorption capacity towards antibiotics pollutants from aqueous solution. *Journal of Hazardous Materials*, 2019. 366: 439–451.

186. Zhang, Y., et al., Humidity sensing properties of $FeCl_3$-NH_2-MIL-125 (Ti) composites. *Sensors and Actuators B: Chemical*, 2014. 201: 281–285.
187. Guo, H., et al., Visible-light photocatalytic activity of Ag@ MIL-125 (Ti) microspheres. *Applied Organometallic Chemistry*, 2015. 29(9): 618–623.
188. Wang, H., et al., Synthesis and applications of novel graphitic carbon nitride/metal-organic frameworks mesoporous photocatalyst for dyes removal. *Applied Catalysis B: Environmental*, 2015. 174: 445–454.
189. Yuan, X., et al., One-pot self-assembly and photoreduction synthesis of silver nanoparticle-decorated reduced graphene oxide/MIL-125 (Ti) photocatalyst with improved visible light photocatalytic activity. *Applied Organometallic Chemistry*, 2016. 30(5): 289–296.
190. Abdelhameed, R.M., D.M. Tobaldi, and M. Karmaoui, Engineering highly effective and stable nano-composite photocatalyst based on NH_2-MIL-125 encirclement with Ag_3PO_4 nanoparticles. *Journal of Photochemistry and Photobiology A: Chemistry*, 2018. 351: 50–58.

9 Photoactive Nanostructured Materials for Antibacterial Action
A Self-Sterilization

Sreejarani Kesavan Pillai
Council for Scientific and Industrial Research

Mabuatsela Virginia Maphoru
Tshwane University of Technology

CONTENTS

9.1   INTRODUCTION

Microbial adherence to several surfaces and the subsequent biofilm development under various ecological circumstances is a typical biological process [1]. The COVID-19 pandemic has deepened the world's focus to the transmission of microorganisms promoted by high-touch exteriors. One problem of high significance is the spread of communicable diseases, which is the capability of germs to endure on different surfaces [2]. Most microorganisms on earth exist in several collective biofilms, for instance, bacteria, fungi, archaea, and viruses [3–6]. Biofilms are tough to eliminate because of their resilience to conservative elimination or prevention strategies such as mechanical cleaning, high energy light radiation, and traditional chemical disinfection agents [1]. Therefore, the hunt for new effective antimicrobial constituents in the form of additives, colloids, or external layers with substantial deactivation rate and stability is a main requisite [7]. Extensive exploration has been directed to study resolutions that avert bacterial accumulation and biofilm establishment by killing and/or decreasing surface adhesion of pathogenic microorganisms [2]. Some innovative strategies are evolving as prospective methods for biofilm eradication and surface sterilization—the usage of

DOI: 10.1201/9781003206385-9

photoresponsive nanomaterials is very promising and has received huge amount of attention due to their outstanding germicidal properties [7,8].

Surfaces play the most important role in microorganisms' circulation, yet they are very often overlooked, with very few available procedures for effective cleaning and maintenance measures and microbiological evaluation [9]. Traditional methods such as the use of chlorine-based surface decontamination are chemical intensive and have many related shortcomings. For instance, chlorine employed for surface sterilization can react with organic material on the respective to generate chloro-organic complexes that are extremely hazardous [10,11]. Additionally, some pathogens such as viruses and specific bacteria have been identified to be impervious to chlorine decontamination [12,13]. Other alternative procedures such as ozonation and radiation of UV lamps have their individual difficulties and constraints, such as the absence of lasting impact [14] and formation of small colony mutations [15] as well as the production of poisonous disinfection by-products [16].

In the fight against biofilm resistance, emerging artificial surfaces, which lower the bonding of germs and offer germicidal activity or combinatorial influences, has surfaced as a major comprehensive approach. Innovations in nanotechnology and biological sciences have made it possible to model intelligent surfaces for reducing contaminations [17]. Surface conversion and/or functionalization to inhibit contamination via self-sterilization is a hot topic of exploration, and numerous distinct methodologies have been established recently. In building a self-sterilizing surface, it is critical not to essentially modify the elements of that surface, such that it stays competent of doing what it was intended to do [18]. Surfaces with anti-adhesive functionality, with integrated disinfectant ingredients or altered with biological active components, are some of the approaches recently recommended [9]. The new methodologies have several benefits over typical sterilization practices. For example, germicidal surfaces are in constant process of activity when compared to standard decontamination procedures. Hence, the microbial load on the surfaces is diminished without delay after exposure, inhibiting its transmission and resulting infection of neighboring surfaces or individuals [19,20]. Many materials exhibit inherent antimicrobial characteristics and hence do not require any form of modification or functionalization to exercise their action as the material in its natural capacity spontaneously destroys microbes [21]. Metals (such as silver [Ag], copper [Cu], gold [Au]) and oxides (such as copper oxide [CuO], zinc oxide [ZnO]) are well known as antimicrobial agents for centuries and applied as antimicrobial materials in various fields. Numerous other ingredients exist where antimicrobial properties are incorporated to a surface to achieve an antimicrobial surface either by surface immobilization or by integration into the bulk material. Another well-illustrated method to keep the surfaces fairly free from germ is to cover the surface using photoactivated self-cleaning materials or films [18]. Light-facilitated microbial deactivation is one such method that takes benefit of the entire wavelength range of light (ultraviolet [UV], visible, and near IR [NIR]) to obliterate a wide spectrum of pathogens [22]. Photoactive antimicrobial materials can be stimulated with light of a suitable wavelength to produce cytotoxic effect that kills infective cells via several mechanisms [23–25]. These mechanisms comprise oxidation of cell-layer lipids and amino acids in protein, cross-linkage of protein chains, and oxidative damage to nucleic acids along with the following interruption of the normal performance of the pathogen [26,27]. There is a wide disparity in the cellular configuration and structure of various microbes, which can impact the efficiency of photoactive materials or surfaces in the inactivation process. For example, Gram-positive bacteria have a solid, permeable peptidoglycan layer, which encloses a cytoplasmic layer, while Gram-negative bacteria hold an external membrane, containing a finer peptidoglycan layer, which is the cytoplasmic layer [28,29]. Therefore, the deactivation of Gram-positive bacteria is much simpler to achieve in comparison with that of Gram-negative strains. The cell walls of fungous cells have got a structure specifically intermediary in porosity between Gram-positive and Gram-negative bacteria. The external portion is a relatively permeable layer of β-glucan and mannan polysaccharides. Fungal strains (and yeasts) are much less vulnerable to photoinactivation than bacteria due to the greater dimension of the fungal (yeast) cell in comparison with the tiny bacterial cell. The quantity of cytotoxic active agent species necessary to kill a yeast cell is considerably larger than that is required to eliminate a bacterial cell [30]. In the case of viruses, outer protein envelope or sheath is a main target for the deactivation [31]. In the presence of a photoactive material, extensive structural and functional modifications occur in the microbial cells, which include structural change, loss of enzymatic activities, protein oxidation and development of protein-protein networks, and reduction in rates of metabolic processes (e.g., DNA synthesis, glucose transportation). Immediate damage to the cell membrane results in the outflow of biological contents of the cells following inactivation of the membrane transport structure and process [32].

9.2 PHOTOACTIVE NANOMATERIALS

Of late, photodisinfection methods count on the contribution of a wide variety of nanoparticles (NPs) and nanostructures that have magnitudes comparable to the wavelength of light [22]. Nanotechnology and related developments have created platforms where it is feasible to exploit nanomaterials as antimicrobial actives, which can be triggered by light radiation starting from ultraviolet (UV) to near-infrared (NIR) region [33]. For photoactive disinfection, one photon (wavelength specific) deactivation was popular earlier. However, multiwavelength or total light sources are encouraged these days as they are observed to be more efficient [34]. Although the shorter-wavelength light offers high energy, equivalent photons with long wavelength have greater penetration power [35]. Photocatalytic inactivation and removal of infective viruses, biofilm-resistant bacteria, and antibiotic resistance genes using nanomaterials can be considered a more effective technique due to its specific properties. Nanomaterials have a size scale ranging between 1 and 100 nm and can be synthesized in various structures and forms. The high specific surface areas due to small size, ample active groups, and large number of functional sites are a few benefits of the nanostructures for photocatalytic activity with high proficiency [36]. Other promising attributes of NPs are stability, uniformity, and broad-spectrum disinfectant activity against different microbes as well as the possibility of being modified with other functional groups, enhancing the disinfectant properties [37,38]. The application of such nanomaterials has presented an excellent possibility in many antibacterial remedies such as food protection, water treatment, medical device, and instruments' sterilization, or as therapeutics for localized microbial infections just to name a few [39–42]. Photoactive nanomaterials include metal oxides such as TiO_2 and ZnO with broader bandgap permitting the excitation simply in the UV light range (<380 nm). The absorption range of these materials can be expanded to visible region by modifications such as non-metal adsorption, doping with metals, or building of heterojunction semiconductors. Additional useful approach is to build new nanomaterials or hybrid structures for photocatalytic disinfection. Carbon materials represented by graphitic carbon nitride (g-C_3N_4), carbon nanotubes (CNTs), and graphene oxide (GO) as well as hybrid metal-carbon or hybrid metal-organic polymers composite nanostructures have been intensely explored in photocatalysis-based sterilization [43].

9.3 MECHANISM OF ANTIMICROBIAL ACTIVITY

Infective microorganisms cause contaminations and are exceptional in acclimatizing to the decontamination approaches that target their biochemical parameters. The increase in amounts of resilient superbugs is frightening, which entices research attention to substitute resolutions. Recently, scientists have dedicated more efforts into designing nanomaterials with antimicrobial responses stimulated by the light irradiation. Understanding of the microbial deactivation mechanisms is vital to improve the effectiveness of nanomaterials and augment the prospects of their application against several germs under diverse milieus [1]. As mentioned earlier, photoinitiated disinfection uses light irradiation at suitable energy, ranging from UV to NIR, to stimulate photoresponsive nanomaterials, which captivates light energy to successfully eliminate pathogens in a short time frame by producing radical oxygen species (ROS) and/or hyperthermic environments [44].

9.3.1 PHOTOCATALYTIC DISINFECTION

Low-energy-state molecule (S0) absorbs light energy to form high energy state (S1) that suffers intersystem crossing to form the relatively stable triplet state (T1, T3). The triplet state undergoes type I (electron transfer) photochemical process to produce superoxide and hydroxyl radicals, and/ or type II (energy transfer) photochemical process to generate singlet oxygen. These ROS can harm and kill various microorganisms (adapted from Kahef et al. [22]).

Semiconductor nanomaterials such as TiO_2, ZnO, or WO_3 are employed as photocatalysts; after exposure to shortwave (high energy) UV light of wavelength corresponding to absorbance maximum (Figure 9.1), molecule is transformed into an excited singlet (S1) state, after which it undertakes a radiation-free process; an intersystem crossing, to a lower energy state, extends the lifetime triplet state (T1, T3) [45,46]. This triplet state can take either of two photochemical pathways, namely, a type I (electron transfer) to form radical ions with surrounding molecules that, on the other hand, react with oxygen to yield cytotoxic species such as superoxide, hydrogen peroxide, and/ or hydroxyl radicals, or a type II (energy transfer) to react with molecular oxygen to generate singlet oxygen [47]. The ROS or singlet oxygen hence produced can afterward initiate a multi-spot attack mechanism against cellular membranes, lipids, intracellular proteins, and DNA of the microbes in the neighborhood. Where one bug is inherently impervious to reactive ROS attack at one location, it may be disposed to an alternative site, and therefore, the killing is still fruitful, plummeting the development of resistance in the microorganisms [48]. The cells could also be photocatalytically deactivated in anaerobic state without ROS involvement. In such cases, the bacterial cell functions as an electron acceptor and the reactive species, namely, photogenerated e^-, can access the bacterial cell straightaway, resulting in the reductive cell inactivation [49,50]. Semiconductor metal oxide nanostructures such as TiO_2 and ZnO, which are known UV photocatalysts, can be modified either by changing the morphologies or by modifying the bandgap (by doping or composite formation with polymer or other metal oxides, metals, etc.) to extend the photoresponse to light region, thus enhancing the continuous generation of hydroxyl radical ($\bullet OH$), superoxide radical ($^\bullet O_2^-$), singlet oxygen ($^1O_2^-$), and in turn antimicrobial property [51]. The relatively less effective antimicrobial non-metal oxide semiconductors like GO and C_3N_4 (g-C_3N_4) (owing to the quick recombination of electron–hole duos) can be modified by composite or heterojunction formation to obtain significantly higher microbial inactivation efficiency [49]. The Bi-based oxides, for instance, $BiVO_4$, Bi_2O_3, and $Bi_2O_2CO_3/Bi_3NbO_7$, as well as ternary mixtures, such as $ZnIn_2S_4$, $CdIn_2S_4$, and $In_2O_3/CaIn_2O_4$, have been intensely explored in the photocatalytic decontamination [33].

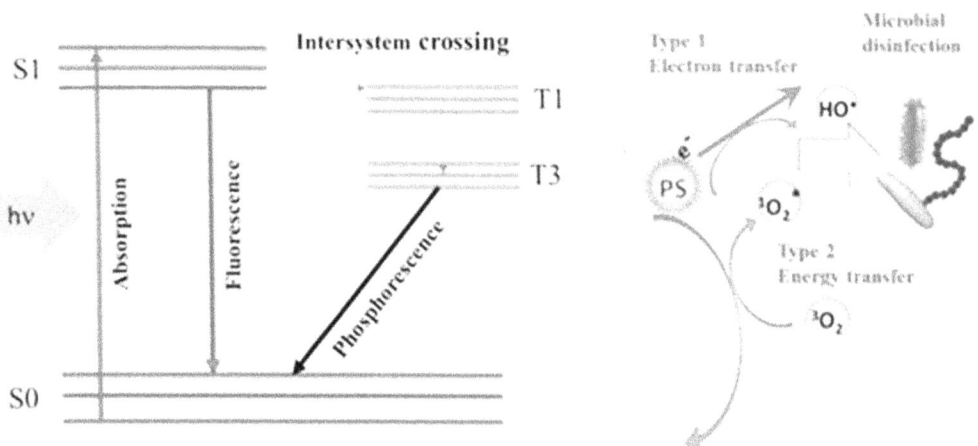

FIGURE 9.1 Jablonski diagram.

In addition to the semiconductors, biodegradable organic nanostructures (such as liposomes, polymeric micelles, nanoemulsions, and polymeric NPs) as well as carbon-based NPs (namely, CNTs), fullerene C_{60}, carbon dots, graphene, and quantum dots) are prospective organic nanovehicles of photosensitizer molecules (like porphyrines, phenothiazines, phthalocyanines, and their derivatives) to augment the photocatalytic disinfection characteristics [52–55]. Covalent or noncovalent alteration of surfaces of metal NPs with photosensitizer molecules as well as changes in the morphology is also employed currently as alternate functional hybrids for the photoinactivation of microorganisms and extends the photoabsorptive property to NIR region [56]. Upconvertion NPs (lanthanide-doped nanocrystals) that can capture light with long wavelength (NIR) and release light with high energy (UV) have enticed interest as antimicrobial nanoagents recently. Metal NPs contribute to the photocatalytic activity of the photosensitizers by surface plasmon resonance (SPR) [57]. In such cases, either a charge transfer mechanism in which the plasmon resonance triggers electron transfer from noble metal to the conduction band (CB) of the photocatalyst [58] or the development of a powerful resident electric field on the surfaces of photocatalysts after the metal NPs exposed to near plasmon resonance frequency leading to, quickening the separation of e^-–h^+ pairs were advocated [59].

9.3.2 Photothermal Disinfection

Photothermal disinfection mechanism comprises hyperthermia damage to the microbial cells with the help of photosensitive agents and light [33]. Under light stimulus, specific nanomaterials can transform solar energy into heat, which can also destroy pathogenic microorganisms [60]. In the photothermal disinfection process, light-engrossing nanostructures are illuminated with appropriate wavelength. On light exposure, the NPs instantly transfer the energy into heat through nonradiative relaxation to the neighboring environment. The produced hyperthermal effect ultimately causes irrevocable harm to bacterial cells, resulting in its death [33]. Ideal attributes of photothermal nanomaterials are strong NIR absorbance, high photothermal exchange, and nontoxic nature [61,62]. Since NIR (700–1,300 nm) generates only minimal damage to healthy cells, compared to UV and visible light, this technique is useful for biologically intense sterilization [63]. Based on the different aspects of the materials, photothermal mechanism can be divided into three groups: (i) the localized surface plasmon resonance (LSPR) effect, (ii) electron-hole generation and relaxation, and (iii) the conjugation or hyperconjugation effect [64].

Noble metal NPs have been extensively used to incapacitate microorganisms, owing to their LSPR effect. Gold nanoparticle, carbon nanomaterials (such as CNTs, fullerenes, GO nanosheets), as well as some other NIR-absorbing nanomaterials (like palladium nanosheets and silicon carbides), have presented great potential in the photothermal destruction of microbes [33,65].

Lately, metal-organic frameworks (MOFs), porous, crystalline materials composed of organic and metal units joined through coordination bonds (e.g., Prussian blue nanotubes), have been used in photothermal disinfection because of their adaptable morphology, structure, size, and changeable compositions [66,67]. Even if single MOF displays low photothermal conversion efficacy, the photothermal performance can be increased by limiting the bandgap of the MOF by metal ion doping, which produces red-shifted LSPR toward the lower-energy region. The modified structure demonstrates a superior germicidal effect, which is assigned to the synergy between photothermal effect of MOF and the intrinsic antimicrobial properties of metal ions [68].

Transition metal sulfides or chalcogenides (CuS, MoS_2, WS_2, TiS_2) hold a robust NIR absorption capacity, which makes them highly effective and capable photothermic agents [44]. Due to innate lattice defects, they are p-type semiconductors with a large number of hole carriers, which can trigger the LSPR phenomenon distinct from that of noble metals. Additionally, the d-d transitions in transition metal can lead to ROS generation, thus significantly boosting the photothermal effect on NIR exposure [69].

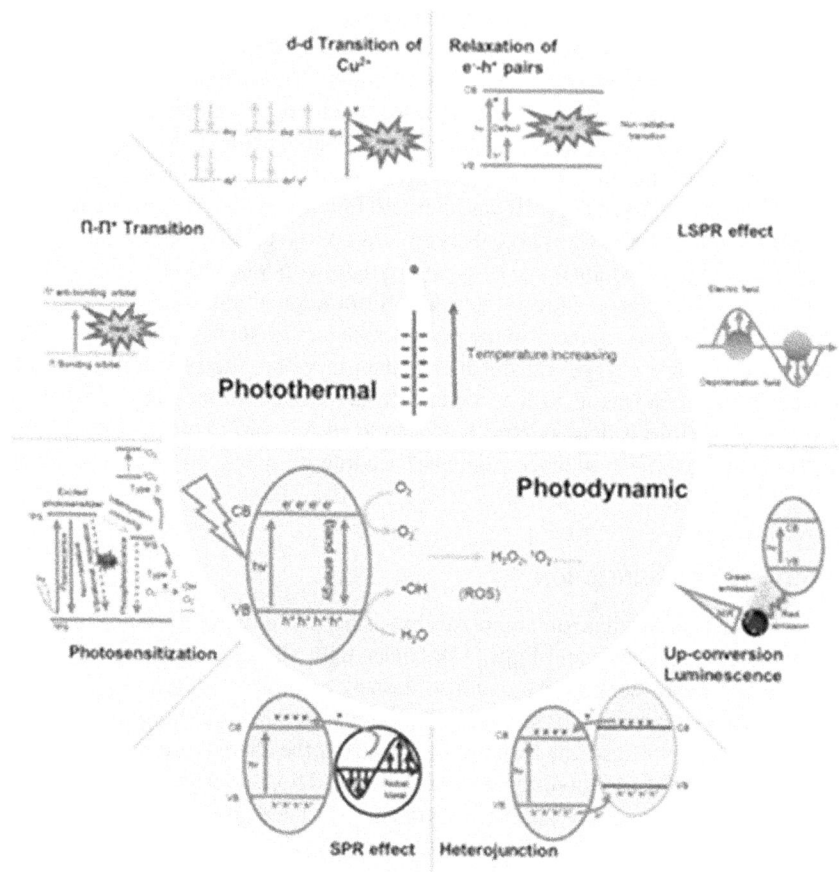

FIGURE 9.2 Schematic illustration of different processes involved in photodynamic and photothermal disinfection mechanisms. (Reproduced from Ren et al. [44].)

NPs of organic polymers, namely, polyaniline (PANI), polypyrrole (PPy), conjugated polyelectrolytes (CPEs), and their composites, show a photothermal antimicrobial activity as a result of their outstanding light-harvesting properties [33,44]. Stability, controllable particle size, and low long-lasting cytotoxicity are emphasized as pluses of these materials. The photothermal effect of polymeric NPs is primarily associated with non-radiative relaxation routes, which necessitates materials to have a low fluorescence quantum yield and singlet oxygen production yield, thus guaranteeing the highest energy-heat transformation. The π conjugate system in the polymer is accountable for causing the photothermal impact following poor singlet oxygen yield and high heat production [70]. Various procedures involved in photocatalytic and photothermal disinfection mechanisms are summarized in Figure 9.2.

9.4 FACTORS AFFECTING KINETICS OF LIGHT-MEDIATED MICROBIAL DISINFECTION

The photocatalytic disinfection is a very complicated process. It entails several factors: let's say pH of the environment, different catalyst loading, light irradiation power, the temperature, and most crucially the intricate structure of the microorganism [71,72].

The influence of physiological properties of the environment such as pH, temperature, and salt concentration in the photocatalytic process is vital. The surface charge of the photoactive material and its microbial cell wall adherence is reliant on the overall pH of the environment. A positive surface charge of the catalyst improves the binding to the negatively charged cell wall of the microbes. This consequently intensifies the decontamination process [73]. The disinfection rate is also impacted by the concentration of photoactive material used. Increase in concentration enhances the disinfection rate primarily and then reaches a steady state, at which the rate of the reaction remains unchanged even with the increase in the amount of catalyst. A higher photoactive material concentration, however, can hinder the light absorption and can be detrimental to the disinfection efficacy [74]. In a photodeactivation reaction, the irradiation intensity and interval of light irradiation are the major factors controlling the proficiency of the process. Higher-intensity continuous irradiation results in heightened ROS generation. Together with the strength of the irradiation, the surface area of the disinfectant is also an element in the successful rate of ROS production [75].

Bandgap of semiconductor photocatalyst also controls the photocatalytic effectiveness. For materials with narrow bandgaps, a feature that diminishes the photocatalytic activity is the fast recombination of photogenerated e^-–h^+ pairs in single semiconductors, and a realistic approach to solve this issue is either doping with other metal or metal oxides or constructing heterojunctions with more than one semiconductors, which can encourage charge separations and transfers [76]. The light-capturing capability of noble metals is related to their dimensions, shapes, homogeneity, and surface chemical states. Once their sizes and forms are suitably varied; they can exhibit substantial localized SPR properties, which can provide them with photothermal characteristics under NIR irradiation, consequently leading to microbial disinfection [77]. In the case of metal sulfides, photothermal conversion productivities can be changed by manipulating the thickness and surface structure of the materials [78].

9.5 PHOTOCATALYTIC ANTIMICROBIAL NANOMATERIALS

For many years, photoactive antibacterial nanomaterials (semiconductors [79–81], carbon-based [82,83], polymer-based [84,85], and metal-based [86–88] nanomaterials) have demonstrated their ability to destroy bacterial cell through multiple bactericidal pathways, thus putting a lot of strain on bacteria to survive [88–90]. The ability of the nanomaterial to generate ROS and free radicals, which raise the oxidative stress that damages the cells of the bacteria through the fragmentation of genomic DNA [91–93], is one of the most important aspects of photoactive antimicrobial activities of the nanomaterials. ROS, singlet oxygen, $O_2^{\bullet-}$, H_2O_2, and OH• originate from redox reactions initiated by electrons (e^-) and holes (h^+) [79,94]. In addition, lots of research have demonstrated that nanomaterials are capable of permeating through the cell membrane as compared to the commercially used antibiotics [95–97]. In many occasions, they have proved to be capable of acting as efflux pump inhibitors and not highly subjected to bacterial resistance [98,99], which is a common problem encountered with currently used antibiotics [96,97].

The antibacterial properties of the nanomaterials, in the absence or presence of light, are related to chemical and physical properties of the materials [100,101]. It is, therefore, of paramount importance that materials are designed and tailored in a way suitable to their intended application. In the past 20 years, nanomaterials of different dimensions and shapes, such as nanoplates [102], nanospheres [88,103–105], nanorods [89,98,106], nanocubes [87,106,107], nanoflowers [108,109], nanodots [110,111], nanofibers [112–115], nanowires [116,117], and nanosheets [118,119], with different sizes were synthesized and proved to have an impact on their antibacterial activities.

Many studies proved that composite materials have high antimicrobial activities as compared to their individual components [108,120,121]. This is attributable to the interaction of components of the composites where their combined physical and chemical properties complement each other and increase their ability to attack the membrane of the bacterial cell [108,120,121]. This can be achieved through the separation of electrons and holes [122], heat conversion [123], physical attack [101,105], exchange of electrons [123], and ability to generate ROS [124]. It is for this reason that heterostructures are the most preferred materials instead of their pure components. However, more efforts should be placed on their design, through their synthetic routes, since their final structures play a very significant role in their antibacterial activities [89,125]. Various preparation methods of photoactive nanomaterials have been studied, and the structures of the resulting nanomaterials and their effect on their photoactive antibacterial properties will be discussed in this section.

9.5.1 METAL OXIDE-BASED NANOMATERIALS

Semiconductor-based materials form a huge class of important photoactive materials with high antibacterial properties due to their ability to generate ROS that play a very important role in their antibacterial activities against Gram-positive and Gram-negative bacteria [126,127]. Common examples of photoactive semiconductors are TiO_2 [95,104,128], Fe_3O_4 [129,130], ZnO [89,100], and CuO [127]. In general, they can generate ROS when irradiated with light, which excite an electron in the valence band (VB) to the CB and leave a positive hole in the VB [79]. Among the generated ROS, OH• is proven to be responsible for the peroxidation of lipid membrane of the bacteria and the fragmentation of the DNA structure in the light-induced antibacterial action of semiconductor-based nanomaterials [79,120]. Semiconductors are often incorporated with each other [121], metal NPs [108,120,122], carbon materials [116,125], and polymers [131,132] to form heterostructures with improved physical and chemical properties, thereby increasing their ability to destroy the bacteria. For example, metal NPs added to the semiconductors act as light harvesters and sensitizers due to their high absorption/scattering cross-sections and increase the ability of the materials to move the electrons into the CB of the semiconductor [93,108].

Hajipour and co-workers [105] synthesized CuO-GO NCs (NCs: nanocomposites, GO: graphene oxide) by chemical bath deposition methods, and CuO-GO-Ag NCs were prepared by the insertion of GO nanosheets and Ag NPs onto copper metal precursor. High dispersion of CuO on CuO-GO was realized in comparison with that of CuO (Figure 9.3a–c), which indicates that the addition of GO inhibited the nucleation of CuO due to its ability to overcome a high super saturation and low electrostatic repulsive barriers of CuO particles. When Cu, GO, and CuO-GO materials were tested for their photoactive properties on Gram-negative *Escherichia coli* and Gram-positive

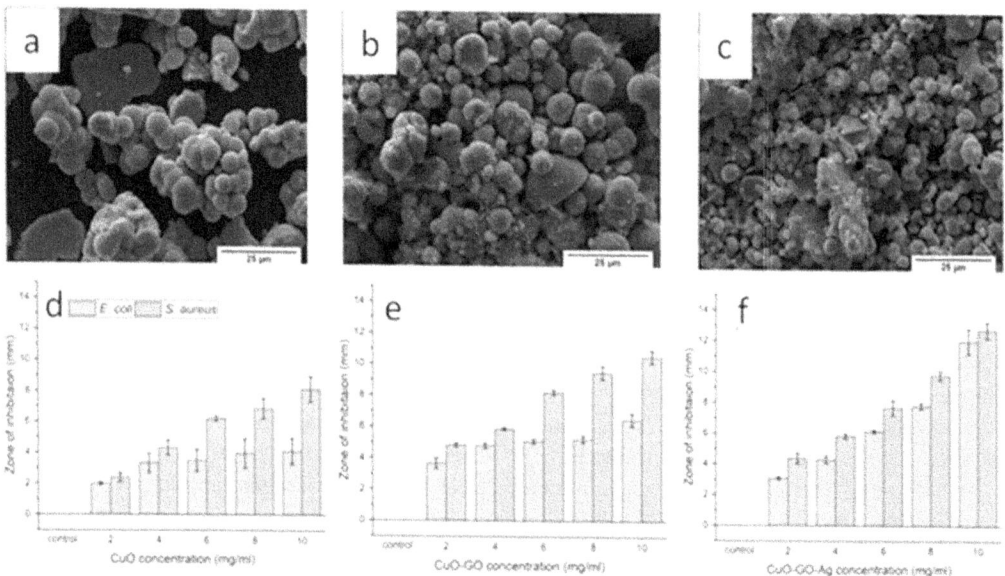

FIGURE 9.3 SEM micrographs of CuO NPs (a), CuO-GO NCs (b), and CuO-GO-Ag NCs (c). Zone of inhibition of CuO NPs (d), CuO-GO NCs (e), and CuO-GO-Ag NCs (f) against *E. coli* and *S. aureus*. (Hajipour [105].)

Staphylococcus aureus bacterial strains using visible light, MIC (minimum inhibition concentration) values of 6.6±0.5, 5.3±0.5, and 2.6±0.5 mg/mL against Gram-negative *E. coli* and 4.6±0.5, 3.3±0.5, and 2.6±0.5 mg/mL against Gram-positive *S. aureus* bacterial strains were realized for CuO, CuO-Go, and CuO-GO-Ag, respectively. However, low MIC values and high inhibition zones were obtained for Gram-positive *S. aureus* bacterial strains as compared to Gram-negative *E. coli* bacteria, and this might be due to the surface of their membranes. Gram-positive bacteria contain amines and carboxyl groups, which interact well with copper, therefore exposing the membrane to Cu toxicity.

In high concentration of CuO-GO-Ag, the difference in the inhibition zones of Gram-positive *S. aureus* from Gram-negative *E. coli* bacteria was very low in comparison with when CuO or CuO-GO was used (Figure 9.3d–f). This might be due to the release of Ag^+ ion from CuO-GO-Ag nanocomposites that is possible at higher concentration of CuO-GO-Ag nanocomposites. The addition of GO and Ag to CuO improved the photoactive antibacterial properties of CuO due to synergic effects of GO nanosheets in charge separation and the SPR effect of Ag NPs that increase the ability of the material to absorb light, leading to the enhancement of ROS formation. In addition, the sharp edges of GO nanosheets are capable of physically cutting the membrane of the bacterial cells.

Mohri and partners [133] investigated the photoactive properties of Sn/SnO_2 foil with lamellar structure against *E. coli* using UV light (366 nm) for 1 hour. The material was prepared by the anodization of Sn at low applied voltages (2–2.2 V) in oxalic acid dihydrate (0.05 wt.-%) and di-potassium oxalate monohydrate buffer solution (4.00 wt.-%). The presence of SnO_2 on Sn was confirmed by CEMS Mossbauer and XPS analysis. Sn/SnO_2 foil inhibited the growth of the bacteria as compared to the polished (non-anodized) foil. This outcome is related to the ability of SnO_2 to generate ROS on Sn/SnO_2 foil surface during its irradiation with UV light, while that is not the case on the surface of non-anodized Sn foil.

Jose and colleagues [97] studied light-induced antimicrobial properties of 2-[(E)-(2-hydroxy naphthalen-1-yl) diazenyl] benzoic acid (dye)-modified lignin encapsulated with zinc oxide NPs (ZONP-lignin-dye) using three Gram-negative strains (*Raoultella ornithinolytica*, *Salmonella typhimurium*, and *S. paratyphi*) and three Gram-positive bacterial strains (*Corynebacterium diphtheriae*, *Bacillus cereus*, and *Staphylococcus haemolyticus*). High inhibition zone was obtained for ZONP-lignin-dye when compared to commercially available antibiotics (tetracycline, vancomycin, penicillin, ampicillin, and chloramphenicol) (Table 9.1). it is noteworthy that low antibacterial activities were obtained in non-irradiated samples as compared to light-irradiated samples, which indicates that light is responsible for high antimicrobial activities of ZONP-lignin-dye. Like in many systems, Gram-positive bacterial strains were more sensitive to the photoactive action than

TABLE 9.1
Zone of Inhibition Obtained from ZONP-Lignin-Dye and Antibiotics Against Different Bacterial Strains

Bacterial Strain	Zone of Inhibition (nm)		
	ZONP-Lignin-Dye (Dark)	ZONP-Lignin-Dye (Light)	Antibiotic
C. diphtheriae	25.66±0.816	27.83±0.752	18.33±0.516 (T)
B. cereus	25.66±0.516	28.66±0.516	19.50±0.836 (V)
S. haemolyticus	30.00±0.894	34.33±1.032	15.66±0.816 (P)
R. ornithinolytica	23.50±0.547	26.33±0.816	18.50±0.547 (A)
S. typhimurium	23.16±0.983	25.83±0.408	9.50±0.547 (C)
S. paratyphi	24.33±0.516	26.16±0.816	10.50±0.547 (C)

Source: Jose [97].

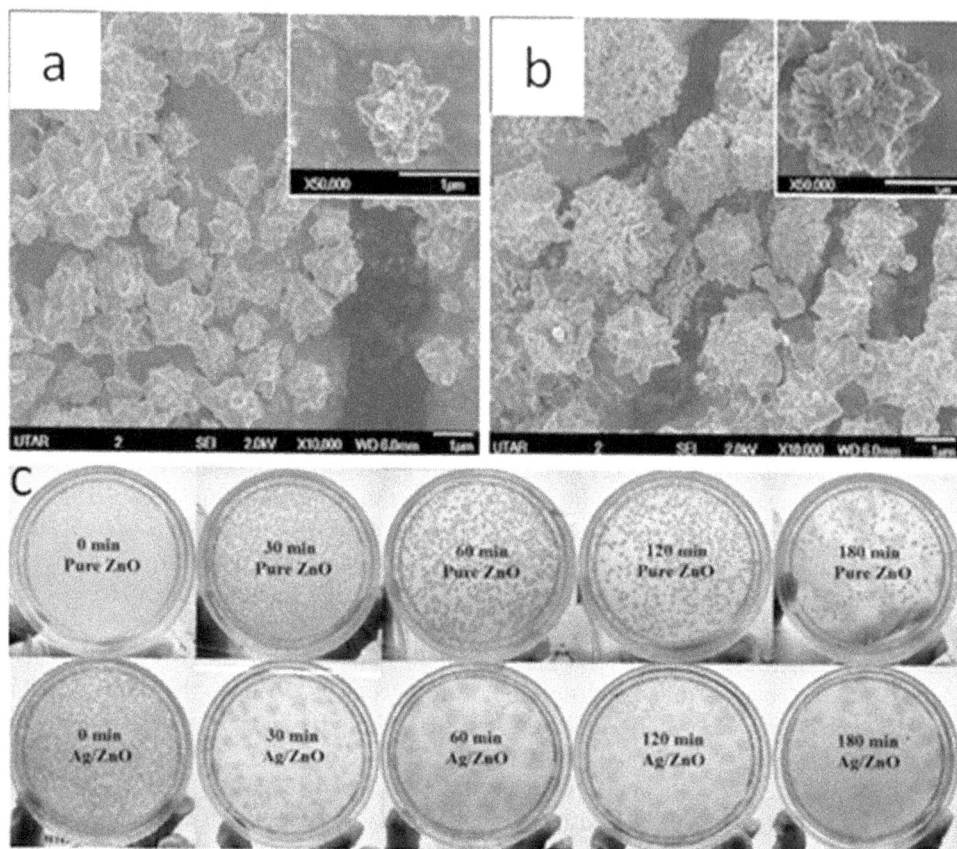

FIGURE 9.4 FESEM micrographs of pure ZnO (a), 5 wt% Ag/ZnO composite (b) and bacterial growth inhibition for *E. coli* on ZnO and 5 wt% Ag/ZnO at various irradiation times (c). (Lam [108].)

Gram-negative strain counterparts. The study proved that the incorporation of photosensitizing agents can destroy the bacterial cell in the presence of light.

Lam and coworkers [108] compared the antimicrobial activities of ZnO nanoflowers, with a broad size range (700 nm–2.4 μm), and 5%Ag/ZnO nanoflowers on *E. coli* in the presence of light (Figure 9.4a and b). The addition of Ag NPs on the surface of ZnO increased the surface roughness of the material. High bacterial activities were obtained on 5%Ag/ZnO as compared to ZnO, which indicates that 5%Ag/ZnO possesses high antimicrobial activities in comparison with ZnO after 180 minutes (Figure 9.4c). This is also visible by a drastic increase in protein concentration obtained from the *E. coli* specimen that was treated with 5%Ag/ZnO (Figure 5a). Combination of light and 5%Ag/ZnO is responsible for these activities as light alone and 5%Ag/ZnO in the dark were not able to achieve these activities. The presence of Ag^+, which is commonly known for increasing the antimicrobial activities, was analyzed by atomic absorption spectroscopy and found to be 0.8 mg/L after 180 minutes. This low concentration excluded Ag^+ toxicity as a possible mechanism for the antibacterial activities of 5%Ag/ZnO, thus leaving the formation of ROS to be responsible for the cell damage of the bacteria. Silver in the composites has the ability to trap the photogenerated e^- after irradiation, thus playing a very important role in the separation of e^- and h^+, which in turn increases its photoactive antibacterial activities (Figure 9.5b).

Sharma and coworkers [89] investigated the antibacterial activities of ZnO nanorods with different sizes on Gram-negative *E. coli* and Gram-positive *S. aureus* in the presence of light. The shape of the rods experienced severe distortion with increasing preparation temperature (275°C, 350°C,

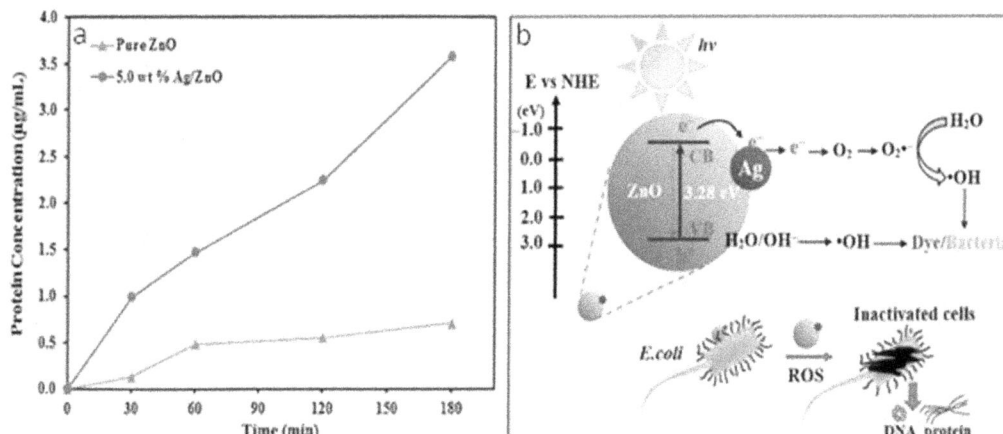

FIGURE 9.5 Concentration of protein released from the bacterial cells (a) and (b) proposed photo-induced antibacterial mechanism for Ag/ZnO composite. (Lam [108].)

425°C, 500°C), i.e., perfect nanorods with the highest aspect ratio of 18.75, with a length×width of 1500 nm×80 nm, were obtained for ZnO prepared at 275°C, while ZnO NPs instead of rods were obtained at 500°C. For both *E. coli* and *S. aureus*, the highest antibacterial activities were obtained from the material with the highest aspect ratio: ZnO, (275°C) due to its highest surface area, allows high generation of ROS and exhibits a reduced rate for charge recombination. Also, its abrasive morphology enables it to physically damage the cell membrane of the bacteria.

Marín-Caba et al. [117] synthesized different TiO_2-based composites, CNT-SNP@AuNR@TiO_2 (CNT: carbon nanotubes, SNP: silica nanospheres, NR: nanorods) and TiNW@AuNR@TiO_2 (NW: nanowires), by sequential assembly of AuNRs and TiO_2 NPs onto the colloidal templates (CNT-SNP and TiNW) following the layer-by-layer strategy. When the composites were tested for their light-assisted antimicrobial properties on *E. coli*, high antimicrobial activities were obtained on CNT-SNP@AuNR@TiO_2 (Figure 6a and b) using simulated sunlight (350–2,400 nm) for 4 hours. The difference in their activities were attributable to the amount of ROS that the two materials were able to produce, where CNT-SNP@AuNR@TiO_2 showed the ability to produce more •OH radicals as compared to TiNW@AuNR@TiO_2 and *E. coli*, which is a Gram-negative bacterium, is prone to •OH radicals' attack. This is evident by severe cell damage obtained in the presence of CNT-SNP@AuNR@TiO_2. In the absence of light, high viable cell count was obtained, thus showing that antimicrobial activities were extremely lower than in irradiated samples.

Joe and group [102] synthesized hexagonal ZnO NPs (NP: nanoplates, 60 nm), ZnO NAs (NAs: nanoassemblies, 200 nm), and ZnO CNs (CNs: conventional NPs, 30 nm) (Figure 9.7a–c) and evaluated their light-assisted (UV-A) antimicrobial activities on *S. aureus* and *K. pneumonia*. The highest antibacterial efficiency was obtained on ZnO NPs as compared to ZnO NAs and CNs (Figure 9.6d and e). Double immunofluorescence analysis with ROS staining assay using calcein-AM showed that more ROS species are produced on ZnO NPs as compared to ZnO NAs and ZnO NCs bacterial specimens, therefore leading to superior antibacterial activities on ZnO NPs. This enhanced ROS generation of ZnO NPs is attributable to the presence of polar facets, which also enhances electron-hole pair separation. When Ag intercalated on ZnO NPs (Ag particle size: 10±5 nm), the bacterium cell viability on ZnO NPs was higher than on Ag/ZnO NPs, which was attributable to the cytotoxicity of Ag NPs and good electron-hole separation brought about the addition of Ag on ZnO.

Wu and coworkers [120] investigated the antibacterial activities of Ag-doped TiO_2 (Ag/TiO_2, rutile) nanofibers on *E. coli* and *S. aureus* in the presence of visible light. There was a visible bacterial growth of *S. aureus* against TiO_2-P25 and pristine TiO_2-treated specimen, while bacterial

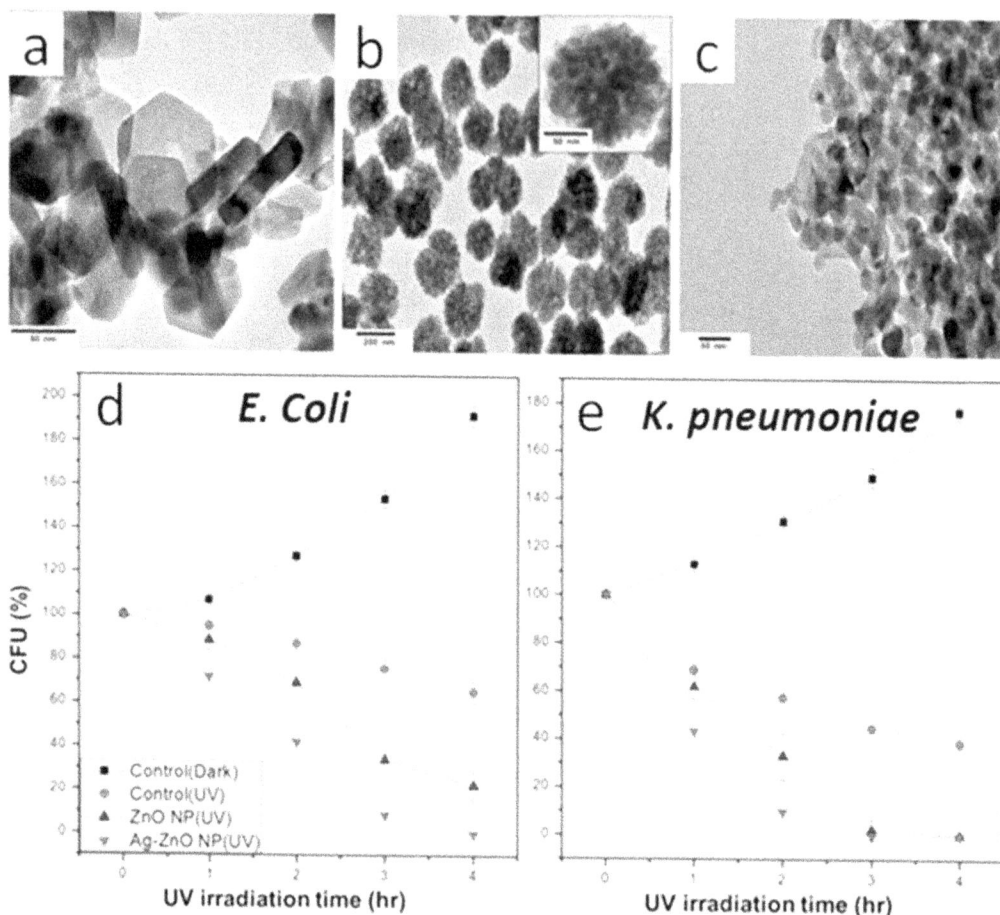

FIGURE 9.6 TEM images of ZnO NPs (a), NAs (b), and CNs (c). CFU (%) variation versus irradiation obtained for *S. aureus* (d) and *K. pneumoniae* (e) on ZnO NPs (0.1 mM and Ag-ZnO NPs (0.05 mM). (Joe [102].)

growth was inhibited on Ag/TiO$_2$-treated *S. aureus* specimen (Figure 9.7). High antibacterial activity of Ag/TiO$_2$ was attributable to the synergic effect of Ag$_2$O, which was found on the composite surface, and ROS (O$_2^{\bullet-}$) species were dominating, which was formed by irradiated Ag/TiO$_2$.

Zhang and co-workers [90] coated brackets for teeth braces with ZnO and ZnO/CQDs composites (CQDs: carbon quantum dots) for the degradation of *S. mutans*, *S. aureus*, and *E. coli* using natural light. The difference in the antibacterial activities of ZnO and ZnO/CQDs is brought about the addition of CQDs, which enhances the ability of the material to absorb visible light. CQDs can also enhance the separation of electrons and holes (Figure 9.8). Superior antibacterial rates, 92.35%, 96.13%, and 90.28%, for *E. coli*, *S. mutans*, and *S. aureus*, respectively, with low CFU were obtained on ZnO/CQD-coated brackets in comparison with 57.14%, 45.31%, and 42.4% for ZnO-coated brackets (Figure 9.9a–c).

In the study carried out by Liu et al. [121], they demonstrated that the addition of AgVO$_3$ QDs on TiO$_2$ NSs (TiO$_2$/AgVO$_3$, 100–300 nm, NSs: nanospheres, QDs: quantum dots) by in situ ultrasonic reaction improves the photoactive antibacterial properties on *E. coli* (Figure 9.9a). The experimental outcome showed that O$_2^-$ and h$^+$ might have contributed significantly to the enhancement of the photoactive properties of the composite. Weaker PL intensity (Figure 9.9b) was obtained for the composite than for TiO$_2$, which indicates that the addition of AgVO$_3$ QDs improves electron-hole separation.

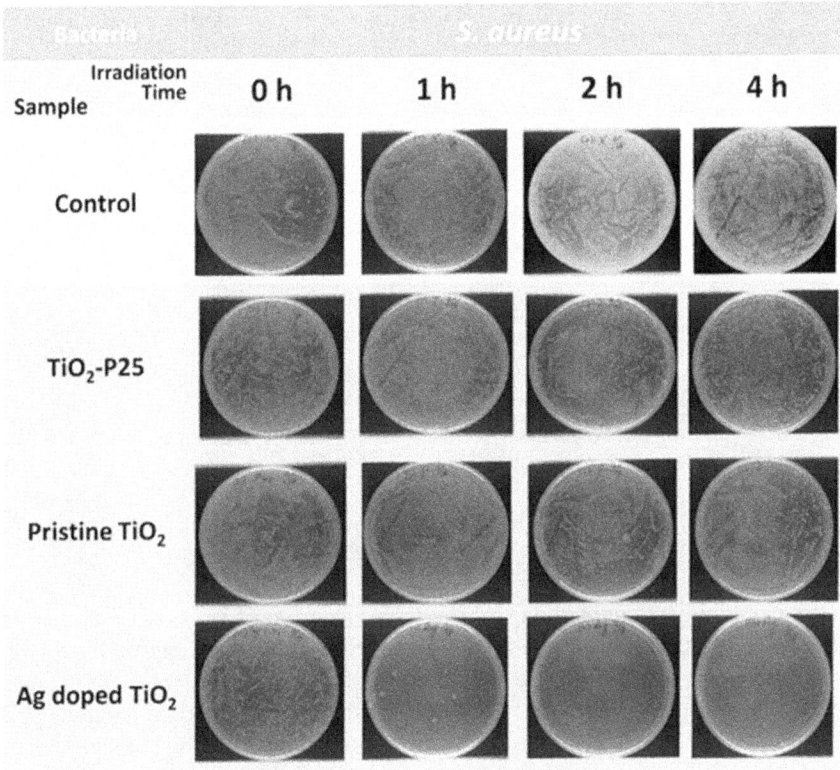

FIGURE 9.7 Bacterial growth inhibition for *E. coli* and *S. aureus* on TiO₂-P25, pristine TiO₂, and Ag/TiO₂ at various irradiation times (c). (Wu [120].)

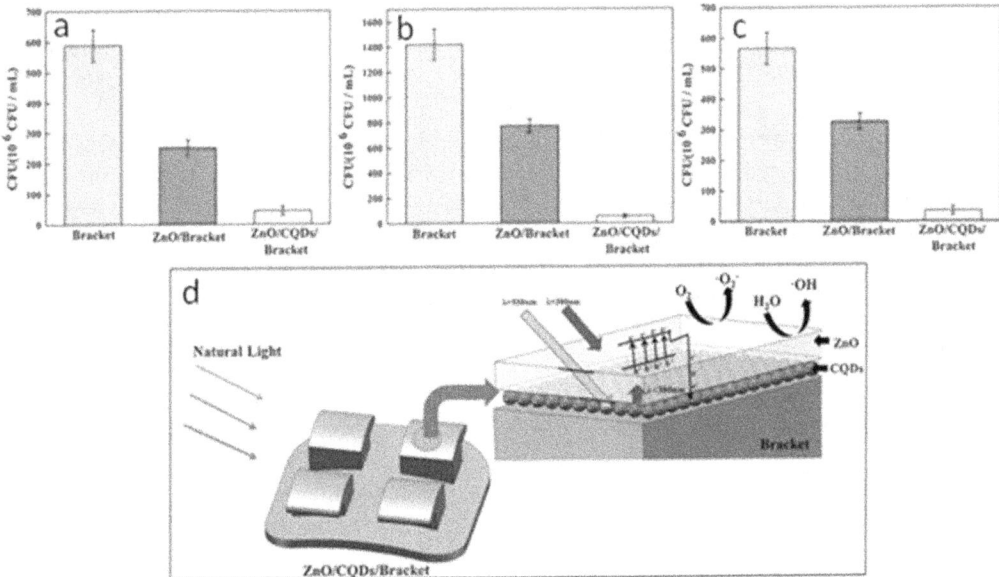

FIGURE 9.8 Colony count for *E. coli* (c), *S. mutans* (b), and *S. aureus* (c). Photo-induced antibacterial mechanism of ZnO- and ZnO/CQD-coated brackets. (Zhang [90].)

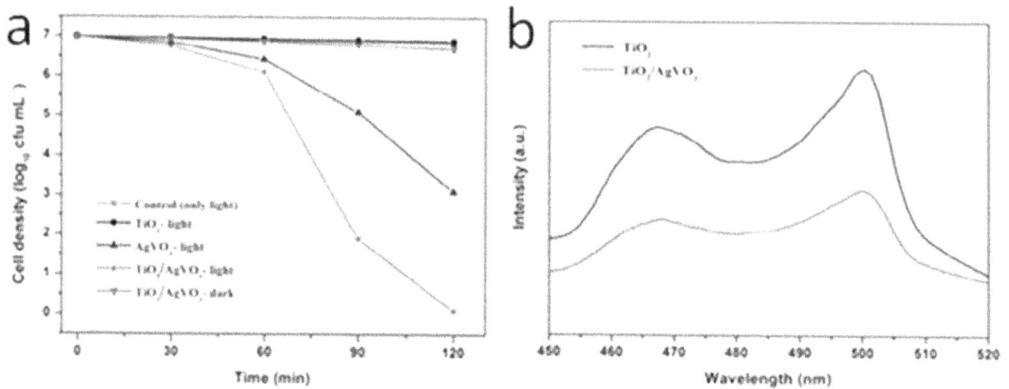

FIGURE 9.9 Light-induced antibacterial activity of the nanocomposites against *E. coli* (a) and PL spectra of TiO$_2$ NSs and TiO$_2$/AgVO$_3$ composite (b). (Liu [121].)

9.5.2 METAL-CARBON-BASED NANOMATERIALS

Carbon-based materials are counted among the wide spectrum of materials that can be used to destroy bacterial strains that developed an ability to resist antibiotics, and its modes of antibacterial action have been widely explored [83,134–136]. Such materials are GO (graphene oxide) [83,123], rGO (reduced graphene) [44,136], SWCNTs (single-walled carbon nanotubes) [134,136], MWCNTs (multi-walled carbon nanotubes) [137,138], and carbon dots [90,110]. Through their functionalization, most carbon materials contain oxygen functional groups, such as epoxy, carboxyl, and hydroxyl, on carbon surface, and these groups promote its dispersion in an aqueous medium, and this property is very important in the removal of biofilms [139–141]. Their ability to generate ROS [111,122,125,137] makes them promising materials as photoactive antibacterial materials. Their structures, sheets and tubes, enable them to physically rupture the cell membrane of the bacteria cell [83]. Incorporating metals and other chemical species on carbon materials brings a lot of improvements on their photoactive antibacterial activities [134–136,142–144]. Silver and gold NPs, due to their plasmonic effect that depends on their concentration, shape, and size, are often used for this purpose [83,122,123]. In this section, various GO-metal-based nanocomposites and their antimicrobial activities will be discussed.

Alayande and co-workers [125] studied light-induced antibacterial properties of rGO-CuO(1) and rGO-CuO(2) (treated with 2 M NaOH) nanocomposite films, synthesized by hydrothermal method, on *Pseudomonas aeruginosa* (PAO1). There is an enhanced antimicrobial activity on both rGO-CuO nanocomposites in comparison with rGO (Figure 9.10a). The authors concluded that the cell damage originated from oxidative stress, which was proven by the oxidation of thiols on glutathione (GSH) to imitate nature (GSH, present on Gram-negative bacteria), where an oxidation of 1%, 73.5, 71.5%, 72%, and 73.5% was obtained for H$_2$O$_2$, rGO-CuO(1), and rGO-CuO(2), respectively. This shows that the rGO-CuO(2) nanocomposites have the ability to generate oxygen species that are responsible for their antibacterial properties. Furthermore, it was observed that the morphology of the nanocomposites plays a significant role in their antibacterial activities, where bacterial cells on rGO-CuO(1) and rGO-CuO(2) films were damaged, while they remained intact on rGO (Figure 9.10b–d). In addition, the cell damage was more prominent on rGO-CuO(2) composite film in comparison with rGO-CuO(1), which was attributable to the presence of the nanorods on rGO-CuO(2), which can physically penetrate the cell membrane of the bacteria, thus increasing its antibacterial activities.

Mallikarjuna and team [122] incorporated Pd-Ag NPs on multi-layered reduced graphene oxide (rGO) (Pd-Ag/rGO) using *stevia* extract and studied their photo-induced antimicrobial activities

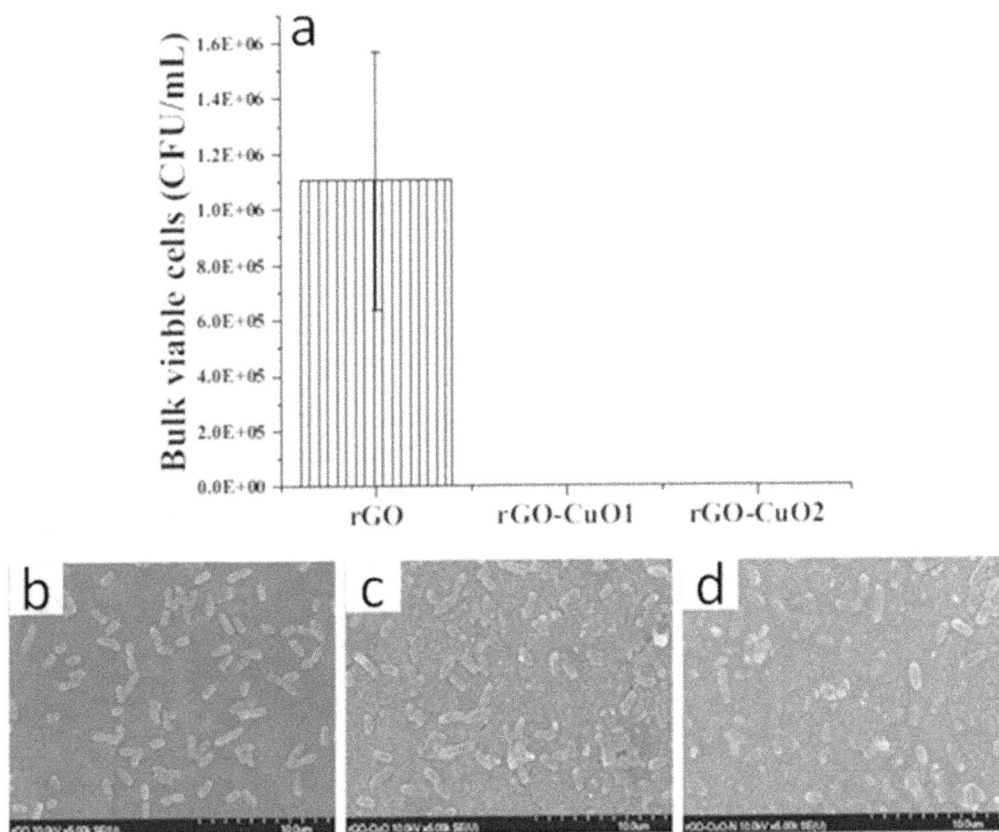

FIGURE 9.10 The photoactive antibacterial activities determined by the total number of colonies on rGO, rGO-CuO(1) and rGO-CuO(2). SEM micrographs of *P. aeruginosa* PAO1 on rGO (a), rGO-CuO(1) (b) and rGO-CuO(2) (c). (Alayande [125].)

against *E. coli*. Combination of Ag and Pd has a synergetic effect and is beneficial to Pd-Ag/rGO nanocomposite since it shifted the SPR peak of Ag from UV to visible region (Figure 9.11a), thus making the composite to be photocatalytically active when it is irradiated with sunlight. Light irradiation increases the efficiency of the antibacterial activity of the nanocomposite (Figure 9.11b and c). Apart from the ROS, rGO sheets are capable of capturing the bacteria to give access to the Ag and Pd NPs that are capable of poisoning the bacteria and separating electrons and holes. The nanocomposite was also active in the absence of light; however, the activity was lower than the one obtained in the presence of light.

Ran and colleagues [123] synthesized photothermal nanocomposite (GO-HA-AgNPs composite) of HA-templated AgNPs incorporated with GO triggered by hyaluronidase (HAase). The function of HAase is to protect AgNPs with HA template during its release to avoid the damage of mammalian cells. GO-HA-AgNPs and GO-HA and cit-AgNPS were tested for the photoactive (NIR irradiation) degradation of *S. aureus*. Although GO-HA and cit-AgNPs showed their ability to photodegrade the *S. aureus*, the highest antibacterial activities were obtained for GO-HA-AgNPs in comparison with GO-HA and cit-AgNPS (Figure 9.12), which were attributable to the synergistic antibacterial effect of both GO and AgNPs, where GO is responsible for heat conversion and AgNPs for the antibacterial activity.

Radwan and coworkers [145] prepared Cu-MNPs-GO nanocomposites by doping magnetite nanoparticles (MNPs) with different amounts of copper (Cu) on GO nanosheets and then crafted

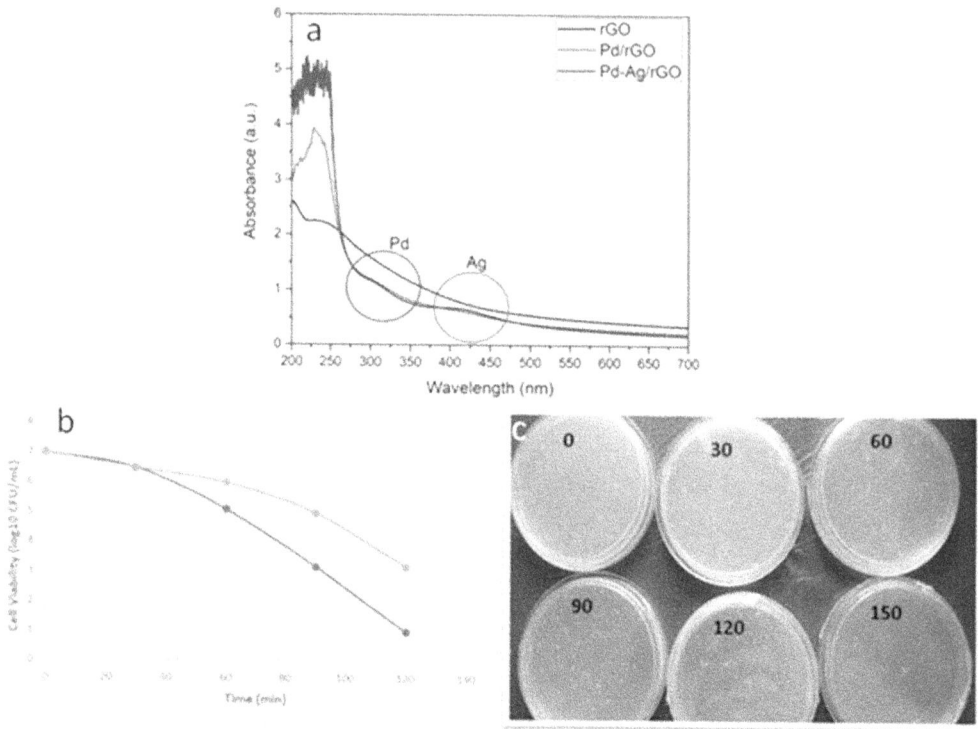

FIGURE 9.11   (a) UV-Vis absorption spectra of rGO, Pd/rGO, Pd-Ag/rGO, (b) bacterial inhibition kinetics of Pd-Ag/rGO against *E. coli* in the presence of light (bottom dark line) and absence of light (top grey line) (C) Bacterial growth inhibition for *E. coli* on Pd-Ag/rGO using light. (Mallikarjuna [122].)

onto electrospun nanofibrous membranes of polycaprolactone (PCL). The composites were evaluated for their photoinduced properties against *E. coli* and *S. aureus*, and the inhibition zone obtained was about 11.4 ± 1.5 and 11.1 ± 1.7 mm in dark conditions and 15.6 ± 2.1 and 13.6 ± 1.8 mm under light, respectively, on the composite with the highest concentration of Cu dopant (08-Cu–MNPs–GO@ PCL). The authors observed that the inhibition zone increases with an increase in the concentration of Cu and also that the presence of light increases the antimicrobial activities of the composites, which is due to the ability of Cu to degrade the DNA material of the bacteria by oxidizing its proteins. GO contributes to the antibacterial activities of the composites by enhancing its ROS generation and by physical destruction of the membrane due to the nature of its structure.

Nagajyothi et al. [137] studied the antibacterial activities of Ag/f-MWCNTs-*E. coli* specimen (f-MWCNTs: functionalized multi-walled carbon nanotubes) under the presence of visible light for 150 minutes. The size of Ag NPs was ranging from 17.1 to 30.5 nm. They observed that the cell density was decreasing with time and the concentration of the proteins, from the degraded cell DNA, was increasing with time due to the presence of ROS. This showed that the material was efficient in degrading the bacterial cells.

Ahmed and Mohammed [138] employed Au/o-MWCNT (o-MWCNTs: oxidized multi-walled carbon nanotubes) and Au_2O_3/o-MWCNTs in the degradation of *E. coli* in the presence of visible light. Peaks at 503 and 508 nm were observed on UV-vis spectra of Au/o-MWCNTs and Au_2O_3/o-MWCNTs hybrids but absent on functionalized MWCNTs. These peaks are related to plasmon resonance bands of Au and Au_2O_3 NPs supported on MWCNTs. The hybrids were found to be very effective in the degradation of bacterial cells in comparison with o-MWCNTs in the presence of light due to the synergic effects of Au-based NPs and o-MWCNTs. o-MWCNTs experienced severe

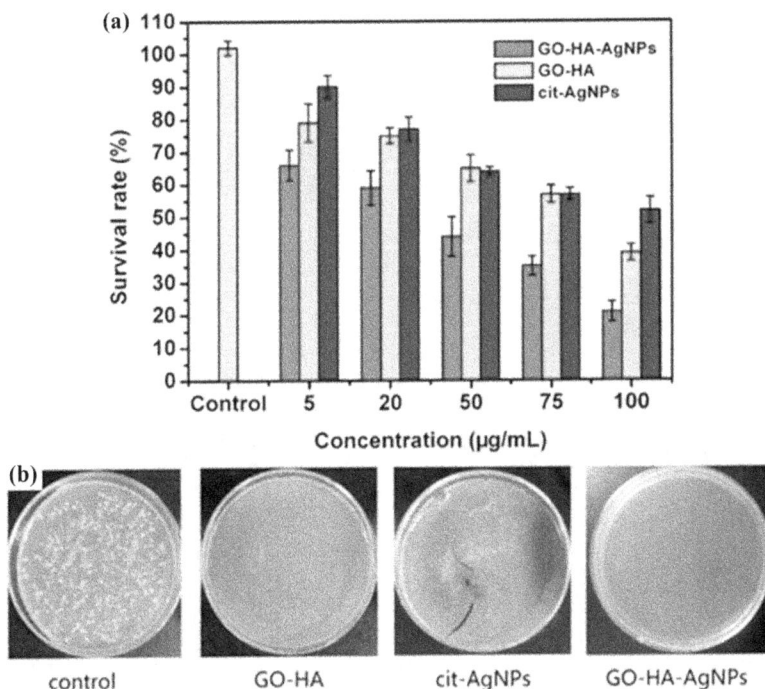

FIGURE 9.12 Survival rate of NIR-irradiated *S. aureus* on GO-HA-AgNPs, GO–HA, and cit-AgNPs with various concentrations for 2 minutes (a) and inhibition of *E. coli* growth on GO-HA, cit-AgNPs, and GO-HA-AgNPs in the presence of light. (b). (Ran [123].)

damage due to their acid oxidation, which resulted in shortening of the tubes that were responsible for the physical damage of the bacteria followed by the oxidation of the bacterial cell membrane by the electrostatic force that exists along the external surface of the bacteria and MWCNTs. Au-based NPs destroyed the bacterial membrane by reaction with the thiol groups, therefore breaking down its protein cell wall.

9.5.3 METAL-ORGANIC POLYMER-BASED NANOMATERIALS

Metal-polymer nanocomposites are promising materials for the removal of bacteria that developed resistance for commercially available drugs [146,147]. In the photoactive antibacterial action of metal-polymer nanocomposites, polymers stabilize the metal NPs and open an access for it to degrade the cell membrane [148,149] through mechanisms that were discussed in this chapter. It is very important that the polymer of choice possesses anti-biofilm and anti-adhesion properties [119] to block the bacteria from adhering on its surface and create a protective bacterial biofilm that will protect it from being destroyed. Polymers are readily available with surfaces that can be easily modified [145], therefore allowing their structures to be tailored based on their intended application.

Xu and co-workers [119] prepared a petal-like nanosheet of β-Bi_2O_3/$Bi_2O_{2.7}$ heterostructures film (thickness: 0.81 µm) on polyvinyl chloride (PVC) polymer matrices by hydrothermal method and incorporated Ag nanospheres on its matrix (140–180 nm) (Figure 9.13). Both PVC film-coated and uncoated PVC exhibited certain photo-induced (visible light) antimicrobial activities on *E. coli* (Figure 9.14) with the highest antibacterial rates of 620.3 ± 34.0 and 823.3 ± 76.9 at 18 hours obtained for β-Bi_2O_3/$Bi_2O_{2.7}$/PVC and Ag-β-Bi_2O_3/$Bi_2O_{2.7}$/PVC within 18 and 24 hours, respectively. It was confirmed by fluorescence analysis that Ag-β-Bi_2O_3/$Bi_2O_{2.7}$/PVC generate more ROS than Ag-β-Bi_2O_3/$Bi_2O_{2.7}$, which are responsible for cell degradation of the bacteria.

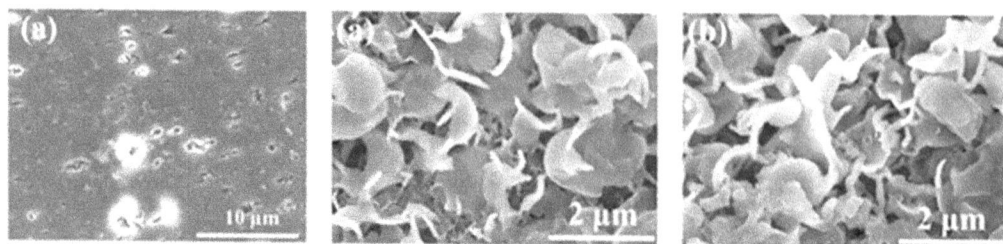

FIGURE 9.13 SEM micrographs of PVC (a), β-Bi$_2$O$_3$/Bi$_2$O$_{2.7}$/PVC (b) and Ag/β-Bi$_2$O$_3$/Bi$_2$O$_{2.7}$/PVC (c) films. (Xu [119].)

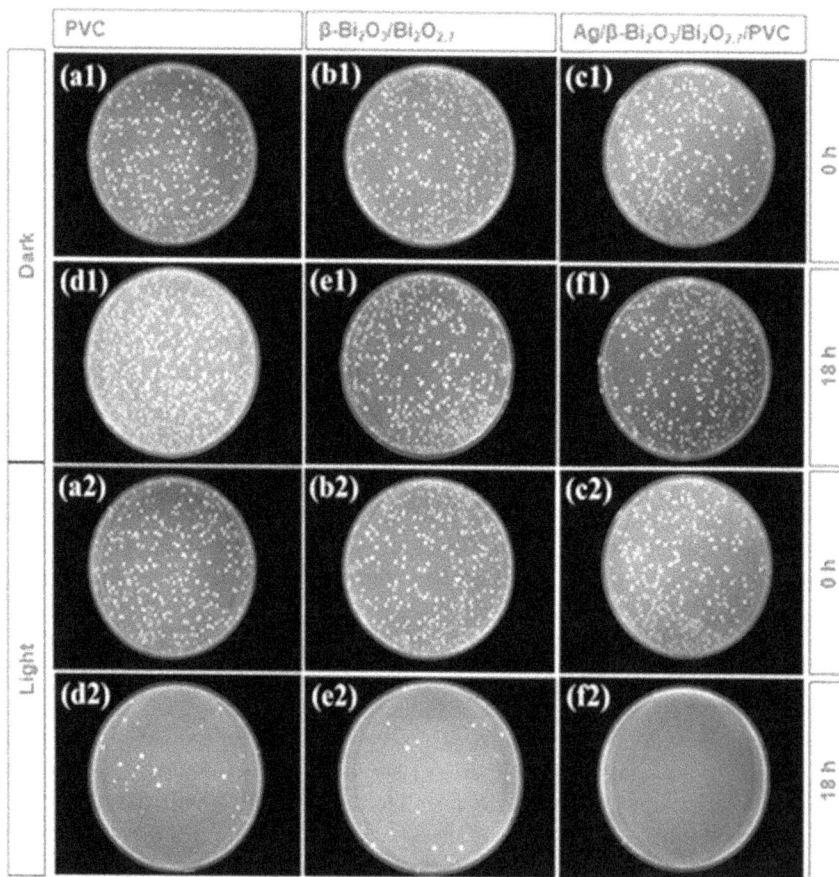

FIGURE 9.14 Bacterial growth inhibition of PVC, β-Bi$_2$O$_3$/Bi$_2$O$_{2.7}$/PVC, and Ag/β-Bi$_2$O$_3$/Bi$_2$O$_{2.7}$/PVC against *E. coli* under dark and photo-irradiated conditions. (Xu [119].)

Rocca and others [150] synthesized well-dispersed lignin-doped Ag NPs on various types of lignin, AgNP@AL (low sulfonate content and no reducing sugars), AgNP@ZHL (depolymerized with 27% sugar content), and AgNP@Alkali (alkali-extracted with 16% sugar content) by thermal and photochemical methods for the photoactive antibacterial properties of *E. coli* and *S. aureus*. All composites showed antibacterial activities to a certain extent in the dark with the exception of AgNP@ZHL. In the presence of light, all nanocomposites were extremely active in the degradation

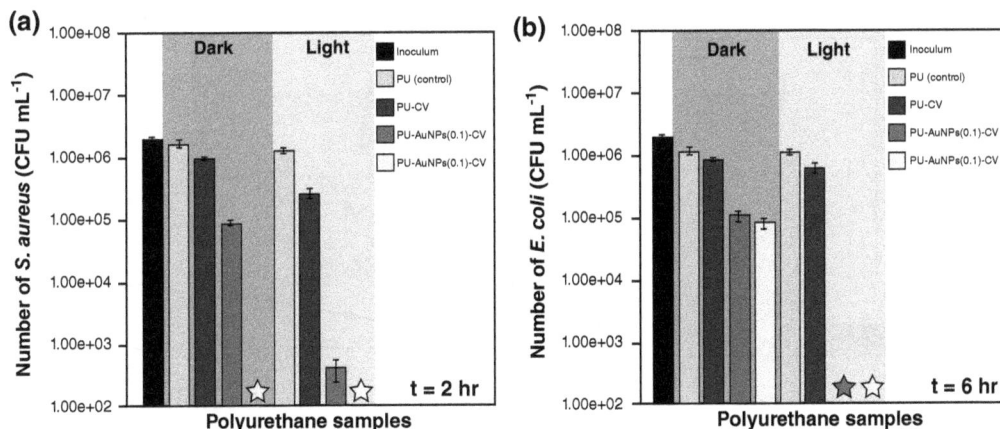

FIGURE 9.15 Antibacterial activities of PV, PU-CV, and PU-AuNPs-CV (2 nm, concentration: 0.1 and 1.0 mg/mL) against *S. aureus* (a) and *E. coli* (b) under dark and light conditions. (Macdonald [85].)

of the bacterial strains except for AgNP@ZHL on *E. coli*. However, AgNP@ZHL showed good selectivity for the degradation of *S. aureus* due to high sugar content on the ZHL, which allows for efficient interaction with the cell walls of *S. aureus*. The authors concluded that the antibacterial activities of these materials happen through photothermal and photochemical mechanisms. Photothermal mechanisms involve the absorption of light by Ag NPs followed by the release of heat to degrade the cell membrane, while the photochemical mechanisms involve the generation of ROS that are responsible for the degradation of the cell membrane. However, it was observed that the antibacterial action of AgNP@ZHL composite does not involve the photothermal mechanism since the amount of ROS did not increase when the sample was irradiated.

Macdonald and group [85] synthesized PU-AuNPs-CV by encapsulation of thiol-capped Au NPs polyurethane (PU) with Ag particle sizes of 2, 3, and 5 nm, and coated it with crystal violet. Au NPs did not show any antibacterial activities under light and dark conditions, and the highest bacterial degradation was obtained for PU-AuNPs-CV in the presence of light for *E. coli* and *S. aureus* specimens (Figure 9.15). Furthermore, PU-AuNPs-CV with Ag NPs size of 2 nm exhibited high antimicrobial activities, which indicates that the antibacterial activity of this material depends on the particle size. This size sensitivity might be due to differing photonics that are found in small AuNPs that promote the formation of ROS, which are responsible for destroying the cell membrane. The composites are also active in the degradation of Gram-positive *S. aureus,* with less complex cell membrane, than Gram-positive *E. coli* under the dark conditions, which is speculated to be due to the redox reactions that generate the reactive oxygen species through electron transfer interaction with crystal violet.

9.6 PHOTOTHERMAL ANTIMICROBIAL NANOMATERIALS

Xu and co-workers [151] prepared gold nanorod graphitic nanocapsules (AuNR@G)-doped poly(vinyl alcohol) (PVA)/chitosan (CS) hydrogels, AG-PC hydrogel, and evaluated them for their photoinduced antibacterial activities against *S. aureus* and *E. coli* bacterial strains. It was observed that irradiation of about 0.8 g AG-PC with 1 and 2 W/cm² lasers at 808 for 10 minutes increases the temperature of the materials by 23.1°C and 43.3°C, respectively, while there is no significant temperature change obtained from PC hydrogel separately under the same conditions. PC hydrogel did not show any ability to kill *S. aureus* and *E. coli*, while almost all *S. aureus* and *E. coli* cells died after 10 minutes of irradiation (2 W/cm²) in the presence of AG-PC hydrogel for 10 minutes. These bacterial cells were killed by the heat generated during their irradiation in the presence of

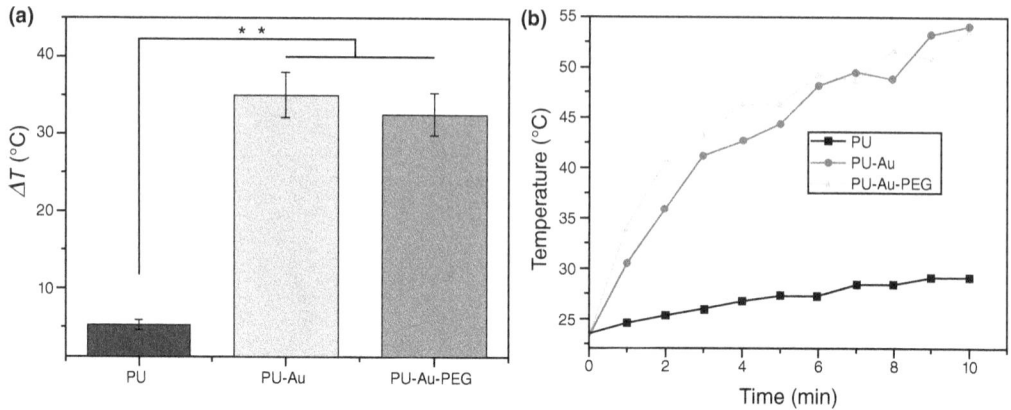

FIGURE 9.16   Temperature change (ΔT) (a) and heating curves (b) of the materials after irradiated for 10 minutes with NIR at 808 nm. (Zhao [64].)

AG-PC hydrogel that proved to increase the temperature of the system. Therefore, the photothermal property of the AG-PC hydrogels was used as an effective tool to destroy the bacterial strains used in this work.

PU-Au-PEG (PU: polyurethane, PEG: polyethylene glycol) hybrid coating was synthesized by Zhao and team [64], and its photothermal antibacterial properties were evaluated against *S. aureus* and *P. aeruginosa* under NIR irradiation (808 nm). The irradiation of PU and Pu-Au for 10 minutes raised the temperature of the materials from 23°C to 28°C and 55°C (Figure 9.16), respectively, due to the strong LSPR absorption of gold nanorods in the NIR region. There was no significant difference in terms of the temperature change for PU-Au and PU-Au-PEG. High bacterial mortality was observed when *S. aureus* and *P. aeruginosa* were irradiated with NIR in the presence of PU-Au for 10 minutes due to the photothermal effect of Pu-Au material. However, dead bacterial cells were deposited on PU-Au material, while deposition of bacteria cells on PU-Au-PEG was at a minimum level, thus showing that the addition of PEG on PU-Au is important in assuring that the material is not easily fouled by the dead bacteria, which can lead to the formation of the second layer of bacterial film, thus making the material ineffective.

Tan and co-workers [135] synthesized rGO/Ag nanocomposite, as shown in Figure 9.17, and determined its antibacterial activities against *E. coli* and *Klebsiella pneumonia* (Kp) using NIR irradiation (808 nm, 0.30 W/cm²) for 10 minutes. It was observed that NIR irradiation increased the temperature on rGO and rGO/Ag materials, while that was not the case on AgNPs and GO. This was due to high ability of rGO to absorb light in the NIR region as compared to AgNPs and GO. However, rGO/Ag was more effective in increasing the temperature than rGO due to the synergistic effects of Ag on rGO (Figure 9.18). rGO/Ag was more effective in killing the bacterial cells in the presence of NIR light due to its high photothermal properties.

Wu and co-workers [152] synthesized silica-coated Au-Ag nanocages (NC) Au-Ag@SiO_2 NCs (Figure 9.18a) with a particle size of 155 nm and evaluated their photothermal antibacterial properties using *S. aureus* and *E. coli* bacterial strains in the presence of NIR irradiation at 808 nm (1 W/cm²). The addition of Au on Ag shifted the SPR peak from 435 to 770 nm and coating with SiO_2 shifted the SPR peak to 804 nm, making the material to be efficient in absorbing in NIR region (Figure 9.18b). NIR irradiation of 50 μg/mL of Au-Ag@SiO_2 NCs for 10 minutes increased the temperature from 20.7°C to 57.4°C (Figure 9.18c). In addition, the photothermal property of the material was effective in increasing the ability of the material to release Ag^+, which is also responsible for the antibacterial properties of the material. High antibacterial activities were obtained on both bacterial strains in the irradiated Au-Ag@SiO_2-bacteria specimens for 0.5 hours, and this is attributable to its photothermal properties rather than the presence of Ag^+.

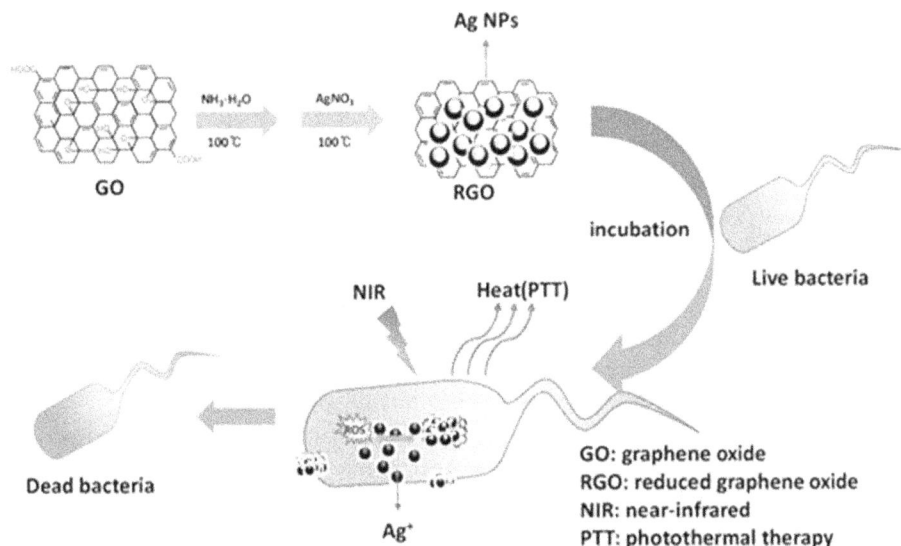

FIGURE 9.17 Synthesis of rGO/Ag and their synergistic antibacterial activity in the presence of NIR irradiation. (Tan [135].)

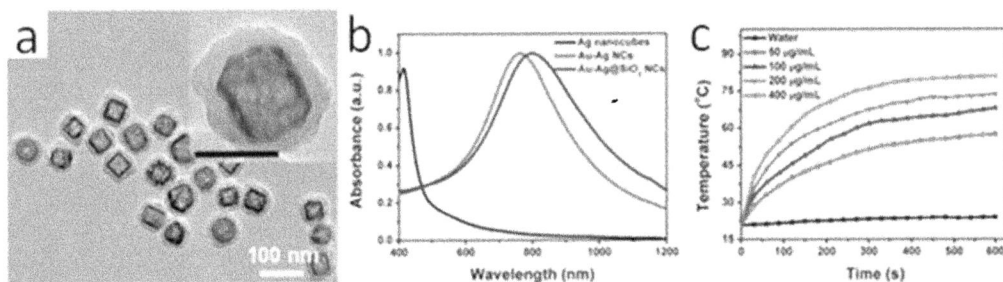

FIGURE 9.18 TEM micrograph of Au-Ag@SiO$_2$ NCs (a). UV-vis-NIR spectra Ag NCs, Au-Ag NCs and Au-Ag@SiO$_2$ NCs (b) and heating curves of Au-Ag@SiO$_2$ NCs in the presence of NIR irradiation at 808 nm (1 W/cm^2) (c). (Wu [152].)

In addition to metal NP-based photothermal nanomaterials, various transition-metal chalcogenides that hold a powerful NIR absorption capacity have been employed as effective photothermal agents. Gargioni et al. [68] observed that the d-d transitions of p-type semiconductor CuS considerably improve antibacterial performance *via* NIR irradiation. Similarly, Zhang et al. [153] revealed that CuS@MoS$_2$ hydrogel had a higher photothermal exchange properties when compared to CuS hydrogel. The presence of MoS$_2$ significantly enhanced hyperthermia and ROS generation, eventually resulting in bacterial cell wall destruction. It was observed that with the exposure of 660- and 808-nm light, 99.3% of *E. coli* and 99.5% of *S. aureus* were destroyed within 15 minutes.

9.7 FUTURE PERSPECTIVE

The progress in photoactive nanomaterial-based disinfection strategies has received great attention from academia and various industries in recent years. Although various nanomaterials have shown acceptable antimicrobial active under solar light, not all single systems are highly efficient.

Moreover, most existing photoresponsive nanomaterials are difficult to degrade and have toxicity issues and other side effects. A lot of research efforts have been diverted to developing photoactive materials with natural components due to their sustainability, environmentally friendly nature, and biocompatibility. Various antimicrobial strategies such as metal or semiconductors, photosensitizers, magnetic and thermal responses of the materials are combined often to design so that the disinfection time is reduced, and the efficacy of sterilization is significantly enhanced. To take benefit of the complete solar spectrum, it is sensible to make use of composite materials that are capable of ROS generation while converting absorbed light to heat. Combining photocatalytic and photothermal systems is also another strategy to improve sterilization efficiency.

Most of the works found in the literature represent laboratory-based preparation using complex, multi-step processes and characterization of prospective materials using various instrumental techniques. Focus of the future studies should be on the production of these materials using simple, practical, cost-effective methods and performance evaluation of the scale-up materials thereof. More initiatives with respect to the phototoxicity of the materials under various solar radiations should be taken, which will help to optimize the concentration and intensity of radiation to find the best possible solution.

REFERENCES

1. C. Liu, J. Guo, X. Yan, Y. Tang, A. Mazumder, S. Wu, Y. Liang, Antimicrobial nanomaterials against biofilms: an alternative strategy, *Environmental Reviews.* 25(2017) 225–244. https://doi.org/10.1139/er-2016-0046
2. S.M. Imani, L. Ladouceur, T. Marshall, R. Maclachlan, L. Soleymani, T.F. Didar. Antimicrobial nanomaterials and coatings: current mechanisms and future perspectives to control the spread of viruses including SARS-CoV-2, *ACS Nano.* 14(2020) 12341–12369. https://doi.org/10.1021/acsnano.0c05937.
3. R.M. Donlan, Biofilms: microbial life on surfaces. *Emerging Infectious Diseases.* 8(2002) 881–890. https://doi.org/10.3201/eid0809.020063
4. S. Skraber, J. Schijven, C. Gantzer, A.M. de Roda Husman, Pathogenic viruses in drinking-water biofilms: a public health risk? *Biofilms.* 2(2005) 105–117. https://doi.org/10.1017/S1479050505001833
5. J. Wingender, H.C. Flemming, Biofilms in drinking water and their role as reservoir for pathogens. *International Journal of Hygiene and Environmental Health.* 214(2011) 417–423. https://doi.org/10.1016/j.ijheh.2011.05.009
6. L. Hall-Stoodley, J.W. Costerton, P. Stoodley, Bacterial biofilms: from the natural environment to infectious diseases, *Nature Reviews Microbiology.* 2(2004) 95–108. https://doi.org/10.1038/nrmicro821
7. P. Ganguly, C. Byrne, A. Breen, S.C. Pillai, Antimicrobial activity of photocatalysts: fundamentals, mechanisms, kinetics and recent advances. *Applied Catalysis B: Environmental.* 225(2018) 51–75. https://doi.org/10.1016/j.apcatb.2017.11.018
8. M.E. de Souza, L.Q.S. Lopes, R. de Almeida Vaucher, R.C.V. Santos, Antibiofilm applications of nanotechnology, *Fungal Genomics Biology.* 4(2014) 1–2. https://doi.org/10.4172/2165-8056.1000e117
9. M.M. Querido, L. Aguiar, P. Neves, C.C. Pereira, J.P. Teixeira, Self-disinfecting surfaces and infection control, *Colloids and Surfaces B: Biointerfaces,* 178(2019) 8–21. https://doi.org/10.1016/j.colsurfb.2019.02.009
10. R.L. Jolley, W.A. Brungs, J.A. Cotruvo, R.B. Cumming, J.S. Mattice, V.A. Jacobs. Water chlorination: environmental impact and health effects, Chemistry and water treatment CONF-811068. Book 1, 4 (1983) Ann Arbor Science Publishers. https://www.osti.gov/biblio/6000538-water-chlorination-environmental-impact-health-effects-volume-book-chemistry-water-treatment
11. P.S.M. Dunlop, J.A. Byrne, N. Manga, B.R. Eggins. The photocatalytic removal of bacterial pollutants from drinking water, *Journal of Photochemistry and Photobiology A: Chemistry.* 148(2002) 355–363.
12. R.D. Letterman, *Water Quality and Treatment: A Handbook of Community Water Supplies.* 1 (1999), American Water Works Association, McGraw-Hill. https://www.worldcat.org/title/water-quality-and-treatment-a-handbook-of-community-water-supplies/oclc/42603208
13. S. Regli, C.E Gilbert, E.J. Calabrese, *Disinfection Requirements to Control for Microbial Contamination, Regulating Drinking Water Quality.* (1992). Lewis, Boca Raton, FL.
14. W.J. Masschelein, R.G. Rice. *Ultraviolet Light in Water and Wastewater Sanitation.* (2016). CRC Press, Boca Raton, Florida.

15. J.M.C. Robertson, P.K.J. Robertson, L.A. Lawton, A comparison of the effectiveness of TiO$_2$ photocatalysis and UVA photolysis for the destruction of three pathogenic micro-organisms, *Journal of Photochemistry and Photobiology A: Chemistry.* 175(2005) 51–56. https://doi.org/10.1016/j.jphotochem.2005.04.033

16. W.-J. Huang, G.-C. Fang, C.-C. Wang, The determination and fate of disinfection by-products from ozonation of polluted raw water, *Science of the Total Environment.* 345(2005) 261–272. https://doi.org/10.1016/j.scitotenv.2004.10.019

17. P. Erkoc, F. Ulucan-Karnak, Nanotechnology-based antimicrobial and antiviral surface coating strategies, *Prosthesis.* 3(2021) 25–52. https://doi.org/10.3390/prosthesis3010005

18. H. Humphreys, Self-disinfecting and microbiocide-impregnated surfaces and fabrics: what potential in interrupting the spread of healthcare-associated infection? *Clinical Infectious Diseases.* 58(2014) 848–853. https://doi.org/10.1093/cid/cit765

19. J. Hasan, R.J. Crawford, E.P. Ivanova, Antibacterial surfaces: the quest for a new generation of biomaterials, *Trends in Biotechnology.* 31(2013) 295–304. https://doi.org/10.1016/j.tibtech.2013.01.017

20. C.D. Sifri, G.H. Burke, K.B. Enfield, Reduced health care-associated infections in an acute care community hospital using a combination of self-disinfecting copper-impregnated composite hard surfaces and linens, *American Journal of Infection Control.* 44(2016) 1565–1571. https://doi.org/10.1016/j.ajic.2016.07.007

21. Tripathy, P. Sen, B. Su, W.H. Briscoe, Natural and bioinspired nanostructured bactericidal surfaces, *Advances in Colloid and Interface Science.* 248(2017) 85–104. https://doi.org/10.1016/j.cis.2017.07.030

22. N. Kashef, Y.-Y. Huang, M.R. Hamblin, Advances in antimicrobial photodynamic inactivation at the nanoscale, *Nanophotonics.* 6(2017) 853–879. https://doi.org/10.1515/nanoph-2016-0189

23. M.R. Hamblin, Antimicrobial photodynamic inactivation: a bright new technique to kill resistant microbes, *Current Opinion in Microbiology.* 33(2016) 67–73. https://doi.org/10.1016/j.mib.2016.06.008

24. M.R. Hamblin, T. Hasan, Photodynamic therapy: a new antimicrobial approach to infectious disease? *Photochemical & Photobiological Sciences.* 3(2004) 436–450. https://doi.org/10.1039/b311900a

25. L. Huang, T. Dai, M.R. Hamblin, Antimicrobial photodynamic inactivation and photodynamic therapy for infections, *Methods in Molecular Biology.* 635(2010) 155–173. https://doi.org/10.1007/978-1-60761-697-9_12

26. M. Ochsner, Photophysical and photobiological processes in the photodynamic therapy of tumours, *Journal of Photochemistry and Photobiology B: Biology.* 39(1997) 1–18. https://doi.org/10.1016/s1011-1344(96)07428-3

27. Z. Luksiene, Photodynamic therapy: mechanism of action and ways to improve the efficiency of treatment, *Medicina (Kaunas, Lithuania).* 39(2003) 1137–1150.

28. L. Bourré, F. Giuntini, I.M. Eggleston, C.A. Mosse, A.J. MacRobert, M. Wilson, Effective photoinactivation of Gram-positive and Gram-negative bacterial strains using an HIV-1 Tat peptide–porphyrin conjugate, *Photochemical & Photobiological Sciences.* 9(2010) 1613–1620. https://doi.org/10.1039/c0pp00146e

29. F. Sperandio, Y.-Y. Huang, M.R. Hamblin, Antimicrobial photodynamic therapy to kill Gram-negative bacteria, *Recent Patents on Anti-Infective Drug Discovery.* 8(2013) 108–120. https://doi.org/10.2174/1574891x113089990012

30. T.N. Demidova, M.R. Hamblin, Effect of cell-photosensitizer binding and cell density on microbial photoinactivation, *Antimicrobial Agents and Chemotherapy.* 49(2005) 2329–2335. https://doi.org/10.1128/AAC.49.6.2329-2335.2005

31. F. Käsermann, C. Kempf, Buckminsterfullerene and photodynamic inactivation of viruses, *Reviews in Medical Virology.* 8(1998) 143–151. https://doi.org/10.1002/(sici)1099-1654(199807/09)8:3<143::aid-rmv214>3.0.co;2-b

32. G. Jori, C. Fabris, M. Soncin, S. Ferro, O. Coppellotti, D. Dei, L. Fantetti, G. Chiti, G. Roncucci, Photodynamic therapy in the treatment of microbial infections: basic principles and perspective applications, *Lasers in Surgery and Medicine: The Official Journal of the American Society for Laser Medicine and Surgery.* 38(2006) 468–481. https://doi.org/10.1002/lsm.20361.

33. Y. Feng, L. Liu, J. Zhang, H. Aslan, M. Dong, Photoactive antimicrobial nanomaterials, *Journal of Materials Chemistry B.* 5(2017) 8631–8652. https://doi.org/10.1039/C7TB01860F

34. A.V. Kachynski, A. Pliss, A.N. Kuzmin, T. Y. Ohulchanskyy, A. Baev, J. Qu, P.N. Prasad, Photodynamic therapy by in situ nonlinear photon conversion, *Nature Photonics.* 8(2014) 455–461. https://doi.org/10.1038/nphoton.2014.90

35. W.-S. Kuo, C.-Y. Chang, H.-H. Chen, C.-L.L. Hsu, J.-Y. Wang, H.-F. Kao, L.C.-S. Chou, Y.-C. Chen, S.-J. Chen, W.-T. Chang, S.-W. Tseng, P.-C. Wu, Y.-C. Pu. Two-photon photoexcited photodynamic therapy and contrast agent with antimicrobial graphene quantum dots, *ACS Applied Materials & Interfaces.* 8(2016) 30467–30474. https://doi.org/10.1021/acsami.6b12014

36. S. Iravani, Nanophotocatalysts against viruses and antibiotic-resistant bacteria: recent advances, *Critical Reviews in Microbiology*. 48(2022) 67–82. https://doi.org/10.1080/1040841X.2021.1944053

37. L.M. Margarucci, V.R. Spica, C. Protano, G. Gianfranceschi, M. Giuliano, V. Di Onofrio, N. Mucci, F. Valeriani, M. Vitali, F. Romano, Potential antimicrobial effects of photocatalytic nanothecnologies in hospital settings, *Annali di Igiene: Medicina Preventiva e di Comunita*. 31(2019) 461–473. https://doi.org/10.7416/ai.2019.2307

38. S. Kim, K. Ghafoor, J. Lee, M. Feng, J. Hong, D.-U. Lee, J. Park, Bacterial inactivation in water, DNA strand breaking, and membrane damage induced by ultraviolet-assisted titanium dioxide photocatalysis, *Water Research*. 47(2013) 4403–4411. https://doi.org/10.1016/j.watres.2013.05.009

39. N. Dragicevic-Curic, A. Fahr, Liposomes in topical photodynamic therapy, *Expert Opinion on Drug Delivery*. 9(2012) 1015–1032. https://doi.org/10.1517/17425247.2012.697894

40. C.-B. Kim, D.K. Yi, P.S.S. Kim, W. Lee, M.J. Kim, Rapid photothermal lysis of the pathogenic bacteria, *Escherichia coli* using synthesis of gold nanorods, *Journal of Nanoscience and Nanotechnology*. 9(2009) 2841–2845. https://doi.org/10.1166/jnn.2009.002

41. M. Pelaez, N.T. Nolan, S.C. Pillai, M.K. Seery, P. Falaras, A.G. Kontos, P.S.M. Dunlop, J.W.J Hamilton, J.A. Byrne, K O'Shea, M.E. Entezari, D.D. Dionysiou, A review on the visible light active titanium dioxide photocatalysts for environmental applications, *Applied Catalysis B: Environmental*. 125(2012) 331–349. https://doi.org/10.1016/j.apcatb.2012.05.036

42. N. Talebian, S.M. Amininezhad, M. Doudi, Controllable synthesis of ZnO nanoparticles and their morphology-dependent antibacterial and optical properties, *Journal of Photochemistry and Photobiology B: Biology*. 120(2013) 66–73. https://doi.org/10.1016/j.jphotobiol.2013.01.004

43. M. Gong, S. Xiao, X. Yu, C. Dong, J. Ji, D. Zhang, M. Xing, Research progress of photocatalytic sterilization over semiconductors, *RSC Advances*. 9(2019) 19278–19284. https://doi.org/10.1039/C9RA01826C

44. Y. Ren, H. Liu, X. Liu, Y. Zheng, Z. Li, C. Li, K.W.K. Yeung, S. Zhu, Y. Liang, Z. Cui, S. Wu. Photoresponsive materials for antibacterial applications, *Cell Reports Physical Science* (2020) 100245. https://doi.org/10.1016/j.xcrp.2020.100245

45. T. Maisch, Anti-microbial photodynamic therapy: useful in the future? *Lasers in Medical Science*. 22(2007) 83–91. https://doi.org/10.1007/s10103-006-0409-7

46. O. Raab, Uber die wirkung fluorescirender stoffe auf infusorien, *Zeitschrift für Biologie* 39(1900) 524–546.

47. I.J. Macdonald, T.J. Dougherty, Basic principles of photodynamic therapy, *Journal of Porphyrins and Phthalocyanines*. 5(2001) 105–129. https://doi.org/10.1002/jpp.328

48. P.A.K. Reddy, P.V.L. Reddy, E. Kwon, K.-H. Kim, T. Akter, S. Kalagara, Recent advances in photocatalytic treatment of pollutants in aqueous media, *Environment International*. 91(2016) 94–103. https://doi.org/10.1016/j.envint.2016.02.012

49. W. Wang, J.C. Yu, D. Xia, P.K. Wong, Y. Li, Graphene and g-C$_3$N$_4$ nanosheets cowrapped elemental α-sulfur as a novel metal-free heterojunction photocatalyst for bacterial inactivation under visible-light, *Environmental Science & Technology*. 47(2013) 8724–8732. https://doi.org/10.1021/es4013504

50. W. Wang, Y. Yu, T. An, G. Li, H.Y. Yip, J.C. Yu, P.K. Wong, Visible-light-driven photocatalytic inactivation of *E. coli* K-12 by bismuth vanadate nanotubes: bactericidal performance and mechanism, *Environmental Science & Technology*. 46(2012) 4599–4606. https://doi.org/10.1021/es2042977

51. A. Tripathy, P. Sen, B. Su, W.H. Briscoe, Natural and bioinspired nanostructured bactericidal surfaces, *Advances in Colloid and Interface Science*. 248(2017) 85–104. https://doi.org/10.1016/j.cis.2017.07.030

52. X. Bai, L. Wang, R. Zong, Y. Zhu, Photocatalytic activity enhanced via g-C$_3$N$_4$ nanoplates to nanorods, *The Journal of Physical Chemistry C*. 117(2013) 9952–9961. https://doi.org/10.1021/jp402062d

53. C. Bombelli, F. Bordi, S. Ferro, L. Giansanti, G. Jori, G. Mancini, C. Mazzuca, D. Monti, F. Ricchelli, S. Sennato, M. Venanzi. New cationic liposomes as vehicles of m-tetrahydroxyphenylchlorin in photodynamic therapy of infectious diseases, *Molecular Pharmaceutics*. 5(2008) 672–679. https://doi.org/10.1021/mp800037d

54. T. Tsai, Y.-T. Yang, T.-H. Wang, H.-F. Chien, C.-T. Chen, Improved photodynamic inactivation of gram-positive bacteria using hematoporphyrin encapsulated in liposomes and micelles, *Lasers in Surgery and Medicine: The Official Journal of the American Society for Laser Medicine and Surgery*. 41(2009) 316–322. https://doi.org/10.1002/lsm.20754

55. C.J.F. Rijcken, J.-W. Hofman, F. van Zeeland, W.E. Hennink, C.F. van Nostrum, Photosensitiser-loaded biodegradable polymeric micelles: preparation, characterisation and in vitro PDT efficacy, *Journal of Controlled Release*. 124(2007) 144–153. https://doi.org/10.1016/j.jconrel.2007.09.002

56. A.A. Tawfik, J. Alsharnoubi, M. Morsy, Photodynamic antibacterial enhanced effect of methylene blue-gold nanoparticles conjugate on *Staphylococcal aureus* isolated from impetigo lesions in vitro study, *Photodiagnosis and Photodynamic Therapy* 12(2015) 215–220. https://doi.org/10.1016/j.pdpdt.2015.03.003

57. J. Zhou, Z. Liu, F. Li, Upconversion nanophosphors for small-animal imaging, *Chemical Society Reviews.* 41(2012) 1323–1349. https://doi.org/10.1039/C1CS15187H

58. Y. Tian, T. Tatsuma, Plasmon-induced photoelectrochemistry at metal nanoparticles supported on nanoporous TiO_2, *Chemical Communications* 16(2004) 1810–1811. https://doi.org/10.1039/B405061D

59. F. Le, D.W. Brandl, Y.A. Urzhumov, H. Wang, J. Kundu, N.J. Halas, J. Aizpurua, P. Nordlander, Metallic nanoparticle arrays: a common substrate for both surface-enhanced Raman scattering and surface-enhanced infrared absorption, *ACS Nano.* 2(2008) 707–718. https://doi.org/10.1021/nn800047e

60. X. Jiang, X. Fan, W. Xu, R. Zhang, G. Wu, Biosynthesis of bimetallic Au-Ag nanoparticles using *Escherichia coli* and its biomedical applications, *ACS Biomaterials Science & Engineering.* 6(2019) 680–689. https://doi.org/10.1021/acsbiomaterials.9b01297

61. O. Khantamat, C.-H. Li, F. Yu, A.C. Jamison, W.-C. Shih, C. Cai, T.R. Lee, Gold nanoshell-decorated silicone surfaces for the near-infrared (NIR) photothermal destruction of the pathogenic bacterium *E. faecalis*, *ACS Applied Materials & Interfaces.* 7(2015) 3981–3993. https://doi.org/10.1021/am506516r

62. J.-W. Xu, K. Yao, Z.-K. Xu, Nanomaterials with a photothermal effect for antibacterial activities: an overview, *Nanoscale.* 11(2019) 8680–8691. https://doi.org/10.1039/C9NR01833F

63. D. Xu, Z. Li, L. Li, J. Wang, Insights into the photothermal conversion of 2D MXene nanomaterials: synthesis, mechanism, and applications, *Advanced Functional Materials.* 30(2020) 2000712. https://doi.org/10.1002/adfm.202000712

64. Y.-Q. Zhao, Y. Sun, Y. Zhang, X. Ding, N. Zhao, B. Yu, H. Zhao, S. Duan, F.-J. Xu, Well-defined gold nanorod/polymer hybrid coating with inherent antifouling and photothermal bactericidal properties for treating an infected hernia, *ACS Nano.* 14(2020) 2265–2275. https://doi.org/10.1021/acsnano.9b09282

65. X. Cai, W. Gao, L. Zhang, M. Ma, T. Liu, W. Du, Y. Zheng, H. Chen, J. Shi, Enabling Prussian blue with tunable localized surface plasmon resonances: simultaneously enhanced dual-mode imaging and tumor photothermal therapy, *ACS Nano.* 10(2016) 11115–11126. https://doi.org/10.1021/acsnano.6b05990

66. H. Maaoui, R. Jijie, G.-H. Pan, D. Drider, D. Caly, J. Bouckaert, N. Dumitrascu, R. Chtourou, S. Szunerits, R. Boukherroub, A 980nm driven photothermal ablation of virulent and antibiotic resistant Gram-positive and Gram-negative bacteria strains using Prussian blue nanoparticles, *Journal of Colloid and Interface Science.* 480(2016) 63–68. https://doi.org/10.1016/j.jcis.2016.07.002

67. J. Li, X. Liu, L. Tan, Z. Cui, X. Yang, Y. Liang, Z. Li, S. Zhu, Y. Zheng, K.W.K. Yeung, X. Wang, S. Wu, Zinc-doped Prussian blue enhances photothermal clearance of *Staphylococcus aureus* and promotes tissue repair in infected wounds, *Nature Communications.* 10(2019) 1–15. https://doi.org/10.1038/s41467-019-12429-6

68. C. Gargioni, M. Borzenkov, L. D'Alfonso, P. Sperandeo, A. Polissi, L. Cucca, G. Dacarro, P. Grisoli, P. Pallavicini, A. D'Agostino, A. Taglietti, Self-assembled monolayers of copper sulfide nanoparticles on glass as antibacterial coatings, *Nanomaterials.* 10(2020) 352. https://doi.org/10.3390/nano10020352

69. Y. Lin, D. Han, Y. Li, L. Tan, X. Liu, Z. Cui, X. Yang, Z. Li, Y. Liang, S. Zhu, S. Wu, $Ag_2S@WS_2$ heterostructure for rapid bacteria-killing using near-infrared light, *ACS Sustainable Chemistry & Engineering.* 7(2019) 14982–14990. https://doi.org/10.1021/acssuschemeng.9b03287

70. T.-W. Wong, S.-Z. Liao, W.-C. Ko, C.-J. Wu, S.B. Wu, Y.-C. Chuang, I. Huang, Indocyanine green-mediated photodynamic therapy reduces methicillin-resistant *Staphylococcus aureus* drug resistance, *Journal of Clinical Medicine.* 8(2019) 411. https://doi.org/10.3390/jcm8030411

71. L.L. Gyürék, G.R. Finch, Modeling water treatment chemical disinfection kinetics, *Journal of Environmental Engineering.* 124(1998) 783–793. https://doi.org/10.1061/(ASCE)0733-9372(1998)124:9(783)

72. J. Marugán, R. Van Grieken, C. Pablos, M.L. Satuf, A.E. Cassano, O.M. Alfano, Rigorous kinetic modelling with explicit radiation absorption effects of the photocatalytic inactivation of bacteria in water using suspended titanium dioxide, *Applied Catalysis B: Environmental.* 102(2011) 404–416. http://hdl.handle.net/10115/4905

73. N. Daneshvar, D. Salari, A.R. Khataee, Photocatalytic degradation of azo dye acid red 14 in water on ZnO as an alternative catalyst to TiO_2, *Journal of Photochemistry and Photobiology A: Chemistry.* 162(2004) 317–322. https://doi.org/10.1016/S1010-6030(03)00378-2

74. H.C. Yatmaz, N. Dizge, M.S. Kurt, Combination of photocatalytic and membrane distillation hybrid processes for reactive dyes treatment, *Environmental Technology.* 38(2017) 2743–2751. https://doi.org/10.1080/09593330.2016.1276222

75. A.-G. Rincón, C. Pulgarin, Bactericidal action of illuminated TiO_2 on pure *Escherichia coli* and natural bacterial consortia: post-irradiation events in the dark and assessment of the effective disinfection time, *Applied Catalysis B: Environmental.* 49(2004) 99–112. https://doi.org/10.1016/j.apcatb.2003.11.013

76. S. Malato, P. Fernández-Ibáñez, M.I. Maldonado, J. Blanco, W. Gernjak, Decontamination and disinfection of water by solar photocatalysis: recent overview and trends, *Catalysis Today.* 147(2009) 1–59. https://doi.org/10.1016/j.cattod.2009.06.018

77. Y. Qiao, F. Ma, C. Liu, B. Zhou, Q. Wei, W. Li, D. Zhong, Y. Li, M. Zhou, Near-infrared laser-excited nanoparticles to eradicate multidrug-resistant bacteria and promote wound healing, *ACS Applied Materials & Interfaces.* 10(2018) 193–206. https://doi.org/10.1021/acsami.7b15251

78. L. Wang, S. Guan, Y. Weng, S.-M. Xu, H. Lu, X. Meng, S. Zhou, Highly efficient vacancy-driven photo-thermal therapy mediated by ultrathin MnO_2 nanosheets, *ACS Applied Materials & Interfaces.* 11(2019) 6267–6275. https://doi.org/10.1021/acsami.8b20639

79. Q.-Y. Lin, Q. Lin, Y.-Q. Zhang, H.-X. Lin, T.-H. Zhou, S.-B. Ning, X.-X. Wang, Effect of modified iodine on defect structure and antibacterial properties of ZnO in visible light, *Research on Chemical Intermediates.* 43(2017) 5067–5081. https://doi.org/10.1007/s11164-017-3053-x

80. B. Janani, A. Syed, A.M. Thomas, S. Al-Rashed, A.M. Elgorban, L.L Raju, S.S. Khan. A Simple approach for the synthesis of bi-functional p-n Type $ZnO@CuFe_2O_4$ heterojunction nanocomposite for photo-catalytic and antimicrobial application, *Physica E: Low-Dimensional Systems and Nanostructures.* 130(2021) 114664. https://doi.org/10.1016/j.physe.2021.114664

81. E. Vélez-Peña, J. Pérez-Obando, D. Pais-Ospina, D.A. Marín-Silva, A. Pinotti, A. Cánneva, J.A. Donadelli, L. Damonte, L.R. Pizzio, P. Osorio-Vargas, J.A. Rengifo-Herrera, Self-cleaning and anti-microbial photo-induced properties under indoor lighting irradiation of chitosan films containing melon/TiO_2 composites, *Applied Surface Science.* 508(2020) 144895. https://doi.org/10.1016/j.apsusc.2019.144895

82. L. Zhang, P. Chen, Y. Xu, W. Nie, Y. Zhou, Enhanced photo-induced antibacterial application of gra-phene oxide modified by sodium anthraquinone-2-sulfonate under visible light, *Applied Catalysis B: Environmental.* 265(2020) 18572. https://doi.org/10.1016/j.apcatb.2019.118572

83. X. Li, S. Li, Q. Bai, N. Sui, Z. Zhu, Gold nanoclusters decorated amine-functionalized graphene oxide nanosheets for capture, oxidative stress, and photothermal destruction of bacteria, *Colloids and Surfaces B: Biointerfaces.* 196(2020) 111313. https://doi.org/10.1016/j.colsurfb.2020.111313

84. J. Wu, S. Hou, D. Ren, P.T. Mather, Antimicrobial properties of nanostructured hydrogel webs contain-ing silver, *Biomacromolecules.* 10(2009) 2686–2693. https://doi.org/10.1021/bm900620w

85. T.J. Macdonald, K. Wu, S.K Sehmi, S. Noimark, W.J. Peveler, H. du Toit, N.H. Voelcker, E. Allan, A.J. MacRobert, A. Gavriilidis, I.P. Parkin, Thiol-capped gold nanoparticles swell-encapsulated into poly-urethane as powerful antibacterial surfaces under dark and light conditions, *Scientific Reports.* 6(2016) 39272. https://doi.org/10.1038/srep39272

86. Y. Kalachyova, A. Olshtrem, O.A. Guselnikova, P.S. Postnikov, R. Elashnikov, P. Ulbrich, S. Rimpelova, V. Švorčík, O. Lyutakov, Synthesis, characterization, and antimicrobial activity of near-IR photoac-tive functionalized gold multibranched nanoparticles, *ChemistryOpen.* 6(2017) 254–260. https://doi.org/10.1002/open.201600159.

87. S. Thangudu, S.S. Kulkarni, R. Vankaya, C.-S. Chiang, K.C. Hwang, Photosensitized reactive chlo-rine species-mediated therapeutic destruction of drug-resistant bacteria using plasmonic core–shell Ag@AgCl nanocubes as an external nanomedicine, *Nanoscale.* 12(2020) 12970–12984. https://doi.org/10.1039/D0NR01300E

88. Z. Xu, C. Zhang, Y. Yu, W. Li, Z. Ma, J. Wang, X. Zhang, H. Gao, D. Liu, Photoactive silver nanoagents for backgroundless monitoring and precision killing of multidrug-resistant bacteria, *Nanotheranostics.* 5(2021) 472–487. https://doi.org/10.7150/ntno.62364

89. R. Sharma, M. Khanuja, S.S. Islam, U. Singhal, A. Varma, Aspect-ratio-dependent photoinduced anti-microbial and photocatalytic organic pollutant degradation efficiency of ZnO nanorods, *Research on Chemical Intermediates.* 43(2017) 5345–5364. https://doi.org/10.1007/s11164-017-2930-7

90. J. Zhang, X. An, X. Li, X. Liao, Y. Nie, F. Zengjie, Enhanced antibacterial properties of the bracket under natural light via decoration with ZnO/carbon quantum dots composite coating, *Chemical Physics Letters.* 706(2018) 702–707. https://doi.org/10.1016/j.cplett.2018.06.029

91. B. Liu, L. Mu, B. Han, J. Zhang, H. Shi, Fabrication of TiO_2/Ag_2O heterostructure with enhanced photo-catalytic and antibacterial activities under visible light irradiation, *Applied Surface Science.* 396(2017) 1596–1603. https://doi.org/10.1016/j.apsusc.2016.11.220

92. K. Mao, Y. Zhu, J. Rong, F. Qiu, H. Chen, J. Xu, D. Yang, T. Zhang, L. Zhong, Rugby-ball like Ag modi-fied zirconium porphyrin metal–organic frameworks nanohybrid for antimicrobial activity: synergistic effect for significantly enhancing photoactivation capacity, *Colloids and Surfaces A: Physicochemical and Engineering Aspects.* 611(2021) 125888. https://doi.org/10.1016/j.colsurfa.2020.125888

93. H. Zhu, N. Liu, Z. Wang, Q. Xue, Q. Wang, X. Wang, Y. Liu, Z. Yin, X. Yuan, Marrying luminescent Au nanoclusters to TiO_2 for visible-light-driven antibacterial application, *Nanoscale.* 13(2021) 18996–19003. https://doi.org/10.1039/D1NR05503H

94. Y. Zhao, R. Lu, X. Wang, X. Huai, C. Wang, Y. Wang, S. Chen, Visible light-induced antibacterial and osteogenic cell proliferation properties of hydrogenated TiO_2 nanotubes/Ti foil composite, *Nanotechnology*. 32(2021) 195101. https://doi.org/10.1088/1361-6528/abe156

95. H. Bai, Z. Liu, L. Liu, D.D. Sun, Large-scale production of hierarchical TiO_2 nanorod spheres for photocatalytic elimination of contaminants and killing bacteria, *Chemistry-A European Journal*. 19(2013) 3061–3070. https://doi.org/10.1002/chem.201204013

96. J.A. Carver, A.L. Simpson, R.P. Rathi, N. Normil, A.G. Lee, M.D. Force, K.A. Fiocca, C.E. Maley, K.M. DiJoseph, A.L. Goldstein, A.A. Attari, H.L. O'Malley, J.G. Zaccaro, N.M. McCampbell, C.A. Wentz, J.E. Long, L.M. McQueen, F.J. Sirch, B.K. Johnson, M.E. Divis, M.L. Chorney, S.L. DiStefano, H.M Yost, B.L. Greyson, E.A. Cid, K. Lee, C.J. Yhap, M.Dong, D.L. Thomas, B.E. Banks, R.B. Newman, J. Rodriguez, A.T. Segil, J.A. Siberski, A.L. Lobo, M.D. Ellison, Functionalized single-walled carbon nanotubes and nanographene oxide to overcome antibiotic resistance in tetracycline-resistant *Escherichia coli*, *ACS Applied Nano Materials*. 3(2020) 3910–3921. https://doi.org/10.1021/acsanm.0c00677

97. L.M. Jose, S. Kuriakose, T. Mathew, Development of photoresponsive zinc oxide nanoparticle-encapsulated lignin functionalized with 2-(E)-(2-hydroxy naphthalen-1-yl) diazenylbenzoic acid: a promising photoactive agent for antimicrobial photodynamic therapy, *Photodiagnosis and Photodynamic Therapy*. 36(2021) 102479. https://doi.org/.1016/j.pdpdt.2021.102479

98. S. Liu, Z. Shen, B. Wu, Y. Yu, H. Hou, X.-X. Zhang, and H.-Q. Ren, Cytotoxicity and efflux pump inhibition induced by molybdenum disulfide and boron nitride nanomaterials with sheetlike structure, *Environmental Science & Technology*. 51(2017) 10834–10842. https://doi.org/10.1021/acs.est.7b02463

99. Z. Shen, J. Wu, Y. Yu, S. Liu, W. Jiang, H. Nurmamat, B. Wu, Comparison of cytotoxicity and membrane efflux pump inhibition in HepG2 cells induced by singlewalled carbon nanotubes with different length and functional groups, *Scientific Reports*. 9(2019) 1–9. https://doi.org/10.1038/s41598-019-43900-5

100. K.-A. Wong, S.-M. Lam, J.-C. Sin, Wet chemically synthesized ZnO structures for photodegradation of pretreated palm oil mill effluent and antibacterial activity, *Ceramics International*. 45(2019) 1868–1880. https://doi.org/10.1016/j.ceramint.2018.10.078

101. G. Sharma, D. Prema, K.S. Venkataprasanna, J. Prakash, S. Sahabuddin, G.D. Venkatasubbu, Photo induced antibacterial activity of CeO_2/GO against wound pathogens, *Arabian Journal of Chemistry*. 13(2020) 7680–7694. https://doi.org/10.1016/j.arabjc.2020.09.004

102. A. Joe, S.-H. Park, D.-J. Kim, Y.-J. Lee, K.-H. Jhee, Y. Sohn, and E.-S. Jang, Antimicrobial activity of ZnO nanoplates and its ag nanocomposites: insight into an ros-mediated antibacterial mechanism under UV light, *Journal of Solid State Chemistry*. 267(2018) 124–133. https://doi.org/10.1016/j.jssc.2018.08.003

103. B. Cui, H. Peng, H. Xia, X. Guo, H. Guo, Magnetically recoverable core–shell Nanocomposites γ-Fe_2O_3@SiO_2@TiO_2-Ag with enhanced photocatalytic activity and antibacterial activity. *Separation and Purification Technology*, 103(2013) 251–257. https://doi.org/10.1016/j.seppur.2012.10.008

104. R. Kaushik, P.V. Daniel, P. Mondal, A. Haldera, Transformation of 2-D TiO_2 to mesoporous hollow 3-D TiO_2 spheres comparative studies on morphology-dependent photocatalytic and antibacterial activity, *Microporous and Mesoporous Materials*. 285(2019) 32–42. https://doi.org/10.1016/j.micromeso.2019.04.068

105. P. Hajipour, A. Bahrami, A. Eslami, A. Hosseini-Abari, H.R.H. Ranjbar, Chemical bath synthesis of CuO-GO-Ag nanocomposites with enhanced antibacterial properties, *Journal of Alloys and Compounds*. 821(2020) 153456. https://doi.org/10.1016/j.jallcom.2019.153456

106. E. Alp, R. Imamoğlu, U. Savacı, S. Turan, M.K. Kazmanlı, A. Genç, Plasmon-enhanced photocatalytic and antibacterial activity of gold nanoparticles-decorated hematite nanostructures, *Journal of Alloys and Compounds*. 852(2021) 157021. https://doi.org/10.1016/j.jallcom.2020.157021

107. C. Zhang, Y. Gu, G. Teng, L. Wang, X. Jin, Z. Qiang, W. Ma, Fabrication of a double-shell Ag/AgCl/G-$ZnFe_2O_4$ nanocube with enhanced light absorption and superior photocatalytic antibacterial activity, *ACS Applied Materials & Interfaces*. 12(2020) 29883–29898. https://doi.org/10.1021/acsami.0c01476

108. S.-M. Lam, J.-A. Quek, J.-C. Sin, Mechanistic investigation of visible light responsive Ag/ZnO micro/nanoflowers for enhanced photocatalytic performance and antibacterial activity, *Journal of Photochemistry and Photobiology A: Chemistry*. 353(2018) 171–184. https://doi.org/10.1016/j.jphotochem.2017.11.021

109. C. Mutalik, D.I. Krisnawati, S.B Patil, M. Khafid, D. Susetiyanto A.P. Santoso, S.-C. Lu, D.-Y. Wang, T.-R. Kuo, Phase-dependent MoS_2 nanoflowers for light-driven antibacterial application, *ACS Sustainable Chem. Eng.* 9(2021) 7904–7912. https://doi.org/10.1021/acssuschemeng.1c01868

110. M. Kováčová, Z.M. Marković, P. Humpolíček, M. Mičušík, H. Švajdlenková, A. Kleinová, M. Danko, P. Kubát, J. Vajďák, Z. Capáková, M. Lehocký, L. Münster, B.M.T. Marković, Z. Špitalský, Carbon quantum dots modified polyurethane nanocomposite as effective photocatalytic and antibacterial agents, *ACS Biomaterials Science & Engineering*. 4(2018) 3983–3993. https://doi.org/10.1021/acsbiomaterials.8b00582
111. X. Wang, Y. Lu, H. Kunwei, D. Yang, Y. Yang, Iodine-doped carbon dots with inherent peroxidase catalytic activity for photocatalytic antibacterial and wound disinfection, *Analytical and Bioanalytical Chemistry*. 413(2021) 1373–1382. https://doi.org/10.1007/s00216-020-03100-x
112. J. Dolanský, P. Henke, P. Kubát, A. Fraix, S. Sortino, J. Mosinger, Polystyrene nanofiber materials for visible-light-driven dual antibacterial action via simultaneous photogeneration of NO and $O_2(1\Delta g)$, *ACS Applied Materials & Interfaces*. 7(2015) 22980–22989. https://doi.org/10.1021/acsami.5b06233
113. E. Preis, T. Anders, J. Širc, R. Hobzova, A.-I. Cocarta, U. Bakowsky, J. Jedelska, Biocompatible indocyanine green loaded PLA nanofibers for in situ antimicrobial photodynamic therapy, *Materials Science & Engineering*. 115(2020) 111068. https://doi.org/10.1016/j.msec.2020.111068
114. S. Yi, Y. Wu, Y. Zhang, Y. Zou, F. Dai, Y. Si, Antibacterial activity of photoactive silk fibroin/cellulose acetate blend nanofibrous membranes against *Escherichia coli*, *ACS Sustainable Chemistry & Engineering*. 8(2020) 16775–16780. https://doi.org/10.1021/acssuschemeng.0c04276
115. F. Wu, P. He, X. Chang, W. Jiao, L. Liu, Y. Si, J. Yu, B. Ding, Visible-light-driven and self-hydrogen-donated nanofibers enable rapid-deployable antimicrobial bioprotection, *Small*. 17(2021) 2100139. https://doi.org/10.1002/smll.202100139
116. F. Kiani, N.A. Astani, R. Rahighi, A. Tayyebi, M. Tayebi, J. Khezri, E. Hashemi, U. Rothlisberger, A. Simchi, Effect of Graphene oxide nanosheets on visible light-assisted antibacterial activity of vertically-aligned copper oxide nanowire arrays, *Journal of Colloid and Interface Science*. 521(2018) 119–131. https://doi.org/10.1016/j.jcis.2018.03.013
117. L. Marín-Caba, G. Bodelón, Y. Negrín-Montecelo, M.A. Correa-Duarte, Sunlight-sensitive plasmonic nanostructured composites as photocatalytic coating with antibacterial properties, *Advanced Functional Materials*. 31(2021) 2105807. https://doi.org/10.1002/adfm.202105807
118. F. Zhang, L. Huang, P. Ding, C. Wang, Q. Wang, H. Wang, Y. Li, H. Xu, H. Li, One-step oxygen vacancy engineering of WO_3-x/2D g-C_3N_4 heterostructure: triple effects for sustaining photoactivity, *Journal of Alloys and Compounds*. 795(2019) 426–435. https://doi.org/10.1016/j.jallcom.2019.04.297
119. X. Xu, Y. Wang, D. Zhang, J. Wang, Z. Yang, In Situ Growth of photocatalytic Ag-decorated β-Bi_2O_3/$Bi_2O_{2.7}$ heterostructure film on PVC polymer matrices with self-cleaning and antibacterial properties, *Chemical Engineering Journal*. 429(2022) 131058. https://doi.org/10.1016/j.cej.2021.131058
120. M.-C. Wu, T.-H. Lin, K.-H. Hsu, J.-F. Hsu, Photo-induced disinfection property and photocatalytic activity based on the synergistic catalytic technique of Ag doped TiO_2 nanofibers, *Applied Surface Science*. 484(2019), 326–334. https://doi.org/10.1016/j.apsusc.2019.04.028
121. B. Liu, X. Han, L. Mu, J. Zhang, H. Shi, TiO_2 nanospheres/$AgVO_3$ quantum dots composite with enhanced visible light photocatalytic antibacterial activity, *Materials Letters*. 253(2019) 148–151. https://doi.org/10.1016/j.matlet.2019.06.057
122. K. Mallikarjuna, O. Nasif, S.A. Alharbi, S.V. Chinni, L.V Reddy, M.R.V. Reddy, S. Sreeramanan, Phytogenic synthesis of Pd-Ag/rGO nanostructures using stevia leaf extract for photocatalytic H_2 production and antibacterial studies, *Biomolecules*. 11(2021) 190. https://doi.org/10.3390/biom11020190
123. X. Ran, Y. Du, Z. Wang, H. Wang, F. Pu, J. Ren, X. Qu, Hyaluronic acid-templated Ag nanoparticles/graphene oxide composites for synergistic therapy of bacteria infection, *ACS Applied Materials & Interfaces*. 9(2017) 19717–19724. https://doi.org/10.1021/acsami.7b05584
124. C. Mao, Y. Xiang, X. Liu, Z. Cui, X. Yang, K.W.K. Yeung, H. Pan, X. Wang, P.K. Chu, S. Wu, Photo-inspired antibacterial activity and wound healing acceleration by hydrogel embedded with Ag/Ag@AgCl/ZnO nanostructures, *ACS Nano*. 11(2017) 9010–9021. https://doi.org/10.1021/acsnano.7b03513
125. A.B. Alayande, M. Obaid, I.S. Kim, Antimicrobial mechanism of reduced graphene oxide-copper oxide (rGO-CuO) nanocomposite films: the case of *Pseudomonas aeruginosa* PAO1, *Materials Science & Engineering C*. 109(2020) 110596. https://doi.org/10.1016/j.msec.2019.110596
126. L. Liu, W. Yang, Q. Li, S. Gao, J.K. Shang, Synthesis of Cu_2O nanospheres decorated with TiO_2 nanoislands, their enhanced photoactivity and stability under visible light illumination, and their post-illumination catalytic memory, *ACS Applied Materials & Interfaces*. 6(2014) 5629–5639.
127. Md. N. Karim, M. Singh, P. Weerathunge, P. Bian, R. Zheng, C. Dekiwadia, T. Ahmed, S. Walia, E. Della Gaspera, S. Singh, R. Ramanathan, V. Bansal, Visible-light-triggered reactive-oxygen-species-mediated antibacterial activity of peroxidase-mimic CuO nanorods, *ACS Applied Nano Materials*. 1(2018) 1694–1704. https://doi.org/10.1021/acsanm.8b00153

128. M.Y. Guo, F. Liu, Y.H. Leung, Y. He, A.M.C. Ng, A.B. Djurisič, H. Li, K. Shih, W.K. Cha, Annealing-induced antibacterial activity in TiO_2 under ambient light, *The Journal of Physical Chemistry C.* 121(2017) 24060–24068. https://doi.org/10.1021/acs.jpcc.7b07325

129. Y. Xu, S. Huang, M. Xie, Y. Li, L. Jing, H. Xu, Q. Zhang, H. Li, Core–shell magnetic $Ag/AgCl@Fe_2O_3$ photocatalysts with enhanced photoactivity for eliminating bisphenol A and microbial contamination, *New Journal of Chemistry.* 40(2016) 3413–3422. https://doi.org/10.1039/C5NJ02898A

130. T.K. Jana, A. Pal, A.K. Mandal, S. Sarwar, P. Chakrabarti, K. Chatterjee, Photocatalytic and antibacterial performance of α-Fe_2O_3 nanostructures, *ChemistrySelect.* 2(2017) 1–11. https://doi.org/10.1002/slct.201700294

131. R. Zhang, Y. Ma, W. Lan, D.E. Sameen, S. Ahmed, J. Dai, W. Qin, S. Li, Y. Liu, Enhanced photocatalytic degradation of organic dyes by ultrasonic-assisted electrospray TiO_2/graphene oxide on polyacrylonitrile/β-cyclodextrin nanofibrous membranes, *Colloids and Surfaces B: Biointerfaces.* 70(2021):105343. https://doi.org/10.1016/j.ultsonch.2020.105343

132. R. Djellabi, J. Ali, X. Zhao, A. Saber, B. Yang, CuO NPs incorporated into electron-rich TCTA@PVP photoactive polymer for the photocatalytic oxidation of dyes and bacteria inactivation, *Journal of Water Process Engineering.* 36(2020) 101238. https://doi.org/10.1016/j.ultsonch.2020.105343

133. N. Mohri, H. Kerschbaumer, T. Link, R. Andre, M. Panthofer, V. Ksenofontov, W. Tremel, Self-organized arrays of SnO_2 microplates with photocatalytic and antimicrobial properties, *European Journal of Inorganic Chemistry.* 2019(2019) 3171–3179. https://doi.org/10.1002/ejic.201900348

134. A. Singh, A. Goswami, S. Nain, Enhanced antibacterial activity and photo-remediation of toxic dyes using Ag/SWCNT/PPy based nanocomposite with core–shell structure, *Applied Nanoscience.* 10(2020), 2255–2268. https://doi.org/10.1007/s13204-020-01394-y

135. S. Tan, X. Wu, Y. Xing, S. Lilak, M. Wu, J.X Zhao, Enhanced synergetic antibacterial activity by a reduce graphene oxide/Ag nanocomposite through the photothermal effect, *Colloids and Surfaces B: Biointerfaces.* 185(2020) 110616. https://doi.org/10.1016/j.colsurfb.2019.110616

136. U. Saha, K. Sharma, N. Chaudhri, M. Sankar, P. Gopinath, Antimicrobial photodynamic therapy: single-walled carbon nanotube (SWCNT)-porphyrin conjugate for visible light mediated inactivation of *Staphylococcus aureus*, *Colloids and Surfaces B: Biointerfaces.* 162(2018) 108–117. https://doi.org/10.1016/j.colsurfb.2017.11.046

137. P.C. Nagajyothi, L.V. Reddy, K.C. Devarayapalli, S.V.P. Vattikuti, Y.J. Wee, J. Shim, Environmentally friendly synthesis: photocatalytic dye degradation and bacteria inactivation using Ag/f-MWCNTs composite, *Journal of Cluster Science.* 32(2021) 711–718. https://doi.org/10.1007/s10876-020-01821-8

138. D.S. Ahmed, M.K.A. Mohammed, Studying the bactericidal ability and biocompatibility of gold and gold oxide nanoparticles decorating on multi-wall carbon nanotubes, *Chemical Papers.* 74(2020) 4033–4046. https://doi.org/10.1007/s11696-020-01223-0

139. S. Sajjad, S.A.K. Leghari, A. Iqbal, Study of graphene oxide structural features for catalytic, antibacterial, gas sensing, and metals decontamination environmental applications, *ACS Applied Materials & Interfaces.* 9(2017) 43393–43414. https://doi.org/10.1021/acsami.7b08232

140. Y. Sun, X. Tang, H. Bao, Z. Yang, F. Ma, The effects of hydroxide and epoxide functional groups on the mechanical properties of graphene oxide and its failure mechanism by molecular dynamics simulations, *RSC Advances.* 10(2020) 29610–29617. https://doi.org/10.1039/D0RA04881J

141. Y.C.G. Kwan, G.M. Ng, C.H.A. Huan, Identification of functional groups and determination of carboxyl formation temperature in graphene oxide using the XPS O 1s spectrum, *Thin Solid Films.* 590(2015) 40–48. https://doi.org/10.1016/j.tsf.2015.07.051

142. Y. Gerasymchuk, A. Lukowiak, A. Wedzynska, A. Kedziora, G. Bugla-Ploskonska, D. Piatek, T. Bachanek, V. Chernii, L. Tomachynski, W. Strek, New photosensitive nanometric graphite oxide composites as antimicrobial material with prolonged action, *Journal of Inorganic Biochemistry.* 159(2016) 142–148. https://doi.org/10.1016/j.jinorgbio.2016.02.019

143. L. Sun, T. Du, C. Hu, J. Chen, J. Lu, Z. Lu, H. Han, Antibacterial activity of graphene oxide/g-C_3N_4 composite through photocatalytic disinfection under visible light, *ACS Sustainable Chemistry & Engineering.* 5(2017) 8693–8701. https://doi.org/10.1021/acssuschemeng.7b01431

144. A. Wanag, P. Rokicka, E. Kusiak-Nejman, J. Kapica-Kozar, R.J Wrobel, A. Markowska-Szczupak, A.W. Morawski, Antibacterial properties of TiO_2 modified with reduced graphene oxide, *Ecotoxicology and Environmental Safety.* 147(2018) 788–793. https://doi.org/10.1016/j.ecoenv.2017.09.039.

145. H. Radwan, R.A. Ismail, S.A. Abdelaal, B.A.A Jahdaly, A. Almahri, M.K. Ahmed, K. Shoueir, Electrospun polycaprolactone nanofibrous webs containing Cu-magnetite/graphene oxide for cell viability, antibacterial performance, and dye decolorization from aqueous solutions, *Arabian Journal for Science and Engineering.* https://doi.org/10.1007/s13369-021-05363-7

146. P. Dallas, V.K. Sharma, R. Zboril, Silver polymeric nanocomposites as advanced antimicrobial agents: classification, synthetic paths, applications, and perspectives, *Advances in Colloid and Interface Science*. 166(2011) 119–135. https://doi.org/10.1016/j.cis.2011.05.008

147. I. Armentano, C.R. Arciola, E. Fortunati, D. Ferrari, S. Mattioli, C.F. Amoroso, J. Rizzo, J.M. Kenny, M. Imbriani, L. Visai, The interaction of bacteria with engineered nanostructured polymeric materials: a review, *Scientific World Journal*. 2014(2014) 1–18. https://doi.org/10.1155/2014/410423

148. M.I.A.A. Maksoud, M.M. Ghobashy, G.S. El-Sayyad, A.M. El-Khawaga, M.A. Elsayed, A.H. Ashour, Gamma irradiation-assisted synthesis of $PANi/Ag/MoS_2/LiCo_{0.5}Fe_2O_4$ nanocomposite: efficiency evaluation of photocatalytic bisphenol A degradation and microbial decontamination from wastewater, *Optical Materials*. 119(2021) 111396. https://doi.org/10.1016/j.optmat.2021.111396

149. W. Wentao, Z. Tao, B. Sheng, T. Zhou, Q. Zhang, W. Fan, Z. Ninglin, S. Jian, Z. Ming, S. Yi, Functionalization of polyvinyl alcohol composite film wrapped in am-ZnO@CuO@Au nanoparticles for antibacterial application and wound healing, *Applied Materials Today*. 17(2019) 36–44. https://doi.org/10.1016/j.apmt.2019.07.001

150. D.M. Rocca, J.P. Vanegas, K. Fournier, M.C. Becerra, J.C. Scaiano, A.E. Lanterna, Biocompatibility and photo-induced antibacterial activity of lignin-stabilized noble metal nanoparticles, *RSC Advances*. 8(2018) 40454–40463. https://doi.org/10.1039/C8RA08169G

151. M.-L. Xu, L.-Y. Guan, S.-K. Li, L. Chen, Z. Chen, Stable gold graphitic nanocapsule doped hydrogels for efficient photothermal antibacterial applications, *Chemical Communications*. 55(2019) 5359–5362. https://doi.org/10.1039/C9CC01933B

152. S. Wu, A. Li, X. Zhao, C. Zhang, B. Yu, N. Zhao, F.-J. Xu, Silica-coated gold–silver nanocages as photothermal antibacterial agents for combined anti-infective therapy, *ACS Applied Materials & Interfaces*. 11(2019) 17177–17183. https://doi.org/10.1021/acsami.9b01149

153. X. Zhang, G. Zhang, H. Zhang, X. Liu, J. Shi, H. Shi, X. Yao, P.K. Chu, X. Zhang, A bifunctional hydrogel incorporated with $CuS@MoS_2$ microspheres for disinfection and improved wound healing, *Chemical Engineering Journal*. 382(2020) 122849. https://doi.org/10.1016/j.cej.2019.122849

10 Advanced Materials for Hydrogen Production and Storage

A New Era of Clean Energy

Muhammad Tahir
UAE University

Sehar Tasleem
Universiti Teknologi Malaysia

CONTENTS

10.1 INTRODUCTION: BACKGROUND

The readily available and cost-effective fossil fuels are recognized to be the main source of producing energy. Approximately, 85% of worldwide consumption of energy is fulfilled from non-renewable energy sources such as fossil fuels [1]. Though, the consumption of fossil fuels leads to the generation of greenhouse gases, including carbon dioxide (CO_2) [2]. So, it is a challenge to explore a clean and carbon-free source of renewable energy. Other than the existence of energy harvesting from sources like wind, heat, geothermal, waves, and biomass, the sunlight reaching the surface of Earth is an ideal renewable source for energy harvesting. Currently, hydrogen (H_2) is known to be an ideal source of clean, long-term, renewable, and storable source of gaining energy [3,4].

Converting solar energy into chemical energy such as H_2 from water is an interesting strategy for efficiently using solar radiation [5,6]. Photogenerated H_2 can be easily stored and used as green fuel. Green technology such as the photocatalytic production of H_2 from water (H_2O) is a promising approach to meet energy needs and does not contribute towards deterioration of environment [7,8].

DOI: 10.1201/9781003206385-10

The motivation for photocatalysis stems from the natural process of photosynthesis, in which plants absorb sunlight and produce chemical energy in the form of glucose with the help of chlorophyll. In the photocatalytic process, irradiations from sun directly provides energy for water splitting, while the use of this H_2 fuel only produces water as a by-product, which is part of the hydrogen cycle. Therefore, the photocatalytic production of H_2 from water is pollution-free technology for power generation [9,10].

Several kinds of semiconductors are utilized by diverse researchers to maximize UV and visible light-induced photocatalytic H_2 evolution. Among the semiconductors for photocatalytic water-splitting processes, titanium dioxide (TiO_2), zinc sulfide (ZnS), graphitic carbon nitride (g-C_3N_4), graphene (GO), cadmium sulfide (CdS), and molybdenum disulfide (MoS_2) are included, while some complex semiconductor structures are also included, such as metal organic frameworks (MOFs), perovskite oxides, and layered double hydroxides (LDH) [10–16]. Among the existing semiconductors, the following parameters should be taken into consideration in designing photocatalysts: valence band (VB) and conduction band (CB) positions, irradiation harvesting ability, transport of electron-hole pairs, reaction area, and stability. The activity of nanomaterials acting as photocatalysts is assisted with their characteristics like porosity, crystallinity, particle shape, and size, which play a role in optical property, electronic features, and surface area [17–19].

Apart from the photocatalytic H_2 generation ability of nanomaterials, also their ability to store H_2 is considered. As compared to the traditional storage methods, material-based approaches are based on the physisorption and chemisorption for restraining and storing H_2 in the solid-state form. Storing H_2 in materials is considered as a safe approach to substitute the liquid- and gas-based H_2 storage because of stable energy states of the H_2 composites and less cost of operation [20,21]. Moreover, it is considered to provide high energy density and display good reversibility, making it applicable in automobile. In material-based H_2 systems, the efficiency of H_2 storage system is dependent on the connection between H_2 and nanomaterials [22].

10.2 CHARACTERISTIC OF HYDROGEN AS A CLEAN ENERGY SOURCE

Since the last 30 years, H_2 has been recognized as a crucial component of decarbonized sustainable energy system, offering less costly and non-toxic energy source. It is known to be alternative to fossil fuels based on its more energy content and lower rate of toxicity towards environment [23]. Moreover, the atom of H_2 constitutes one proton and one electron and has a density that is less than the density of air, while the H_2 fuel constitutes sevenfold increased gravimetric density than other fossil fuels [24]. H_2 can be stored easily after compressing, as cryogenic liquid H_2 by the solid-state storing approach. Mehrizi et al. explored that H_2 offer cost-effective storage options along with providing storage at normal temperature and also offer elevated storing volume [25]. As compared to other fuels, H_2 constitutes better features, including diffusion coefficient, kinematic viscosity, thermal conductivity, low density, and luminosity that induces good diffusion and heat transfer properties. Also, the flammable range of H_2 in air provides rich mixtures to support combustion and needs less energy to ignite [26]. The only product formed from directly combusting H_2 is water that is easily manageable. However, the initiation of H_2-dependent sustainable economy needs the use of H_2 fuel generated from renewable energy sources instead of other highly costly and environmental deteriorating methods such as steam reforming of methane, resulting in large amount of CO_2. H_2 as an energy carrier offers storage capacity at large scale and can be easily transformed into electricity in the form of fuel cells [27]. Pivovar et al. recognized the chemical-to-electrical energy conversion ability of H_2 along with excellent energy density and good chemical bonding for storage over extended time span. There exists great flexibility regarding the integration of H_2 in various systems dealing with energy, usefulness in form of dispatchable load and usefulness as a basis for power production. This makes it possible for processes involving H_2 to be combined with the energy systems for maximum harvesting of benefits [28]. Figure 10.1 illustrates the major characteristics of H_2 as a clean source of energy.

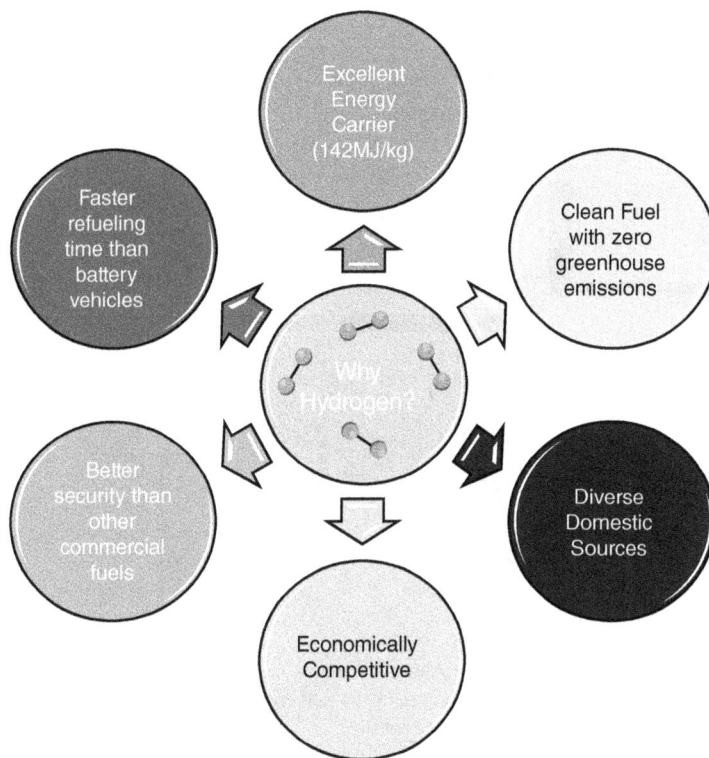

FIGURE 10.1   Characteristics of H_2 as source of clean energy. (Reprinted with permission from [29]. Copyright 2019 Springer nature.)

10.3   UTILITY OF HYDROGEN PRODUCTION AND STORAGE

Currently, the usage of fuel cells and battery systems based on H_2 is being highly encouraged to promote the reduction in greenhouse gas emissions. These fuel cells promote the production of electrical energy along with heat from chemical energy [30]. The combination of fuel cells with batteries can derive the utilization of fuel cells in transportation. Normally, internal combustion engines require combustion process for energy harvesting, while the fuel cells are based on electrocatalytic process. In comparison with the batteries, fuel cells offer increased energy conversion efficiency and also provide high energy density and power density required for transportation and other industrial applications. Fuel cells consist of electrons (anode, cathode) containing coated catalyst and an electrolyte. Moreover, they have less storage space, which is 8–14 times less in weight, and offer quick refueling capacity. Fuel cells have been estimated to produce 1 kW of electric energy and 2 kW of heat [31]. Also, they produce less greenhouse gas emissions and require low maintenance cost [32]. Thus, as shown in Figure 10.2a, H_2 is produced from renewable sources by water splitting in combination with fuel cells, and the conversion efficiency is increased with zero pollution for the sustainable production of energy.

In case of direct fuel cells, the electrodes are used to carry out electrochemical oxidative reaction, while in case of indirect system, the electrochemical oxidative reaction takes place on electrodes after feeding them fuels containing H_2. In case of transportation, proton exchange membrane fuel cells operating at 80°C temperature are used as they offer increased efficiencies [36]. Figure 10.2b shows the proton exchange membrane fuel cell, which uses H_2 to operate for energy generation. Moreover, H_2 is recognized to be a future fuel based on its high content of energy, less carbon emissions, and less atomic mass. Its more mass energy as compared to oil enables it to be used in

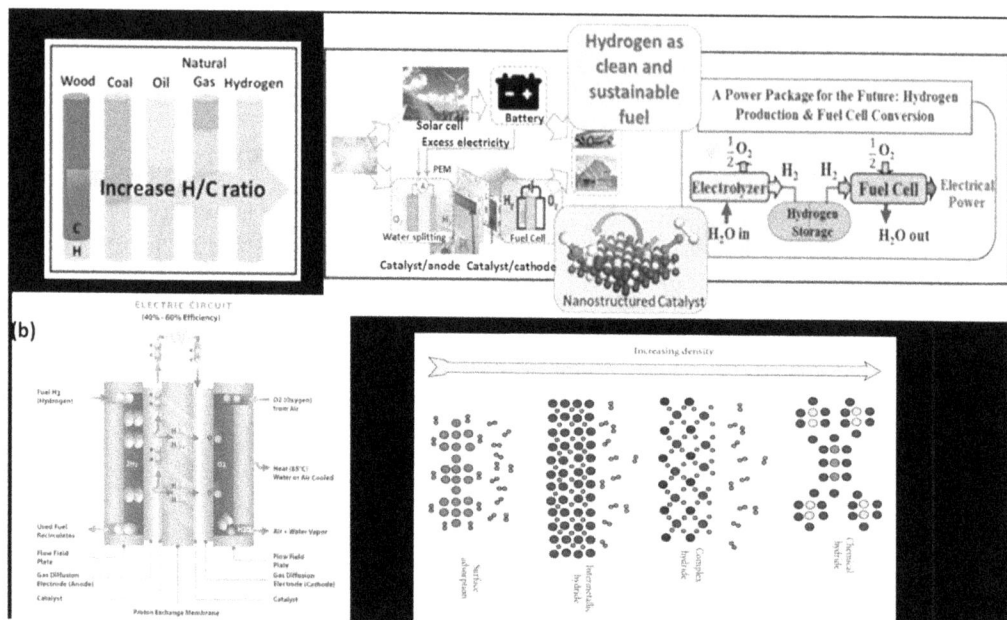

FIGURE 10.2 (a) Representation of H_2-to-C ratio of various fuels and the illustration of a complete sustainable system of dual fuel cell having solar to H_2 generation and H_2 to electric energy generation. Reprinted with permission from [33]. (Copyright 2021 Springer Nature); (b) Display of basic fuel cell. (Reprinted with permission from [34]. Copyright 2022 Elsevier; (c) Storage of H_2 with respect to density in metal hybrids, materials, and chemical hybrids. Reprinted with permission from [35]. Copyright 2014 American Chemical Society.)

transportation sector and electronics. The electric energy generation from fuel cells is pollutants free. H_2-fed transportation requires around 10 minutes to recharge and is a lightweight fuel alternative for heavy transport [37].

Nanostructured materials have the ability to store H_2 due to their outstanding surface and bulk absorption, and intragrain boundary. Nanostructured and nanoscale materials are likely to have a strong effect on the thermodynamics and kinetics towards H_2 absorption and dissociation. This happens due to more rate of diffusion and less diffusion length. Additionally, the materials at the nanoscale offer the possibility of controlling material tailoring parameters independently as compared to their bulk counterparts. They also lead to the design of lightweight H_2 storage systems with better H_2 storage characteristics [35]. As shown in Figure 10.2c, there are various storage systems for H_2, including metal, chemical, and complex hydrides, nanomaterials, polymer-based composites, adsorbents, clathrate hydrates, and MOFs, etc. [38–40]. The characteristics of materials for H_2 storage should include more than 6.0 wt.% content of H_2, good thermodynamics, i.e., 30–55 kJ/mol H_2, low operating temperature, onboard refueling, and about 1,000 cycle-based reversibility.

10.4 OVERVIEW OF PHOTOCATALYTIC H_2 GENERATION

Hydrogen generation from photocatalysis is similar to that from photosynthesis, where chemical energy is acquired from solar energy. Photocatalysis occurs under solar light utilization for the generation of photogenerated electron at the VB of semiconductor photocatalyst, which later jumps to CB, leaving behind holes in the VB, and redox reaction takes place. The utilization of photocatalyst with a wide or narrow bandgap influences the absorption of light as wide bandgap photocatalyst can absorb visible light in the range of 400–800 nm and wider bandgap can only absorb wavelength ranging from 200 to 400 nm [41]. Moreover, the usage of sacrificial agent having a varying capacity

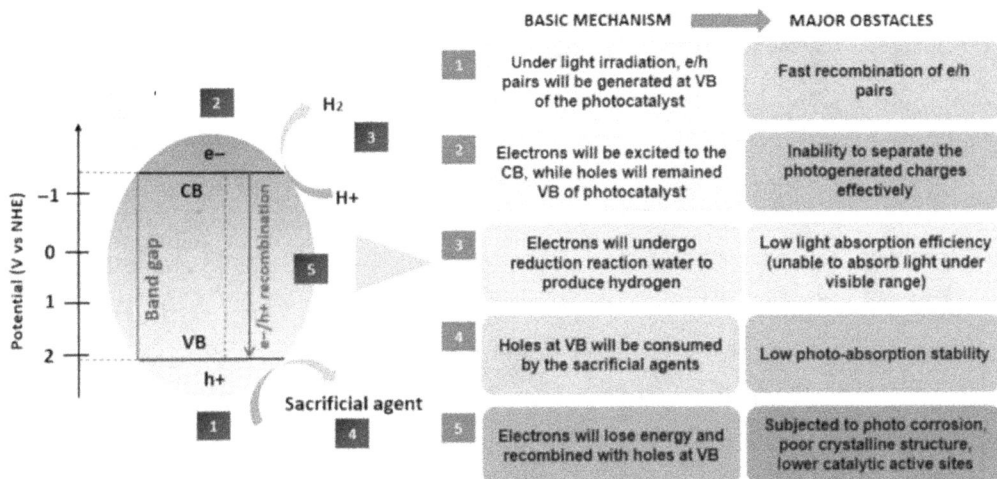

FIGURE 10.3 Illustration of mechanism of water splitting over the semiconductor photocatalyst. (Reprinted with permission from [44]. Copyright John Wiley and Sons.)

of adsorption for carrying out improved redox reaction is another necessary element of photocatalytic system [42]. The photoreactor offers varying efficiencies of utilizing concentration of irradiation, which affects the overall photocatalytic process [43].

Photocatalytic H_2 generation initiates by light absorption having an energy greater than or equal to the bandgap of photocatalyst. After the generation of electron and hole pairs in the CB of photocatalyst, the charge carrier's separation takes place and electrons are transferred to CB of photocatalyst and holes are left in the VB of photocatalyst (Equation 10.1). Recombination of electron-hole pairs occurs with the release of heat energy, which minimizes the activity of the process (Equation 10.2). The electrons at CB undergo a reduction reaction, while holes in VB undergo oxidation (Equations 10.3 and 10.4). The following main processes take place in photocatalytic processes: (i) absorption of light, (ii) charge carriers' separation, (iii) migration of charge carriers, and (iv) redox reactions at the surface of photocatalyst. Figure 10.3 displays the illustration of steps involved in the photocatalytic process for H_2 generation [44]. In a work on photocatalysis, Ida et al., explored the role of energy and time needed by nanomaterials to carry out water splitting. Four electrons under reduction process for the generation of H_2 require four photons having a suitable energy at a short span of time. The solar photon flux density is estimated to be $2,000 \, \mu mol/s/m^2$. The time needed for the photons to absorb on the nanomaterials is around 4 ms, while 1 μs is the approximate lifetime of charges [45].

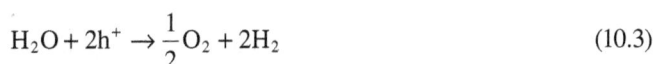

$$\text{Photocatalyst} \xrightarrow{h\nu} e^- + h^+ \tag{10.1}$$

$$\text{Photocatalyst } (e^- + h^+) \rightarrow \text{photocatalyst} \tag{10.2}$$

$$H_2O + 2h^+ \rightarrow \frac{1}{2}O_2 + 2H_2 \tag{10.3}$$

10.5 CHARACTERISTICS OF NANOMATERIALS FOR PHOTOCATALYTIC H_2 GENERATION AND STORAGE

Since the pioneering of H_2 production through photocatalysis in 1972, various nanomaterials working as semiconductor photocatalysts have been explored; however, the requirements for

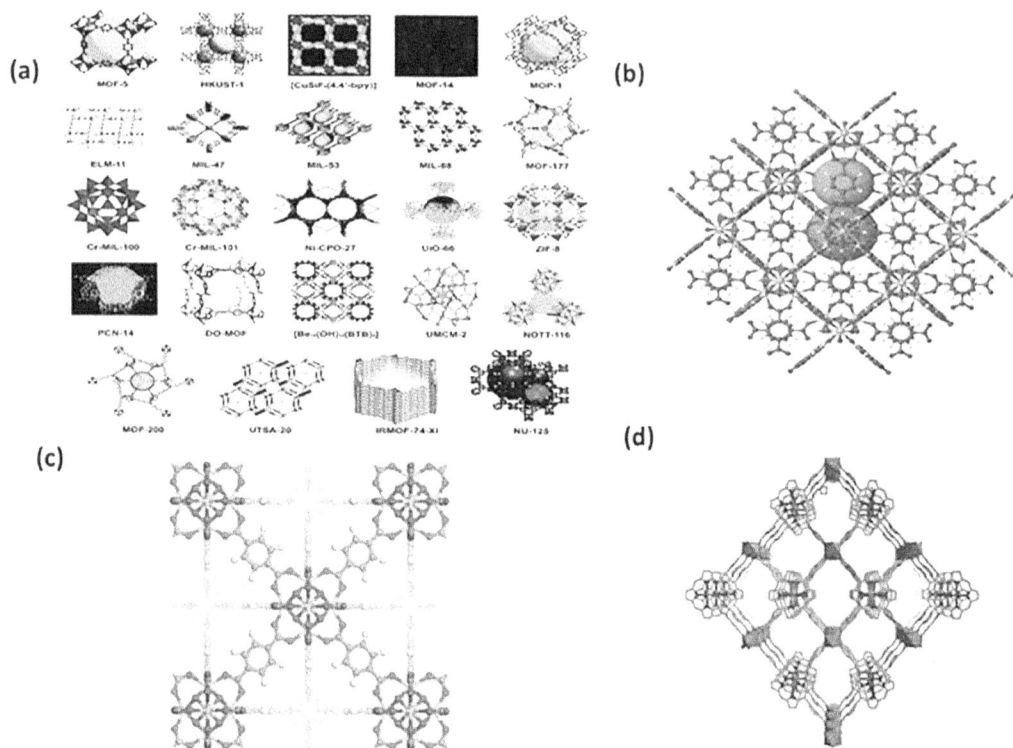

FIGURE 10.4   (a) Illustration of various structures of MOFs. (Reprinted with permission from [49]. Copyright Royal Society of Chemistry); (b) Structural illustration for UiO-66 MOF with organic linkers. (Reprinted with permission from [50]. Copyright Royal Society of Chemistry); (c) Illustration of Zr-MOFs with 1,4-benzene-dicarboxylate (BDC) linker, UiO-66. (Reprinted with permission from [51]. Copyright American Chemical Society); (d) 3D network of USTC-8(In) along an axis. (Reprinted with permission from [52]. Copyright 2014 American Chemical Society).

photocatalyst range from appropriate bandgap for charge separation, photostability, and stability of structure for large-scale applications. Other than this, the morphological properties also play an important role. Following are the properties in various nanomaterials, which play a key role in photocatalytic activity.

10.5.1   METAL ORGANIC FRAMEWORKS

MOFs are recognized as a category of 3D porous materials having link between organic linkers and metal ions via covalent bonds. The excellent stability, flexibility in structure, and high surface area ranging from 1,000 to 10,000 m^2/g make them a good photocatalyst [46]. Moreover, MOFs play the role of utile semiconductor photocatalyst for acting as support for developing extremely distinct photocatalysts for photocatalytic H$_2$ evolution. Apart from this, the MOF provides a porous structural property, which offers excellent mass transfer for photoreaction in the nanometric spaces [47]. Around 20,000 MOFs have been already explored and studied [48]. MOFs provide the following properties: (i) The organic linkers and nodes of metal oxide and sulfide clusters in the MOFs are lightly active and MOFs also work under near-infrared (NIR) irradiation, (ii) the ability of MOFs to be flexible for modification, (iii) more surface area for increased photoactivity, (iv) photostability, (v) stability in varying pH and in the presence of harsh solvents, (vi) recyclable and, (vii) cost-effectiveness. Figure 10.4a shows the structural variation of MOFs due to the existence of widely present metals and linkers.

In this context, zirconium (Zr)-based MOFs show higher VB potential and also constitute oxophilic nature and excellent photophysical properties, leading to a stable structure [53,54]. Countless linear polyiodide chains and π–π^* transition inside organic ligands of Zr-based MOFs make them to act as semiconductor. Figure 10.4b shows the porous framework of $[Ti_8O_8(OH)_4(ATA)_6]$; UiO-66,$[Zr_6(\mu_3\text{-}O)_4(\mu_3\text{-}OH)_4(BDC)_6]$ having organic linkers along with metal clusters [50]. In a study on MOFs, UiO-66 MOFs acquired from three linker ligands H_2N–H_2BDC, O_2N–H_2BDC, and Br–H_2BDC have been reported to show excellent thermally and chemically stable nature, beneficial for photocatalytic processes. Figure 10.4c displays the Zr-MOFs with 1,4-benzene-dicarboxylate (BDC) linker, i.e., UiO-66 [51]. In another similar study, $[Zr_6(\mu_3\text{-}O)_4(\mu_3\text{-}OH)_4(ATA)_6]$ ($H_2ATA = 2$-aminoterephthalic acid) (UiO-66(NH_2)) and $[Zr_6O_4(OH)_4]$ octahedron clusters and BDC ligands (UiO-66) were studied for photoactivity towards H_2 evolution showing a quantum yield of 3.5 and <0.1, respectively, under UV light [55]. Titanium (Ti)-based MOFs are constituted of Ti-O clusters, inducing photocatalytic capacity similar to the photoactivity of titanium and offering excellent boosted surface area with high porous structure. In this regard, Ti-based $[Ti_8O_8(OH)_4(BDC)_6]$ (MIL-125(Ti)) MOFs were recognized to have highly crystalline structure composing of octameric $Ti_8O_8(OH)_4$ rings interrelated with 1,4-benzenedicarboxylic acid (H_2BDC), where the clusters of Ti-O make a bonding with BDC ligands, inducing stability to the entire structure of MOF [53]. MIL-125-$(SCH_3)_2$ MOF composed of cyclic octamers of TiO_2 octahedra and terephthalic acid linker as well as methylthio functional group on the terephthalic acid linker (H_2BDC-$(SCH_3)_2$) was utilized to carry out visible light-induced H_2 production, attributing to the -SCH_3, which gave out electrons (S 3p) to aromatic linking unit and contributed towards visible light absorption [56]. Cu-based MOFs were discussed for photocatalytic H_2 generation activity based on their modes of copper acting as open coordination sites. Moreover, they depict efficient light utilization and redox reactions [57]. In another study, Ni-based MOFs were recognized as efficient photocatalysts based on their enlarged surface area with thin-sheeted structure [58]. Another Ni-metal organic framework named UNiMOF provided unsaturated sites containing Ni, inducing an efficient reduction process for H_2 generation [59]. Apart from this, iron-based (Fe) MOFs containing Fe oxo-clusters were also studied for H_2 generation processes, making the MOF capable of inducing activity under visible irradiation [60]. Also, the oxo-clusters depicted a ligand-to-metal charge transfer for visible light absorption. As shown in Figure 10.4d, the photocatalytic activity was carried out using in-based MOF (USTC-8) having central metal and porphyrin ligand, which offered the migration of electrons to it to carry out water reduction process [52]. A highly stable Al-based MOF (AlTCS-1) towards aqua regia acted as a photocatalyst for H_2 generation under irradiations [61]. Similarly, 2D $[Ni(phen)(oba)]_n \cdot 0.5nH_2O$ MOF acted as a photocatalyst for visible light-induced photoactivity because of the enlarged specific surface area, reaction sites, acceptable band structure, and synergistic phenomenon [62].

10.5.2 Perovskite Oxides

A Russian scientist was the pioneer of exploring perovskite materials since he first discovered calcium titanate ($CaTiO_3$) in 1839, after which the entire group of materials having structural similarity to $CaTiO_3$ were known as perovskite having a formula of ABX_3. The research on the utilization of perovskite for the application of solar energy conversion expanded rapidly after Miyasaka in 2012 reported solar cells having perovskite materials. Perovskite has a general formula of ANX_3, where A site contains monovalent and divalent bigger radius cations like cesium (Cs), barium (Ba), rubidium (Rb), serium (Sr), sodium (Na), Nd, and lanthanum (La), whereas the B site denotes transition metals like gold (Au), niobium (Nb), cobalt (Co), and ruthenium (Ru). The X site in the perovskite contains non-metals like bromine (Br), nitrogen (N), oxygen (O), and chlorine (Cl) and is categorized into oxide perovskite, halide perovskite, and their derivatives.

The semiconductor-like properties in perovskites arise from the outer p orbitals of B-site, which form the CB, and the orbitals of X-site, which form the VB of perovskites. The energy band is not affected by the atoms at A site; however, perovskites help to stabilize their lattice.

On the other hand, by partially replacing the atoms at B site and the change of X site ions, the tuning of band gap and edges can be achieved. This can lead to good light utilization and good catalytic properties of perovskites [63]. The ability of perovskite to contain many electrons of the VB states makes perovskite oxides to show a higher adsorption coefficient as compared to other oxide materials. Also, provision of longer diffusion distances for electron-hole pairs is another structural property of these oxides. There exists a wide compositional and structural variation in perovskites based on defects in their lattice and tolerance factor. Perovskites with desired energy bands and specific surface areas can be obtained as the bulk components and surface components of these materials are flexible; thus, the excitation of electrons for water splitting is favored [64].

Generally, electrocatalysis is the major field of catalysis utilizing perovskite oxides as catalysts based on their high catalytic capacity, electronic conductivity, and O_2 vacancies. However, a few perovskites have also been reported for photocatalytic H_2 generation. The limitations such as wide band gap and slow migration of charges make them less active for photocatalysis. Likewise, halide perovskites constitute optoelectronic features, visible light activity, good coefficient for adsorption, and larger path for diffusion but, at the same time, show instability in water environment, restricting their uses [65,66]. Various structural and morphological approaches, including encapsulation approaches have been done.

In the context of utilizing and harvesting the properties of perovskite towards high-efficiency photocatalytic H_2 generation, a current study demonstrated single-atom Pt–I_3 sites anchored on cesium tin iodide double perovskite (Cs_2SnI_6) and inorganic halide perovskite ($PtSA/Cs_2SnI_6$) for improved H_2 generation from hydrogen iodide (HI) splitting. The synthesized Cs_2SnI_6 depicted an excellent stability at higher temperature, i.e., 350°C in air, as shown in Figure 10.5a. Moreover, Cs_2SnI_6 showed zero solubility at room temperature. As indicated in Figure 10.5b, Cs_2SnI_6 depicted an absorption over a wide wavelength range with 1.22 eV band gap, which is smaller than the

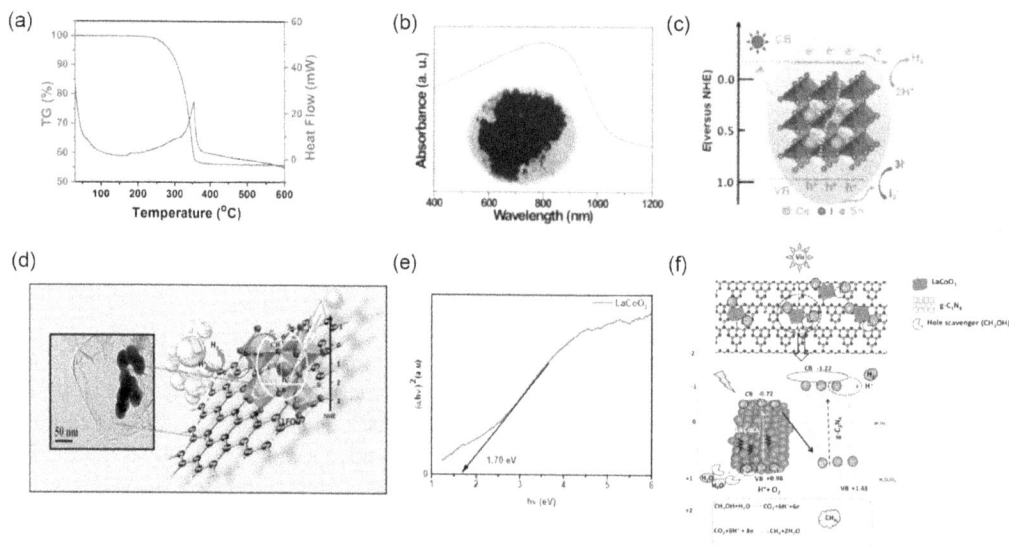

FIGURE 10.5 (a) Thermogravimetry and differential scanning calorimetry of Cs_2SnI_6; (b) UV-visible absorption spectrum of Cs_2SnI_6; (c) Energy band structure of Cs_2SnI_6. Reprinted with permission from [67]. Copyright 2019 Springer Nature); (d) Illustration of photocatalytic H_2 evolution in $LaFeO_3/RGO$. (Reprinted with permission from [68]. Copyright Elsevier); (e) Tauc plot of $LaCoO_3$; (f) Illustration of photocatalytic H_2 evolution in $LaCoO_3/g$-C_3N_4 Z-scheme. (Reprinted with permission from [69]. Copyright, American Chemical Society.)

commonly existing methylammonium lead triiodide (MAPbI$_3$) halide perovskite and also depicted a good conductivity. The energy band levels of Cs$_2$SnI$_6$ offered an efficient splitting of HI into H$_2$ and I$_3^-$, as shown in Figure 10.5c. The Pt–I$_3$ species in PtSA/Cs$_2$SnI$_6$ generated an exceptional coordination structure, which caused a strong metal-support interaction for improved migration of change carriers from perovskite to Pt atoms along with a requirement of less Gibbs free energy and more kinetics for efficient H$_2$ evolution [67].

Lanthanum ferrite (LaFeO$_3$) constitutes an appropriate band gap and band edges for reduction and oxidation photocatalytic processes and normally has a bandgap of 2–2.66 eV, which is achieved through inducing changes in morphological features and particle size [70]. The ability of LaFeO$_3$ to carry out both oxidation and reduction makes it an efficient photocatalysts for water splitting H$_2$ evolution and other photodegradation reactions. For instance, the water-splitting ability of LaFeO$_3$ was utilized, where it was modified on the surface of RGO. As shown in Figure 10.5d, it supported the photogenerated electron-hole separation and extended the lifetime of charge carriers as well as improved the light absorption by facilitating the migration of electrons from LaFeO$_3$ to RGO [68].

Likewise, in another work, as shown by Tauc plot in Figure 10.5e, a narrow band gap (1.70 eV) of lanthanum cobaltite perovskite (LaCoO$_3$) was synthesized by a combination of co-precipitation and hydrothermal methods. LaCoO$_3$ formed a Z-scheme with g-C$_3$N$_4$ for improved charge separation and light absorption, as indicated in Figure 10.5f. The improved photocatalytic H$_2$ generation due to visible light activity was dependent on the efficient light utilization of LaCoO$_3$. Moreover, the improvement in efficient light utilization was attributable to the synergistic effect of La and Co on the structure of perovskite [69]. Silver-based perovskites are known for their efficient photocatalytic activity towards H$_2$ generation based on their absorption and photoactivity under light. Silver tantalate (AgTaO$_3$) has been studied as an excellent photocatalyst as it offers Ta-O-Ta bond angle of 164° and is near to the bond angle of the most efficient tantalate perovskite, i.e., sodium tantalate (NaTaO$_3$). AgTaO$_3$ was studied for improved photocatalytic H$_2$ generation based on the light activation property of AgTaO$_3$ and due to the surface plasmonic resonance (SPR) effect of metals [71].

Moreover, layered perovskites, including titanates and niobates, also exhibit outstanding photocatalytic characteristics for water-splitting processes in comparison with bulk structural materials. They constitute tuneable structural properties and efficient charge migration and conductivity [72]. For instance, bismuth tungstate (Bi$_2$WO$_6$) is the simplest layered perovskite based on structural and chemical characteristics arising from the presence of alternating WO$_4^{2-}$- and fluorite-like Bi$_2$O$_2^{2+}$ blocks [73]. Bi$_2$WO$_6$ shows ferro-/piezoelectric response and is lead-free with no toxicity [74]. The ionic conductivity along with the migration of O$_2$ ions makes it applicable towards photocatalytic processes. Recently, the piezoelectric feature of Bi$_2$WO$_6$ was utilized to achieve catalytic H$_2$ generation. The minimum CB of Bi$_2$WO$_6$ enables an effective reduction of water to yield H$_2$ from triethanolamine, as shown in Figure 10.6a and b [75]. Bismuth-based oxyhalide material Bi$_4$MO$_8$X, where X=Cl, Br; M=Nb, Ta, was studied for photocatalytic H$_2$ generation. The Raman analysis, shown in Figure 10.6c, indicated that Bi$_4$NbO$_8$Br showed good migration of charge carriers due to the layers of [Bi$_2$O$_2$]$^{2+}$ and also because of its lesser twisting degree. Figure 10.6d indicates that bismuth-based perovskites also showed a narrow band gap, offering an efficient photocatalytic H$_2$ generation under visible light [76]. Hu et al., declared layered perovskites to be promising for photocatalytic H$_2$ generation due to the existence of slabs, which consist of cations of metals and tuneable interlayered spaces. This offers the modification of layered perovskites for improved activity, more stable structure, and lesser cost. Also, they show improved surface area and charge migration [77]. In another work, the photocatalytic properties of 2D lead-free double perovskite Cs$_2$AgBiBr$_6$ were studied to depict its narrow band gap and visible- and UV-induced light absorption. The reduced thickness of Cs$_2$AgBiBr$_6$ induced direct band gap, outstanding migration of charges, and maximized exciton-binding energy. Moreover, the water splitting was favored by the ability of adsorption of water and desorption of H$_2$ [78].

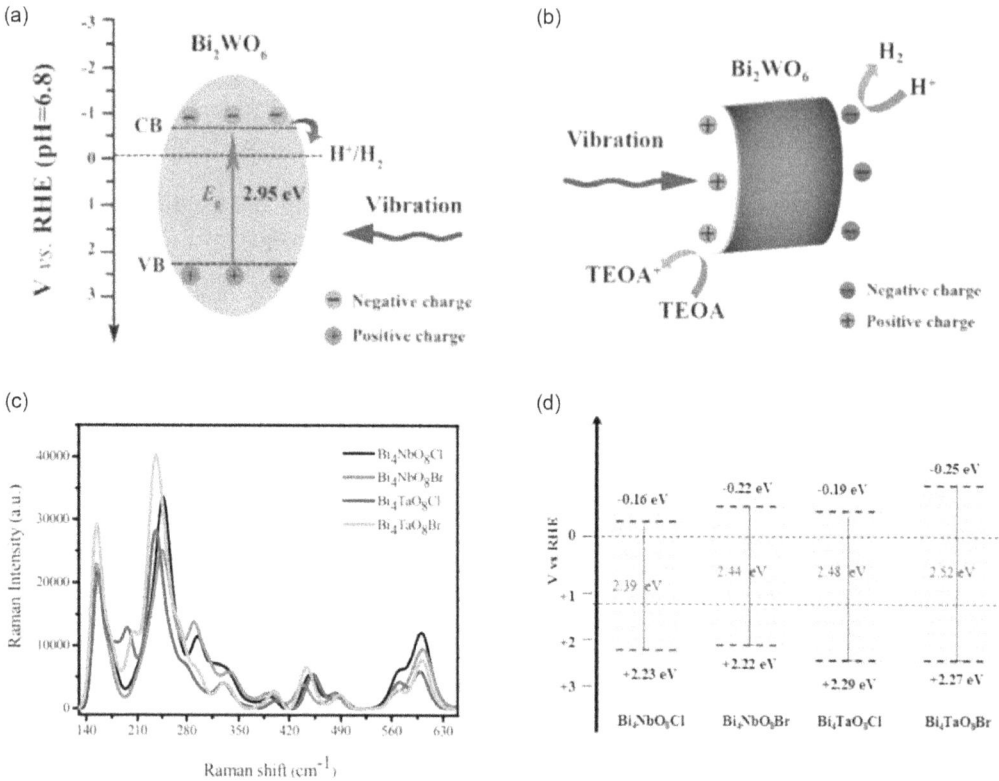

FIGURE 10.6   (a) The energy band level of Bi_2WO_6; (b) Schematic of the piezo-catalytic hydrogen production. Reprinted with permission from [75]. Copyright Elsevier; (c) Raman spectra of Bi_4MO_8X (X=Cl, Br; M=Nb, Ta); (d) Schematic depiction of the calculated VBM and CBM positions as well as band gap values of Bi_4MO_8X (M=Nb, Ta; X=Cl, Br). Reprinted with permission from [76]. Copyright Royal Society of Chemistry.)

10.5.3   LAYERED DOUBLE HYDROXIDES

Currently, LDH has been recognized as a unique group of anionic clays having layered structure showing a prominent contribution in photocatalytic activity [79], electrochemistry [80], and adsorption applications [81]. There exists no significant existence of LDH naturally; however, they can be economically synthesized. LDH constitutes outstanding light absorption property and a layered structural feature, which makes it applicable for photocatalytic processes [82]. The photocatalytic H_2 generation is dependent on the electronic and structural characteristics of LDH.

LDHs are recognized as hydrotalcites, and commonly occurring LDHs contain Mg^{2+} and Al^{3+} having the formula $Mg_6Al_2·(OH)_{16}·CO_3·4H_2O$. As hydrotalcite is the prominent mineral in LDHs, LDHs are known as hydrotalcite-like compounds having the formula $[M(II)_{1-x}M(III)_x (OH)_2]^{x+}[A_{x/n} H_2O]^{x-}$, where M(II) and M(III) denote metal cations, and A denotes the intercalated anions like CO_3^{2-}, NO_3^-, SO_4^{2-}, F^-, Cl^- [83].

In context of the structural properties of LDH, there exists a structural similarity in hydrotalcite- and brucite-like $Mg(OH)_2$, containing ions of Mg^{2+} in a octahedral coordination with OH ions, as shown in Figure 10.7a and b. These octahedral units consist of infinite layers by edge-shared octahedrons of MO_6, whereas the MO_6 octahedron is responsible for inducing photophysical and photocatalytic characteristics in LDHs [82]. Also, the highly dispersal feature of LDH makes it

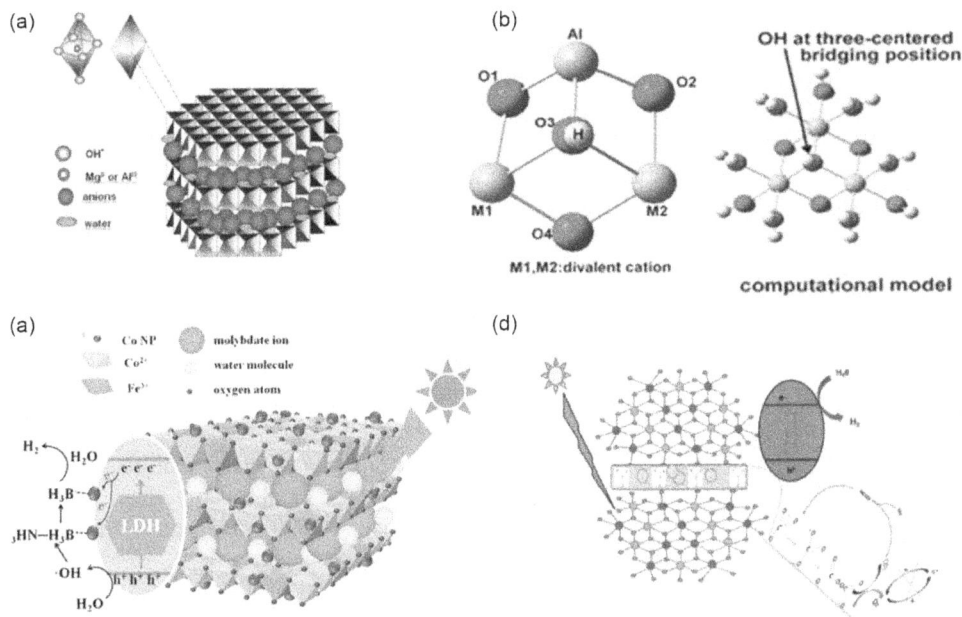

FIGURE 10.7 (a) Illustration of LDH; (b) Computational model of $M_2Al(OH_2)_9(OH)_4]^{3+}$ clusters. (Reprinted with permission from [82]. Copyright Royal Society of Chemistry); (c) Photocatalytic H_2 generation from molybdate-intercalated Co/CoFe layered double hydroxide. (Reprinted with permission from [89]. Copyright John Wiley and Sons); (d) Photocatalytic H_2 generation of carbonate (CO_3)-intercalated Zn/Cr LDH. (Reprinted with permission from [90]. Copyright Royal Society of Chemistry.)

applicable towards photoactivity, providing an efficient migration of electrons for minimized charge recombination [84]. In comparison with the bulk materials, the layered morphology of LDH benefits them to offer more events of diffusion and better charge separation due to MO_6 octahedron [85]. Moreover, LDHs are capable of absorbing water which is accessible to the inter-gallery gaps and the external surface, increasing the catalyst surface for photoactivity. This maximizes the surface of the catalyst and helps in the photocatalysis [86]. The movement of charges in the lamellar morphology of LDH induces the generation of photoactive defects, beneficial for photocatalytic processes. There exists an electrostatic host–guest contact in the anions of the layered structure due to the uniformly dispersed positive charges on layered LDH structure, and it contribute in the stable nature of the anions as the water and H_3O^+ ions in the layered spaces of LDH are advantageous for the separation of charge carriers in photo-processes [87]. Furthermore, the interlayered structure of LDH offers anion intercalation for overcoming the issues of charge recombination [88]. Apart from this, LDH consists of metal-oxygen-metal oxo bridge links, which connect the metal octahedra and behave like a redox center. This redox center is visible light active and offers metal-to-metal charge transition in photocatalytic processes [82].

In this context, Figure 10.7c illustrates the molybdate-intercalated Co/CoFe LDH as a H_2-evolving photocatalyst from ammonia borane hydrolysis. The charge transport was improved by tuning of the Co–O–Fe oxo-bridges, while the utilization of irradiation and charge recombination was minimized due to the improved intercalated molybdate. In another work, the photo-catalytic H_2 generation of carbonate (CO_3)-intercalated zinc-chromium Zn/Cr LDH under visible light was reported. The octahedral site containing Cr in the LDH helped the electron excitation after splitting of Cr-$3dt_{2g}$–Cr-$3d_{eg}$ orbital. As observed in Figure 10.7d, the oxidation of CO_3 took place by the help of holes and the CO_3 was transformed into CO_3 radicals, helpful for charge carriers' separation. Moreover, the LDH was easy to synthesize and was reusable [90]. In another work, the photocatalytic activity of cyanide-intercalated CoBi LDH was attributed to the 1.32 eV narrow band gap of LDH, its octahedral assembly of Co and Bi, and water molecule diffusion in the interlayered structure with spaces of 0.76 nm in layers [91]. Sherryna et al., reported the photocatalytic ability of nickel (Ni)-based LDH, making up its brucite layer. This led to excellent conductivity linked to the band gap for more light utilization and greater work function of 5.3 eV [44]. LDH constituting Zn in brucite layer showed higher absorption of irradiation for photoactivity [92]. The incorporation of Zn is capable of minimizing the electron-hole pairs recombination and maximizing thermal activation and efficient charge migration [93]. Apart from this, crystal lattice was increased due to the higher ionic radii of Zn^{2+} [94]. The Al-based LDH has been reported to induce properties of more concentration of charges and also supply the LDH with more reaction sites for catalytic activity [95]. A study on photocatalytic reaction for cobalt-aluminum (CoAl) LDH showed good photoactivity along with excellent photostability based on the ability of Al to get activated under irradiation. It also depicted enhanced surface area, supplying more sites for redox reactions [96]. The Cr-based LDH has been studied to produce ligand-to-metal charge transfer in the CrO_6 octahedral units. This led to the enhanced light absorption from 340- to 700-nm wavelength range as the ligand-to-metal charge transfer offered a photosensitizing effect for improved photocatalytic H_2 generation activity [97].

10.5.4 CARBON MATERIALS

Polymeric materials with semiconducting property and constituting more interactive area with H_2O are known to be good substitutes for catalysts containing metals. GO has been recognized to be similar in structure to graphene having an apparent bandgap due to its linkage with the materials containing O_2, enabling its easy dispersion in water. The structural, chemical, and physical characteristics make them favorable to be used in field of photocatalysis [98]. GO is known as a transitional state between graphene and graphite [17–23] but, dissimilar to graphite, the O_2 as functional group of GO makes it to undergo exfoliation and dispersion in water. The sp3 hybridization supports the modification of the structure and electronic features of GO [99]. The change in level of oxidation is reported to widen the bandgap of GO, whereas the oxidation of GO makes it to act as an insulator, and partial oxidation leads to GO that is a semiconductor [100], offering tuneable oxidation level for the modification of electronic properties. In this context as shown in Figure 10.8a, Pt/GO-ZnS was studied for photocatalytic H_2 generation, where GO acted as a mediator for accepting electrons from ZnS, transferring them to Pt, and also offered a site for reduction

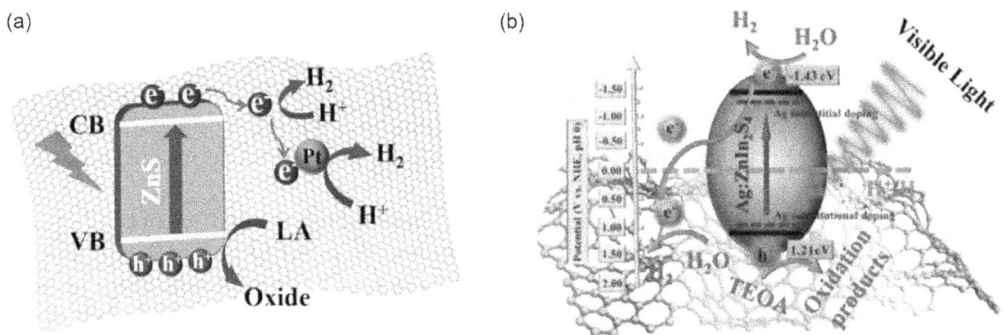

FIGURE 10.8 Photocatalytic H_2 generation mechanism for composite involving GO and RGO (a) Pt/GO-ZnS. (Reprinted with permission from [101]. Copyright Elsevier); (b) Ag:ZnIn$_2$S$_4$/RGO. (Reprinted with permission from [102]. Copyright Elsevier.)

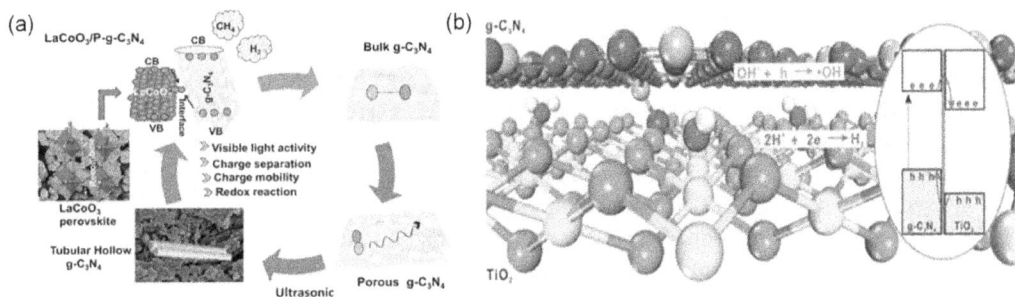

FIGURE 10.9 (a) Illustrative description for properties of hollow tubes of porous g-C$_3$N$_4$ for H$_2$ generation. (Reprinted with permission from [69]. Copyright American Chemical Society); (b) Interface bonding of TiO$_2$ with g-C$_3$N$_4$ via (Ti)$_2$-N-C bond. (Reprinted with permission from [108]. Copyright American Chemical Society.)

reaction [101]. In another work, indicated in Figure 10.8b, the photoactivity of Ag:ZnIn$_2$S$_4$/RGO was reported, where an interaction between RGO and ZnIn$_2$S$_4$ was established for electron accepting and migration, contributing towards the extended lifetime of electron-hole pairs [102]. Long et al. studied photochemical H$_2$ generation in ZnAgInSe quantum dots (QDs) and hybrid TiO$_2$/GO film. The role of GO was vital in the hybrid as it offered minimized resistance to the movement of charges, less charge recombination, and enhanced migration of electrons in the QDs-TiO$_2$ photoanodes [103]. The charge separation ability of RGO was explored for TiO$_2$/In$_{0.5}$WO$_3$ composite forming an S-scheme for photocatalytic H$_2$ generation [104].

Metal free g-C$_3$N$_4$ semiconductor is known to be applicable for the large-scale utilization for H$_2$ generation because of its stable structure, absorption in visible range, and cost-effectiveness [105]. Moreover, ultrathin conjugated g-C$_3$N$_4$ offers a narrow bandgap, i.e., 2.7 eV, is chemically and thermally stable in air at less than 600°C under air and in acidic and basic solutions, as well as shows an optical wavelength of around 460 nm. A unique curly g-C$_3$N$_4$ was reported to be highly efficient towards visible light-induced H$_2$ generation with 10.8% quantum efficiency based on the unique structure, large surface area, less defects, rapid charge migration, and strong reduction ability [106]. In another current study, as shown in Figure 10.9a, perovskite with g-C$_3$N$_4$ was explored for photocatalytic H$_2$ generation, where the hollow tubes of porous g-C$_3$N$_4$ were studied to contribute towards photoactivity, attributing to its porous nature, more reaction sites, and good charge migration [69]. Chen et al. studied nanosheets of g-C$_3$N$_4$ through copolymerization of urea and diaminodiphenyl sulfone, showing good kinetics for HER activity, light utilization, good rate of transmission, and charge separation [107]. Figure 10.9b shows TiO$_2$/g-C$_3$N$_4$-depicted H$_2$ generation based on the existence of a strong interaction formed due to (Ti)$_2$-N-C bond for charge separation and also the interface-supported H$_2$O molecules, leading to improved reduction reaction [108]. In another study, the property of g-C$_3$N$_4$ supports the dispersion of Au for developing suitable interface between ligand-free Au and g-C$_3$N$_4$ having an ultrathin structural morphology [109].

10.5.5 Metal Sulfides

Semiconductors having a visible light activity like metal sulfides such as ZnS, CdS, and MoS$_2$ are known to be highly efficient for photocatalytic activity. Various metal sulfides having varying compositional and morphological features, crystalline structure, and valence states have been explored for photoactivity [110]. Zinc sulfide (ZnS) belongs to II-VI group of transition metal sulfide and is known for its high ability of generation of electron-hole pairs, which is greater than TiO$_2$ [111]. ZnS consists of a relatively negative CB, leading to the generation of electrons for the reduction reaction of H_2 generation. It possesses a wide bandgap of 3.6–3.8 eV and works under UV irradiation, and the visible light activity is achieved through the modification for achieving high-efficiency system [112]. In a current study, defect-rich ZnS along with Ni-P alloy was used for photocatalytic H_2 generation. The defect containing Zn led to the creation of a defect energy level in the bandgap of ZnS for improving the visible light absorption. Figure 10.10a shows that under visible light irradiation, based on the efficient visible light activity of defect-rich ZnS, the electrons were transferred from ZnS to Ni-P for reduction reaction [113]. UV-treated ZnS films were studied for photocatalytic performance based on their property to absorb VIS-NIR irradiations and also due to the generation of Zn and S vacancies as they played role of trapping centers in delaying the recombination of charge carriers [114]. Hollow-structured ZnS having a large surface area in heterojunction with ZnO has been used to carry out photoactivity towards H_2 generation due to the efficient charge separation [115].

MoS_2 is known as a cocatalyst based on its properties like cost-effectiveness and high reactivity for H_2 generation reaction and is known to be the simplest material to replace Pt. Significantly, MoS_2 features a typical 2D structure, and its combination with other 2D structures creates contact interfaces, which give ample channels for photogenerated charge transfer. For instance, Figure 10.10b indicates the 2D/2D face-to-face interaction of MoS_2 with black phosphorous (BP), which contributed towards photocatalytic activity due to efficient charge transfer [116]. In another work, 2D/2D MoS_2/g-C_3N_4 heterojunction was reported, in which MoS_2 acted as a center for reduction of H^+ to H_2 after accepting electrons from g-C_3N_4 [117]. $Cd_{0.5}Zn_{0.5}S$ QDs with flaked MoS_2 was used to study HER photocatalytic process, attributing to the property of MoS_2 to act as charge mediator, light absorption, and provision of active sites [118]. Zhuge et al., reported the dual function of MoS_2 in MoS_2/CdS, including (i) efficient charge separation by acting as an electron acceptor and (ii) the role

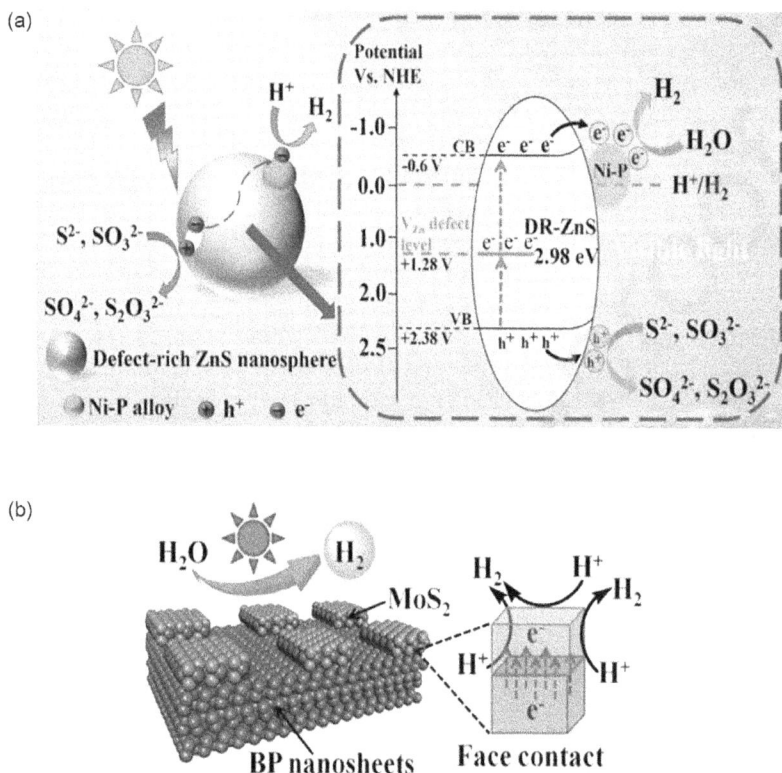

FIGURE 10.10 (a) Photocatalytic H_2 generation mechanism involving defect-rich ZnS/Ni-P alloy. (Reprinted with permission from [113]. Copyright Elsevier); (b) 2D/2D face-to-face interaction of MoS_2 with black phosphorous (BP) for photocatalytic H_2 generation. (Reprinted with permission from [116]. Copyright Elsevier.)

of Mo-S bonds in reducing the H_2 generation barrier [119]. The photocatalytic and electrocatalytic H_2 production was reported for the monolayer MoS_2 having 9–42 nm lateral sizes. MoS_2 monolayer showed characteristics contributing towards photoactivity as the edges of MoS_2 had active sites and also showed reduced distance of transfer for electrons [120].

10.5.6 METAL OXIDES

Titanium dioxide (TiO_2) is a readily available metal oxide because of its outstanding properties, including appropriate position of CB and VB, good stability, efficient migration and separation of electron-hole pairs, and non-toxic behavior. Figure 10.11 shows that TiO_2 can be found in anatase, rutile, and brookite crystalline forms, and the anatase is recognized to be more efficient towards photocatalytic water-splitting application. Moreover, the indirect band gap of anatase as compared to rutile and brookite having a direct band gap overcomes the issue of charge carriers' recombination, and also, the existence of less average effective mass of electrons generated after light absorption leads to better electron-hole pair separation in anatase phase of TiO_2 [121]. However, the anatase phase constitutes a band gap, i.e., 3.2 eV, which is slightly larger than the band gap of rutile

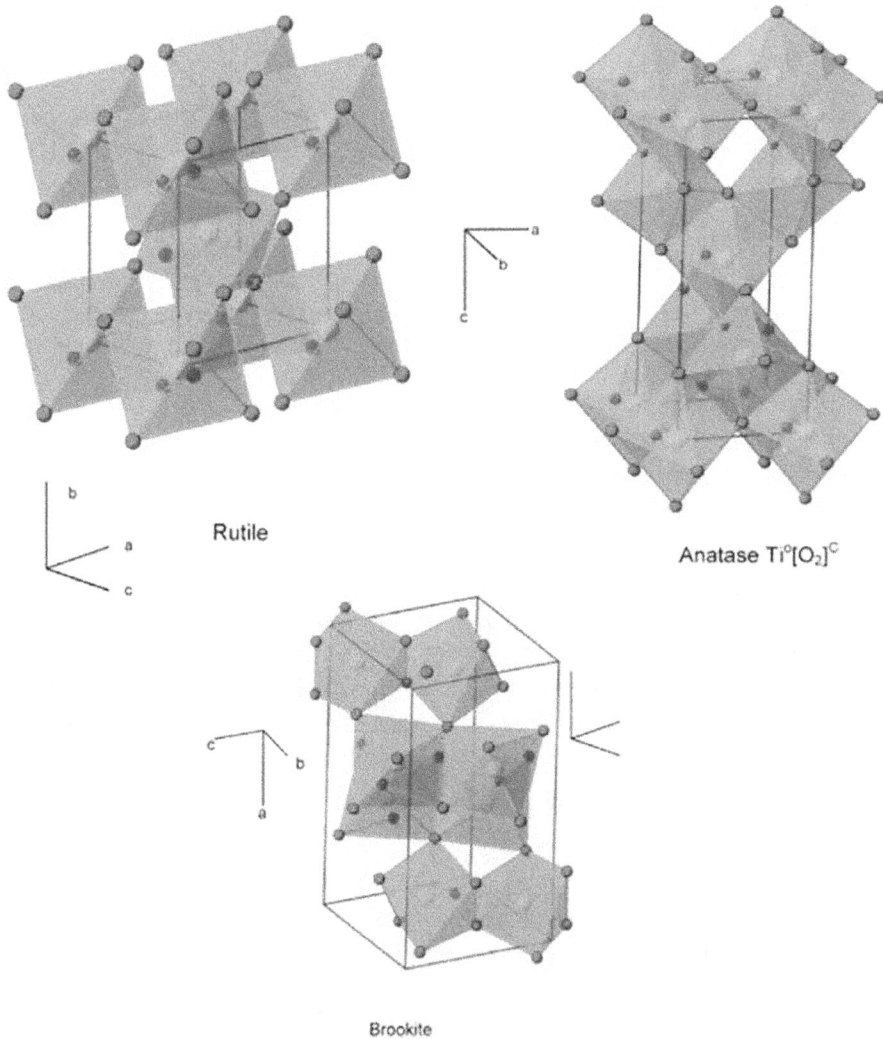

FIGURE 10.11 Rutile, anatase, and brookite phases of TiO_2 [125].

phase (3.0 eV), hindering and negatively affecting its light utilization [122]. Furthermore, in general, TiO_2 faces the limitation of wide band gap, leading to only absorption of 4% of sunlight. Also, the efficient anatase phase of TiO_2 also offers thermal instability where it is converted into rutile phase at temperatures higher than 600°C [123]. The excellent structural and chemical characteristics of TiO_2 can be utilized fully as they offer the option of tuning the band gap for altering the light utilization ability and improving charge recombination. This is achievable through boosting active reaction sites on TiO_2 along with enhancing the electric conductivity [124].

Kumar et al. studied that the photocatalytic ability of TiO_2 was highly dependent on the crystalline size and shape of crystal. Figure 10.12a–c displays the HR-TEM images of TiO_2 nanosquares (TNS) having agglomerated nature, 1-D nanotubular structure (TNT), and nanorods (TNR), respectively. The property like high surface area was dominant in TiO_2 nanotubes (186 m²/g) and was significantly greater than in nanosquares and nanorods. As shown in Figure 10.12d, improved photocatalytic H_2 generation was depicted by nanotubes of TiO_2 under UV-A irradiations based on the existence of both anatase and rutile phases of TiO_2; a wider band gap boosted specific surface area and hollow structural property. These properties led to electron-hole pair separation due to the photocurrent density existing because of one-directional movement of charges [126].

In another work, TiO_2-CTAB-300°C NPs were synthesized after decentralizing Ti(IV) polymer and cetrimonium bromide (CTAB). After the water and ethanol were evaporated, CTAB was left in higher concentration, and as its concentration became higher than the concentration of micelle, the CTAB self-assembled in form of ball micelle. Next, calcination occurred at 300°C to form TiO_2 NPs. Figure 10.13a shows the illustration for the synthesis process of TiO_2-CTAB-300°C. The synthesized TiO_2-CTAB-300°C depicted a boosted activity towards H_2 generation under visible as the CTAB used formed homogeneous pore size distribution and well-dispersed NPs, leading to easy dispersion of NPs in solution for boosted light absorption [127]. Ultrathin TiO_2 nanosheets obtained with the assistance of Cl ions were studied for photocatalytic H_2 generation, leading to nanosheets with good crystalline structure, boosted surface area, and more Ti^{3+} defects for improved charge transfer resistance for efficient charge separation as well as for introducing electronic states [128]. Currently, density functional tight-binding approach revealed that the brookite and rutile phases of TiO_2 showed visible light activity due to their reduced band gap from 3.59 to 2.62 eV and 2.75 to 2.60 eV, respectively, because of temperature-dependent nature [129]. In another work, the thin-deposited films of TiO_2 having various deposit times of 30, 60, 90, and 120 minutes by DC magnetron sputtering were studied. The thin film of TiO_2 showed good optical properties. As indicated in Figure 10.13b, the efficiency activity can be attributed to the mid-gap states having Ti^{3+} in-between the VB and CB, responsible for efficient H_2

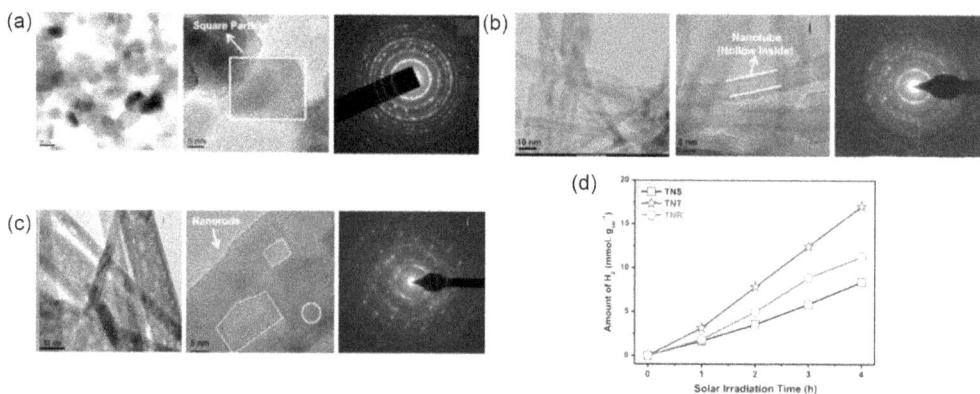

FIGURE 10.12 HR-TEM images of (a) TiO_2 nanosquares; (b) TiO_2 nanotubes;(c) TiO_2 nanorods; (d) photocatalytic H_2 generation of TiO_2 nanosquares, nanotubes, and nanorods. (Reprinted with permission from [126]. Copyright Elsevier.)

(a)

(b)

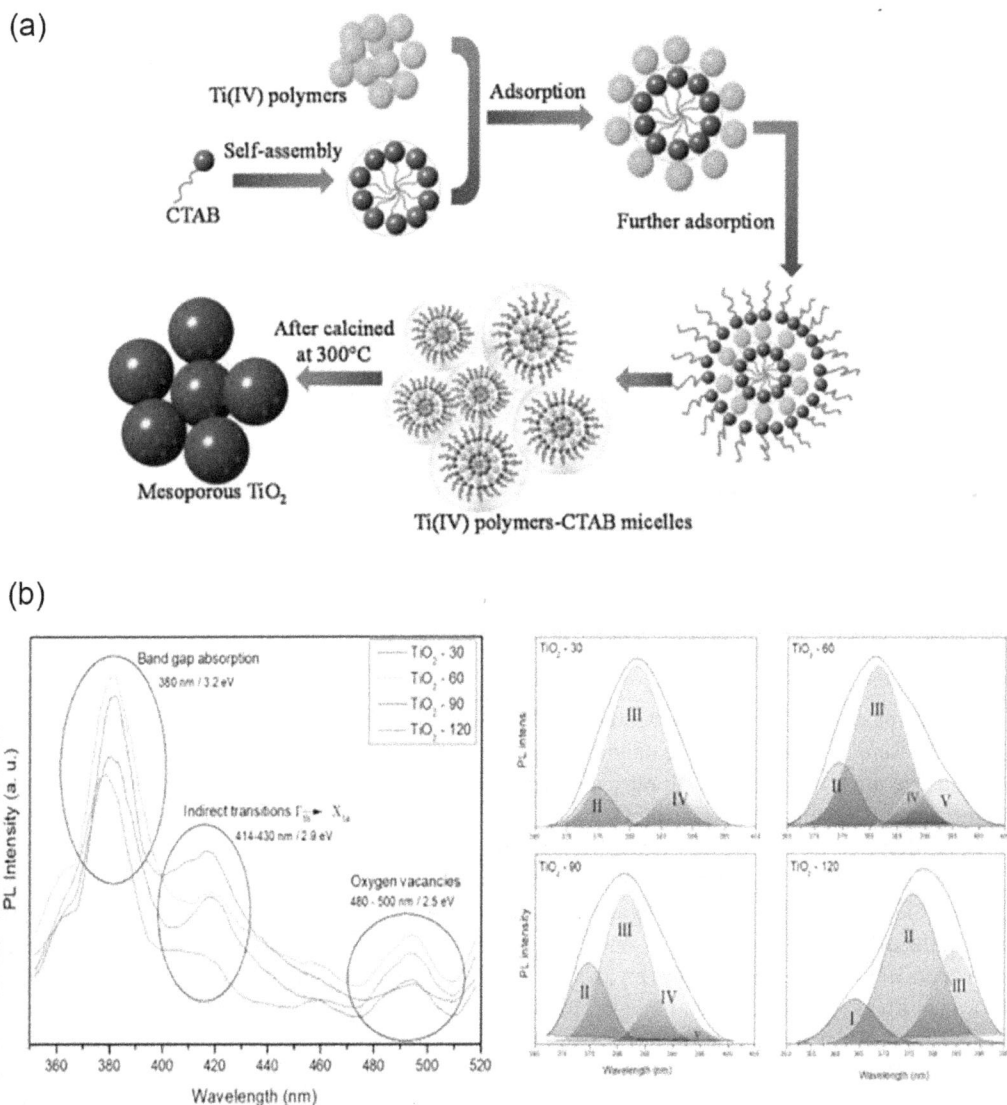

FIGURE 10.13 (a) Illustration of synthesis of TiO$_2$-CTAB-300°C NPs. (Reprinted with permission from [127]. Copyright Elsevier); (b) PL spectra for thin films at different deposit times, and the principal emission of each film. (Printed with permission from [130]. Copyright Elsevier.)

production. The electrons were present in the lower energy and not in the band gap. Moreover, the thin films also showed refractive index and the extinction coefficient value [130].

Zinc oxides (ZnO) are known to be photocatalyst nanomaterials constituting photoelectric properties, are chemically stable, have unique physical features, vast availability, are less costly and nontoxic in nature [131]. Moreover, ZnO also shows a faster rate of hole transportation, and increased light absorption coefficient, and is highly stable in water [132]. ZnO synthesized by the sol-gel method followed by ultrasonication at high temperature was studied for photocatalytic H$_2$ generation. Increased surface areas, small-sized particles, porous structure, and less O$_2$ vacancies were the physical characteristics responsible for the photocatalytic H$_2$ generation of ZnO [133]. Figure 10.14a shows another work in which hierarchical flower-like ZnO showed UV and visible optically active

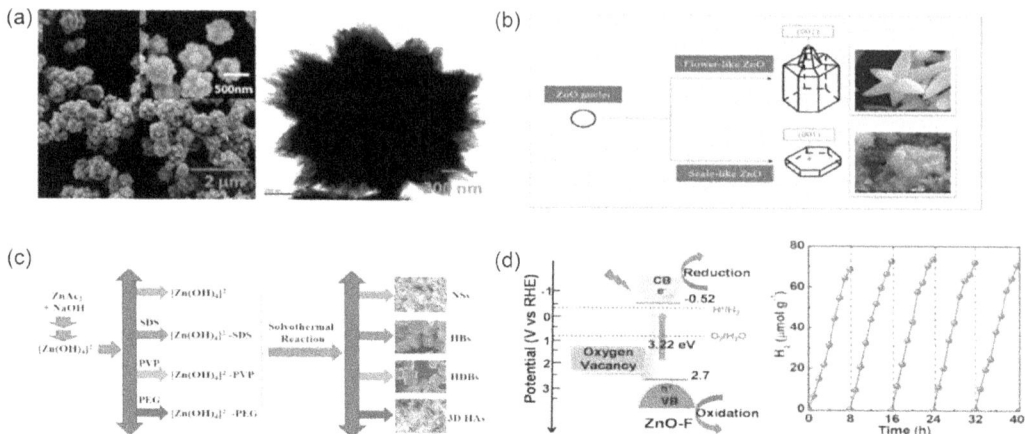

FIGURE 10.14 (a) SEM and HR-TEM of flower-like hierarchical ZnO. (Reprinted with permission from [134]. Copyright Royal Society of Chemistry); (b) SEM images of 13D flower-like and 1D scale-like ZnO. (Reprinted with permission from [136]. Copyright Elsevier); (c) Illustration for the synthesis of ZnO having various morphologies by using different surfactants. (Reprinted with permission from [137]. Copyright Elsevier); (d) Band gap diagram of F-ZnO and; (e) Stability of F-ZnO for photocatalytic H_2 generation. (Reprinted with permission from [140]. Copyright Elsevier.)

defects for improved photoactivity. Moreover, it also showed a stability of up to 50 cycles [134]. The ZnO QDs were studied for photocatalytic activity based on their characteristics such as large specific surface area, high adsorption rate, efficient separation of charge carriers, and reduced rate of charge recombination. Moreover, the QDs depicted boosted kinetic rate constants for photoactivity [135]. As depicted in Figure 10.14b, hydrogen-treated 3D flower-like ZnO and 1D scale ZnO showed a visible light-induced photoactivity, attributing to the existence of O_2 vacancies on ZnO [136]. Mao et al., studied the visible light activity and reduced charge recombination due to less existence of O_2 defects; the different morphologies of ZnO synthesized from the use of different surfactants, including sodium dodecyl sulfate (SDS), polyvinylpyrrolidone (PVP), and polyethylene glycol (PEG), to obtain hexagonal disks (HDs), hexagonal bilayer disk-like structures (HBDs), and 3D flower-like hierarchitectures (HAs) structures as shown in Figure 10.14c [137]. Rod-like calcined ZnO was reported to constitute efficient optical, morphological, and structural properties for efficient photocatalytic activity. The presence of more adsorbed O_2 on ZnO from calcination instead of O_2 vacancies influenced the trapping of photoelectrons in the CB of the ZnO [138]. Electrodeposition followed by heat treatment was observed to be highly efficient towards photocatalytic ability based on the large specific surface area for radical-organic contact as well as for good migration of charge across the interface [139]. Currently, Wang et al., reported that fructose regulated ZnO (F-ZnO) for photocatalytic H_2 generation and showed the characteristics of strong oxidation ability due to the shifted VB of ZnO, as shown in Figure 10.14d, and also showed photostability over five cyclic runs (Figure 10.14e) [140].

10.6 CONCLUSION

Photocatalytic production of H_2 from water is a promising approach to meet energy needs and does not contribute towards the deterioration of environment. The characteristics offered by H_2 as a clean source of renewable energy include cost-effective storage, elevated storing volume, good diffusion coefficient, kinematic viscosity, thermal conductivity, low density and luminosity, good diffusion and heat transfer properties, requiring less energy to ignite, and easy transformation in the form of fuel cells. The photocatalytic H_2 generation process is achieved through the use of various nanomaterials, which can also play a role in designing of lightweight H_2 storage systems with better H_2

storage characteristics. Moreover, the photocatalytic H_2 can be utilized in fuel cells for energy-gaining purposes. Various advanced nanomaterials showing a photocatalytic ability, including metal oxides, metal sulfides, carbon materials, perovskites, MOFs, and LDH, have been explored based on their following main characteristics contributing towards H_2 production:

- MOFs have organic linkers and nodes of metal oxide and sulfide clusters for activity under NIR irradiation, flexible modification, more surface area, photostability, recyclable, and cost-effectiveness.
- Perovskite shows high catalytic capacity, electronic conductivity, and O_2 vacancies. Halide perovskites depict optoelectronic features, visible light activity, good coefficient for adsorption, and larger path for diffusion.
- LDH constitutes outstanding light absorption property and a layered structural feature.
- GO acts a mediator, contains O_2 as functional group, which makes it to undergo exfoliation and dispersion in water.
- g-C_3N_4 offers stable structure, absorption in visible range, cost-effectiveness, narrow band gap, chemical and thermal stabilities.
- ZnS constitutes high ability of generation of electron-hole pairs and has a relatively negative CB, for the reduction reaction.
- MoS_2 is cost-effective, highly reactive for H_2 generation reaction, and has a typical 2D structure.
- TiO_2 is a readily available metal oxide because of its outstanding properties, including appropriate position of CB and VB, good stability, efficient migration, separation of electron-hole pairs and non-toxic behavior.
- ZnO shows photoelectric properties, chemical stability, unique physical features, vast availability, is less costly and non-toxic.

REFERENCES

[1] M. Umer, M. Tahir, M. Usman Azam, S. Tasleem, T. Abbas, and A. Muhammad, Synergistic effects of single/multi-walls carbon nanotubes in TiO_2 and process optimization using response surface methodology for photo-catalytic H_2 evolution, *J Environ Chem Eng* 7 (2019), p. 103361.

[2] M. Tahir, Hierarchical 3D VO_2/ZnV_2O_4 microspheres as an excellent visible light photocatalyst for CO_2 reduction to solar fuels, *Appl Surf Sci* 467–468 (2019), pp. 1170–1180.

[3] M.J. Rivero, O. Iglesias, P. Ribao, and I. Ortiz, Kinetic performance of TiO_2/Pt/reduced graphene oxide composites in the photocatalytic hydrogen production, *Int J Hydrog Energy* 44 (2019), pp. 101–109.

[4] T. Lv, M. Wu, M. Guo, Q. Liu, and L. Jia, Self-assembly photocatalytic reduction synthesis of graphene-encapusulated $LaNiO_3$ nanoreactor with high efficiency and stability for photocatalytic water splitting to hydrogen, *Chem Eng J* 356 (2019), pp. 580–591.

[5] N. Shehzad, M. Tahir, K. Johari, T. Murugesan, and M. Hussain, Fabrication of highly efficient and stable indirect Z-scheme assembly of AgBr/TiO_2 via graphene as a solid-state electron mediator for visible light induced enhanced photocatalytic H_2 production, *Appl Surf Sci* 463 (2019), pp. 445–455.

[6] X. Qian, J. Zhang, Z. Guo, S. Liu, J. Liu, and J. Lin, Facile ultrasound-driven formation and deposition of few-layered MoS_2 nanosheets on CdS for highly enhanced photocatalytic hydrogen evolution, *Appl Surf Sci* 481 (2019), pp. 795–801.

[7] Y. Zou, D. Ma, D. Sun, S. Mao, C. He, Z. Wang, X. Ji, and J.-W. Shi, Carbon nanosheet facilitated charge separation and transfer between molybdenum carbide and graphitic carbon nitride toward efficient photocatalytic H_2 production, *Appl Surf Sci* 473 (2019), pp. 91–101.

[8] K. Wang, T. Peng, Z. Wang, H. Wang, X. Chen, W. Dai, and X. Fu, Correlation between the H_2 response and its oxidation over TiO_2 and N doped TiO_2 under UV irradiation induced by Fermi level, *Appl Catal B Environ* 250 (2019), pp. 89–98.

[9] Y.-J. Yuan, Z. Shen, S. Wu, Y. Su, L. Pei, Z. Ji, M. Ding, W. Bai, Y. Chen, Z.-T. Yu, and Z. Zou, Liquid exfoliation of g-C_3N_4 nanosheets to construct 2D-2D MoS_2/g-C_3N_4 photocatalyst for enhanced photocatalytic H_2 production activity, *Appl Catal B Environ* 246 (2019), pp. 120–128.

[10] N. Fajrina, and M. Tahir, A critical review in strategies to improve photocatalytic water splitting towards hydrogen production, *Int J Hydrog Energy* 44 (2019), pp. 540–577.

[11] P. Zhou, F. Lv, N. Li, Y. Zhang, Z. Mu, Y. Tang, J. Lai, Y. Chao, M. Luo, F. Lin, J. Zhou, D. Su, and S. Guo, Strengthening reactive metal-support interaction to stabilize high-density Pt single atoms on electron-deficient g-C_3N_4 for boosting photocatalytic H_2 production, *Nano Energy* 56 (2019), pp. 127–137.

[12] L. Ye, and Z. Wen, $ZnIn_2S_4$ nanosheets decorating WO_3 nanorods core-shell hybrids for boosting visible-light photocatalysis hydrogen generation, *Int J Hydrog Energy* 44 (2019), pp. 3751–3759.

[13] S. Wang, B. Zhu, M. Liu, L. Zhang, J. Yu, and M. Zhou, Direct Z-scheme ZnO/CdS hierarchical photocatalyst for enhanced photocatalytic H_2-production activity, *Appl Catal B: Environ* 243 (2019), pp. 19–26.

[14] P. Wang, H. Li, Y. Sheng, and F. Chen, Inhibited photocorrosion and improved photocatalytic H_2-evolution activity of CdS photocatalyst by molybdate ions, *Appl Surf Sci* 463 (2019), pp. 27–33.

[15] T. Di, B. Cheng, W. Ho, J. Yu, and H. Tang, Hierarchically CdS–Ag_2S nanocomposites for efficient photocatalytic H_2 production, *Appl Surf Sci* 470 (2019), pp. 196–204.

[16] C.-J. Chang, Y.-G. Lin, P.-Y. Chao, and J.-K. Chen, AgI-BiOI-graphene composite photocatalysts with enhanced interfacial charge transfer and photocatalytic H_2 production activity, *Appl Surf Sci* 469 (2019), pp. 703–712.

[17] N. Serpone, D. Lawless, and E. Pelizzetti, Subnanosecond characteristics and photophysics of nanosized TiO_2 particulates from R_{part} = 10A to 34 A: Meaning for heterogeneous photocatalysis, in *Fine Particles Science and Technology*, E. Pelizzetti (ed), Springer, Dordrecht, 1996.

[18] W. Cui, L. Liu, S. Ma, Y. Liang, and Z. Zhang, CdS-sensitized $K_2La_2Ti_3O_{10}$ composite: A new photocatalyst for hydrogen evolution under visible light irradiation, *Catal Today* 207 (2013), pp. 44–49.

[19] W. Yao, C. Huang, N. Muradov, and A. T-Raissi, A novel Pd–Cr_2O_3/CdS photocatalyst for solar hydrogen production using a regenerable sacrificial donor, *Int J Hydrog Energy* 36 (2011), pp. 4710–4715.

[20] X. Yu, Z. Tang, D. Sun, L. Ouyang, and M. Zhu, Recent advances and remaining challenges of nanostructured materials for hydrogen storage applications, *Prog Mater Sci* 88 (2017), pp. 1–48.

[21] M. Hirscher, V.A. Yartys, M. Baricco, J.B. von Colbe, D. Blanchard, R.C. Bowman Jr, D.P. Broom, C.E. Buckley, F. Chang, and P. Chen, Materials for hydrogen-based energy storage–past, recent progress and future outlook, *J Alloys Compd* 827 (2020), p. 153548.

[22] J. Zheng, C.-G. Wang, H. Zhou, E. Ye, J. Xu, Z. Li, and X.J. Loh, Current research trends and perspectives on solid-state nanomaterials in hydrogen storage, *Research* 2021 (2021), p. 3750689.

[23] R. Singh, A. Altaee, and S. Gautam, Nanomaterials in the advancement of hydrogen energy storage, *Heliyon* 6 (2020), p. e04487.

[24] J. Yang, A. Sudik, C. Wolverton, and D.J. Siegel, High capacity hydrogen storage materials: Attributes for automotive applications and techniques for materials discovery, *Chem Soc Rev* 39 (2010), pp. 656–675.

[25] M.Z. Mehrizi, J. Abdi, M. Rezakazemi, and E. Salehi, A review on recent advances in hollow spheres for hydrogen storage, *Int J Hydrog Energy* 45 (2020), pp. 17583–17604.

[26] G.A. Karim, Hydrogen as a spark ignition engine fuel, *Int J Hydrog Energy* 28 (2003), pp. 569–577.

[27] J. Nowotny, J. Dodson, S. Fiechter, T.M. Gür, B. Kennedy, W. Macyk, T. Bak, W. Sigmund, M. Yamawaki, and K.A. Rahman, Towards global sustainability: Education on environmentally clean energy technologies, *Renew Sustain Energy Rev* 81 (2018), pp. 2541–2551.

[28] B. Pivovar, N. Rustagi, and S. Satyapal, Hydrogen at scale (H_2@ Scale): Key to a clean, economic, and sustainable energy system, *Electrochem Soc Interface* 27 (2018), p. 47.

[29] V. Jain, and B. Kandasubramanian, Functionalized graphene materials for hydrogen storage, *J Mater Sci* 55 (2020), pp. 1865–1903.

[30] P.H. Cyril, and G. Saravanan, Development of advanced materials for cleaner energy generation through fuel cells, *New J Chem* 44 (2020), pp. 19977–19995.

[31] S.M.M. Ehteshami, and S. Chan, The role of hydrogen and fuel cells to store renewable energy in the future energy network–potentials and challenges, *Energy Policy* 73 (2014), pp. 103–109.

[32] E. Fleury, J. Jayaraj, Y. Kim, H. Seok, K. Kim, and K. Kim, Fe-based amorphous alloys as bipolar plates for PEM fuel cell, *J Power Sources* 159 (2006), pp. 34–37.

[33] S. Wang, A. Lu, and C.-J. Zhong, Hydrogen production from water electrolysis: Role of catalysts, *Nano Converg* 8 (2021), p. 4.

[34] V. Raj, Direct methanol fuel cells in portable applications: Materials, designs, operating parameters, and practical steps toward commercialization, in *Direct Methanol Fuel Cell Technology*, Elsevier, 2020, pp. 495–525.

[35] M.U. Niemann, S.S. Srinivasan, A.R. Phani, A. Kumar, D.Y. Goswami, and E.K. Stefanakos, Nanomaterials for hydrogen storage applications: A review, *J Nanomater* 2008 (2008), p. 950967.

[36] A. Dalvi, and M. Guay, Control and real-time optimization of an automotive hybrid fuel cell power system, *Control Eng Pract* 17 (2009), pp. 924–938.

[37] M.K. Singla, P. Nijhawan, and A.S. Oberoi, Hydrogen fuel and fuel cell technology for cleaner future: A review, *Environ Sci Pollut Res* 28 (2021), pp. 15607–15626.

[38] R. Sule, A.K. Mishra, and T.T. Nkambule, Recent advancement in consolidation of MOFs as absorbents for hydrogen storage, *Int J Energy Res* 45 (2021), pp. 12481–12499.

[39] A.H. Majeed, D.H. Hussain, E.T.B. Al-Tikrity, and M.A. Alheety, Poly (o-Phenylenediamine-GO-TiO$_2$) nanocomposite: Modulation, characterization and thermodynamic calculations on its H$_2$ storage capacity, *Chem Data Collect* 28 (2020), p. 100450.

[40] B. Chakraborty, and G. Sanyal, Green hydrogen energy: Storage in carbon nanomaterials and polymers, in *Applied Biopolymer Technology and Bioplastics*, Apple Academic Press, New York, 2021, pp. 183–239.

[41] S. Tasleem, and M. Tahir, Recent progress in structural development and band engineering of perovskites materials for photocatalytic solar hydrogen production: A review, *Int J Hydrog Energy* 45 (2020), pp. 19078–19111.

[42] W.-T. Chen, Y. Dong, P. Yadav, R.D. Aughterson, D. Sun-Waterhouse, and G.I. Waterhouse, Effect of alcohol sacrificial agent on the performance of Cu/TiO$_2$ photocatalysts for UV-driven hydrogen production, *Appl Catal A: General* 602 (2020), p. 117703.

[43] W.K. Fan, and M. Tahir, Recent developments in photothermal reactors with understanding on the role of light/heat for CO$_2$ hydrogenation to fuels: A review, *Chem Eng J* 427 (2021), p. 131617.

[44] A. Sherryna, and M. Tahir, Recent developments in layered double hydroxide structures with their role in promoting photocatalytic hydrogen production: A comprehensive review, *Int J Energy Res* 46 (2022), pp. 2093–2140.

[45] S. Ida, and T. Ishihara, Recent progress in two-dimensional oxide photocatalysts for water splitting, *J Phys Chem Lett* 5 (2014), pp. 2533–2542.

[46] A.W. Peters, Z. Li, O.K. Farha, and J.T. Hupp, Toward inexpensive photocatalytic hydrogen evolution: A nickel sulfide catalyst supported on a high-stability metal–organic framework, *ACS Appl Mater Interfaces* 8 (2016), pp. 20675–20681.

[47] X. Li, J.-L. Shi, H. Hao, and X. Lang, Visible light-induced selective oxidation of alcohols with air by dye-sensitized TiO$_2$ photocatalysis, *Appl Catal B: Environ* 232 (2018), pp. 260–267.

[48] X.S. Wang, L. Li, D. Li, and J. Ye, Recent progress on exploring stable metal–organic frameworks for photocatalytic solar fuel production, *Solar RRL* 4 (2020), p. 1900547.

[49] P. Silva, S.M.F. Vilela, J.P.C. Tomé, and F.A. Almeida Paz, Multifunctional metal–organic frameworks: From academia to industrial applications, *Chem Soc Rev* 44 (2015), pp. 6774–6803.

[50] H.N. Abdelhamid, UiO-66 as a catalyst for hydrogen production via the hydrolysis of sodium borohydride, *Dalton Trans* 49 (2020), pp. 10851–10857.

[51] M. Kandiah, M.H. Nilsen, S. Usseglio, S. Jakobsen, U. Olsbye, M. Tilset, C. Larabi, E.A. Quadrelli, F. Bonino, and K.P. Lillerud, Synthesis and Stability of Tagged UiO-66 Zr-MOFs, *Chem Mater* 22 (2010), pp. 6632–6640.

[52] F. Leng, H. Liu, M. Ding, Q.-P. Lin, and H.-L. Jiang, Boosting photocatalytic hydrogen production of porphyrinic MOFs: The metal location in metalloporphyrin matters, *ACS Catal* 8 (2018), pp. 4583–4590.

[53] D. Sun, and Z. Li, Robust Ti-and Zr-based metal-organic frameworks for photocatalysis, *Chin J Chem* 35 (2017), pp. 135–147.

[54] T. Tachikawa, J.R. Choi, M. Fujitsuka, and T. Majima, Photoinduced charge-transfer processes on MOF-5 nanoparticles: Elucidating differences between metal-organic frameworks and semiconductor metal oxides, *J Phys Chem C* 112 (2008), pp. 14090–14101.

[55] C.G. Silva, I. Luz, F.X.L. i Xamena, A. Corma, and H. García, Water stable Zr–benzenedicarboxylate metal–organic frameworks as photocatalysts for hydrogen generation, *Chem Eur J* 16 (2010), pp. 11133–11138.

[56] S.Y. Han, D.L. Pan, H. Chen, X.B. Bu, Y.X. Gao, H. Gao, Y. Tian, G.S. Li, G. Wang, and S.L. Cao, A methylthio-functionalized-MOF photocatalyst with high performance for visible-light-driven H$_2$ evolution, *Angew Chem Int Ed* 57 (2018), pp. 9864–9869.

[57] X.Y. Dong, M. Zhang, R.B. Pei, Q. Wang, D.H. Wei, S.Q. Zang, Y.T. Fan, and T.C. Mak, A crystalline copper (II) coordination polymer for the efficient visible-light-driven generation of hydrogen, *Angew Chem Int Ed* 55 (2016), pp. 2073–2077.

[58] X. Liu, X. Lv, H. Lai, G. Peng, Z. Yi, and J. Li, A simple Ni-based metal–organic framework as catalyst for dye-sensitized photocatalytic H_2 evolution from water reduction, *Photochem Photobiol* 96 (2020), pp. 1169–1175.

[59] A. Cao, L. Zhang, Y. Wang, H. Zhao, H. Deng, X. Liu, Z. Lin, X. Su, and F. Yue, 2D–2D heterostructured UNiMOF/g-C_3N_4 for enhanced photocatalytic H_2 production under visible-light irradiation, *ACS Sustain Chem Eng* 7 (2019), pp. 2492–2499.

[60] S. Bauer, C. Serre, T. Devic, P. Horcajada, J. Marrot, G. Férey, and N. Stock, High-throughput assisted rationalization of the formation of metal organic frameworks in the iron (III) aminoterephthalate solvothermal system, *Inorg Chem* 47 (2008), pp. 7568–7576.

[61] Y. Guo, J. Zhang, L.Z. Dong, Y. Xu, W. Han, M. Fang, H.K. Liu, Y. Wu, and Y.Q. Lan, Syntheses of exceptionally stable aluminum (III) Metal–organic frameworks: How to grow high-quality, large, single crystals, *Chem Eur J* 23 (2017), pp. 15518–15528.

[62] J. Ran, J. Qu, H. Zhang, T. Wen, H. Wang, S. Chen, L. Song, X. Zhang, L. Jing, and R. Zheng, 2D metal organic framework nanosheet: A universal platform promoting highly efficient visible-light-induced hydrogen production, *Adv Energy Mater* 9 (2019), p. 1803402.

[63] H. Tanaka, and M. Misono, Advances in designing perovskite catalysts, *Curr Opin Solid State Mater Sci* 5 (2001), pp. 381–387.

[64] H. Bian, D. Li, J. Yan, and S. Liu, Perovskite – A wonder catalyst for solar hydrogen production, *J Energy Chem* 57 (2021), pp. 325–340.

[65] Q. Dong, Y. Fang, Y. Shao, P. Mulligan, J. Qiu, L. Cao, and J. Huang, Electron-hole diffusion lengths > 175 µm in solution-grown $CH_3NH_3PbI_3$ single crystals, *Science* 347 (2015), pp. 967–970.

[66] G.W. Adhyaksa, L.W. Veldhuizen, Y. Kuang, S. Brittman, R.E. Schropp, and E.C. Garnett, Carrier diffusion lengths in hybrid perovskites: Processing, composition, aging, and surface passivation effects, *Chem Mater* 28 (2016), pp. 5259–5263.

[67] P. Zhou, H. Chen, Y. Chao, Q. Zhang, W. Zhang, F. Lv, L. Gu, Q. Zhao, N. Wang, J. Wang, and S. Guo, Single-atom Pt-I_3 sites on all-inorganic Cs_2SnI_6 perovskite for efficient photocatalytic hydrogen production, *Nature Commun* 12 (2021), p. 4412.

[68] S. Acharya, D. Padhi, and K. Parida, Visible light driven $LaFeO_3$ nano sphere/RGO composite photocatalysts for efficient water decomposition reaction, *Catal Today* 353 (2020), pp. 220–231.

[69] S. Tasleem, and M. Tahir, Constructing $La_xCo_yO_3$ perovskite anchored 3D g-C_3N_4 hollow tube heterojunction with proficient interface charge separation for stimulating photocatalytic H_2 production, *Energy & Fuels* 35 (2021), pp. 9727–9746.

[70] T. Vijayaraghavan, R. Sivasubramanian, S. Hussain, and A. Ashok, A facile synthesis of $LaFeO_3$-based perovskites and their application towards sensing of neurotransmitters, *ChemistrySelect* 2 (2017), pp. 5570–5577.

[71] O.A. Carrasco-Jaim, A.M. Huerta-Flores, L.M. Torres-Martínez, and E. Moctezuma, Fast in-situ photodeposition of Ag and Cu nanoparticles onto $AgTaO_3$ perovskite for an enhanced photocatalytic hydrogen generation, *Int J Hydrog Energy* 45 (2020), pp. 9744–9757.

[72] L. Zhang, and W. Liang, How the structures and properties of two-dimensional layered perovskites $MAPbI_3$ and $CsPbI_3$ vary with the number of layers, *J Phys Chem Lett* 8 (2017), pp. 1517–1523.

[73] L. Zhang, H. Wang, Z. Chen, P.K. Wong, and J. Liu, Bi_2WO_6 micro/nano-structures: Synthesis, modifications and visible-light-driven photocatalytic applications, *Appl Catal B: Environ* 106 (2011), pp. 1–13.

[74] S. Adhikari, S. Selvaraj, and D.-H. Kim, Construction of heterojunction photoelectrode via atomic layer deposition of Fe_2O_3 on Bi_2WO_6 for highly efficient photoelectrochemical sensing and degradation of tetracycline, *Appl Catal B: Environ* 244 (2019), pp. 11–24.

[75] X. Xu, L. Xiao, Z. Wu, Y. Jia, X. Ye, F. Wang, B. Yuan, Y. Yu, H. Huang, and G. Zou, Harvesting vibration energy to piezo-catalytically generate hydrogen through Bi_2WO_6 layered-perovskite, *Nano Energy* 78 (2020), p. 105351.

[76] Z. Wei, J. Liu, W. Fang, Z. Qin, Z. Jiang, and W. Shangguan, A visible-light driven novel layered perovskite oxyhalide Bi_4MO_8X (M=Nb, Ta; X=Cl, Br) constructed using BiOX (X=Cl, Br) for enhanced photocatalytic hydrogen evolution, *Catal Sci Technol* 8 (2018), pp. 3774–3784.

[77] Y. Hu, L. Mao, X. Guan, K.A. Tucker, H. Xie, X. Wu, and J. Shi, Layered perovskite oxides and their derivative nanosheets adopting different modification strategies towards better photocatalytic performance of water splitting, *Renew Sustain Energy Rev* 119 (2020), p. 109527.

[78] B.-H. Wang, B. Gao, J.-R. Zhang, L. Chen, G. Junkang, S. Shen, C.-T. Au, K. Li, M.-Q. Cai, and S.-F. Yin, Thickness-induced band-gap engineering in lead-free double perovskite $Cs_2AgBiBr_6$ for highly efficient photocatalysis, *Phys Chem Chem Phys* 23 (2021), pp. 12439–12448.

[79] J. Tao, X. Yu, Q. Liu, G. Liu, and H. Tang, Internal electric field induced S–scheme heterojunction MoS$_2$/CoAl LDH for enhanced photocatalytic hydrogen evolution, *J Colloid Interface Sci* 585 (2021), pp. 470–479.

[80] P.M. Bodhankar, P.B. Sarawade, G. Singh, A. Vinu, and D.S. Dhawale, Recent advances in highly active nanostructured NiFe LDH catalyst for electrochemical water splitting, *J Mater Chem A* 9 (2021), pp. 3180–3208.

[81] H. Xu, S. Zhu, M. Xia, and F. Wang, Rapid and efficient removal of diclofenac sodium from aqueous solution via ternary core-shell CS@ PANI@ LDH composite: Experimental and adsorption mechanism study, *J Hazard Mater* 402 (2021), p. 123815.

[82] L. Mohapatra, and K. Parida, A review on the recent progress, challenges and perspective of layered double hydroxides as promising photocatalysts, *J Mater Chem A* 4 (2016), pp. 10744–10766.

[83] S. Velu, K. Suzuki, and T. Osaki, A comparative study of reactions of methanol over catalysts derived from NiAl- and CoAl-layered double hydroxides and their Sn-containing analogues, *Catal Lett* 69 (2000), pp. 43–50.

[84] X. Wang, C.J. Summers, and Z.L. Wang, Mesoporous single-crystal ZnO nanowires epitaxially sheathed with Zn$_2$SiO$_4$, *Adv Mater* 16 (2004), pp. 1215–1218.

[85] S. Ali, M. Asif, A. Razzaq, and S.-I. In, Layered double hydroxide (LDH) based photocatalysts: An outstanding strategy for efficient photocatalytic CO$_2$ conversion, *Catalysts* 10 (2020), p. 1185.

[86] K. Parida, L. Mohapatra, and N. Baliarsingh, Effect of Co^{2+} substitution in the framework of carbonate intercalated Cu/Cr LDH on structural, electronic, optical, and photocatalytic properties, *J Phys Chem C* 116 (2012), pp. 22417–22424.

[87] X. Chen, S. Shen, L. Guo, and S.S. Mao, Semiconductor-based photocatalytic hydrogen generation, *Chem Rev* 110 (2010), pp. 6503–6570.

[88] X. Zong, and L. Wang, Ion-exchangeable semiconductor materials for visible light-induced photocatalysis, *J Photochem Photobiol C: Photochem Rev* 18 (2014), pp. 32–49.

[89] S. Zhang, J. Xu, H. Cheng, C. Zang, F. Bian, B. Sun, Y. Shen, and H. Jiang, Photocatalytic H$_2$ evolution from ammonia borane: Improvement of charge separation and directional charge transmission, *ChemSusChem* 13 (2020), pp. 5264–5272.

[90] K. Parida, and L. Mohapatra, Recent progress in the development of carbonate-intercalated Zn/Cr LDH as a novel photocatalyst for hydrogen evolution aimed at the utilization of solar light, *Dalton Trans* 41 (2012), pp. 1173–1178.

[91] M.S. Mostafa, L. Chen, M.S. Selim, M.A. Betiha, R. Zhang, Y. Gao, S. Zhang, and G. Ge, Novel cyanate intercalated CoBi layered double hydroxide for ultimate charge separation and superior water splitting, *J Clean Prod* 313 (2021), p. 127868.

[92] S. Pang, J.G. Huang, Y. Su, B. Geng, S.Y. Lei, Y.T. Huang, C. Lyu, and X.J. Liu, Synthesis and modification of Zn-doped TiO$_2$ nanoparticles for the photocatalytic degradation of tetracycline, *Photochem Photobiol* 92 (2016), pp. 651–657.

[93] M. Cao, K. Wang, I. Tudela, and X. Fan, Synthesis of Zn doped g-C$_3$N$_4$ in KCl-ZnCl$_2$ molten salts: The temperature window for promoting the photocatalytic activity, *Appl Surf Sci* 533 (2020), p. 147429.

[94] C. Foschini, L. Perazolli, and J.A. Varela, Sintering of tin oxide using zinc oxide as a densification aid, *J Mater Sci* 39 (2004), pp. 5825–5830.

[95] M. Thambidurai, S. Foo, K.M. Salim, P. Harikesh, A. Bruno, N.F. Jamaludin, S. Lie, N. Mathews, and C. Dang, Improved photovoltaic performance of triple-cation mixed-halide perovskite solar cells with binary trivalent metals incorporated into the titanium dioxide electron transport layer, *J Mater Chem C* 7 (2019), pp. 5028–5036.

[96] H. Cheng, J. Wang, Y. Zhao, and X. Han, Effect of phase composition, morphology, and specific surface area on the photocatalytic activity of TiO$_2$ nanomaterials, *RSC Adv* 4 (2014), pp. 47031–47038.

[97] G. Zhang, G. Kim, and W. Choi, Visible light driven photocatalysis mediated via ligand-to-metal charge transfer (LMCT): An alternative approach to solar activation of titania, *Energy Environ Sci* 7 (2014), pp. 954–966.

[98] W.-C. Hou, and Y.-S. Wang, Photocatalytic generation of H$_2$O$_2$ by graphene oxide in organic electron donor-free condition under sunlight, *ACS Sustain Chem Eng* 5 (2017), pp. 2994–3001.

[99] H.J. Shin, K.K. Kim, A. Benayad, S.M. Yoon, H.K. Park, I.S. Jung, M.H. Jin, H.K. Jeong, J.M. Kim, and J.Y. Choi, Efficient reduction of graphite oxide by sodium borohydride and its effect on electrical conductance, *Adv Func Mater* 19 (2009), pp. 1987–1992.

[100] R.J.W.E. Lahaye, H.K. Jeong, C.Y. Park, and Y.H. Lee, Density functional theory study of graphite oxide for different oxidation levels, *Phys Revp B* 79 (2009), p. 125435.

[101] L. Tie, Y. Liu, S. Shen, C. Yu, C. Mao, J. Sun, and J. Sun, In-situ construction of graphene oxide in microsphere ZnS photocatalyst for high-performance photochemical hydrogen generation, *Int J Hydrog Energy* 45 (2020), pp. 16606–16613.

[102] Y. Gao, B. Xu, M. Cherif, H. Yu, Q. Zhang, F. Vidal, X. Wang, F. Ding, Y. Sun, and D. Ma, Atomic insights for Ag Interstitial/Substitutional doping into $ZnIn_2S_4$ nanoplates and intimate coupling with reduced graphene oxide for enhanced photocatalytic hydrogen production by water splitting, *Appl Catal B: Environ* 279 (2020), p. 119403.

[103] Z. Long, X. Tong, C. Liu, A.I. Channa, R. Wang, X. Li, F. Lin, A. Vomiero, and Z.M. Wang, Near-infrared, eco-friendly ZnAgInSe quantum dots-sensitized graphene oxide-TiO_2 hybrid photoanode for high performance photoelectrochemical hydrogen generation, *Chem Eng J* 426 (2021), p. 131298.

[104] C.B. Carter and M.G. Norton, Sols, gels, and organic chemistry, in *Ceramic Materials: Science and Engineering*, C. B. Carter and M. G. Norton eds., Springer, New York, 2007, pp. 400–411.

[105] Q. Zhu, Z. Xu, B. Qiu, M. Xing, and J. Zhang, Emerging cocatalysts on g-C_3N_4 for photocatalytic hydrogen evolution, *Small* 17 (2021), p. 2101070.

[106] Y. Hong, L. Wang, E. Liu, J. Chen, Z. Wang, S. Zhang, X. Lin, X. Duan, and J. Shi, A curly architectured graphitic carbon nitride (g-C_3N_4) towards efficient visible-light photocatalytic H_2 evolution, *Inorg Chem Front* 7 (2020), pp. 347–355.

[107] H. Chen, Y. Fan, Z. Fan, H. Xu, D. Cui, C. Xue, and W. Zhang, Electronic tuning of g-C_3N_4 via competitive coordination to stimulate high-efficiently photocatalytic for hydrogen evolution, *J Alloys Compd* 891 (2022), p. 162027.

[108] N. Feng, H. Lin, F. Deng, and J. Ye, Interfacial-bonding Ti–N–C boosts efficient photocatalytic H_2 evolution in close coupling g-C_3N_4/TiO_2, *J Phys Chem C* 125 (2021), pp. 12012–12018.

[109] S. Yang, S. Ding, C. Zhao, S. Huo, F. Yu, J. Fang, and Y. Yang, Ligand-free Au nanoclusters/g-C_3N_4 ultra-thin nanosheets composite photocatalysts for efficient visible-light-driven photocatalytic H_2 generation, *J Mater Sci* 56 (2021), pp. 13736–13751.

[110] Z. Wang, C. Li, and K. Domen, Recent developments in heterogeneous photocatalysts for solar-driven overall water splitting, *Chem Soc Rev* 48 (2019), pp. 2109–2125.

[111] Y. Hong, J. Zhang, X. Wang, Y. Wang, Z. Lin, J. Yu, and F. Huang, Influence of lattice integrity and phase composition on the photocatalytic hydrogen production efficiency of ZnS nanomaterials, *Nanoscale* 4 (2012), pp. 2859–2862.

[112] I. Tsuji, H. Kato, and A. Kudo, Photocatalytic hydrogen evolution on ZnS– $CuInS_2$– $AgInS_2$ solid solution photocatalysts with wide visible light absorption bands, *Chem Mater* 18 (2006), pp. 1969–1975.

[113] S. Zhu, X. Qian, D. Lan, Z. Yu, X. Wang, and W. Su, Accelerating charge transfer for highly efficient visible-light-driven photocatalytic H_2 production: In-situ constructing Schottky junction via anchoring Ni-P alloy onto defect-rich ZnS, *Appl Catal B: Environ* 269 (2020), p. 118806.

[114] E. Puentes-Prado, C. Garcia, J. Oliva, R. Galindo, J. Bernal-Alvarado, L. Diaz-Torres, and C. Gomez-Solis, Enhancing the solar photocatalytic hydrogen generation of ZnS films by UV radiation treatment, *Int J Hydrog Energy* 45 (2020), pp. 12308–12317.

[115] X. Yang, H. Liu, T. Li, B. Huang, W. Hu, Z. Jiang, J. Chen, and Q. Niu, Preparation of flower-like ZnO@ZnS core-shell structure enhances photocatalytic hydrogen production, *Int J Hydrog Energy* 45 (2020), pp. 26967–26978.

[116] Y.-J. Yuan, P. Wang, Z. Li, Y. Wu, W. Bai, Y. Su, J. Guan, S. Wu, J. Zhong, and Z.-T. Yu, The role of bandgap and interface in enhancing photocatalytic H_2 generation activity of 2D-2D black phosphorus/MoS_2 photocatalyst, *Appl Cataly B: Environ* 242 (2019), pp. 1–8.

[117] Y.-J. Yuan, Z. Shen, S. Wu, Y. Su, L. Pei, Z. Ji, M. Ding, W. Bai, Y. Chen, and Z.-T. Yu, Liquid exfoliation of g-C_3N_4 nanosheets to construct 2D-2D MoS_2/g-C_3N_4 photocatalyst for enhanced photocatalytic H_2 production activity, *Appl Catal B: Environ* 246 (2019), pp. 120–128.

[118] T. Su, L. Xiao, Y. Gao, T. Liu, X. Peng, H. Yuan, Y. Han, S. Ji, and X. Wang, Multifunctional MoS_2 ultrathin nanoflakes loaded by $Cd_{0.5}Zn_{0.5}S$ QDs for enhanced photocatalytic H_2 production, *Int J Energy Res* 43 (2019), pp. 5678–5686.

[119] K. Zhuge, Z. Chen, Y. Yang, J. Wang, Y. Shi, and Z. Li, In-suit photodeposition of MoS_2 onto CdS quantum dots for efficient photocatalytic H_2 evolution, *Appl Surf Sci* 539 (2021), p. 148234.

[120] L. Yin, X. Hai, K. Chang, F. Ichihara, and J. Ye, Synergetic exfoliation and lateral size engineering of MoS_2 for enhanced photocatalytic hydrogen generation, *Small* 14 (2018), p. 1704153.

[121] J. Zhang, P. Zhou, J. Liu, and J. Yu, New understanding of the difference of photocatalytic activity among anatase, rutile and brookite TiO_2, *Phys Chem Chem Phys* 16 (2014), pp. 20382–20386.

[122] Y.-X. Pan, T. Zhou, J. Han, J. Hong, Y. Wang, W. Zhang, and R. Xu, CdS quantum dots and tungsten carbide supported on anatase–rutile composite TiO_2 for highly efficient visible-light-driven photocatalytic H_2 evolution from water, *Catal Sci Technol* 6 (2016), pp. 2206–2213.

[123] V. Etacheri, C. Di Valentin, J. Schneider, D. Bahnemann, and S.C. Pillai, Visible-light activation of TiO_2 photocatalysts: Advances in theory and experiments, *J Photochem Photobiol C: Photochem Rev* 25 (2015), pp. 1–29.

[124] X. Chen, and S.S. Mao, Titanium dioxide nanomaterials: Synthesis, properties, modifications, and applications, *Chem Rev* 107 (2007), pp. 2891–2959.

[125] Y.-H. Wang, K.H. Rahman, C.-C. Wu, and K.-C. Chen, A review on the pathways of the improved structural characteristics and photocatalytic performance of titanium dioxide (TiO_2) thin films fabricated by the magnetron-sputtering technique, *Catalysts* 10 (2020), p. 598.

[126] D.P. Kumar, V.D. Kumari, M. Karthik, M. Sathish, and M.V. Shankar, Shape dependence structural, optical and photocatalytic properties of TiO_2 nanocrystals for enhanced hydrogen production via glycerol reforming, *Sol Energy Mater Sol Cells* 163 (2017), pp. 113–119.

[127] T. Wang, W. Li, D. Xu, X. Wu, L. Cao, and J. Meng, A novel and facile synthesis of black TiO_2 with improved visible-light photocatalytic H_2 generation: Impact of surface modification with CTAB on morphology, structure and property, *Appl Surf Sci* 426 (2017), pp. 325–332.

[128] H. Li, S. Wu, Z.D. Hood, J. Sun, B. Hu, C. Liang, S. Yang, Y. Xu, and B. Jiang, Atomic defects in ultrathin mesoporous TiO_2 enhance photocatalytic hydrogen evolution from water splitting, *Appl Surf Sci* 513 (2020), p. 145723.

[129] H. Kurban, M. Dalkilic, S. Temiz, and M. Kurban, Tailoring the structural properties and electronic structure of anatase, brookite and rutile phase TiO_2 nanoparticles: DFTB calculations, *Comput Mater Sci* 183 (2020), p. 109843.

[130] M.R.A. Cruz, D. Sanchez-Martinez, and L.M. Torres-Martínez, Optical properties of TiO_2 thin films deposited by DC sputtering and their photocatalytic performance in photoinduced process, *Int J Hydrog Energy* 44 (2019), pp. 20017–20028.

[131] S. Huo, and C. Chen, One-step synthesis CdS/single crystal ZnO nanorod heterostructures with high photocatalytic H_2 production ability, *Inorg Chem Commun* 132 (2021), p. 108841.

[132] M. Baek, D. Kim, and K. Yong, Simple but effective way to enhance photoelectrochemical solar-water-splitting performance of ZnO nanorod arrays: Charge-trapping $Z(OH)_2$ annihilation and oxygen vacancy generation by vacuum annealing, *ACS Appl Mater Interfaces* 9 (2017), pp. 2317–2325.

[133] E. Luévano-Hipólito, and L.M. Torres-Martínez, Sonochemical synthesis of ZnO nanoparticles and its use as photocatalyst in H_2 generation, *Mate Sci Eng B* 226 (2017), pp. 223–233.

[134] K. Ranjith, and R.R. Kumar, Regeneration of an efficient, solar active hierarchical ZnO flower photocatalyst for repeatable usage: Controlled desorption of poisoned species from active catalytic sites, *RSC Adv* 7 (2017), pp. 4983–4992.

[135] I.J. Peter, E. Praveen, G. Vignesh, and P. Nithiananthi, ZnO nanostructures with different morphology for enhanced photocatalytic activity, *Mater Res Express* 4 (2017), p. 124003.

[136] Y. Lin, H. Hu, and Y.H. Hu, Role of ZnO morphology in its reduction and photocatalysis, *Appl Surf Sci* 502 (2020), p. 144202.

[137] Y. Mao, Y. Li, Y. Zou, X. Shen, L. Zhu, and G. Liao, Solvothermal synthesis and photocatalytic properties of ZnO micro/nanostructures, *Ceram Int* 45 (2019), pp. 1724–1729.

[138] L. He, Z. Tong, Z. Wang, M. Chen, N. Huang, and W. Zhang, Effects of calcination temperature and heating rate on the photocatalytic properties of ZnO prepared by pyrolysis, *J Colloid Interface Sci* 509 (2018), pp. 448–456.

[139] T. Wanotayan, J. Panpranot, J. Qin, and Y. Boonyongmaneerat, Microstructures and photocatalytic properties of ZnO films fabricated by Zn electrodeposition and heat treatment, *Mater Sci Semicond Process* 74 (2018), pp. 232–237.

[140] X. Wang, H. Xu, Y. Zhang, X. Ji, and R. Zhang, Fructose-regulated ZnO single-crystal nanosheets with oxygen vacancies for photodegradation of high concentration pollutants and photocatalytic hydrogen evolution, *Ceram Int* 47 (2021), pp. 16170–16177.

11 Advancement in Biofuels Production

Sustainable Perception towards Green Energy and Environment

Jyoti Kataria, Seema Devi, and Pooja Devi
Guru Jambheshwar University of Science & Technology

CONTENTS

DOI: 10.1201/9781003206385-11

11.1 INTRODUCTION

Developing secure and sustainable energy resources is the most pressing need for human beings. Energy resources are vital for the overall stability and economic growth of our society and nation. The population on earth is increasing day by day along with standard of their living, which enhances the energy demand and results in unexpected environmental fluctuations [1,2]. Also, more consumption of fossil fuels, which are non-renewable, non-sustainable, and expensive, creates negative impacts on ecosystem, such as global climate change, retreating of glaciers, sea-level rise, biodiversity loss, pollution, etc. [3,4]. Therefore, there is a need for an alternative resource of energy, such as biofuels. Biofuels are the fuels that can be produced from biomass of plants, algae, animal waste, used cooking oil, etc. [5,6]. Since, the feedstock material used for biofuels synthesis can be restocked readily; therefore, they are known as renewable source of energy [2]. Biofuels are inexhaustible, cost-effective, environment-friendly (reduce greenhouse gases emission), non-toxic, and sulfur-free [7]. However, there are few obstructions in the utilization of biofuels like production cost, their physicochemical behavior, and less adequacy for engines as per their configurations and requirements [8]. So, it is inevitable to develop the different biofuels with variable chemical and physical properties, and also, the study of their blends is necessary [9]. As per the Government enticements to make fuels economical and according to the requirements of industries, vehicles, and the fields for cultivation, are the driving factors boosting the production of biofuels from crop wastes generated feedstock [10,11].

On the basis of feedstock materials, biofuels are classed into two categories: primary and secondary biofuels [6]. Secondary biofuels are further classified into first, second, third, and fourth generation, which are shown in Figure 11.1, and their description is given in proceeding sections [12,13]. First-generation biofuels are formed from extracted oil of plants, carbohydrates, etc., which contributes to the nutrition-related problems, whereas biofuel synthesized using lignocellulosic biomass of second generation minimizes the food issues [14]. Algae-based third-generation biofuels can be generated in bulk, reduce environmental pollution, and easy to purify [15,16]. Over the past years, researchers focused on third-generation biofuels from microalgae that can grow vigorously under divergent conditions and produce various types of inexhaustible and viable fuels like biohydrogen, biodiesel, and biogas [17,18]. The fourth-generation biofuels utilized designed cyanobacterial algae, which is a novel and advanced technology for the production of biofuels [19]. Fourth-generation

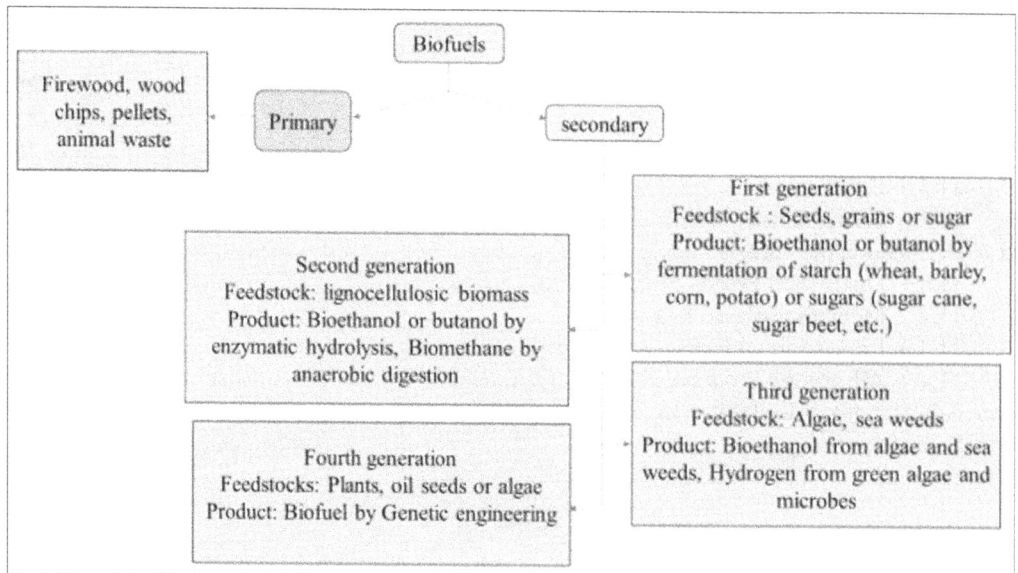

FIGURE 11.1 Classification of biofuels.

biofuels amalgamate both feedstock biology and carbon capture and storage (CCS) processes, which provide a new pathway for manufacturing high-quality biofuels with negligible content of carbon and maximum engine performance efficiencies [2,20]. These advanced biofuels used novel synthetic biology tools and are in their initial stage of development and research [21].

The goal of this chapter is to summarize recent advancement in biofuels production and their possible applications. This chapter provides an idea of different generations of biofuels, including bioethanol, biodiesel, biogas, and biobutanol, using biochemical and thermochemical approaches. Further, the study focused on the energy-saving purification methods to improve the biofuel yield and make them commercially feasible and ecologically benign. This chapter also reviewed the application of biofuels as gasoline in the industrial and transport sector. The biofuel production also concerns with environmental, economic, and societal realm. Therefore, switching to biofuels appears as a prominent way to save money, reduce the dependency on fossil fuels, and keep our economy secure.

11.2 CLASSIFICATION OF BIOFUELS ON THE BASIS OF THEIR FEEDSTOCK

Biofuels are generally classified as primary and secondary biofuels on the basis of their feedstock [16]. Primary biofuels are natural and utilized for cooking and in energy generation. Secondary biofuels, such as ethanol, biodiesel, and dimethyl ether (DME), are made from biomass and can be utilized in cars and industrial processes [22]. Biofuels can be solid, liquid, or gaseous. Solid biofuels include fuel wood, wood pellets, and charcoal; liquid biofuels include pyrolysis oils, bioethanol, and biodiesel; and gaseous biofuels include biogas (methane) [23].

11.2.1 OIL EXTRACTION METHODS FOR FIRST-GENERATION BIOFUELS

Biofuels of first-generation are made from arable food crops, food wastes, and animal fats by simple methods. The biofuels are ultimately derived from grain crops, pulse crops, forage crops, and oil seed crops [24]. As per the literature, initially vegetable oils were used for the synthesis of first-generation biofuels. Bioethanol, biodiesel, and biogas are first-generation biofuels. Bioethanol is most frequently used biofuel due to its high-octane number and ability to improve the air quality by reducing hazardous gas emissions [22]. Generally, bioethanol is synthesized from carbohydrate polymers using hydrolysis and sugar fermentation processes [25]. In hydrolysis process, cellulosic part of biomass is broken down into sugars with the help of enzymes or dilute acids, which further fermented into ethanol [26]. Biodiesel is another renowned first-generation biofuel formed from oils of sebaceous plants by transesterification or cracking methods [27]. Basically, two methods (mechanical and solvent extraction) are applied to extract highly pure oils from arable crops. Mechanical extraction methods involve hydraulic or screw pressing of oleaginous materials and are more effective than solvent extraction methods due to their advantageous features such as low equipment cost and lack of volatile organic chemicals [28]. In solvent extraction methods, the extraction of oil from feedstock is carried out by using organic solvents such as hexane, acetone, chloroform, etc., which are chosen based on their solubility (like dissolves like) and biocompatibility [29]. It was reported by Steinbock that a combination of mechanical pressing and solvent extraction provides stronger outcomes than either of the process used separately [30]. The first-generation biofuels production is still questionable, since it is formed by using food stuffs [31]. Moreover, certain limitations like reduced land efficiency, increased crop prices, low return on investment, and some environmental issues downgrade the economic feasibility of biofuel production. Therefore, in "food versus fuel" conflict, these feedstocks are not sustainable [32,33]. These constraints encourage the researchers for biofuels production from the non-edible resources [23,34]. Accordingly, the current study focuses on successive generations of biofuels by considering the sustainability factors [35].

11.2.2 Oil Extraction Methods for Second-Generation Biofuels

The second-generation biofuels have emerged as a sustainable response to the growing controversy over the first-generation biofuels [36]. As stated by Larson, the basic characteristics of raw materials should have the potential to provide low prices, substantial environment, and energy benefits for most of the second-generation biofuels [23]. The feedstocks for the second-generation biofuels are non-edible lignocellulosic sources such as agricultural residues, jatropha, bagasse, tree biomass, demolition wood, grass, straw, etc., which therefore impose a limit between direct food and fuel competition. Three distinct approaches, i.e., biochemical, physicochemical, and thermochemical processing methods, are used for the production of the second-generation biofuels from non-edible wastes or whole plant [14]. The majority of second-generation fuels, including methanol, refined Fischer Tropsch liquids (FTL), DME, etc., are developed thermochemically, but some second-generation biofuels such as biobutanol and bioethanol are produced by biochemical processes [37]. Pretreatment, hydrolysis, and fermentation are the three phases in which the biochemical conversion of cellulose into ethanol occurs [38]. In case of thermochemical processes, cellulosic part needs to undergo extreme conditions of temperatures and pressures, i.e., gasification or pyrolysis (discussed later in Section 11.3) [39]. In comparison with the first-generation biofuels, the utilization of cheaply available non-edible biomass for the production of second-generation biofuels provides a low-scale land acquisition, which promotes further studies for the second-generation biofuels [23]. The following methods are the commonly used extraction methods for second-generation feedstocks [28].

1. Conventional solvent extraction (CSE)
2. Physical-supported solvent extraction (PSSE)
3. Supercritical fluid extraction (SFE)
4. Novel extractions

11.2.2.1 Conventional Solvent Extraction (CSE)

CSE has been used in various domains, including food and pharmaceutical industry. Chemical solvents used in CSE are usually in liquid state at room temperature and ambient pressure and have maximum solubility and selectivity for the composition of interest. The premise of CSE is that the extraction of oil from feedstocks is carried out by using organic solvent. Soxhlet extraction and Bligh-Dyer methods based on CSE principle are still applied for the separation of precursors from oleaginous materials [29]. A wide array of solvents such as methanol, hexane, and chloroform or their blends can be used for the Soxhlet technique [40]. Roschat et al. extracted rubber seed oil by using different types of solvents, including acetone, hexane, ethyl acetate, and dichloromethane. They successfully obtained 24 wt. % of oil by using hexane as an extractant solvent [41]. In the Bligh-Dyer method, the mixture of organic solvents is used for the extraction of oil, and the function of solvents can be activated by accelerating their polarity by using non-polar-, aprotic polar-, and protic polar-based solvents [42,43]. However, the most disappointing point of CSE is that all the organic solvents are volatile in nature, which are detrimental to human health and cause environmental pollution. Furthermore, in order to extract the accessible compounds in CSE, a pretreatment is usually necessary, which adds to the overall expense burden [29].

11.2.2.2 Physical-Supported Solvent Extraction (PSSE)

PSSE is an integrated version of CSEs, including microwave and ultrasound sonication methods for oil extraction. In microwave extraction, electromagnetic radiations of frequency of 0.3–300 GHz are penetrating into the biomaterial, interact with polar entities, which in turn disrupt the plant cells [44]. The factors like temperature, quality of feedstocks, amount and physical properties of solvent, and power of microwave determine the efficiency of microwave-assisted extraction. The advantageous features of this technique are high yield, high purity, fast process, low cost, and low solvent consumption. Sonication method is based on phenomenon of cavitation, in which cellular structure of oleaginous substances gets ruptured due to the formation of high- and low-pressure cycles. Sonication method is more effective in the situation of simultaneous extraction of oils and production of biofuels from different feedstocks. Furthermore, it operates at lower frequency (18–50 kHz) and under medium conditions of temperature and pressure. Rajendran et al. extracted oil from *Calophyllum inophyllum* seeds using different combinations of solvent (chloroform:methanol, diethyl ether:ethanol, and isopropanol: methanol) in ultrasonic-assisted technique to maintain the oil quality and property. The results revealed that maximum oil extraction was achieved in an equimolar ratio of diethyl ether:ethanol in assistance with ultrasonic waves [45]. However, these methods produce high vibrating waves that may induce shock into working biomass and cause to destroy the whole. Therefore, it is proposed to represent a viable approach for the extraction of oils with necessary components [46].

11.2.2.3 Supercritical Fluid Extraction (SFE)

SFE was introduced in 1980 for successfully extraction of oil without its degradation and oxidation. It has a distinct edge over CSE in terms of their high selectivity, high reaction rate, use of non-toxic solvents, and neat product formation (no solvent residue). SFE uses fluid at temperature and pressure above its critical point used as an extractant, and shows mass transfer properties as gases and solvation properties like liquids. Also, their extraction properties can be significantly modified by subtle changes in pressure and temperature. Carbon dioxide (CO_2) and water are primarily used as supercritical fluids for biofuel production. Supercritical CO_2 ($scCO_2$) can efficiently solubilize non-polar molecules due to its low polarity, and in case of polar compounds, their solubilizing power can be enhanced by using co-solvents of different polarities. Couto et al. studied the SFE of oils from spent coffee grounds using $scCO_2$ at different temperature and pressure [47]. Further, the experimental parameters, including reaction temperature, type of biomass and solvent, biomass/solvent ratio, and extraction time, can significantly affect the product quality and yield. Current SFE methods also have some limitations, including safety issues, need for specialized equipment, and high running cost, thereby suggesting their use near- or subcritical solvents [48].

11.2.2.4 Novel Methods

Novel methods are assigned to those extraction techniques, which do not follow the pre-described categories of extraction methods. Nowadays, ionic liquids-covalent molecule co-solvent systems are significantly used to extract essential oils from different biomass [49]. Further, liquified gases were introduced to reduce moisture content and extract the oil from different vegetal biomasses. Li et al. used liquified DME for dewatering up to 81.3%–88.7% and extraction of bio-crude with a yield of 5.8 wt.%–16.8 wt.% from total dried feedstocks [50]. Generally, all the industrial chemical processes involve more than one steps and require different solvents for each step, as stated by Murphy's Law of Solvents, i.e., "The best solvent for any process step is bad for the next step." Therefore, scientists introduced the concept of switchable solvent to meet the need of first step to dewater the crude oil before its extraction into pure oil by using supercritical fluid modified CO_2 [51]. These are simple liquids, which can be reversibly switched from one form to another. Phan et al. used switchable solvents for the extraction of oil from soybean flakes and comparatively required less time, small amount of solvent, and low energy consumption [52].

11.2.3 Oil Extraction Methods for Third-Generation Biofuels

Second-generation biofuels are more advantageous but still there is major concern over land acquisition, food security, complex methods for production processes, and time-consuming growth period, which is directing the attention of researchers toward third-generation biofuels. Therefore, based on current scientific knowledge and technological predictions, third-generation biofuels developed from microalgae and microorganisms are considered as feasible substitute energy sources that reduce the key limitations of first and second-generation biofuels [23]. Third-generation biofuels are also known as algae biofuels because they are specifically derived from photosynthetic microbes, such as algae, cyanobacteria, and microalgae. Accordingly, algae can produce several distinct renewable biofuels (e.g., biohydrogen, hydrocarbons, and biodiesel) [28]. Algae has a number of advantages as biofuel feedstock, including low area needs, high growing rate, high endurance for harsh circumstances, and potential to reduce greenhouse gas emission. Furthermore, algae can develop in a variety of environment and water, including fresh water, marine, and even industrial wastewaters [36]. Similar to terrestrial biomass, thermochemical and biochemical conversion technologies are used for micro-algae biomass. Thermochemical conversion involves thermal degradation of organic compounds into biofuels using either thermochemical liquefaction, direct combustion, gasification, or pyrolysis methods. However, biological process includes alcoholic fermentation, anaerobic digestion (AD), and photobiological hydrogen production for energy conversion of biomass into other fuels [53]. The methods (CSE, PSSE, SFE, and novel) are also used for the extraction of lipids from algal biomass.

11.2.3.1 Extraction of Lipids from Algal Biomass Using CSE Method

A variety of pure solvents and co-solvent systems (hexane/ethanol, methanol/chloroform, dichloromethane/ethanol, hexane/isopropanol, etc.) have been examined and tested for algae biomass extraction. In co-solvent systems, the polar molecules disrupt the intermolecular forces (hydrogen bonding and electrostatic) between polar lipids, whereas non-polar solvents break the hydrophobic non-polar/neutral lipids. Long and Abdelkader evaluated the efficiency of some commonly used solvents for lipid extraction from *Nannochloropsis* microalgae [54]. They observed cyclohexane/1-butanol co-solvent as a most effective solvent system in Soxhlet extraction. Shin et al. found that hexane and methanol (1:1) mixture acts as a promising solvent for the large-scale lipid extraction from algae [55]. Pretreatment methods involve drying and cell distraction, which are often necessary prior to CSE for the removal of the water from algae cells and recognition of target site in the cells, which is one of the downsides of CSE for algae biofuel production. Such operations consume a lot of energy, and the use of organic solvents has a negative impact on the environment.

11.2.3.2 Extraction of Lipids from Algal Biomass Using PSSE Method

Physical methods (microwaving and sonication, grinding, autoclaving, osmotic shock, or bead beating) are normally integrated with CSE for lipid extraction from algae biomasses [56]. PSSE methods are applied to mechanically break thick walls of microalgae to release intralipids; they also reduce the extraction time by tenfold. Adam et al. used ultrasound-assisted method for the extraction of lipids from fresh *Nannochloropsis oculata* biomass [57]. They also compare oil recovery with Bligh Dyer's method, and optimize various experimental parameters like extraction time, water content of biomass, and ultrasonic power using response surface methodology (RSM) [57]. Zhang et al. evaluated the efficiency of lipid extraction from *Trichosporon oleaginosus* by using chloroform/methanol and different polarity solvents (water, methanol, and hexane) under ultrasonication [58]. They found the highest lipid recovery in a very short time (15 minutes) under low temperature conditions in chloroform/methanol mixture at an ultrasonication frequency of 50 Hz [58]. The study reported by Lee et al. showed that the use of microwave provides most easy, simple, and effective way of lipid extraction from algae biomasses [59]. In their study, they evaluated lipid recovery by using different methods from *Botryococcus sp.*, *Scenedesmus sp.*, and *Chlorella vulgaris*, and found *Botryococcus sp.*, yields highest lipid content using microwave oven method of extraction [59]. Zhou et al. showed lipid extraction using microwave-assisted process from *Scenedesmus obliquus* microalgae grown on municipal wastewater [60]. The optimum parameters temperature, extraction time, solvent/solvent ratio, and co-solvent/biomass ratio were 130°C, 0.25 hour, 3:2 (V: V), and 50:1 (mL:g), respectively, for 88.25% extraction rate of lipid. However, their highly intensive energy demands promote researchers towards alternate efficient pathway of lipid extraction [60].

11.2.3.3   Extraction of Lipids from Algal Biomass Using SFE Method

SFE has potential to replace CSE in terms of selectivity, solvation power, and toxicity. Soh and Zimmerman reported lipid extraction from *Scenedesmus dimorphus* using supercritical conditions and compared results with CSE [61]. They observed that the maximum solubility of triglycerides in $scCO_2$ can be achieved at moderate temperatures (70°C–80°C) and high pressures (above 7,000 psi) [61]. In a recent investigation, $scCO_2$ extraction of lipids from algae biomasses using azeotropic co-solvents (hexane+ethanol; 1:1) has been performed [62]. In $scCO_2$-assisted conditions with process parameters such as reaction time (60 minutes), reaction pressure (340 bar), reaction temperature (80°C), and algae/solid ratio (12:1), the maximum obtained yield of algal lipid was 31.37% [62]. It was revealed in a study by Postma et al. that bead milling significantly improves the extraction yield obtained from oleaginous algae using $scCO_2$ [63]. This technique offers many benefits over CSE, including product without solvent residue, less extraction time, and improved quality and yield. However, high operational expenses and infrastructure linked with $ScCO_2$ are their primary drawbacks [63].

11.2.3.4   Extraction of Lipids from Algal Biomass Using Novel Method

For algae biofuel production, several innovative extraction techniques have been recently investigated. A study by Kamaroddin et al. showed lipid extraction by direct ozonation, integrated with microbubble-driven photobioreactor system, which eliminates or reduces contaminants [64]. Du et al. used CO_2 switchable solvents for lipid extraction from wet algal biomass to produce biofuels or food ingredients, pharmaceutical and cosmetics products [65]. They also compared efficiency and energy consumption with CSE and SFE techniques and found that switchable solvent exhibits a maximum product yield with less energy consumption [65]. Further, a report came with CO_2 switchable ionic liquids (S-ILs), which switched back and forth between a hydrophobic and hydrophilic nature. S-ILs provide a sustainable approach for the separation and extraction of lipids from algae with less energy-intensive steps [66]. Also, ionic liquids (ILs) combined with other solvent can act as an extraction solvent system for lipid production from algal biomasses. In addition to second-generation feedstocks, the liquefied gases are also used to extract the biofuel precursor from third-generation feedstocks. Goto et al. applied subcritical DME to extract components from algae without drying and cell disruption as shown in Figure 11.2 [67]. Moreover, it is easy, simple, and low energy-consuming process, which also eliminates the step of solvent evaporation performed usually in all the techniques.

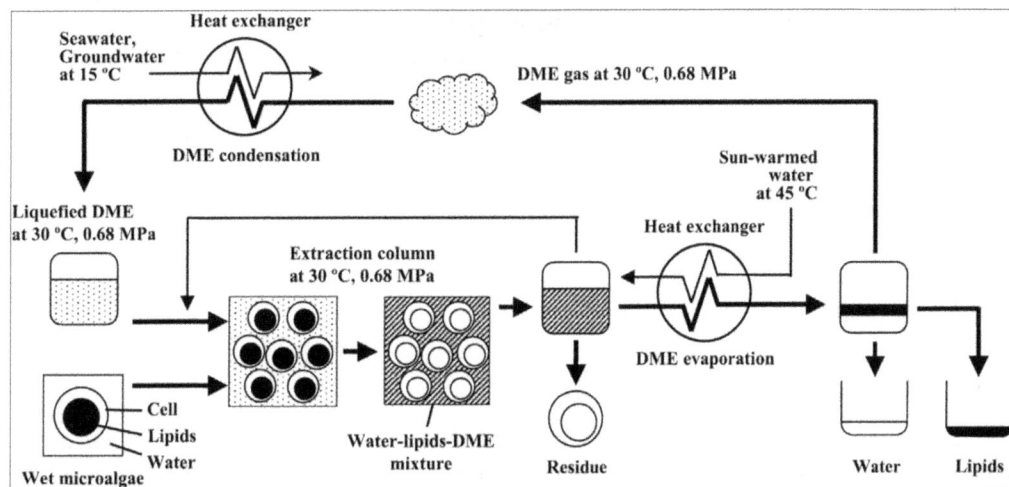

FIGURE 11.2   Extraction of lipids from microalgae using subcritical DME. (Reprinted with permission from Goto et al. [67]. Copyright 2015 Elsevier B. V.)

11.2.4 FOURTH-GENERATION BIOFUELS

Fourth-generation biofuels are different in manner of feedstocks and processing technologies adopted for the above-described generation of biofuels. In the production of these fuels, feedstocks are tailored genetically to improve carbon capture and the processing efficiency. In this sense, fourth-generation biofuels embody the concept of "Bioenergy with Carbon Storage (BECS)." They involve the direct conversion of sunlight into useful fuels using readily available and inexhaustible raw materials (water, CO_2, and sunlight) and are expected to be produced via engineered photosynthetic microorganisms. There are many microorganisms, including yeast, bacteria, cyanobacteria, and microalgae, but cyanobacteria and microalgae play a vital role in global carbon balance and oxygen production. It is based on genetically engineered DNA formed by recombinant fragments of DNA from different kind of organisms, or involves other bioengineering and biological techniques to adjust the cellular metabolism and characteristics of algal metabolic networks. Yoshikawa et al. produced bioethanol from *Synechocystis sp. PCC 6803 strain* using metabolic engineering [68]. The usual process stages required for biofuel production employing fourth-generation methods are substantially shorter than preceding approaches, resulting in lower and capital operational cost. Kruse et al. reviewed a wide range of organisms for biohydrogen production from solar energy using molecular genetic tools [69]. Deng and Coleman demonstrated the production of bioethanol in cyanobacteria through *Zymomonas mobilis pdc/adh* gene transformation [70]. Though, the study of these advanced biofuels is still in earlier stage of development and experiment. Further, certain limitations like high cost and deficiency of research on its practical performance from an economic and technical perspective necessitate the new technologies for the development of final product.

11.3 TECHNIQUES USED IN PRODUCTION OF BIOFUELS

The most important characteristics of the advanced biofuels resulting from organic waste are their superior production and ability to decrease the environment pollution, although the most specific technique for biofuel production continues to evolve. Biofuels include both solid and liquid fuels as well as gaseous fuels. The major popular techniques include fermentation of compounds from carbohydrates for the production of ethanol and transesterification of waste cooking oil or animal fat to produce biodiesel. Manufacture of liquid biofuels is quite modern compared to the manufacture of solid or gaseous biofuels, due to greater transformation capacity, less waste production, usage of low area, and water compared with gas-based biofuel manufacturing methods [71]. The techniques used for the production of biofuels from biomass are biochemical, physicochemical, and thermochemical. The most well-established technologies for the conversion of organic waste into biofuels are as follows (shown in Figure 11.3) [2].

    i. Hydrolysis and fermentation
   ii. Pyrolysis
  iii. Hydrothermal liquefaction
   iv. Anaerobic digestion
    v. Gasification
   vi. Transesterification

11.3.1 HYDROLYSIS AND FERMENTATION

Bioethanol is an alternate source of renewable energy that can be produced from carbohydrates, cellulose, and hemicellulose using hydrolysis and fermentation processes [23]. Hydrolysis process mainly includes conversion or breaking of carbohydrates into the corresponding sugars, which are further converted into ethanol biofuels [72]. Abo et al. state that three steps are required for the production of biofuel from lignocellulose biomass, as shown in Figure 11.4 [38]. Pretreatment was used

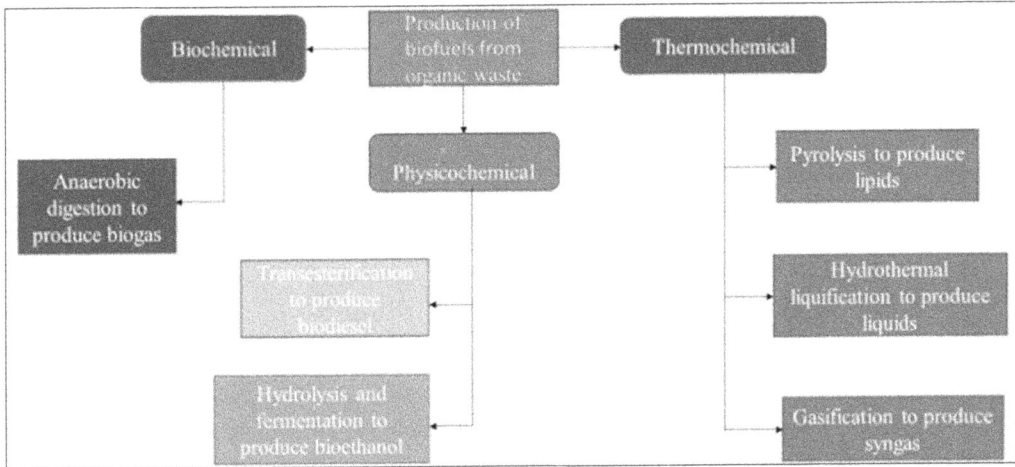

FIGURE 11.3 Production of biofuels from organic waste.

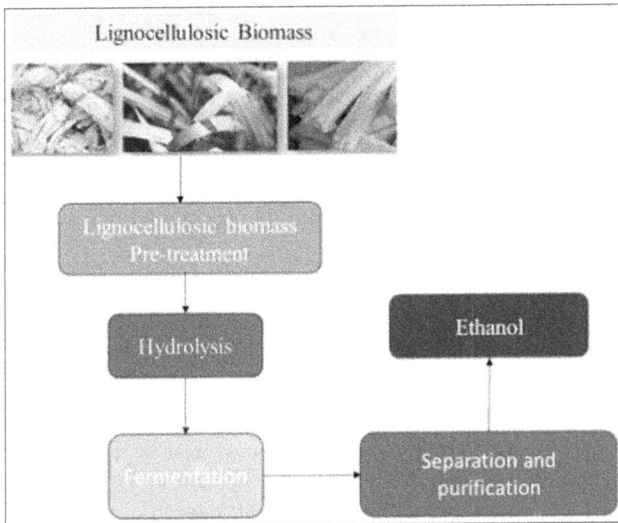

FIGURE 11.4 Production of bioethanol by hydrolysis and fermentation [38].

to destroy the crystalline structure of the lignocellulosic material. Then, the hydrolysis stage was designed to break down complex polymer chains into simple sugar. Finally, sugars were converted into ethanol using fermentation [38]. Talebnia et al. reviewed the most recent advances in wheat straw pretreatment before hydrolysis and fermentation for ethanol production [73]. After enzymatic hydrolysis of wheat straw, a maximum theoretical sugar yield was 74%–99.6% depending on the type of pretreatment method (physical, physicochemical, chemical, and biological) [73]. Kumar et al. reported the conversion of lignocellulose biomass into ethanol biofuel via hydrolysis process [74]. The cellulose hydrolysis processes mainly consist of three types of hydrolysis, i.e., (i) dilute acid hydrolysis, (ii) concentrated acid hydrolysis, and (iii) an enzymatic hydrolysis. Conversion of cellulose biomass into biofuels via dilute acid-catalyzed hydrolysis process requires high temperature (160°C–230°C) and pressure (~10 atm) conditions. Yield obtained in this process is very low because dilute acid-catalyzed hydrolysis does not lead to complete conversion of cellulose into glucose. However, concentrated acid-catalyzed hydrolysis requires low temperature (50°C) and pressure conditions, higher retention time, and results in the 100% conversion of cellulose into glucose, which is further used to produce ethanol biofuel. Obtained quantity of bioethanol in concentrated acid-catalyzed process is much more than that in the dilute acid-catalyzed process [74]. Beside this acid-catalyzed conversion of biomass into biofuels, there are many other processes, out of which enzyme-catalyzed process is most promising and common. Retention time of this enzyme-catalyzed process is higher. The production of bioethanol from sugar-rich feedstock is a well-established conversion method than the starch-rich crops as extra step is necessary to hydrolyze starch to its sugar monosaccharide units (glucose), which often cost more energy and money. Furthermore, bioethanol produced from fermentation is normally purified and dewatered before being used instead of gasoline or combined at various percentages [75].

11.3.2 Pyrolysis

Pyrolysis is a thermochemical method, in which biomass is heated under a temperature range of 300°C–700°C and at atmospheric pressure to crack the polymer structure (large chain hydrocarbon into small), then further converted to biofuel without the need of oxygen [76]. Solid (char), an organic liquid (crude bio-oil), and pyrogas (having low calorific value) are the main products of pyrolysis. The product yield in pyrolysis process depends upon the factors, including feedstock type, rate of heating, vapor residence time, range of temperature, pressure, type of catalysts, and presence of hydrogen gas [77]. On the basis of these parameters, the pyrolysis process is further divided into slow and fast pyrolysis [76]. Slow pyrolysis is primarily used for char and described by its low reactor temperatures (<400°C), long residence time (60 minutes), and slow heating rate (45–50°C/min). However, fast pyrolysis is employed to maximize the liquid product yield and characterized by its high heating rate (120–127°C/min), high reactor temperature (>400°C), and short vapor residence time (3 seconds). Further, slow pyrolysis consists of a tubular furnace and fixed bed (in which materials were heated by internal electrical furnace) and nitrogen gas (carrier gas) blown to reactor from the top. Flow rate of nitrogen gas was controlled by flow meter, and the products were removed by nitrogen gas. However in case of fast pyrolysis, fluidized bed was used and the materials were heated by external electrical furnace. Also, the carrier gas is supplied to the reactor from the bottom [78]. Duman et al. studied the effect of reactor (fixed-bed and fluidized bed) and temperature on the composition of product produced by cherry seeds (CWS) and cherry seeds shells (CSS) using slow and fast pyrolysis. The results showed that the maximum bio-oil yield was 44 wt. % at a temperature of 500°C for both CWS and CSS in fast pyrolysis, whereas in slow pyrolysis, 21 wt.% and 15 wt. % yields were obtained at 500°C from CWS and CSS, respectively [79]. Karaosmanoglu and Gollu investigated the slow pyrolysis of straw and stalk of the rapeseed plant under different temperatures (350°C, 450°C, 550°C, and 650°C) and heating rates (10 and 30°C/min) [80]. The maximum product yield was obtained with 650°C temperature and 30°C/min heating rate. Miao and Wu discussed a method of fast pyrolysis to improve the yield of bio-oil by modifying the metabolism of microalgae through heterotrophic growth [81]. They found that the quantity of bio-oil produced by heterotopic cells was 3.4 times greater than that produced by autotropic cells.

11.3.3 Hydrothermal Liquefaction (HTL)

HTL is a process for the conversion of biomass into partially oxygenated liquid hydrocarbon fuel (bio-oil) at low temperature and high pressure (10–25 MPa) in the presence of catalyst and hydrogen [82]. HTL is similar to aqueous pyrolysis but more stable than pyrolysis because bio-oils produced by HTL have less content of moisture and oxygen. Biller and Ross studied that the product of hydrothermal process was solid (char), liquid, and gaseous biofuels (at temperature less than 200°C and pressure lower than 2 MPa using hydrothermal carbonization), at temperature range 280°C–370°C; and high pressure 10–12 MPa by hydrothermal liquefaction) and temperature above 375 and pressure 25–30 MPa using hydrothermal gasification, respectively [83]. Milledge et al. used the process of HTL for the conversion of seaweed biomass into crude bio-oil [84]. Because of its capacity to handle wet

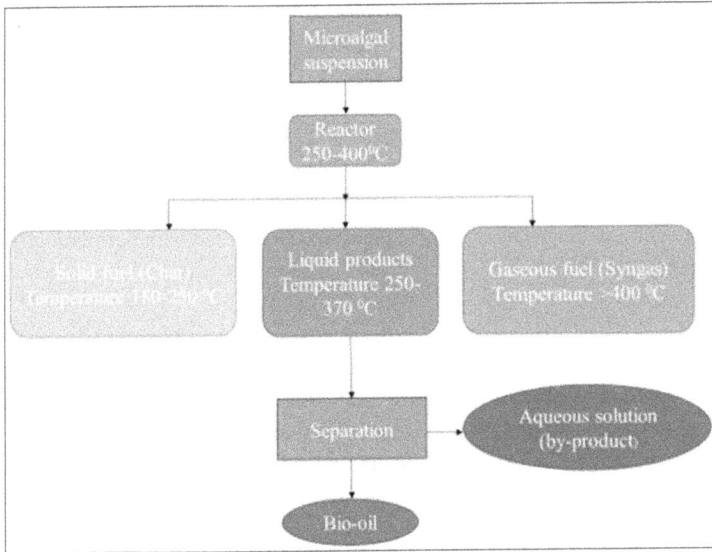

FIGURE 11.5 Hydrothermal liquefaction process for the production of biofuels.

biomass, HTL is one of the most stimulating techniques for the production of biofuel from algae [84]. Vlaskin et al. performed hydrothermal process using microalgae in the presence of moisture at a temperature of 100°C and above. Vlaskin et al. stated that the target product of hydrothermal process was various types of fuels at a definite temperature as shown in Figure 11.5 [85]. In that process, a reactor was used (autoclave), in which aqueous microalgae was suspended and the reactor was pressurized with nitrogen, then the reactor was further heated to the desired temperature. The reactor was allowed to cool after heating, the gases were removed, and the liquid and solid products were discharged from the reactor. The solid residue was separated from the liquid by decanting and bio-oil can be separated from the remaining liquid fraction with the help of solvents like hexane, acetone, dichloromethane, etc. The main advantages of microalgae HTL were that there is no need to pre-drying the feedstock, and confirms a comparative increase in the yield of product owing to all microalgae components [85].

Another advantage of HTL was that not only lipids but also carbohydrates and proteins can be directly converted into biofuel by thermochemical process as stated by Elliot et al. [86]. This provides an additional probability of regenerating the heat from the already-heated products to the feedstock, which enhances the energy efficiency of this processing technique [85]. As per literature, thermal treatments for the production of biofuel represent that profit-oriented interest in liquefaction is low due to the complex feed systems and higher prices as compared to pyrolysis and gasification.

11.3.4 ANAEROBIC DIGESTION

AD is gaining popularity as a solution to environmental concerns as well as an energy source for today's energy-intensive lifestyle [87]. Adekunle and Okolie stated that AD is a biochemical method in which decomposition of organic matter takes place by using specialized anaerobic bacteria for the production of biogas, containing methane (55%–75%) and CO_2 (25%–45%) [87]. The obtained yield of methane depends on microalgae species, algae biomass pretreatment, and inhibitors of methanogenesis. Jankowska et al. studied that AD consists of multiple steps such as hydrolysis, fermentation, acetogenesis, and methanogenesis [88]. Peter Weiland stated that hydrolyzing and fermenting microorganisms are responsible for the initial attack on polymers and monomers, producing primarily acetate and hydrogen as well as variable amounts of volatile

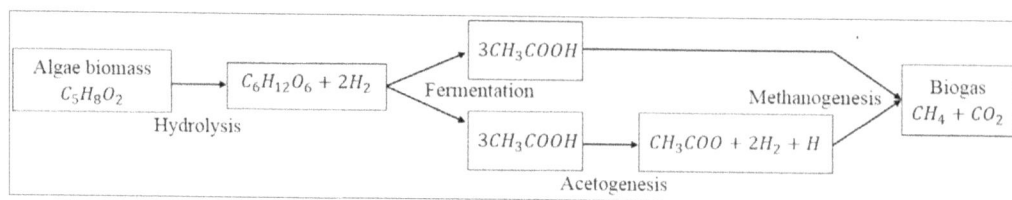

FIGURE 11.6   Production of biogas by anaerobic digestion. (Reprinted with permission from Adiniyi et al. [90]. Copyright 2018 Elsevier B. V.)

fatty acids like propionate and butyrate [89]. Adeniyi et al. highlighted all the pathways involved in AD conversion process as shown in Figure 11.6 [90]. Hydrolysis includes breaking of organic substrate of algae into soluble sugars, acids, ketones, and amino acids. The soluble sugar is then converted to CO_2 and hydrogen in second stage (fermentation) by acidogenic bacteria. In the next stage, acetogenesis demonstrates the oxidation of fermented compounds that are converted into acetate. In the final stage, methanogenesis, acetate is combined with CO_2 and hydrogen to form methane by a group of bacteria known as methanogens [90]. Ward et al. recorded that the production of biogas from microalgae biomass was lowest ~70 mL/g VS (volatile solids) for untreated *Microcystis sp* [91]. Key advantage of the microalgae AD is very high energy efficacy as compared to other biofuel due to the fact that there was no need for the extraction of oil and lipids, and the main product methane was captured [78]. Schamphelaire and Verstraete recorded a biogas production of 600 mL/g VS, which was the highest methane production for a mixed undetermined freshwater microalgae consortium [92]. AD has become an interesting alternative to produce biogas with low cost as compared to other techniques but still new research effort is needed on pretreatment processes in order to maximize biogas production [93].

11.3.5   GASIFICATION

The production of biofuels from sustainable resources is a most vital task in research, and gasification is one of the most promising thermochemical processes for the conversion of biomass to biofuels [94]. In this process, syngas was produced by using biomass, which mainly comprises carbon monoxide, hydrogen, CO_2, and methane with traces of some other elements [95]. Obtained syngas is further used to produce biofuels like biomethanol, bioethanol, biohydrogen, and synthetic natural gas (SNG) [94]. Fischer Tropsch (FT) is one of the techniques used for the conversion of syngas into methanol with a conversion rate of more than 99%, in the presence of an alkali metal catalyst. The FT process is a method capable for the manufacture of liquid hydrocarbon fuels from bio-syngas [96]. In FT process, methanol is obtained when syngas is fed to a methanol reactor, in the presence of alkali metal catalyst and water. Obtained methanol is not pure, and it contains many chemical species like alcohols, ketones, etc. that can be separated easily by direct distillation, and non-reacted syngas is sent back to the reactor. Molino et al. [94] studied the production of various types of biofuels (ethanol, methanol, biohydrogen) from syngas (as shown in Figure 11.7).

11.3.6   TRANSESTERIFICATION

Transesterification is a chemical process that converts triglycerides (fats) found in oils into usable biofuel. In this process, one alcohol is displaced by another alcohol from an ester so that this process is also called alcoholysis. Transesterification is similar to hydrolysis but the difference is that an alcohol is used in place of water [97]. This is frequently used method for lowering the viscosity of triglycerides. The transesterification is expressed by the following reaction:

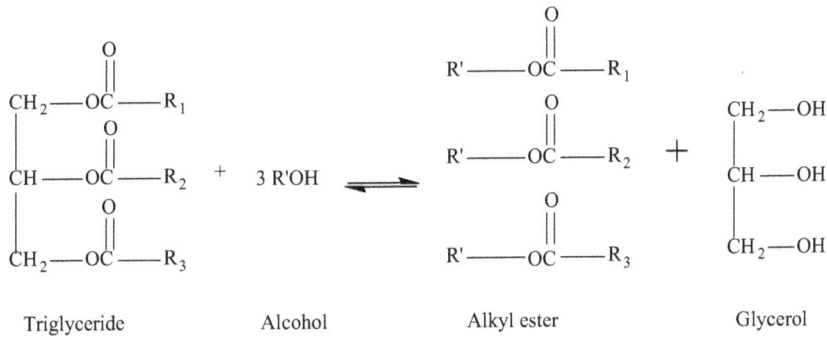

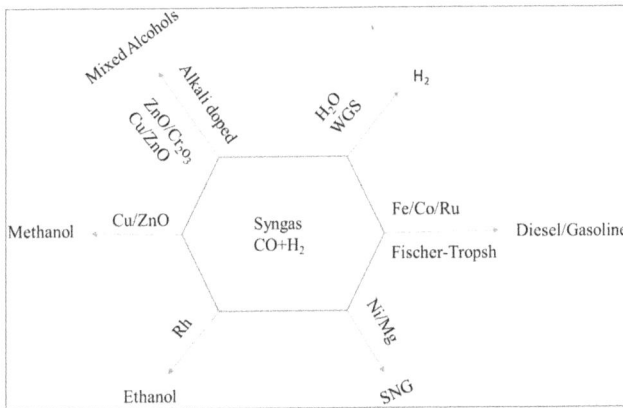

FIGURE 11.7 Gasification process to produce biofuels.

Transesterification, which involves the addition of a catalyst, can be used to make biodiesel from vegetable oils (methanol or ethanol and alkali such as sodium hydroxide, and acidic catalyst). Used cooking oil, oil crop residue and animal waste were used as feedstock in this process [75]. The key factors that affect the yield of the transesterification reaction were free fatty acid (FFA) and the moisture content, catalyst type, reaction time, molar ratio of methanol to oil, and reaction temperature as reported by Leung and Guo [98]. Akubude et al. stated that biodiesel is a liquid biofuel made up of monoalkyl esters (methyl or ethyl) of long-chain fatty acids derived from vegetable oils and animal fats [99].

Homogeneous acid catalysts like sulfuric acid and p-toluene sulfonic acid were used for the esterification of FFA to fatty acid methyl ester (FAME). But these acid catalysts are non-renewable, resulting in drawbacks such as increase in by-products, corrosion of apparatus, time-consuming process, and environmental issues. So, there is need to go search for those catalysts that are active, stable, biodegradable, and recyclable from the reaction mixture of transesterification reaction. ILs are catalysts and solvents that are regarded to be ecologically benign. Because of their variable with physicochemical properties like non-volatile nature, design possibility, extensive fluid range, good chemically and thermally stable, good recoverability, possible catalytic activity, and the ability to be easily separated from reactant products attracted the interest of researchers from academia and industries [100,101]. Lately, the flexibility of ILs is enhanced through the acidic functional group in which both the reagent and medium are attached. Out of all ILs catalysts, sulfonic acid group-functionalized ILs are extremely valuable, since they are very effective in a wide range of acidic-catalyzed reactions and have recently been employed in biodiesel productions [102]. The most effective way to increase their acidities is to increase the number of acidic sites in the structure of ILs. The whole scientific and industrial community is interested towards microwave irradiation, due to numerous advantages such as improvement in the rate of reaction, eco-friendly, decrease in reaction time and enhancement in the product yield and quality over conventional heating [103].

11.4 PURIFICATION OF BIOFUELS

Despite the advances in biofuel technology, they still include oxygenated chemicals, in particular phenolics, which results in negative consequences such as decreased engine performance and higher hazardous gas emissions [104]. Therefore, biofuel purification is a difficult problem that must be solved before it can be universally acknowledged as a viable substitute for fossil fuels. Biofuels have been purified using a variety of techniques, including

1. Distillation process
2. Membrane-based process
3. Liquid-liquid extraction process
4. Adsorption process

11.4.1 DISTILLATION PROCESS

Distillation is a method of separating chemical substances based on vapor pressure differences [105]. For separating and purifying biofuels, distillation is often regarded as the best approach. Bioethanol is one of the most widely utilized biofuels due to its high-quality octane booster and positive impact on the environment. The main issue in the processing of bioethanol production is the formation of homogeneous azeotrope of ethanol and water, which further necessitates to obtain high-purity ethanol [106]. There are different terms described in literature for the separation of pure ethanol, including azeotropic distillation (AD), extractive distillation (ED), adsorption, pervaporation, and pressure-swing distillation [107,108]. Among them, ED is most frequently used method for the production of bioethanol at large scale. Typically, ED is performed in two columns: first one is extractive distillation column (EDC), which distillates ethanol with desired purity, and then solvent recovery column (SRC), which recovers the solvent, cools and recycles back to EDC [109]. Li and Bai modified two-column process to a three-column process by introducing a concentrator column after the recovery column [110]. Moreover, a pre-concentrator was added before the extractive column by Errico et al. in their study [111]. Luo et al. proposed a new heat pump-assisted extractive distillation unit for the purification of bioethanol [112]. They observed that the specific energy requirement was reduced from 1.24 to 2.07 kW.h/kg (classic sequence) and led to overall energy saving up to 40%. Pacheco-Basulto et al. produced high-purity ethanol from azeotropic mixture (ethanol+water) using batch and semi-batch distillation with ethylene glycol and ILs as entrainers. The separation was recommended through ethylene glycol rather than IL due to its higher cost [113]. However, vinasse is the distillation by-product, which can be an issue for the environment, because 1 L of ethanol produces roughly 15 L of vinasse [114]. Also, the quantity of vinasse is greatly influenced by the rate of distillation and substrates used for biofuel production [115]. Therefore, there is a need to focus on new energy-efficient distillation processes.

11.4.2 MEMBRANE-BASED PROCESS

Today, membrane-based separation technology has achieved specific attention due to its high separation efficiency, low energy and operating expense, and absence of extra chemicals [116,117]. Therefore, it is economic and effective method, and can also be used to separate temperature-sensitive components [118]. Membrane creates a physical barrier in terms of shape, size, or properties, and subsequently enhances their selectivity towards specific components [119]. Reverse osmosis, microfiltration, nanofiltration, ultrafiltration, pervaporation, and membrane distillation are conventionally applied membrane techniques for bioethanol production [120,121]. Pervaporation is widely accepted membrane-based technique among all the available membrane techniques. They consist of a non-porous membrane, which separates the liquid mixture through partial vaporization [122]. Pervaporation offers many beneficial features like simplicity, flexibility, small floor area, high efficiency, low energy consumption, and lower cost compared to conventional separation technologies [123,124].

Mass transport through pervaporation technique generally follows the three steps: (i) components from feed side are selectively adsorbed onto the membrane, (ii) diffusion of sorbed molecules through the membrane, and (iii) desorption of the components as vapor phase into other side of membrane [72,125]. Polydimethylsiloxane, poly-1-(trimethylsilyl)-1-propyne, porous polytetrafluoroethylene, and polyether block amide membranes are the type of polymeric membranes that have been used to separate ethanol from the fermentation broth [126]. In addition to the forgoing, inorganic zeolite membranes (hydrophobic) have also been employed. Zhang and Liu prepared an A4-type zeolite membrane through secondary hydrothermal growth method and further tested their separation efficiency by using water/ethanol mixture [127]. Wu et al. modified MFI zeolite membranes by adopting dopamine as a modification material for the ethanol/water separation by pervaporation. This modification provides the long-term pervaporation stability by reducing or eliminating Si-OH from the zeolite, thereby avoiding the chances of chemical reaction taking place between Si-OH groups and components' membrane surface as shown in Figure 11.8 [128]. Kaewkannetra et al. analyzed the effect of different operating conditions (operating time, temperature, and ethanol concentration) on the permeate fluxes, selectivity, and percent separation of ethanol from ethanol/water binary mixture [129]. They observed a significant reduction in their selectivity with rising the concentration and temperature of ethanol. This may be attributed to swelling in membrane structures in high ethanol concentration. Marjani et al. synthesized and characterized polyamide carbon nanotube composite membrane

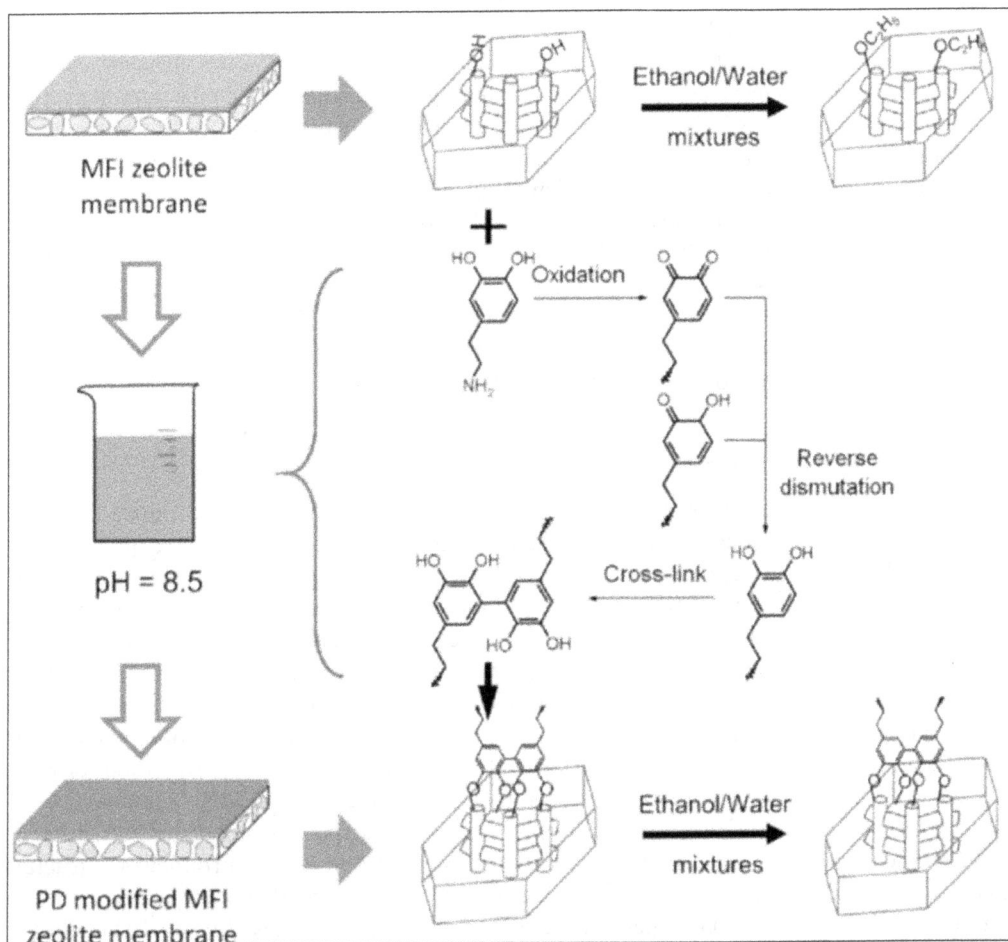

FIGURE 11.8 Modification of MFI zeolite membrane by removing surface Si-OH groups using dopamine. (Reprinted with permission from Wu et al. [128]. Copyright 2018 American Chemical Society.)

for the purification of alcohol, and results demonstrated that the composite membrane was best fitted to the mixtures that contain ethanol more than 50 wt. % [130].

11.4.3 LIQUID-LIQUID EXTRACTION PROCESS

Liquid-liquid extraction (LLE) has widely used for bioethanol separation from fermentation broth. LLE is a mass transfer process that involves the diffusion of a solute between two immiscible liquids, usually an organic liquid and an aqueous organic liquid. The solvent usually has a preference for one or more of the liquid mixture's constituents. When the two phases come into contact, the product of interest, which is usually an aqueous solution, begins to partition between the two phases. The separated material then establishes an equilibrium distribution between the immiscible solvents, allowing the two phases to be separated due to their immiscibility [131,132]. The selected solvent for this purpose should follow some standards, including reasonable chemical stability and extraction efficiency, water insolvability, non-toxicity, environmental friendliness, and affordability [72]. The most appealing classes of chemicals are ketones, esters, alcohols, ethers, and acetates because of their high distribution coefficients and low reactivity. The toxicity issues of several interesting solvents can be resolved through natural organic molecules like fatty acids, carboxylic acids, and β-alcohols [133]. As a result, numerous fatty acids were investigated as solvents for the extraction of ethanol from water. Valeric acid is a low-molecular-weight fatty acid, exhibited a maximum extraction efficiency for ethanol, but it also extracts water. Though, low-molecular-weight fatty acids were able to extract a small amount of bioethanol. The best solvents were octanoic and nonanoic acid; however, nonanoic acid was the most appropriate solvent since it evaporated least during flash distillation. For the same quantity of bioethanol, this process uses 38% less energy than fractional distillation [134]. Nowadays, the mixed solvent system is preferentially used in extraction process instead of single-component system. Habaki et al. employed mixed solvents for the extraction of ethanol from fermentation liquor, and maximum separation selectivity and distribution coefficient was observed with furfural + m-xylene mixture [135]. Further, Chafer et al. used two ILs (1-ethyl-3-methylimidazolium bis(trifluoromethylsulfonyl)imide ([emim][Tf$_2$N]), 1-hexyl-3-methylimidazolium bis(trifluoromethylsulfonyl)imide ([hmim][Tf$_2$N])) for the separation of azeotropic mixture comprising water and ethanol by LLE, but it is easier to extract ethanol by using [hmim][Tf$_2$N]. Also, it was found that temperature significantly affected the distribution coefficient as compared to cationic structure of ILs [136].

11.4.4 ADSORPTION PROCESS

Adsorption is a separation technique with a more efficient energy level as compared to other [137]. The phenomenon of adsorption relies on intrinsic characteristics (pore size, particle size, adsorption phase, and adsorption flow rate) of adsorbent, at which gases or solutes are adsorbed. The adsorbent is a stable crystal-like solid, and exhibits specific selectivity and absorption capacity for water and ethanol [138]. A series of materials can be used as adsorbents such as zeolites, polymeric resins, activated carbon molecular sieves, and activated carbon [120,137,139]. Among them, zeolites have received a lot of interest as adsorbent for the dehydration of ethanol. Zeolites are crystalline, hydrated aluminosilicate of groups IA and IIA (sodium, potassium, magnesium, and calcium), exhibited three-dimensional structures, which can be occupied by cations or water molecules to balance the electrostatic charge of tetrahedral units [140,141]. Yamamoto et al. examined the adsorption of water on zeolites having different sporous properties and observed that a zeolite exchanged with a monovalent cation species and possessed a framework structure having small Si/Al ratio exhibited the strongest affinity to water in ethanol [142]. Sudibandriyo and Putri analyzed the adsorption performance of two synthetic zeolites (3A and 4A) and found that adsorbent 3A had better adsorption capacity, longer saturation time, and improved purity, due to larger surface area and pore diameter of 3A [143]. The adsorption capacity is significantly affected by the operational time, temperature, and material of adsorbent as shown by Laksmono et al. in their research [106]. They tested polyvinyl alcohol, zeolite, and activated carbon as adsorbent materials and observed that activated carbon had shown higher adsorption capacity as compared to zeolite and polyvinyl alcohol, but zeolites had better selectivity. Jones et al. used activated carbon in two different separation modes as an adsorbent for ethanol dehydration [144]. In the first scheme, activated carbon was directly added to fermentation flask, whereas in the second scheme, externally located activated carbon columns were used. The second separation mode proved to be much more efficient in enhancing the production of ethanol [144]. The use of activated molecular sieving carbon (MSC) in adsorption process composes a relatively new area in surface chemistry. The pore diameters of MSC are comparable with the molecular diameter of adsorbates. Seo et al. used two activated MSC (MSC 4A has a pore volume of 0.24 cm^3/g and MSC 5A has a pore volume of 0.17 cm^3/g) for the dehydration of ethanol under various conditions of temperature and initial composition of ethanol [145]. It was confirmed that as compared MSC 4A, MSC 5A showed the better ethanol adsorption capacity [145]. Now, based on previous discussion, it is apparent that the most of research continue to employ two-component ethanol and water systems, ignoring the presence of other components in the fermentation broth. As a result, much studies are required to check the effect of other substances on adsorption, or to look into other methods that can be used in conjunction with adsorption to make purification and separation of ethanol easier.

11.5 APPLICATIONS OF BIOFUELS

In the past few years, the most significant environmental problems are the pollution caused by traffic and transport due to increasing use of petroleum fuels. There is no doubt that emissions from engine particularly from automobiles and industrial sectors have been related to serious environmental and human health consequences. Therefore, an alternate way to replace these conventional fuels like gasoline and diesel by biofuels is considered to reduce pollution and support sustainable cultivation [20,146]. Biofuels are one of the most promising fuels of sustainable energy with many applications. The majority of biofuel applications are centered towards industrial sectors and in automobiles [147]. Transportation and agriculture are major consumers of fossil fuels and major contributors to environmental pollution, which can be reduced by substituting bio-origin renewable fuels by replacing mineral-based fuels [148]. Bioethanol, butanol, and biodiesel are three liquid biofuels that have lately gained popularity. Bioethanol is environmentally favorable biofuel, which can mix with gasoline or can be used in its pure form for transportation. Bioethanol also enhances the combustion of gasoline in automobiles, thus lowering the emission of carcinogenic gases. On the other hand, butanol can be transported or distributed by using pipelines and filling station because it is less corrosive than ethanol. Biobutanol is mainly used as a solvent but it could replace gasoline and can be used in conventional automobiles according to experts [23]. Biodiesel is a synthetic fuel that can be used directly or mixed with petroleum diesel but requires some engine modifications [149]. Ahmad et al. stated that biohydrogen is a zero-carbon emission fuel that is considered as the cleanest fuel among all other fuels [150]. Hydrogen reacts with oxygen on combustion to give water. The water produced was radiant and the energy obtained lesser than that which was spent in production, so that it can be used in vehicles to replace gasoline. Internal engine combustion can generate power and electricity that can be used for beneficial purposes. Because hydrogen has the highest heating value, it can be used as a fuel in spacecraft propulsion [150]. To sum up, the application of biofuel in the automotive sector is still considered one of the most viable options to reduce the carbon emission of this sector significantly [147].

Kilic et al. conducted an experiment to investigate the characteristics of combustion by using different types of blends (diesel/butanol/biodiesel) in a flame tube boiler [151]. Mediavilla et al. investigated the application of broom plant clearing for the production of solid biofuels in addition to its use in industrial and commercial boilers [152]. Shemfe et al. looked at the economy of using biofuel in the electrical generation system [153]. This study discovered that electrical generation from biodiesel could be economically viable, sustainable, and profitable. But the industrial sector as well as the combined industrial and power generation sector contributed little share towards general use of biofuel, which accounts for 50% of global renewable energy use [154]. Furthermore, increased study and implementation of waste-derived biofuel in the industrial sector has the possible way to advance biofuel production. No doubt that biofuels have better environmental performance than that of fossil fuels but still some demerits like high production and supply cost make biofuels non-attractive [155].

11.6   SUMMARY

A review on the recent advancement used in the production of biofuels and their applications has been presented in the preceding sections. The growing demand for non-renewable energy has prompted scientists and analysts around the world to consider biofuel, as it is environmentally friendly, less polluting, non-toxic, and more appealing power source. Further, techniques are systematically evaluated based on the feedstocks of various generations. Also, the techniques and purification methods of biofuels have been summarized in this chapter. Nowadays, the fourth-generation biofuel production is in its initial phase of advancement to overcome the limitations associated with first-, second-, and third-generation biofuels. The major benefit of the fourth generation is to reduce the number of process steps. The long-term economic feasibility of fourth-generation biofuels appears to be superior, but still there are some technical risks that must be overcome in the establishment of supporting infrastructure. A better understanding of algal growth and metabolism is required for fourth-generation biofuels, which is not up to level that exists for other biofuel production methods. Also, an excellent energy balance is needed for the provision of cost-effective setup and biofuel separation technologies. Other barriers faced during the production of various generation biofuels are labor costs, lack of production infrastructure, increased demand for land and agricultural, competing food and fuel dilemma, water scarcity, especially in dry climates, which in turn disturbs the normal life cycle. Therefore, it is necessary to conduct further research in order to eradicate these bottlenecks of biofuel trades in different regions. Moreover, the viscous nature of biofuel can significantly affect performance of engines. Generally, low-viscous biofuel has maximum cetane number. Thus, it is obligatory to produce biofuels with satisfactory and constant quality by monitoring fuel quality standards during the biofuel production processes. Also, keep in mind that regardless of how many novel fuels have been produced, in order to make difference in the fuel market, technologies must be economically viable. Therefore, a cost-effective method for producing and processing biofuel should be investigated. Thus, we can conclude that biofuel with advanced techniques can easily eliminate fossil fuel dependency in the future and become a renewable and safe source of energy.

REFERENCES

[1] P.T. Sekoai, C.N.M. Ouma, S.P. du Preez, P. Modisha, N. Engelbrecht, D.G. Bessarabov, A. Ghimire, Application of nanoparticles in biofuels: An overview, *Fuel*. 237 (2019) 380–397. https://doi.org/10.1016/j.fuel.2018.10.030.

[2] J.L. Stephen, B. Periyasamy, Innovative developments in biofuels production from organic waste materials: A review, *Fuel*. 214 (2018) 623–633. https://doi.org/10.1016/j.fuel.2017.11.042.

[3] S. Vasistha, A. Khanra, M. Clifford, M.P. Rai, Current advances in microalgae harvesting and lipid extraction processes for improved biodiesel production: A review, *Renew. Sustain. Energy Rev*. 137 (2021) 110498. https://doi.org/10.1016/j.rser.2020.110498.

[4] R. Shan, L. Lu, Y. Shi, H. Yuan, J. Shi, Catalysts from renewable resources for biodiesel production, *Energy Convers. Manag*. 178 (2018) 277–289. https://doi.org/10.1016/j.enconman.2018.10.032.

[5] Z. Ullah, M.A. Bustam, Z. Man, Biodiesel production from waste cooking oil by acidic ionic liquid as a catalyst, *Renew. Energy*. 77 (2015) 521–526. https://doi.org/10.1016/j.renene.2014.12.040.

[6] M.V. Rodionova, R.S. Poudyal, I. Tiwari, R.A. Voloshin, S.K. Zharmukhamedov, H.G. Nam, B.K. Zayadan, B.D. Bruce, H.J.M. Hou, S.I. Allakhverdiev, Biofuel production: Challenges and opportunities, *Int. J. Hydrogen Energy*. 42 (2017) 8450–8461. https://doi.org/10.1016/j.ijhydene.2016.11.125.

[7] M.N. Nabi, M.S. Akhter, M.M.Z. Shahadat, Improvement of engine emissions with conventional diesel fuel and diesel-biodiesel blends, *Bioresour. Technol.* 97 (2006) 372–378. https://doi.org/10.1016/j.biortech.2005.03.013.

[8] M. Abdul Hakim Shaah, M.S. Hossain, F.A. Salem Allafi, A. Alsaedi, N. Ismail, M.O. Ab Kadir, M.I. Ahmad, A review on non-edible oil as a potential feedstock for biodiesel: Physicochemical properties and production technologies, *RSC Adv.* 11 (2021) 25018–25037. https://doi.org/10.1039/d1ra04311k.

[9] D.H. Qi, H. Chen, L.M. Geng, Y.Z. Bian, Experimental studies on the combustion characteristics and performance of a direct injection engine fueled with biodiesel/diesel blends, *Energy Convers. Manag.* 51 (2010) 2985–2992. https://doi.org/10.1016/j.enconman.2010.06.042.

[10] P.V Naikwade, R.P. Bansode, S.T. Sankpal, Biofuels: Potential, current issues and future trends, *J. Today's Biol. Sci. Res. Rev.* 1 (2012) 186–198.

[11] G. Francis, R. Edinger, K. Becker, A concept for simultaneous wasteland reclamation, fuel production, and socio-economic development in degraded areas in India: Need, potential and perspectives of Jatropha plantations, *Nat. Resour. Forum.* 29 (2005) 12–24. https://doi.org/10.1111/j.1477-8947.2005.00109.x.

[12] B.K. Highina, I.M. Bugaje, B. State, B. Umar, B. State, A review on second generation biofuel: A comparison of its carbon footprints, *Eur. J. Eng. Technol.* 2 (2014) 117–125.

[13] M.A. Carriquiry, X. Du, G.R. Timilsina, Second generation biofuels: Economics and policies, *Energy Policy.* 39 (2011) 4222–4234. https://doi.org/10.1016/j.enpol.2011.04.036.

[14] A. Singh, S.I. Olsen, P.S. Nigam, A viable technology to generate third-generation biofuel, *J. Chem. Technol. Biotechnol.* 86 (2011) 1349–1353. https://doi.org/10.1002/jctb.2666.

[15] G. Dragone, B. Fernandes, A.A. Vicente, J.A. Teixeira, Third generation biofuels from microalgae, (2010) pp. 1355–1366. http://hdl.handle.net/1822/16807.

[16] D. Singh, D. Sharma, S.L. Soni, S. Sharma, P. Kumar Sharma, A. Jhalani, A review on feedstocks, production processes, and yield for different generations of biodiesel, *Fuel.* 262 (2020) 116553. https://doi.org/10.1016/j.fuel.2019.116553.

[17] H.L. Chum, F.E.B. Nigro, R. Mccormick, G.T. Beckham, J.E.A. Seabera, J. Saddler, L. Tao, E. Warner, R. Overend, Conversion technologies for biofuels and their use, *Bioenergy Sustain. Bridg. Gaps.* (2015) 374–467. http://bioenfapesp.org/scopebioenergy/index.php/project-overview.

[18] L.M. González-González, D.F. Correa, S. Ryan, P.D. Jensen, S. Pratt, P.M. Schenk, Integrated biodiesel and biogas production from microalgae: Towards a sustainable closed loop through nutrient recycling, *Renew. Sustain. Energy Rev.* 82 (2018) 1137–1148. https://doi.org/10.1016/j.rser.2017.09.091.

[19] B. Abdullah, S.A.F. ad Syed Muhammad, Z. Shokravi, S. Ismail, K.A. Kassim, A.N. Mahmood, M.M.A. Aziz, Fourth generation biofuel: A review on risks and mitigation strategies, *Renew. Sustain. Energy Rev.* 107 (2019) 37–50. https://doi.org/10.1016/j.rser.2019.02.018.

[20] A. Datta, A. Hossain, S. Roy, An overview on biofuels and their advantages and disadvantages, *Asian J. Chem.* 8 (2019) 1851–1858. https://doi.org/10.14233/ajchem.2019.22098.

[21] E.M. Aro, From first generation biofuels to advanced solar biofuels, *Ambio.* 45 (2016) 24–31. https://doi.org/10.1007/s13280-015-0730-0.

[22] S.N. Naik, V.V. Goud, P.K. Rout, A.K. Dalai, Production of first and second generation biofuels : A comprehensive review, *Renew. Sust. Energ. Rev.* 14 (2010) 578–597. https://doi.org/10.1016/j.rser.2009.10.003.

[23] P.S. Nigam, A. Singh, Production of liquid biofuels from renewable resources, *Prog. Energy Combust. Sci.* 37 (2011) 52–68. https://doi.org/10.1016/j.pecs.2010.01.003.

[24] J. Popp, Z. Lakner, M. Harangi-Rákos, M. Fári, The effect of bioenergy expansion: Food, energy, and environment, *Renew. Sustain. Energy Rev.* 32 (2014) 559–578. https://doi.org/10.1016/j.rser.2014.01.056.

[25] Z. Xu, F. Huang, Pretreatment methods for bioethanol production, *Appl. Biochem. Biotechnol.* 174 (2014) 43–62. https://doi.org/10.1007/s12010-014-1015-y.

[26] A. Demirbaş, Bioethanol from cellulosic materials: A renewable motor fuel from biomass, *Energy Sources.* 27 (2005) 327–337. https://doi.org/10.1080/00908310390266643.

[27] M.R. Javed, M. Noman, M. Shahid, T. Ahmed, M. Khurshid, M.H. Rashid, M. Ismail, M. Sadaf, F. Khan, Current situation of biofuel production and its enhancement by CRISPR/Cas9-mediated genome engineering of microbial cells, *Microbiol. Res.* 219 (2019) 1–11. https://doi.org/10.1016/j.micres.2018.10.010.

[28] P. Li, K. Sakuragi, H. Makino, Extraction techniques in sustainable biofuel production: A concise review, *Fuel Process. Technol.* 193 (2019) 295–303. https://doi.org/10.1016/j.fuproc.2019.05.009.

[29] A. del P. Sánchez-Camargo, M. Bueno, F. Parada-Alfonso, A. Cifuentes, E. Ibáñez, Hansen solubility parameters for selection of green extraction solvents, *TrAC - Trends Anal. Chem.* 118 (2019) 227–237. https://doi.org/10.1016/j.trac.2019.05.046.

[30] L.M. Khan, M.A. Hanna, Expression of oil from oilseeds-A review, *J. Agric. Eng. Res.* 28 (1983) 495–503. https://doi.org/10.1016/0021-8634(83)90113-0.

[31] V. Patil, K.Q. Tran, H.R. Giselrød, Towards sustainable production of biofuels from microalgae, *Int. J. Mol. Sci.* 9 (2008) 1188–1195. https://doi.org/10.3390/ijms9071188.

[32] W.H. Liew, M.H. Hassim, D.K.S. Ng, Review of evolution, technology and sustainability assessments of biofuel production, *J. Clean. Prod.* 71 (2014) 11–29. https://doi.org/10.1016/j.jclepro.2014.01.006.

[33] Q. Nguyen, J. Bowyer, J. Howe, S. Bratkovich, H. Groot, E. Pepke, K. Fernholz, Global production of second-generation biofuels: Trends and influences. *Biofuels/Biorefinery Dev. Rep. Card.* (2017) 1–15. http://www.annualreviews.org/doi/10.1146/annurev-environ-101813-013253.

[34] A. Kumar, S. Sharma, Potential non-edible oil resources as biodiesel feedstock: An Indian perspective, *Renew. Sustain. Energy Rev.* 15 (2011) 1791–1800. https://doi.org/10.1016/j.rser.2010.11.020.

[35] A. Mohr, S. Raman, Lessons from first generation biofuels and implications for the sustainability appraisal of second generation biofuels, *Effic. Sustain. Biofuel. Prod. Environ. Land-Use Res.* 63 (2013) 114–122. https://doi.org/10.1016/j.enpol.2013.08.033.

[36] S. Anto, S.S. Mukherjee, R. Muthappa, T. Mathimani, G. Deviram, S.S. Kumar, T.N. Verma, A. Pugazhendhi, Algae as green energy reserve: Technological outlook on biofuel production, *Chemosphere.* 242 (2020) 125079 https://doi.org/10.1016/j.chemosphere.2019.125079.

[37] V.K. Sharma, G. Braccio, P.B.L. Chaurasia, V. Lomonaco, Innovation for production of next generation biofuel from lignocellulosic wastes, *Innovation.* 1 (2018) 20–44.

[38] B.O. Abo, M. Gao, Y. Wang, C. Wu, H. Ma, Q. Wang, Lignocellulosic biomass for bioethanol: An overview on pretreatment, hydrolysis and fermentation processes, *Rev. Environ. Health.* 34 (2019) 57–68. https://doi.org/10.1515/reveh-2018-0054.

[39] A. Demirbas, Products from lignocellulosic materials via degradation processes, *Energy Sources A Recover. Util. Environ. Eff.* 30 (2008) 27–37. https://doi.org/10.1080/00908310600626705.

[40] K. Ramluckan, K.G. Moodley, F. Bux, An evaluation of the efficacy of using selected solvents for the extraction of lipids from algal biomass by the soxhlet extraction method, *Fuel.* 116 (2014) 103–108. https://doi.org/10.1016/j.fuel.2013.07.118.

[41] W. Roschat, T. Siritanon, B. Yoosuk, T. Sudyoadsuk, V. Promarak, Rubber seed oil as potential non-edible feedstock for biodiesel production using heterogeneous catalyst in Thailand, *Renew. Energy.* 101 (2017) 937–944. https://doi.org/10.1016/j.renene.2016.09.057.

[42] E.G. Bligh, W.J. Dyer, A rapid method of total lipid extraction and purification, *Can. J. Biochem. Physiol.* 37 (1959) 911–917. https://doi.org/10.1139/o59-099.

[43] W. Herchi, H. Kallel, S. Boukhchina, Physicochemical properties and antioxidant activity of Tunisian date palm (*Phoenix dactylifera* L.) oil as affected by different extraction methods, *Food Sci. Technol.* 34 (2014) 464–470. http://dx.doi.org/10.1590/1678-457X.6360.

[44] J.L. Luque-García, M.D. Luque De Castro, Focused microwave-assisted Soxhlet extraction: Devices and applications, *Talanta.* 64 (2004) 571–577. https://doi.org/10.1016/j.talanta.2004.03.054.

[45] N. Rajendran, B. Gurunathan, I.A.E. Selvakumari, Optimization and technoeconomic analysis of biooil extraction from *Calophyllum inophyllum* L. seeds by ultrasonic assisted solvent oil extraction, *Ind. Crop. Prod.* 162 (2021) 113273. https://doi.org/10.1016/j.indcrop.2021.113273.

[46] N. Pragya, K.K. Pandey, P.K. Sahoo, A review on harvesting, oil extraction and biofuels production technologies from microalgae, *Renew. Sustain. Energy Rev.* 24 (2013) 159–171. https://doi.org/10.1016/j.rser.2013.03.034.

[47] R.M. Couto, J. Fernandes, M.D.R.G. da Silva, P.C. Simões, Supercritical fluid extraction of lipids from spent coffee grounds, *J. Supercrit. Fluids.* 51 (2009) 159–166. https://doi.org/10.1016/j.supflu.2009.09.009.

[48] A. Depeursinge, D. Racoceanu, J. Iavindrasana, G. Cohen, A. Platon, P.-A. Poletti, H. Muller, Fusing visual and clinical information for lung tissue classification in HRCT data, *Artif. Intell. Med.* 50 (2010) 13–21. https://doi.org/10.1016/j.artmed.2010.04.006.

[49] L.D. Simoni, A. Chapeaux, J.F. Brennecke, M.A. Stadtherr, Extraction of biofuels and biofeedstocks from aqueous solutions using ionic liquids, *Comput. Chem. Eng.* 34 (2010) 1406–1412. https://doi.org/10.1016/j.compchemeng.2010.02.020.

[50] P. Li, H. Kanda, H. Makino, Simultaneous production of bio-solid fuel and bio-crude from vegetal biomass using liquefied dimethyl ether, *Fuel.* 116 (2014) 370–376. https://doi.org/10.1016/j.fuel.2013.08.020.

[51] P.G. Jessop, S.M. Mercer, D.J. Heldebrant, CO_2-triggered switchable solvents, surfactants, and other materials, *Energy Environ. Sci.* 5 (2012) 7240–7253. https://doi.org/10.1039/c2ee02912j.

[52] L. Phan, H. Brown, J. White, A. Hodgson, P.G. Jessop, Soybean oil extraction and separation using switchable or expanded solvents, *Green Chem.* 11 (2009) 53–59. https://doi.org/10.1039/b810423a.

[53] S.Y. Lee, R. Sankaran, K.W. Chew, C.H. Tan, R. Krishnamoorthy, D.-T. Chu, P.-L. Show, Waste to bioenergy: A review on the recent conversion technologies, *BMC Energy.* 1 (2019) 1–22. https://doi.org/10.1186/s42500-019-0004-7.

[54] R.D. Long, E. Abdelkader, Mixed-polarity azeotropic solvents for efficient extraction of lipids from nannochloropsis microalgae, *Am. J. Biochem. Biotechnol.* 7 (2011) 70–73. https://doi.org/10.3844/ajbbsp.2011.70.73.

[55] H. Shin, S. Shim, Y. Ryu, J. Yang, S. Lim, C. Lee, Lipid extraction from *Tetraselmis* sp. microalgae for biodiesel production using hexane-based solvent mixtures, *Biotechnol. Bioprocess Eng.* 23 (2018) 16–22. https://doi.org/10.1007/s12257-017-0392-9.

[56] M. Gong, A. Bassi, Carotenoids from microalgae: A review of recent developments, *Biotechnol. Adv.* 34 (2016) 1396–1412. https://doi.org/10.1016/j.biotechadv.2016.10.005.

[57] F. Adam, M. Abert-Vian, G. Peltier, F. Chemat, "Solvent-free" ultrasound-assisted extraction of lipids from fresh microalgae cells: A green, clean and scalable process, *Bioresour. Technol.* 114 (2012) 457–465. https://doi.org/10.1016/j.biortech.2012.02.096.

[58] X. Zhang, S. Yan, R.D. Tyagi, P. Drogui, R.Y. Surampalli, Ultrasonication assisted lipid extraction from oleaginous microorganisms, *Bioresour. Technol.* 158 (2014) 253–261. https://doi.org/10.1016/j.biortech.2014.01.132.

[59] J.Y. Lee, C. Yoo, S.Y. Jun, C.Y. Ahn, H.M. Oh, Comparison of several methods for effective lipid extraction from microalgae, *Bioresour. Technol.* 101 (2010) S75–S77. https://doi.org/10.1016/j.biortech.2009.03.058.

[60] X. Zhou, W. Jin, R. Tu, Q. Guo, S. Fang Han, C. Chen, Q. Wang, W. Liu, P.D. Jensen, Q. Wang, Optimization of microwave assisted lipid extraction from microalga *Scenedesmus obliquus* grown on municipal wastewater, *J. Clean. Prod.* 221 (2019) 502–508. https://doi.org/10.1016/j.jclepro.2019.02.260.

[61] L. Soh, J. Zimmerman, Biodiesel production: The potential of algal lipids extracted with supercritical carbon dioxide, *Green Chem.* 13 (2011) 1422–1429. https://doi.org/10.1039/c1gc15068e.

[62] P.D. Patil, K. Phani, R. Dandamudi, J. Wang, Q. Deng, S. Deng, Extraction of bio-oils from algae with supercritical carbon dioxide and co-solvents, *J. Supercrit. Fluids.* 135 (2018) 60–68 https://doi.org/10.1016/j.supflu.2017.12.019.

[63] P.R. Postma, T.L. Miron, G. Olivieri, M.J. Barbosa, R.H. Wijffels, M.H.M. Eppink, Mild disintegration of the green microalgae *Chlorella vulgaris* using bead milling, *Bioresour. Technol.* 184 (2015) 297–304. https://doi.org/10.1016/j.biortech.2014.09.033.

[64] M.F. Kamaroddin, J. Hanotu, D.J. Gilmour, W.B. Zimmerman, In-situ disinfection and a new downstream processing scheme from algal harvesting to lipid extraction using ozone-rich microbubbles for biofuel production, *Algal Res.* 17 (2016) 217–226. https://doi.org/10.1016/j.algal.2016.05.006.

[65] Y. Du, B. Schuur, S.R.A. Kersten, D.W.F. Brilman, Opportunities for switchable solvents for lipid extraction from wet algal biomass: An energy evaluation, *Algal Res.* 11 (2015) 271–283. https://doi.org/10.1016/j.algal.2015.07.004.

[66] W. Tang, K. Ho Row, Evaluation of CO_2-induced azole-based switchable ionic liquid with hydrophobic/hydrophilic reversible transition as single solvent system for coupling lipid extraction and separation from wet microalgae, *Bioresour. Technol.* 296 (2020) 122309. https://doi.org/10.1016/j.biortech.2019.122309.

[67] M. Goto, H. Kanda, S. Machmudah, Extraction of carotenoids and lipids from algae by supercritical CO_2 and subcritical dimethyl ether, *J. Supercrit. Fluids.* 96 (2015) 245–251. https://doi.org/10.1016/j.supflu.2014.10.003.

[68] K. Yoshikawa, Y. Toya, H. Shimizu, Metabolic engineering of *Synechocystis* sp. PCC 6803 for enhanced ethanol production based on flux balance analysis, *Bioprocess Biosyst. Eng.* 40 (2017) 791–796. https://doi.org/10.1007/s00449-017-1744-8.

[69] O. Kruse, J. Rupprecht, J.H. Mussgnug, G.C. Dismukes, B. Hankamer, Photosynthesis: A blueprint for solar energy capture and biohydrogen production technologies, *Photochem. Photobiol. Sci.* 4 (2005) 957–970. https://doi.org/10.1039/b506923h.

[70] M. De Deng, J.R. Coleman, Ethanol synthesis by genetic engineering in cyanobacteria, *Appl. Environ. Microbiol.* 65 (1999) 523–528. https://doi.org/10.1128/aem.65.2.523-528.1999.

[71] S.J. Malode, K.K. Prabhu, R.J. Mascarenhas, N.P. Shetti, T.M. Aminabhavi, Recent advances and viability in biofuel production, *Energy Convers. Manag. X.* 10 (2021) 100070. https://doi.org/10.1016/j.ecmx.2020.100070.

[72] A. Bušić, N. Mardetko, S. Kundas, G. Morzak, H. Belskaya, M.I. Šantek, D. Komes, S. Novak, B. Šantek, Bioethanol production from renewable raw materials and its separation and purification: A review, *Food Technol. Biotechnol.* 56 (2018) 289–311. https://doi.org/10.17113/ftb.56.03.18.5546.

[73] F. Talebnia, D. Karakashev, I. Angelidaki, Production of bioethanol from wheat straw: An overview on pretreatment, hydrolysis and fermentation, *Bioresour. Technol.* 101 (2010) 4744–4753. https://doi.org/10.1016/j.biortech.2009.11.080.

[74] P. Kumar, D.M. Barrett, M.J. Delwiche, P. Stroeve, Methods for pretreatment of lignocellulosic biomass for efficient hydrolysis and biofuel production, *Ind. Eng. Chem. Res.* 48 (2009) 3713–3729. https://doi.org/10.1021/ie801542g

[75] W. Stafford, A. Lotter, A. Brent, G. Von Maltitz, WIDER Working Paper 2017/87 Biofuels technology A look forward, (2017). https:// doi:10.35188/UNU-WIDER/2017/311-0.

[76] A.T. Sipra, N. Gao, H. Sarwar, Municipal solid waste (MSW) pyrolysis for bio-fuel production: A review of effects of MSW components and catalysts, *Fuel Process. Technol.* 175 (2018) 131–147. https://doi.org/10.1016/j.fuproc.2018.02.012.

[77] D. Chiaramonti, M. Prussi, M. Buffi, A.M. Rizzo, L. Pari, Review and experimental study on pyrolysis and hydrothermal liquefaction of microalgae for biofuel production, *Appl. Energy.* 185 (2017) 963–972. https://doi.org/10.1016/j.apenergy.2015.12.001.

[78] K. Azizi, M. K.Moraveji, H. A. Najafabadi, A review on bio-fuel production from microalgal biomass by using pyrolysis method, *Renew. Sustain. Energy Rev.* 82 (2018) 3046–3059. https://doi.org/10.1016/j.rser.2017.10.033.

[79] G. Duman, C. Okutucu, S. Ucar, R. Stahl, J. Yanik, The slow and fast pyrolysis of cherry seed, *Bioresour. Technol.* 102 (2011) 1869–1878. https://doi.org/10.1016/j.biortech.2010.07.051.

[80] F. Karaosmanoglu, E. Tetik, E. Göllü, Biofuel production using slow pyrolysis of the straw and stalk of the rapeseed plant, *Fuel Process. Technol.* 59 (1999) 1–12. https://doi.org/10.1016/S0378-3820(99)00004-1.

[81] X. Miao, Q. Wu, High yield bio-oil production from fast pyrolysis by metabolic controlling of *Chlorella protothecoides*, *J. Biotechnol.* 110 (2004) 85–93. https://doi.org/10.1016/j.jbiotec.2004.01.013.

[82] S. Nanda, F. Pattnaik, V. B. Borugadda, A. K. Dalai, J.A. Kozinski, S. Naik, Catalytic and noncatalytic upgrading of bio - oil to synthetic Fuels : An introductory review, *ACS Symp. Ser.* 1379 (2021) 1–28. https ://doi.org/10.1021/bk-2021-1379.ch001.

[83] P. Biller, A.B. Ross, Hydrothermal processing of algal biomass for the production of biofuels and chemicals, *Biofuel Res. J.* 3 (2012) 603–623. https://doi.org/10.4155/bfs.12.42.

[84] J.J. Milledge, B. Smith, P.W. Dyer, P. Harvey, Macroalgae-derived biofuel: A review of methods of energy extraction from seaweed biomass, *Energies.* 7 (2014) 7194–7222. https://doi.org/10.3390/en7117194.

[85] M.S. Vlaskin, N.I. Chernova, S. V Kiseleva, O.S. Popel, A.Z. Zhuk, Hydrothermal liquefaction of microalgae to produce biofuels : State of the art and future prospects, *Therm. Eng.* 64 (2017) 627–636. https://doi.org/10.1134/S0040601517090105.

[86] D.C. Elliott, Review of recent reports on process technology for thermochemical conversion of whole algae to liquid fuels, *Algal Res.* 13 (2016) 255–263. https://doi.org/10.1016/j.algal.2015.12.002.

[87] K.F. Adekunle, J.A. Okolie, A review of biochemical process of anaerobic digestion, *Adv. Biosci. Biotechnol.* 06 (2015) 205–212. https://doi.org/10.4236/abb.2015.63020.

[88] E. Jankowska, A.K. Sahu, P. Oleskowicz-Popiel, Biogas from microalgae: Review on microalgae's cultivation, harvesting and pretreatment for anaerobic digestion, *Renew. Sustain. Energy Rev.* 75 (2017) 692–709. https://doi.org/10.1016/j.rser.2016.11.045.

[89] P. Weiland, Biogas production: Current state and perspectives, *Appl. Microbiol. Biotechnol.* 85 (2010) 849–860. https://doi.org/10.1007/s00253-009-2246-7.

[90] O.M. Adeniyi, U. Azimov, A. Burluka, Algae biofuel: Current status and future applications, *Renew. Sustain. Energy Rev.* 90 (2018) 316–335. https://doi.org/10.1016/j.rser.2018.03.067.

[91] A.J. Ward, D.M. Lewis, F.B. Green, anaerobic digestion of algae biomass: A review, *Algal Res.* 5 (2014) 204–214. https://doi.org/10.1016/j.algal.2014.02.001.

[92] L. De Schamphelaire, W. Verstraete, Revival of the biological sunlight-to-biogas energy conversion system, *Biotechnol. Bioeng.* 103 (2009) 296–304. https://doi.org/10.1002/bit.22257.

[93] I.S. Horváth, M. Tabatabaei, K. Karimi, R. Kumar, Recent updates on biogas production - A review, *Biofuel Res. J.* 3 (2016) 394–402. https://doi.org/10.18331/BRJ2016.3.2.4.

[94] A. Molino, V. Larocca, S. Chianese, D. Musmarra, Biofuels production by biomass gasification: A review, *Energies.* 11 (2018) 1–31. https://doi.org/10.3390/en11040811.

[95] V.S. Sikarwar, M. Zhao, P.S. Fennell, N. Shah, E.J. Anthony, Progress in biofuel production from gasification, *Prog. Energy Combust. Sci.* 61 (2017) 189–248. https://doi.org/10.1016/j.pecs.2017.04.001.

[96] A. Demirbas, Progress and recent trends in biofuels, *Prog. Energy Combust. Sci.* 33 (2007) 1–18. https://doi.org/10.1016/j.pecs.2006.06.001.

[97] Z.M. Hasib, J. Hossain, S. Biswas, A. Islam, Bio-diesel from mustard oil: A renewable alternative fuel for small diesel engines, *Mod. Mech. Eng.* 1 (2011) 77–83. https://doi.org/10.4236/mme.2011.12010.

[98] D.Y.C. Leung, Y. Guo, Transesterification of neat and used frying oil: Optimization for biodiesel production, *Fuel Process. Technol.* 87 (2006) 883–890. https://doi.org/10.1016/j.fuproc.2006.06.003.

[99] V.C. Akubude, K.N. Nwaigwe, E. Dintwa, Production of biodiesel from microalgae via nanocatalyzed transesterification process: A review, *Mater. Sci. Energy Technol.* 2 (2019) 216–225. https://doi.org/10.1016/j.mset.2018.12.006.

[100] R.L. Vekariya, A review of ionic liquids: Applications towards catalytic organic transformations, *J. Mol. Liq.* 227 (2017) 44–60. https://doi.org/10.1016/j.molliq.2016.11.123.

[101] N. Nasirpour, M. Mohammadpourfard, S. Z. Heris, Ionic liquids: Promising compounds for sustainable chemical processes and applications, *Chem. Eng. Res. Des.* 160 (2020) 264–300. https://doi.org/10.1016/j.cherd.2020.06.006.

[102] B. Karimi, H.M. Mirzaei, A. Mobaraki, Periodic mesoporous organosilica functionalized sulfonic acids as highly efficient and recyclable catalysts in biodiesel production, *Catal. Sci. Technol.* 2 (2012) 828–834. https://doi.org/10.1039/c2cy00444e.

[103] A. Breccia, B. Esposito, G. Breccia Fratadocchi, A. Fini, Reaction between methanol and commercial seed oils under microwave irradiation, *J. Microw. Power Electromagn. Energy.* 34 (1999) 3–8. https://doi.org/10.1080/08327823.1999.11688383.

[104] D.B. Kaymak, Design and control of an alternative bioethanol purification process via reactive distillation from fermentation broth, *Ind. Eng. Chem. Res.* 58 (2019) 1675–1685. https://doi.org/10.1021/acs.iecr.8b04832.

[105] R.I. Canales, J.F. Brennecke, Comparison of ionic liquids to conventional organic solvents for extraction of aromatics from aliphatics, *J. Chem. Eng. Data.* 61 (2016) 1685–1699. https://doi.org/10.1021/acs.jced.6b00077.

[106] J.A. Laksmono, U.A. Pangesti, M. Sudibandriyo, A. Haryono, A.H. Saputra, Adsorption capacity study of ethanol-water mixture for zeolite, activated carbon, and polyvinyl alcohol, *IOP Conf. Ser. Earth Environ. Sci.* 105 (2018) 012025. https://doi.org/10.1088/1755-1315/105/1/012025.

[107] A.A. Kiss, D.J.P.C. Suszwalak, Enhanced bioethanol dehydration by extractive and azeotropic distillation in dividing-wall columns, *Sep. Purif. Technol.* 86 (2012) 70–78. https://doi.org/10.1016/j.seppur.2011.10.022.

[108] A.A. Kiss, R.M. Ignat, Innovative single step bioethanol dehydration in an extractive dividing-wall column, *Sep. Purif. Technol.* 98 (2012) 290–297. https://doi.org/10.1016/j.seppur.2012.06.029.

[109] G.W. Meindersma, E. Quijada-Maldonado, T.A.M. Aelmans, J.P.G. Hernandez, A.B. De Haan, Ionic liquids in extractive distillation of ethanol/water: From laboratory to pilot plant, *ACS Symp. Ser.* 1117 (2012) 239–257. https://doi.org/10.1021/bk-2012-1117.ch011.

[110] G. Li, P. Bai, New operation strategy for separation of ethanol-water by extractive distillation, *Ind. Eng. Chem. Res.* 51 (2012) 2723–2729. https://doi.org/10.1021/ie2026579.

[111] M. Errico, B.G. Rong, G. Tola, M. Spano, Optimal synthesis of distillation systems for bioethanol separation. Part 1: Extractive distillation with simple columns, *Ind. Eng. Chem. Res.* 52 (2013) 1612–1619. https://doi.org/10.1021/ie301828d.

[112] H. Luo, C.S. Bildea, A.A. Kiss, Novel heat-pump-assisted extractive distillation for bioethanol purification, *Ind. Eng. Chem. Res.* 54 (2015) 2208–2213. https://doi.org/10.1021/ie504459c.

[113] J. A. Pacheco-Basulto, D. Hernández-McConville, F.O. Barroso-Muñoz, S. Hernández, J.G. Segovia-Hernández, A.J. Castro-Montoya, A. Bonilla-Petriciolet, Purification of bioethanol using extractive batch distillation: Simulation and experimental studies, *Chem. Eng. Process. Process Intensif.* 61 (2012) 30–35. https://doi.org/10.1016/j.cep.2012.06.015.

[114] R.B. Nair, M.J. Taherzadeh, Valorization of sugar-to-ethanol process waste vinasse: A novel biorefinery approach using edible ascomycetes filamentous fungi, *Bioresour. Technol.* 221 (2016) 469–476. https://doi.org/10.1016/j.biortech.2016.09.074.

[115] A.R. Navarro, M. Del C. Sepulveda, M.C. Rubio, Bio-concentration of vinasse from the alcoholic fermentation of sugar cane molasses, *Waste Manag.* 20 (2000) 581–585. https://doi.org/10.1016/S0956-053X(00)00026-X.

[116] X. Kong, J. Ma, P. Le-Clech, Z. Wang, C.Y. Tang, T.D. Waite, Management of concentrate and waste streams for membrane-based algal separation in water treatment: A review, *Water Res.* 183 (2020) 115969. https://doi.org/10.1016/j.watres.2020.115969.

[117] S. Basu, A.L. Khan, A. Cano-Odena, C. Liu, I.F.J. Vankelecom, Membrane-based technologies for biogas separations, *Chem. Soc. Rev.* 39 (2010) 750–768. https://doi.org/10.1039/b817050a.

[118] E. Drioli, A.I. Stankiewicz, F. Macedonio, Membrane engineering in process intensification-An overview, *J. Memb. Sci.* 380 (2011) 1–8. https://doi.org/10.1016/j.memsci.2011.06.043.

[119] A. Alkhudhiri, N. Darwish, N. Hilal, Membrane distillation: A comprehensive review, *Desalination.* 287 (2012) 2–18. https://doi.org/10.1016/j.desal.2011.08.027.

[120] P. Wei, L. H. Cheng, L. Zhang, X. Hua. Xu, H. L. Chen, C. J. Gao, A review of membrane technology for bioethanol production, *Renew. Sustain. Energy Rev.* 30 (2014) 388–400. https://doi.org/10.1016/j.rser.2013.10.017.

[121] N. Hajilary, M. Rezakazemi, S. Shirazian, Biofuel types and membrane separation, *Environ. Chem. Lett.* 17 (2019) 1–18. https://doi.org/10.1007/s10311-018-0777-9.

[122] M.B. Patil, T.M. Aminabhavi, Pervaporation separation of toluene/alcohol mixtures using silicalite zeolite embedded chitosan mixed matrix membranes, *Sep. Purif. Technol.* 62 (2008) 128–136. https://doi.org/10.1016/j.seppur.2008.01.013.

[123] L. Wang, Y. Wang, L. Wu, G. Wei, Fabrication, properties, performances, and separation application of polymeric pervaporation membranes: A review, *Polymers.* 12 (2020) 1–30. https://doi.org/10.3390/polym12071466.

[124] D. Mitra, Desulfurization of gasoline by pervaporation, *Sep. Purif. Rev.* 41 (2012) 97–125. https://doi.org/10.1080/15422119.2011.573044.

[125] A. Khalid, M. Aslam, M.A. Qyyum, A. Faisal, A.L. Khan, F. Ahmed, M. Lee, J. Kim, N. Jang, I.S. Chang, A.A. Bazmi, M. Yasin, Membrane separation processes for dehydration of bioethanol from fermentation broths: Recent developments, challenges, and prospects, *Renew. Sustain. Energy Rev.* 105 (2019) 427–443. https://doi.org/10.1016/j.rser.2019.02.002.

[126] F. Liu, L. Liu, X. Feng, Separation of acetone-butanol-ethanol (ABE) from dilute aqueous solutions by pervaporation, *Sep. Purif. Technol.* 42 (2005) 273–282. https://doi.org/10.1016/j.seppur.2004.08.005.

[127] J. Zhang, W. Liu, Thin porous metal sheet-supported NaA zeolite membrane for water/ethanol separation, *J. Memb. Sci.* 371 (2011) 197–210. https://doi.org/10.1016/j.memsci.2011.01.032.

[128] Z. Wu, C. Zhang, L. Peng, X. Wang, Q. Kong, X. Gu, Enhanced stability of MFI zeolite membranes for separation of ethanol/water by eliminating Surface Si-OH Groups, *ACS Appl. Mater. Interfaces.* 10 (2018) 3175–3180. https://doi.org/10.1021/acsami.7b17191.

[129] P. Kaewkannetra, N. Chutinate, S. Moonamart, T. Kamsan, T.Y. Chiu, Separation of ethanol from ethanol-water mixture and fermented sweet sorghum juice using pervaporation membrane reactor, *Desalination.* 271 (2011) 88–91. https://doi.org/10.1016/j.desal.2010.12.012.

[130] J. Diani, Y. Liu, K. Gall, Finite strain 3D thermoviscoelastic constitutive model for shape memory polymers, *Polym. Eng. Sci.* 46 (2006) 486–492. https://doi.org/10.1002/pen.20497.

[131] A. Bokhary, M. Leitch, B.Q. Liao, Liquid – liquid extraction technology for resource recovery: Applications, potential, and perspectives, *J. Water Process Eng.* 40 (2021) 101762. https://doi.org/10.1016/j.jwpe.2020.101762.

[132] H. Huang, S. Ramaswamy, B.V. Ramarao, Overview of biomass conversion processes and separation and purification technologies in biorefineries, in *Separation and Purification Technologies in Biorefineries* (2013) pp. 1–35. https://doi.org/10.1002/9781118493441.ch1

[133] C. Weilnhammer, E. Blass, Continuous fermentation with product recovery by in-situ extraction, *Chem. Eng. Technol.* 17 (1994) 365–373. https://doi.org/10.1002/ceat.270170602

[134] T.M. Boudreau, G.A. Hill, Improved ethanol – water separation using fatty acids, *Process Biochem.* 41 (2006) 980–983. https://doi.org/10.1016/j.procbio.2005.11.006.

[135] H. Habaki, H. Hu, R. Egashira, Liquid – liquid equilibrium extraction of ethanol with mixed solvent for bioethanol concentration, *Chin. J. Chem. Eng.* 24 (2016) 235–258. https://doi.org/10.1016/j.cjche.2015.07.022.

[136] A. Cháfer, J. De Torre, S. Loras, J.B. Montón, Study of liquid – liquid extraction of ethanol + water azeotropic mixtures using two imidazolium-based ionic liquids, *J. Chem. Thermodyn.* 118 (2017) 92–99. https://doi.org/10.1016/j.jct.2017.11.006.

[137] J.A. Laksmono, M. Sudibandriyo, A.H. Saputra, A. Haryono, Structured polyvinyl alcohol/zeolite/carbon composites prepared using supercritical fluid extraction techniques as adsorbent for bioethanol dehydration, *Int. J. Chem. Eng.* (2019) 6036479 https://doi.org/10.1155/2019/6036479

[138] S. Karimi, M. T. Yaraki, R. R. Karri, A comprehensive review of the adsorption mechanisms and factors influencing the adsorption process from the perspective of bioethanol dehydration, *Renew. Sustain. Energy Rev.* 107 (2019) 535–553. https://doi.org/10.1016/j.rser.2019.03.025.

[139] J.C. Saint Remi, R.G. Baron, J. Denayer, Adsorptive separations for the recovery and purification of biobutanol, *Adsorption*. 18 (2012) 367–373. https://doi.org/10.1007/s10450-012-9415-1.

[140] E.M. Flanigen, Zeolites and molecular sieves: An historical perspective, *Stud. Surf. Sci. Catal.* 137 (2001) 11–35. https://doi.org/10.1016/S0167-2991(01)80243-3.

[141] V. Krsti, B. Bhanvase, S. Sonawane, V. Pawade, A. Pandit, Role of zeolite adsorbent in water treatment, in *Handbook of Nanomaterials for Wastewater Treatment*. (2021) pp. 417–481. https://doi.org/10.1016/B978-0-12-821496-1.00024-6.

[142] T. Yamamoto, Y.H. Kim, B. Chul Kim, A. Endo, N. Thongprachan, Adsorption characteristics of zeolites for dehydration of ethanol : Evaluation of diffusivity of water in porous structure, *Chem. Eng. J.* 181–182 (2012) 443–448. https://doi.org/10.1016/j.cej.2011.11.110.

[143] M. Sudibandriyo, F.A. Putri, The Effect of various zeolites as an adsorbent for bioethanol purification using a fixed bed adsorption column, *Int. J. Technol.* 11 (2020) 1300–1308. https://doi.org/10.14716/ijtech.v11i7.4469.

[144] R.A. Jones, J.A. Gandier, J. Thibault, F.H. Tezel, Enhanced ethanol production through selective adsorption in bacterial fermentation, *Biotechnol Biopricess. Eng.* 16 (2011) 531–541. https://doi.org/10.1007/s12257-010-0299-1.

[145] D. Seo, A. Takenaka, H. Fujita, K. Mochidzuki, A. Sakoda, Practical considerations for a simple ethanol concentration from a fermentation broth via a single adsorptive process using molecular-sieving carbon, *Renew. Energy*. 118 (2017) 257–264. https://doi.org/10.1016/j.renene.2017.11.019.

[146] E.A. Viornery-Portillo, B. Bravo-Díaz, V.Y. Mena-Cervantes, Life cycle assessment and emission analysis of waste cooking oil biodiesel blend and fossil diesel used in a power generator, *Fuel*. 281 (2020) 118739. https://doi.org/10.1016/j.fuel.2020.118739.

[147] J. Sentanuhady, G.P.S.G. Atmaja, M.A. Muflikhun, Challenges of biofuel applications in industrial and automotive: A Review, *J. Eng. Sci. Technol. Rev*. 14 (2021) 119–134. https://doi.org/10.25103/jestr.144.16.

[148] A.K. Agarwal, Biofuels (alcohols and biodiesel) applications as fuels for internal combustion engines, *Prog. Energy Combust. Sci.* 33 (2007) 233–271. https://doi.org/10.1016/j.pecs.2006.08.003.

[149] A. Demirbas, Political, economic and environmental impacts of biofuels: A review, *Appl. Energy*. 86 (2009) S108–S117. https://doi.org/10.1016/j.apenergy.2009.04.036.

[150] M.I. Ahmad, M. Ismail, S. Riffat, *Renewable Energy and Sustainable Technologies for Building and Environmental Applications: Options for a Greener Future*, 2016. https://doi.org/10.1007/978-3-319-31840-0.

[151] G. Kilic, B. Sungur, B. Topaloglu, H. Ozcan, Experimental analysis on the performance and emissions of diesel/butanol/biodiesel blended fuels in a flame tube boiler, *Appl. Therm. Eng.* 130 (2018) 195–202. https://doi.org/10.1016/j.applthermaleng.2017.11.006.

[152] I. Mediavilla, E. Borjabad, M.J. Fernández, R. Ramos, P. Pérez, R. Bados, J.E. Carrasco, L.S. Esteban, Biofuels from broom clearings: Production and combustion in commercial boilers, *Energy*. 141 (2017) 1845–1856. https://doi.org/10.1016/j.energy.2017.11.112.

[153] M.B. Shemfe, S. Gu, P. Ranganathan, Techno-economic performance analysis of biofuel production and miniature electric power generation from biomass fast pyrolysis and bio-oil upgrading, *Fuel*. 143 (2015) 361–372. https://doi.org/10.1016/j.fuel.2014.11.078.

[154] B.C. McLellan, G.D. Corder, D.P. Giurco, K.N. Ishihara, Renewable energy in the minerals industry: A review of global potential, *J. Clean. Prod.* 32 (2012) 32–44. https://doi.org/10.1016/j.jclepro.2012.03.016.

[155] E.C. Petrou, C.P. Pappis, Biofuels: A survey on pros and cons, *Energy Fuels*. 23 (2009) 1055–1066. https://doi.org/10.1021/ef800806g.

12 Advanced Fluids in Chemical Absorption of CO_2

Development in CO_2 Capture Technology

Fernando Vega and Luz Marina Gallego-Fernández
University of Seville

David Abad-Correa
Centro Nacional del Hidrógeno

Francisco Manuel Baena-Moreno
University of Seville

CONTENTS

DOI: 10.1201/9781003206385-12

GLOSSARY CHEMISTRY

AEEA	2-((2-aminoethyl)amino)ethanol
AEP	N-(2-aminoethyl) piperazine
AMP	2-amino-2-methyl-1-propanol
DEA	diethanolamine
DEEA	N, N-diethylethanolamine
DETA	diethylenetriamine
DGA	diglycolamine
DMCA	N, N-dimethylcyclohexylamine
DMF	N, N-dimethylformamide
DPA	dipropylamine
MAPA	3-(methylamino)-propylamine
[Bmin][Acetate]	1-butyl-3-methylimidazolium - acetate
[Bmin][Cl]	1-butyl-3-methylimidazolium - chloride
[Bmin][Gly]	1-butyl-3-methylimidazolium - glycine
[Bmin][i-but]	1-butyl-3-methylimidazolium - isobutylene
[Bmin][NTF2]	1-butyl-3-methylimidazolium - bis(trifluoromethylsulfonyl)imide
[Bmin][PF6]	1-butyl-3-methylimidazolium - hexafluorophosphate
[Bmin][Pro]	1-butyl-3-methylimidazolium - prolinate
[Emin][Cl]	1-etyl-3-methylimidazolium - chloride
[Emin][DCN]	1-etyl-3-methylimidazolium - dicyanamide
[Emin][NTF2]	1-etyl-3-methylimidazolium - bis(trifluoromethylsulfonyl)imide
[3MEPY][NTF2]	3-methyl-1-etyl pyridinium - bis(trifluoromethylsulfonyl)imide
[N1111][Gly]	tetramethylammonium - glycinate
[P66614][CNPyr]	trihexyl (tetradecyl)phosphonium – 2-cyanopyrrole
[P2228][CNPyr]	triethyl (octyl)phosponium – 2-cyanopyrrole
[TETA]Br-PMDETA	Triethylenetetramine hydrobromide-5N-pentamethyldiethylenetriamine
[TETAH][Lys]	Triethylenetetramine - L-lysine
MCA	N-methylcyclohexylamine
MAE	methylaminoethanol
MDEA	N-methyldiethanolamine
MEA	monoethanolamine
PDMS	poly dimethylsiloxane
PMDETA 5N-	pentamethyldiethylenetriamine
PPGDME	polypropyleneglycoldimethylether
PZ	piperazine
S1N	N-cyclohexylpropane-1,3-diamine
TEA	triethanolamine

12.1 INTRODUCTION

Despite the ongoing development on renewables, energy storage and efficiency improvements at industrial processes, more efforts must be done to avoid greenhouse gas emissions (GHG) derived from the use of fossil fuel in the current energy scenario and the foreseeable future. Carbon capture and storage (CCS) technologies are one of the most feasible and practical options for long-term CO_2 sequestration and/or its reuse and, hence, GHG emission reduction. This technology allows a sustainable and responsible use of fossil fuels through a permanent CO_2 storage.

Among all the CCS alternatives, post-combustion capture based on chemical absorption is a mature technology ready for its implementation not only power plants – first-priority application – but also, and currently priority, other carbon-intensive industries such as cement, iron, and steel manufacturing. However, the elevated CO_2 capture cost compared with the current CO_2 emission trading market and the prohibitive energy requirements and environmental issues are the major barriers that constrain its deployment at large scale.

Chemical absorption separates CO_2 from the exhaust gas after both combustion and, particularly in cement plants, $CaCO_3$ calcination. Typically, the CO_2 capture is located downstream and continues the use of a specific abatement unit applied for a particular pollutant, being CO_2 at the current case. In most of the industrial processes that use fossil fuels, chemical absorption for CO_2 capture can be applied directly as retrofitting but, in some cases, additional pollutants removal is required to ensure solvent functionality and stability. This capture approach offers feasibility and versatility in terms of operating conditions and process integration.

A conventional chemical absorption scheme is shown in Figure 12.1 [1]. A flue gas with low CO_2 concentration is introduced in the absorber in countercurrent with lean solvent coming from the regeneration unit at temperatures ranging from 40°C to 60°C. In some cases, further flue gas quench is required due to volatility constrains, as it occurs using ammonia (chilled ammonia™ process). Once CO_2 is absorbed, the solvent containing CO_2 – rich amine – from the absorber bottom is pumped into a cross-exchanger in countercurrent with hot lean amine from stripper before it is introduced into the stripper in order to pre-heat the rich amine fed to the stripper. The stripping temperature ranges between 120°C and 150°C, and the operating pressure can vary between 2 and 5 bar. The solvent resistance to be thermally degraded sets the operational temperature at the stripper [2]. The pressure is recommended to be as higher as the stripping temperature allows in order to reduce the energy requirements of the regeneration stage [3]. The CO_2 leaves the stripper from the top. In this stream, water vapor is removed by cooling prior to purification, transport, and storage, while lean amine at the bottom of the stripper is pumped back into the absorber. CO_2 concentration in the flue gas from a combustion process varies from 12% to 14%v/v in coal-fueled power plants to 4%–8%v/v in combined cycle and boilers using natural gas. Other industries such as cement, iron and steel, and petrochemicals produce a higher CO_2 concentrated flue gas ranging between 14% and 33%v/v, mainly depending on the operating

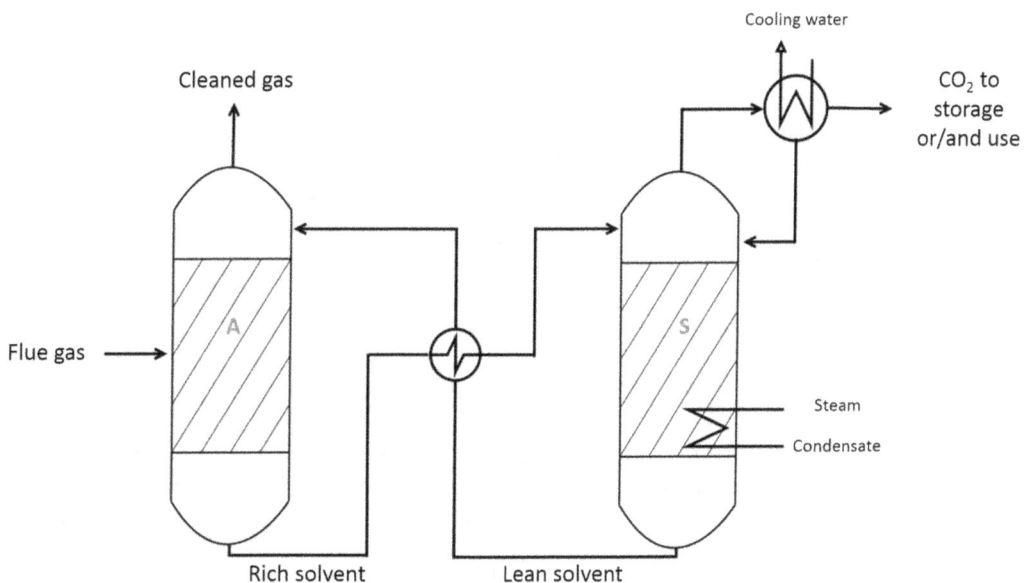

FIGURE 12.1 Conventional configuration of CO_2 chemical absorption unit: (a) absorption section and (b) stripping section. (Adapted from [1].)

conditions and the air leakages [4]. The key drawbacks constraining the implementation at large scale of this technology rely on the large mass flow of exhaust gas that should be treated and the low CO_2 concentration of the gas stream, which implies a very low CO_2 partial pressure and, consequently, a low driving force for the CO_2 separation stage. In addition, the current solvents used industrially for CO_2 separation require high energy penalties, mainly linked to the CO_2 desorption stage. The presence of high concentrations of solid particles, O_2, SO_x, NO_x, and other trace compounds such as Hg and the relatively high temperature of the flue gas, typically between 120°C and 180°C, are also two design challenges that must be also addressed for the deployment of this technology [4]. Based on the available post-combustion technologies, those disadvantages have not been addressed yet and have a significant impact on the electricity and capture costs.

Finally, it should be noted that the energy requirement associated with the CO_2 desorption process and the oxidative and thermal degradation of amine-based solvents are identified as the key challenges limiting its deployment at the commercial scale of this technology. Solvent regeneration is a high-intensive energy process that requires temperatures over 100°C in order to break the CO_2-amine bonds. Novel solvents and blends have been proposed to diminish the energy demand and solvent flow rates in comparison with the benchmark – MEA 5M.

This work summarized the traditional solvents applied to CO_2 chemical absorption applications, from amine-based solvents to alkali aqueous solutions, and the most recent insights and innovation in this field, such as ionic liquids (ILs) and biphasic behavior solvents. Their evaluation and comparison help to conclude a comprehensive dissertation on the current state of art of solvents for carbon capture applications based on post-combustion.

12.2 CONVENTIONAL SOLVENTS

12.2.1 AMINE-BASED SOLVENTS

Aqueous amine solutions having at least one $-NH_2$ group in the chemical structure are the most extended and industrially developed solvents for CCS technology. The chemical absorption process using amines has been applied for CO_2 and H_2S removal in natural gas-upgrading plants since the 1950s [5]. CO_2 is absorbed typically using amines due to their alkaline character. In fact, amines can form weak bonds with CO_2 rapidly, selectively, and reversibly, and hence, they can be applied for low CO_2 partial pressure flue gases, as it occurs in most of the industrial processes that require to reduce their CO_2 emissions. Amines exhibit an elevated volatility and are cheap and safe to handle compounds. Their most significant issues are related to the fact that they are corrosive. For this reason, the construction material should be adequately selected to resist the corrosion environment. Furthermore, amines form stable salts in the presence of O_2, SO_x, and other impurities such as dust, HCl, HF, and organic and inorganic Hg trace compounds that severely constrain the content of those compounds in the treated gas.

5M aqueous solution of MEA is established as the most widely amine-based solvent used in this kind of applications. MEA is considered the benchmark due to its high cyclic capacity, significant absorption-stripping kinetic rates at low CO_2 concentration, and high solubility in water. Some other amine-based absorbents have also traditionally been utilized, such as DEA, TEA, DGA, MDEA, PZ, AMP, and AEP [5–7].

In general, the absorption process links the CO_2 with amines, which occurs in two different ways. As the acid pathway, amines behave as a base and react with aqueous CO_2 from flue gas absorbed as carbonic acid. This acid-base equilibrium is governed by pH and CO_2 concentration in the liquid bulk, being bicarbonate (HCO_3^-) the predominant species at absorber conditions [4]. The second way is based on the carbamate formation pathway – zwitterion mechanism – which is described in detail below. The zwitterion mechanism was originally proposed by Caplow and later by Danckwerts [8]. It consists of a two-step mechanism. The first step suggests that the reaction of CO_2 with amine occurs through the zwitterion formation (Eq. 12.1). In the second step, the zwitterion undergoes deprotonation in the presence of a base or bases to form the carbamate ion (Eq. 12.2). The sum of all the bases presented in the aqueous solution, such as H_2O, OH-, and the combination of all the amines in the aqueous solution, must be taken into account in the reaction rate:

- **Step 1: Carbamate formation (rate-limiting step):**

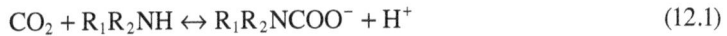

$$CO_2 + R_1R_2NH \leftrightarrow R_1R_2NCOO^- + H^+ \tag{12.1}$$

- **Step 2: Protonated alkanolamine formation:**

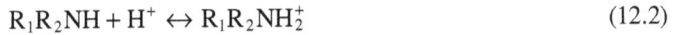

$$R_1R_2NH + H^+ \leftrightarrow R_1R_2NH_2^+ \tag{12.2}$$

- **Overall reaction:**

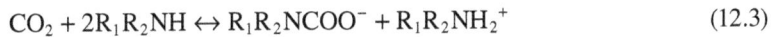

$$CO_2 + 2R_1R_2NH \leftrightarrow R_1R_2NCOO^- + R_1R_2NH_2^+ \tag{12.3}$$

The carbamate formation pathway predominates as a mechanism in the absorber, but the amine ratio is limited to 0.5, which results in a low CO_2 absorption capacity in comparison with the hydrolysis mechanism, which can be considered a "pseudo"-physical absorption. Carbamate formation shows a high formation enthalpy, resulting in higher energy requirements if CO_2 should be released (regeneration stage).

Several mechanisms have been proposed for CO_2-amine reactions. In general, the reaction of CO_2 with primary, secondary, and sterically hindered amines is governed either by the zwitterion mechanism or by the termolecular mechanism, whereas the reaction that involves tertiary amines is described by the base-catalyzed hydration of CO_2.

The termolecular mechanism suggests that a molecule of amine reacts simultaneously with one molecule of both CO_2 and a base. In this case, the mechanism consists of a one-step process based on the loosely bound encounter complex formation, which acts as intermediate [8].

The base-catalyzed hydration mechanism was first proposed by Donaldson and Nguyen to explain the CO_2 absorption process involving tertiary amines. In this case, tertiary amines cannot react with CO_2 by carbamate formation. On the contrary, CO_2 is absorbed by a hydration mechanism that is catalyzed in the presence of a base as tertiary amine [8].

The zwitterion mechanism fits accurately with the kinetic phenomena observed when fresh or low CO_2 loading solvent is used. According to Aboudheir, the termolecular reaction mechanism must be considered in high CO_2-loaded and highly concentrated MEA aqueous solution cases [9].

In summary, Table 12.1 describes the advantages and disadvantages regarding chemical absorption using amine-based solvents.

12.2.2 Aqueous Ammonia

Aqueous ammonia is feasible to simultaneously remove SOx and NOx along with CO_2 from flue gases. In this case, the absorption process occurs at low temperature, typically between 15°C and 27°C, due to the high volatility of the solvent. Ammonia reacts with CO_2 to form ammonium carbonate and bicarbonate, which are solid species, as given in reactions (12.4) and (12.5):

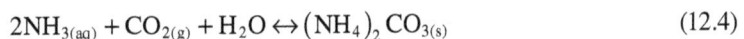

$$2NH_{3(aq)} + CO_{2(g)} + H_2O \leftrightarrow (NH_4)_2 CO_{3(s)} \tag{12.4}$$

$$NH_{3(aq)} + CO_{2(g)} + H_2O \leftrightarrow NH_4HCO_{3(s)} \tag{12.5}$$

TABLE 12.1

List of Advantages and Disadvantages of CCS Based on Chemical Absorption Using Amine-Based Solvents

Advantages	Disadvantages
• Feasible for low CO_2 concentrated flue gas	• Low CO_2 capacity
• Mature technology	• Corrosion issues
	• Amine degradation and waste management
	• High energy requirements for CO_2 stripping
	• Scale-up from actual (800 t/day) to required (8,000 t/day) CO_2 capacity

Source: Adapted from [10].

For the removal of SOx and NOx, a pretreatment oxidation unit is required to produce the oxidized form of those compounds, SO_3 and NO_2. Ammonia absorbs those to form ammonium nitrate and sulfate, which are both well-known fertilizers and can be sold to generate profits. Once CO_2 is captured, the rich CO_2 solution is sent to the stripper column where CO_2 is released and solvent is regenerated. The stripping temperature varies from 27°C to 92°C, producing the ammonium compound's decomposition. Stripping temperature is constrained at higher values due to the ammonia losses during the desorption process, which can substantially decrease the solvent capacity of the overall CO_2 capture unit. According to Spigarelly, ammonium bicarbonate has a lower enthalpy of reaction than ammonium carbonate, which means a lower energy requirement at the stripping stage [10]. The bicarbonate-carbonate ratio can be controlled by monitoring pH. Low pH values are desirable to shift the equilibrium towards bicarbonate formation. Table 12.2 summarizes the most relevant insights extracted from chemical absorption using amines.

12.2.3 DUAL ALKALI PROCESS

Dual alkali absorption is a novel CO_2 capture approach derived from the Solvay process, which converts CO_2 into Na_2CO_3 in two stages. A brief description of the Solvay process is shown below. Equation 12.6 represents the first step of the Solvay process, where CO_2 reacts with sodium chloride (NaCl) to obtain sodium bicarbonate ($NaHCO_3$). The reaction is catalyzed in the presence of ammonia as a primary alkali:

$$CO_{2(g)} + NaCl_{(aq)} + NH_{3(aq)} \leftrightarrow NaHCO_{3(s)} + NH_4Cl_{(aq)} \qquad (12.6)$$

TABLE 12.2

List of Advantages and Disadvantages of CCS Based on Chemical Absorption Using Aqueous Ammonia

Advantages	Disadvantages
• High CO_2 loading	• High cooling power requirements to provide a low temperature of the flue gas
• Multi-pollutant removal	
• Avoiding oxidative degradation	• Plugging issues due to the presence of solids during the capture process
• No corrosion issues	
• Possible by-products (fertilizers)	• Large ammonia losses mainly during stripping

Source: Adapted from [10].

In the second step (12.7), the formed sodium bicarbonate is filtered and heated to produce commercial-grade sodium carbonate ($NaCO_3$). Finally, ammonia is recovered from ammonium chloride (NH_4Cl) using $Ca(OH)_2$ as secondary alkali (12.8).

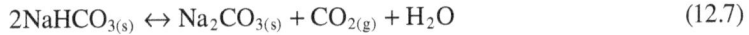

$$2NaHCO_{3(s)} \leftrightarrow Na_2CO_{3(s)} + CO_{2(g)} + H_2O \tag{12.7}$$

$$NH_4Cl_{(aq)} + Ca(OH)_{2(aq)} \leftrightarrow 2NH_{3(aq)} + CaCl_{2(aq)} + 2H_2O \tag{12.8}$$

The Solvay process is not suitable as CO_2 capture approach is mainly due to the use of calcium hydroxide ($Ca(OH)_2$) for ammonia regeneration. Calcium hydroxide is produced from lime, which is derived from limestone calcination. Therefore, calcium hydroxide production is provided from a high-intensive energy process and CO_2 is produced during the calcination. Those drawbacks make the Solvay process inefficient and not suitable for CO_2 capture on a large scale.

Dual alkali absorption proposes modifications to the Solvay process to overcome its drawbacks and to make it feasible for large-scale applications. MEA [4] and MAE [11] have been proposed as primary alkalis instead of ammonia. The reaction of the first step related to both amines is given below:

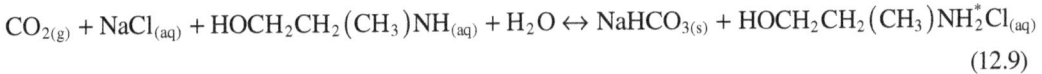

$$CO_{2(g)} + NaCl_{(aq)} + HOCH_2CH_2(CH_3)NH_{(aq)} + H_2O \leftrightarrow NaHCO_{3(s)} + HOCH_2CH_2(CH_3)NH_2^*Cl_{(aq)} \tag{12.9}$$

MEA and MAE can replace the primary alkali and, consequently, avoid the use of calcium hydroxide as a secondary alkali in the ammonia regeneration process. On the contrary, further NOx and SOx removal is required to avoid solvent degradation and flue gas must be cooled below 25°C for a proper performance of the removal process. Researchers are focused on developing an efficient secondary alkali that could recover amine compounds from the first step. This secondary alkali has not been identified yet. Other authors have also proposed an alternative method for ammonia regeneration using active carbon instead of calcium hydroxide, and therefore, a second alkali is not further required [12]. Table 12.3 resumes the main characteristics of the dual alkali process.

12.2.4 SODIUM CARBONATE

Sodium carbonate (Na_2CO_3) has been proposed as an alternative solvent for CO_2 chemical absorption. The mechanism associated with CO_2 absorption by sodium carbonate is summarized in 12.10 and 12.11. A 30wt% sodium carbonate slurry is used to provide a basic environment in which CO_2 is absorbed as bicarbonate followed by sodium bicarbonate formation [13]:

TABLE 12.3
List of Advantages and Disadvantages of Dual Alkali Absorption

Advantages	Disadvantages
• High CO_2 capacity	• Use of $Ca(OH)_2$ for ammonia recovery
• Commercial-grade $NaCO_3$ production	• An alkali must be found to replace $Ca(OH)_2$ in the process
• Low energy requirements of the process	• Higher pollutant removal and gas cooling if amine compounds were used as primary alkali

Source: Adapted from [10].

Two reactions form the $Na_2CO_3 - NaHCO_3$ formation pathway, which is described in detail below:

- **Step 1: Bicarbonate formation:**

$$CO_3^{2-}{}_{(aq)} + H_2O + CO_2 \leftrightarrow 2HCO_3^-{}_{(aq)} \tag{12.10}$$

- **Step 2: Sodium bicarbonate formation:**

$$Na^+{}_{(aq)} + HCO_3^-{}_{(aq)} \leftrightarrow NaHCO_{3(s)} \tag{12.11}$$

The $NaHCO_3$ precipitation represented in 12.11 enhances the bicarbonate formation expressed in 12.10 and, hence, the CO_2 capture capacity of the solvent. Sodium carbonate has shown a high performance in CO_2 capture in comparison with the MEA benchmark. It produces a high CO_2 loading capacity (0.73 kg CO_2/kg CO_3^{2-}) and a reboiler duty of 3.2 MJ/kg CO_2 rather than 0.5 kg CO_2/kg MEA and 3.5–4.2 MJ/kg CO_2 reported for MEA [13]. It should be noted that sodium carbonate can react with pollutants such as SO_2, enabling its cyclic capacity for CO_2 absorption. Despite those advantages, sodium carbonate obtains low absorption rates, which lead to higher absorption columns. It assumes that sodium carbonate needs the use of additives such as amines to enhance its CO_2 absorption rates [13]. The advantages and disadvantages are listed in Table 12.4.

12.2.5 GAS ABSORPTION MEMBRANE

Membranes are widely utilized in industry for natural gas treatment. They consist of semi-permeable barriers that allow selective separation of a gas component from a mixture based on the permeation velocity of each component through the membrane. The membrane can be considered as a filter for one or more gas components and produces a component-enriched permeate stream. The differential pressure between the feed side and the permeate side is the main driving force to promote component separation from the flue gas.

Gas absorption membrane aims to enhance the gas component separation using higher differential pressure as a driving force along the membrane due to the use of a CO_2 solvent in the permeate side. In this case, the membrane is combined with a liquid solvent, typically an amine, which can immediately remove CO_2 from the flue gas. CO_2 is first diffused into the membrane and then is chemically absorbed by the use of a solvent in the permeate side. The high CO_2 removal rates provide smaller contactor devices using this principle than those from simple gas separation membrane. Moreover, using a membrane prior to CO_2 removal by chemical absorption reduces the volume of gas to be treated in the absorber column and consequently decreases the equipment size and the operating cost associated with the overall process.

TABLE 12.4
List of Advantages and Disadvantages of CCS Based on Chemical Absorption Using Sodium Carbonate Absorption

Advantages	Disadvantages
• Multi-pollutant capture system	• Slow absorption rate, meaning that the solvent should be promoted with an increasing rate of additives
• Use of a non-hazardous and non-volatile solvent	• Solid and slurry management
• Lower fouling and corrosion issues than amine compounds	• High pollutant removal

Source: Adapted from [10].

TABLE 12.5

List of Advantages and Disadvantages of CCS Based on Gas Absorption Membrane

Advantages	Disadvantages
• Avoidance of regeneration stages and operational issues related to absorption technology	• High performance at high pressure (15–20 bar)
• Lower operational costs	• High CO_2 selectivity requirements for post-combustion applications (12%–15%v/v CO_2 flue gas)
• Compact design	• Membrane degrades at temperatures over 100°C
• No corrosion issues	• Impurities cause damage on the membrane surface and limit CO_2 permeability
	• Multistage operation mode is suitable for post-combustion capture applications

Source: Adapted from [10].

Table 12.5 summarizes the advantages and disadvantages of the use of membrane in CO_2 separation from flue gas. As indicated below, the driving force along the membrane decreases as CO_2 is diffused into the membranes. This phenomena impacts on the flexibility of the process, limiting its application in post-combustion due to the poor CO_2 recovery and purity. Barbieri et al. indicate that the use of membrane in CO_2 separation from flue gas containing less than 20% CO_2 is inefficient. Membrane also requires a cooling stage to ensure the gas stays at a temperature below 100°C as higher temperatures produce a further degradation of the membrane materials [14].

Despite the low energy requirements (0.5–6 MJ/kg CO_2), the low CO_2 removal capacity and the low purity of the CO_2 stream make its implementation difficult on a large scale. Moreover, membranes do not provide a high separation level, so multi-stage and recycle configuration must be implemented to achieve an affordable degree of CO_2 separation. It seems that membrane technology cannot tackle the energy penalties related to CO_2 separation in the post-combustion capture process, but its combination with existing processes, such as absorption membrane separation, may reduce the penalties of the overall CO_2 capture process. Some promising alternatives such as facilitated transport membranes and mixed matrix membranes have also been proposed in the literature [10].

12.3 IONIC LIQUIDS

Traditional amine-based solvents containing, at least, a $-NH_2$ group are widely considered for chemical absorption in CCS applications at industrial scale. In general, they show high operational and capital costs, in which solvent loss due to volatility and degradation, corrosion, and prohibitive energy requirements for regeneration are the main drawbacks [15]. Recent research in CO_2 solvents currently focuses on lowering both energy consumption in the regeneration stage and volatile emissions – more stable solvents [16].

ILs are considered more environmentally acceptable solvents that can become economically feasible at mid-term in comparison with conventional amine-based solvents. ILs are organic salts composed of a large organic cation such as imidazolium, pyridinium, pyrrolidinium, piperidinium or phosphonium cation – weakly coordinated to small anions, either organic anions such as HSO_4^-, $FeCl_4^-$, BF_4^- and PF_6^-, or an organic anion, i.e., RCO_2^- and $CF_3SO_3^-$ [15,17]. Among them, imidazole salts are the most extended IL for CCS applications. The strong ion-ion interactions in combination with a low molecular symmetry reduce both the lattice energy and the crystalline structure, thus decreasing their melting point. Their state is liquid at ambient conditions in spite of their strong ionic nature [15,18]. ILs are also classified based on their molecular structure in proton-donating and non-proton-donating solvents [15].

Several authors have identified the key properties in which the potential advantages of ILs relied on for CO_2 capture applications [19,20]:

- Low regeneration energy – ILs present a high CO_2 uptake capacity in the absorption stage and low energy penalty in the solvent regeneration step, although it occurs under vacuum pressure (Table 12.6).
- Tunable solvent capacity – ILs can be synthetized in a large number of combinations. In particular, they are easily tailored by using different cation-anion configurations, and also, additional functional groups can be attached. Their high capacity to be modified allows ILs to be specifically designed for a particular application such as low CO_2-concentrated exhaust gas derived from fossil fuel combustion processes [18].
- Negligible vapor pressure – ILs are non-volatile compounds compared with conventional volatile organic amine-based solvents. In fact, they are considered as "green solvents" because they prevent solvent emissions, avoid solvent loss during operation, and hence mitigate environmental pollution [15,17,18]. The extremely low vapor pressure has impact on the energy requirements in the regeneration step, which can be reduced up to 15% for MEA-functionalized ILs compared to 5M aqueous MEA solution (Orhan [18]; Vega et al. [21]).
- High thermal stability, non-flammability, and low corrosiveness – these properties in combination with the zero solvent loss during operation lead to a high recycling and reuse capacity that strengthen the economic benefits of their use instead of volatile and moderate stable amine-based solvents [20].

The CO_2 absorption process involved in conventional ILs should be defined as physical absorption instead of chemical, being not suitable for most of the energy-intensive industrial processes – low CO_2 concentrated flue gas at ambient pressure. For this kind of CCS applications, ILs should be enhanced using specifically designed configurations, namely, functionalized ILs or task-specific ILs. They add either primary or secondary amino groups into ILs structure to increase the CO_2 absorption kinetics by means of zwitterion mechanisms. In fact, amine solvents and ILs blends have been also proposed in order to trade off the ILs limitations in terms of absorption kinetics and viscosity. Table 12.7 reports the advantages and disadvantages of each IL family under CCS applications [15].

TABLE 12.6
Summary of ILs' Properties Reported from the Literature

IL	MW (g/mol)	ρ (kg/m³)	μ (mPa*s)	ΔHr (kJ/mol)	Max Solubility (mol CO_2/mol IL)	Ref.
[P₂₂₂₈][CNPyr]	323	940	163.5	−48		[19]
[P₆₆₆₁₄][CNPyr]	575	900	166.4	−40		
[Bmin][Acetate]	198	1,030–1,060	179.8–393.3	−35	0.89	[18], [19]
[Bmin][i-but]	226	1,010	198.0	−20		
[Bmin][GLY]	213	1,030	424.9	−23		
[Bmin][PRO]	253	1,060	891.3	−14		
[Bmin][Cl]	175	1,080	1,534.0		0.55	[18]
[Bmin][PF₆]	-	1,366	36.9			[15]
[Bmin][NTF₂]	419	1,430	69.0		0.77	[18]
[Emin][Cl]	147	1,090	36.0		0.52	
[Emin][NTF₂]	391	1,510	32.6		0.67	
[CH₃CH₂NH₃][NO₃]	-	1,210	35.9			[15]
TCMI-IL	-	1,155	1,956.3			[22]

TABLE 12.7
List of Advantages and Disadvantages of Several ILs Families

IL Family	Advantages	Disadvantages
Conventional ILs	CO_2 selectivity	Viscosity
		Price
Functionalized ILs	Kinetic enhancement	Synthesis
	Increase of CO_2 absorption capacity	Viscosity
		Price
Amine-ILs blends	Kinetic enhancement	Volatile solvent
	Low viscosity	Solvent loss

In addition to the limited CO_2 absorption rate, the high viscosity and the price are the two main drawbacks that constrain their competitiveness against conventional amine-based solvents. All the conventional ILs show high viscosity, which is increased once CO_2 is absorbed, limiting their mass transfer capacities and producing operational issues (Table 12.6). This limitation in the CO_2 mass transfer requires an increase in the solvent demand during the CO_2 capture process to balance the low CO_2-ILs interactions due to high viscosity. The higher viscosity observed as CO_2 is absorbed is due to the formation of hydrogen-bonded networks between carbamate species and those derived from dication mechanism [23]. Some authors proposed the use of either non-amine groups or ether oxygen atoms into the functionalized ILs structure to avoid the production of hydrogen-bonded networks [24,25]. Other works proposed ILs diluted in organic amines instead of water to decrease the viscosity without affecting the CO_2 absorption capacity of the amine-IL blend [26].

The combination of the intense solvent demand and the cost of IL purchase makes ILs currently not economically feasible for CCS applications. Table 12.8 summarizes the characteristics and properties of conventional ILs compared with other licensed commercial solvents. Efforts have to be made in researching novel less viscous and thermally stable ILs that can reduce the intensive solvent demand with losing CO_2 absorption selectivity and capacity [16].

The CO_2 absorption performance of ILs strongly depends on the CO_2 solubility. An increase in the CO_2 solubility implies a further reduction in the solvent rate demanded for a fixed CO_2 removal efficiency [19]. According to Freitas et al. [27], the energy of vaporization and the molar

TABLE 12.8
Main Properties of Conventional IL Versus Amine-Based Commercial Solvents [15,17]

	Econoamine FG+	Ionic Liquid
Solvent	Aqueous solution MEA	Salts
Licensor	Fluor	-
Price (€/kg)	1.5	30
Density (kg/m^3)	1,017	800–1,500
Molecular Weight (g/mole)	61	200–750
Viscosity (mPa*s)	2,2	20–1,000
Vapor Pressure (mmHg)	0,36	1×10^{-6}
CO_2 Absorption Enthalpy (kJ/mole CO_2)	−85	−15
Vaporization Enthalpy (kJ/mole solvent)	49,8	>120

volume are the key properties that have an impact on the CO_2 solubility. Most of ILs exhibit high energy of vaporization. Therefore, the molar volume is the differential parameter [15]. Their highly tunable characteristic allows ILs to be designed in order to maximize the free space available to accept CO_2 molecules and then the amount of CO_2 that can be absorbed [28]. In this sense, several authors demonstrated that anions have more impact on the solvent performance during the absorption process for conventional ILs, being the influence of cations considerably lower [15,18]. The use of large anions increases the molar volume, resulting in a significant enhancement of the CO_2 solubility. The size of cations has also influence on the CO_2 solubility. Higher alkyl chain length of the cations results in a higher CO_2 solubility [18] but also increases the viscosity [15]. Kazmi and co-workers recommend increasing the moiety on the cations to produce more molecules' free volume, hence enhancing the CO_2 absorption [20]. It should be noted that the nature of the anions has also influence on the IL absorption behavior. The presence of fluoro-alkyl groups in anions, such as $[NTf_2]^-$, $[BF_4]^-$, and $[PF_6]^-$, causes a strong interaction with CO_2, improving the CO_2 solubility [18,20].

From the operational point of view, viscosity plays a key role in the CO_2 absorption performance of ILs. The high values of viscosity provided by ILs constrain the mass transfer between both gas phase (CO_2) and liquid phase (IL). Hospital-Benito and co-workers identified three alternatives to overcome the mass transfer limitations observed in ILs [19]:

- The use of specific anions that can enhance the CO_2 diffusivity, such as $[DCN]^-$ and $[TCM]^-$
- Synthetize methods such as supporting or encapsulating
- Increasing the enthalpy of the CO_2-ILs absorption reaction

All the above-mentioned options result in a further decrease in the viscosity and hence an enhancement of the mass transfer rates, resulting in a higher CO_2 solubility. Current research on ILs is focused on developing novel IL configurations that achieve moderate viscosity, such as aprotic-heterocyclic-based anions (AHA-ILs). Works from Hospital-Benito et al. identified $[P_{2228}]$ [CNPyr] as a promising IL for CCS applications due to its moderate viscosity and elevated both enthalpy of reaction and CO_2 cyclic capacity. The use of solvents with higher enthalpy of reaction produces an increase in the temperature profile along with the absorber as CO_2 reacts with ILs. Under these conditions, the viscosity of the IL is reduced, and hence, the mass transfer limitation observed in conventional IL is partially mitigated, enhancing the CO_2 solubility and the absorption kinetics [16,19].

Other authors proposed ILs supported by membranes. The combination of IL with membrane decreases the viscosity of the CO_2-IL liquid solution during the CO_2 absorption process and also improves the CO_2 permeability and selectivity [28].

Several works related to the techno-economic evaluation involving ILs at different CCS applications such as exhaust flue gas cleaning and biogas upgrading are reported in Table 12.9. These studies are supported on simulations and laboratory characterization of ILs, presenting a lack of experience from pilot plant campaigns. In general, the CO_2 capture costs regarding IL solvents are ranging between 100 and 180 $/t CO_2 from flue gas treatment (at ambient pressure). Increasing the pressure of the absorption process can reduce this CO_2 capture cost window below 100 $/t CO_2 as occurs in biogas upgrading applications [16]. Some works established non-convincing values further below, around 60 and up to 25 $/t CO_2. However, the most realistic CO_2 capture costs are significantly far from 40 $/t CO_2 reported from DOE as the target value to promote a solvent at industrial scale [29,30]. Most of these studies indicate that the elevated operating costs are associated with the electric consumption of either compressors in the high-pressure absorbers or vacuum operation in the stripping section. From the capital costs, the mass transfer and kinetics limitations shown by ILs lead to prohibitive designs of the absorption and stripping columns that increase drastically the equipment investment.

TABLE 12.9

Economic Evaluation of ILs Used for Carbon Capture Applications

	Post			
IL	[P$_{2228}$][CNPyr]	[P$_{66614}$][CNPyr]	[Bmin][Acetate]	[Emin][NTf$_2$]
Reference		[16,19]		[30]
Year		2020–2021		2017
CO_2 removal capacity (t/h)	3.96	3.96	3.96	3.48
Gas temperature (°C)	40	40	40	50
Gas pressure (bar)	1	1	1	1 (20)[a]
Cyclic capacity (mol/kg)	0.94	0.54	0.49	-
Regeneration requirement (GJ/t CO_2)	3.8	6.3	7.0	1.4
Operating costs ($/t CO_2)	63.2	69.9	71.8	73.3
Annualized capital costs ($/t CO_2)	45.9	50.2	47.0	-
Total annualized costs ($/t CO_2)	109.1	120.1	118.8	-

	Post			
IL	[Emin][DCN]	[Emin][NTf$_2$]	[Bmin][Acetate]	[P$_{66614}$][CNPyr]
Reference		[31]	[32]	[29]
Year		2018	2010	2014
CO_2 removal capacity (t/h)	437.4	437.4	3.7	451.3
Gas temperature (°C)	30	30	49	40
Gas pressure (bar)	1 (20)[a]	1 (20)[a]	1 (7.9)[a]	1
Cyclic capacity (mol/kg)	-	-	-	0.19
Regeneration requirement (GJ/t CO_2)	2.8	-	-	-
Operating costs ($/t CO_2)	40.0	60.0	117.0	16.4
Annualized capital costs ($/t CO_2)	50.0	120.0	23.0	46.0
Total annualized costs ($/t CO_2)	90.0	180.0	140.0	62.4

	Biogas Upgrading			
IL	[P$_{2228}$][CNPyr]	[P$_{66614}$][CNPyr]	[Bmin][Acetate]	[3MEPY][NTF$_2$]
Reference		[16,19]		[20]
Year		2020–2021		2021
CO_2 removal capacity (t/h)	3.96	3.96	3.96	8
Gas temperature (°C)	40	40	40	30
Gas pressure (bar)	3.9	3.9	3.9	40
Cyclic capacity (mol/kg)	1.82	1.18	1.09	1.40
Regeneration requirement (GJ/t CO_2)	1.4	1.6	3.4	1.1
Operating costs ($/t CO_2)	57.8	60.5	61.1	13.9
Annualized capital costs ($/t CO_2)	39.1	40.9	39.4	11.7
Total annualized costs ($/t CO_2)	96.9	101.4	100.5	25.6

[a] Absorption occurs at pressure in brackets.

12.4 CUTTING-EDGE SOLVENTS

12.4.1 PHASE-CHANGE SOLVENTS

Phase-change solvents show a dual behavior depending on either their CO_2 loading or their temperature. Under variations on either the amount of CO_2 absorbed or temperature, a physical phase separation appears in this type of solvents, producing two different phases: one highly concentrated in

CO_2 and another poorly concentrated in CO_2, namely, CO_2-rich phase and CO_2-lean phase, respectively [33]. They are also called biphasic solvents.

The CO_2-rich phase is able to contain more than 95% of the total absorbed CO_2 within the 50%–80% of the solution volume flow [34,35]. This phase separation feature allows to potentially reduce the energy requirements for the regeneration stage in which only a CO_2-enriched fraction of the total solvent flow used in the absorption section is sent to the stripper. Therefore, lower amount of solvent should be regenerated. The use of phase-change solvents affects all the terms that the reboiler duty consists of. The diminishment of the solvent flow rate sent to the stripper directly reduces the sensible heat required to reach the stripping temperature [35,36]. In addition, the CO_2-rich phase separation can be considered as a high CO_2 cyclic capacity system. The higher CO_2 loading of the CO_2-rich phase solvent leads to higher CO_2 partial pressure profiles along the stripper, resulting in a higher CO_2-vapor ratio and hence a lower energy consumption associated with the water vaporization [33,37]. It should be also noted that the use of a highly CO_2 concentrated rich solvent enables stripping operations at both higher pressure and temperature, which can produce a further energy reduction in the regeneration stage. Finally, the phase-change solvents exhibit lower enthalpy of CO_2 absorption than conventional amines [33]. The sum of all the above-mentioned advantages is translated to a lower energy consumption during the solvent regeneration.

12.4.1.1 CO_2-Loading-Dependent Biphasic Solvents

This type of phase-change solvents exhibits phase separation once CO_2 is absorbed [35,36]. Both non-aqueous and aqueous solvents have been proposed for carbon capture applications.

12.4.1.1.1 Amine Blends

Amines exhibit phase-change features as specific types are blending. The phase-change amine blends consist of mixing either a primary or secondary amine, acting as absorption promoter, combined with a tertiary amine, acting as CO_2 capacity promoter. Primary and secondary amines containing multiple protonatable sites are recommended as diamines and triamines, whereas high-hydrophobicity-capacity tertiary amines promote the formation of two phases [36]. DEEA-based amine blends achieve up to 95% of absorbed CO_2, which is contained in the rich phase, and the specific energy consumption of the solvent regeneration was set in the window 2.3–2.4 Gt/CO_2 as is blended with MAPA [36]. Experiments with AEEA blended with PMDETA in dimethyl sulfoxide can reduce further the energy requirements up to 1.66 GJ/t CO_2 [38].

Kazepidis et al. proposed a novel solvent blend combining S1N and DMCA. This solvent blend demonstrated its high performance for CO_2 absorption applications at the laboratory scale: It exhibited low viscosity, 1.35 mol CO_2 per mole solvent, and lowering the volatile loss up to 10% compared with other phase-change amine blends [39]. An experimental campaign was run in a pilot plant at Rolincap research facility in which a 3M S_1N/DMCA obtained a reboiler duty ranging between 2.1 and 2.3 Gt/CO_2 for flue gas from a conventional flue gas (14%v/v CO_2) and gas-fired flue gas (3.5%v/v CO_2), respectively [39]. According to Kazepidis et al., this solvent can lead to further reductions on the operating costs – 47% respect to MEA – with only 1.7% of increase in the investment cost, mainly regarding a decanter required prior to the regeneration stage [39].

Other authors proposed the use of non-aqueous amine blends based on alcohols. In these alcohol-solubilizer amine blends, long-chain alcohols such as isooctanol and polyethers promote the liquid-liquid phase separation. Once CO_2 is absorbed, alcohol is in the lean CO_2 phase, whereas the carbamate and protonated amines are mainly distributed in the rich phase. 80% of CO_2 is found in the rich CO_2 phase. On the other hand, short-chain alcohols such as ethanol and 1-propanol promote the solid-liquid phase separation [36]. The alcohol participates in the CO_2 absorption process and enhances the absorption capacity compared to the aqueous amine blend. Wang et al. tested DETA combined with 1-propanol. The 30%wt DETA and 50%wt 1-propanol exhibited an excellent result. In particular, the rich-phase volume ratio was 41.7%, which leads to a 2.14 GJ/t CO_2 reboiler duty.

The CO_2 loading in the rich phase was 6.54 mol/L. However, most of the amine-alcohol blends showed high volatility that conduces to elevate solvent loss compared to aqueous amine solutions.

12.4.1.1.2   Amino Acid Salts

Amino acid salts have been recently proposed as an eco-solvent for carbon capture applications [40]. They are considered environmentally friendly solvents due to their negligible vapor pressure (null volatile solvent loss), high resistance to oxidation, and low toxicity [40,41]. Aqueous amino acid solvents precipitate under the presence of high CO_2 concentration into the solution by means of either the formation of amino acid zwitterion carbamate or salting-out effect. According to Zhang et al., amino acid concentrations in aqueous solution over 3M facilitate the solid-phase formation, producing the solid precipitation under lower CO_2 loadings [36]. The high energy penalty regarding solvent regeneration and solid management during operations, particularly blockages and foaming production, have been reported as the main drawbacks related to their use in CCS [41].

Numerous amino acid salts were proposed from linear amino acid salts, such as glycine, taurine, sarcosine, to polyamino acid salts such as asparagine and glutamine [36]. PAS6, a phase-change solvent, was studied for capturing CO_2 from a 416-MW NGCC power plant [42]. Results extracted from this work report an 11% reduction of energy penalty of the overall CO_2 capture process in comparison with MEA. Works from Aronu et al. [42] proposed a novel solvent, namely, SARMAPA. Although it exhibited lower CO_2 absorption rates than MEA 5M and high energy demand for regeneration (7.5 GJ/t CO_2), the reboiler duty decreases around 50% due to the volume flow reduction to be treated at the stripping section. An amino acid-based solvent proposed by TNO, namely, DECAB plus, is mainly composed of potassium-taurine and can be regenerated at 2.6 GJ/t CO_2 energy requirement.

To strengthen the CO_2 absorption features of amino acids, several authors proposed the use of non-aqueous amino acid solvents. Alcohols can enhance the absorption performance of amino acid-based solvents in terms of CO_2 capacity, absorption rates, and energy requirements. Prolinate was mixed with ethanol to increase its CO_2 solubility and absorption rates compared to aqueous prolinate solvent up to 3.5 and 4.2 times, respectively. This solvent presents solid-liquid phase features in which prolinate carbamate, bicarbonate, and ethyl carbonate form the solid phase. Unfortunately, only 55%–60% of the absorbed CO_2 was placed at the solid phase, and hence, the liquid phase should be sent to the strip instead to the absorber. Short-chain alcohols showed high volatility and solvent loss. To minimize these issues, DMF was proposed as diluent for amino acid salts [40]. Allivand and co-workers optimized the solvent blends and proposed the use of potassium glycinate combined with a 60:40 volume ratio of DMF-water solution. The solvent blend exhibits excellent results, producing a 63% of CO_2-free phase volume, which can be directly used to absorb CO_2. The rest of highly-loaded solvent solution requires 6.32 GJ/t CO_2, a 59.1% lower the reboiler duty than aqueous glycine. The elevated energy demand remains as a main drawback associated with amino acid salts for CCS applications.

12.4.1.1.3   Ionic Liquids

ILs can feature a CO_2-loading phase-change behavior as they are blended with either amines or alcohols, mostly solid-liquid phase separation. The combination of alkanolamines with IL can produce the carbamate precipitation in the presence of CO_2, leading to an increase in the CO_2 capacity and lower energy requirements related to the minor amount of CO_2-rich phase that has to be stripped. It seems that the IL hydrophobicity causes the segregation of solid particles during the absorption process and enhances the separation of the solid phase from the liquid phase [36]. According to Hasib-ur-Rahman et al., alkanolamines-IL blends avoid the formation of oxidizing products that are responsible for the main corrosion issues in CO_2 capture facilities [43].

Zhou and co-workers proposed [TETA]Br-PMDETA as a phase-change IL-alkanolamine blend. The optimal concentration was set at 4M with a 3:7 ratio between [TETA]-Br and PMDETA. Under conventional flue gas conditions −323 K and ambient pressure, the solvent blend can absorb 0.658 mol CO_2/mol solvent and a liquid-liquid separation with a 46% volume ratio related to the lower phase was achieved. Almost all the CO_2-bonded species were found in the lower phase (>99%), which makes [TETA]Br-PMDETA a promising candidate for further energy reductions in the regeneration stage [34]. Non-aqueous IL solvents have also been proposed. Jiang et al. evaluated a 1:2 molar ratio [N_{1111}][Gly] with ethanol instead of water as diluent. The presence of ethanol reduces the IL viscosity and promotes the phase change when CO_2 is absorbed. A 0.85 mol CO_2/mol IL was achieved under conventional flue gas conditions [44]. Ethanol was tested in combination with [TETAH][Lys]. The volume ratio of the lower phase was 33% of the total volume, containing 93% of the absorbed CO_2 [44,45].

12.4.1.2 Temperature-Dependent Biphasic Solvents

Thermomorphic solvents experience liquid-liquid phase-change behavior under temperatures ranging between 60°C and 90°C once CO_2 is absorbed. Therefore, the phase separation occurs upon heating the loaded aqueous solvent solution. This type of compounds are formed by molecules combining hydrophobic and hydrophilic functional groups, namely, lipophilic amine solvents, which are partially miscible in aqueous solution. The key parameter is the lower critical solution temperature (LCST), which identifies the minimal temperature that produces phase separation of the loaded aqueous solvent solution [33,46]. Lipophilic amines with elevated LCST lead to higher stripping temperatures, whereas lower LCST solvents conduce to extremely low absorption temperatures. Blends of both types of lipophilic amines should be recommended for CCS applications. As it occurs with conventional amines, primary and secondary lipophilic amines contribute to enhance the absorption performance of the aqueous amine blend, whereas tertiary lipophilic amines increase the CO_2 absorption capacity of the solution. The advantages regarding their use in CCS rely on the reduction in the amount of CO_2-loaded solvent to be regenerated and the low stripping temperature, which allows the use of both waste heat and low-quality steam for heating [36].

Zhang and co-workers proposed DPA-DMCA blends as a temperature-induced phase-change solvent. This solvent blend strips at 70°C, showing a good resistance to degrade. The steam requirements were 30% lower compared with benchmark – MEA 5M. However, the low CO_2 absorption rates and the presence of precipitates derived from protonated DPA ions and bicarbonates constrained its use. To solve these drawbacks, DPA was substituted by MCA. The 1:1 molar ratio (4M) MCA-DMCA blend provided higher CO_2 absorption rates with CO_2 loadings up to 3.34 mol CO_2/L at 30°C but its low LCST produced phase-change separation during the CO_2 absorption process. Other operational issues such as foaming and volatile solvent loss were reported. The authors proposed the use of AMP to increase the LCST at 40°C. They found the addition up to 9%wt of AMP as the optimal AMP amount. AMP concentrations above 20%wt should lead to excessive solvent degradation [36].

A novel lipophilic solvent blend licensed by IFPEN was developed for CO_2 capture in gas-fired power plants, namely, DMX™ [36,41]. DMX™ has high LCST, which allows the absorption of CO_2 that occurs at 40°C. From laboratory analysis of DMX™, the enthalpy of CO_2 absorption is around 29% lower compared with MEA and it doubles its CO_2 capacity with similar values of CO_2 absorption rates. DMX™ shows high thermal stability at temperatures over 150°C. At pilot plant campaign, DMX™ requires only 25%–30% lower liquid flow rate to be sent to the stripper, which can further reduce the energy penalty during the regeneration process and the CO_2 avoided costs – up to 26% and 23% compared to MEA 5M, respectively [36,41].

12.4.1.3 Hydrate-Based Separation Solvents

This separation approach is based on the hydrate formation from high pressure water in contact with the flue gas containing CO_2. Hydrates are crystalline and consist of water and gas under suitable low temperature and high pressure conditions. The CO_2 is captured as hydrate is formed. A pure CO_2 stream is then obtained since CO_2 is released from the hydrates, achieving up to 99% of CO_2 recovery.

The most positive characteristic for CO_2 capture of hydrates is that hydrates can store large amounts of CO_2. Most of the studies are focused on adding an additive to enhance the CO_2 capture kinetic rate. High pressure required for hydrate formation is also relaxed by the use of additives.

Tetrahydrofuran and tetra-n-butyl salts such as tetra-n-butyl ammonium bromide and tetra-n-butyl ammonium fluoride are the most attractive additives reported in the literature. The slow hydrate formation rate and the very low temperature required are two of the current advantages of this approach.

12.4.2 SOLID-SUPPORTED LIQUID SOLVENTS

Recent researches on the field of solvents for CO_2 chemical absorption are focused on combining high molecular weight solvents with porous solid as support base, which can overcome the main drawbacks related to conventional solvents [44,45].

12.4.2.1 Polymeric Solvents

Polymeric solvents in non-aqueous solutions have been recently proposed for CO_2 absorption. Polyesters, PDMS, PPGDME, and polyethers are the most common polymeric solvents studied. They show low viscosity, resistance to thermal degradation, and low volatility as main properties. Liquid surfactants have been also proposed due to their high commercial availability and non-toxicity. Their use in combination with a superbase can enhance the reaction surfactant-CO_2 and hence the CO_2 absorption capacity [47]. Other studies demonstrate the performance of amine-infused hydrogels in high internal phase emulsion. Although the high CO_2 absorption performance, they exhibit operational issues such as foaming and high pressure drop that constrain its use only to CO_2/H_2 separation [48]. Polymerized IL solvents have been used to produce porous polymeric solid materials [49].

12.4.2.2 Nanosolvents

This type of solvents combine the advantages of both liquid solvents such as high selectivity, kinetics, and CO_2 absorption capacity and solid solvents, in terms of high resistance, low volatility, and mass transfer area. Typically, the nanosolvents are formed by a solvent coated with a permeable polymer producing spherical-shaped nanoparticles. The CO_2 absorption process is similar to conventional amine-based solvents. The unique modification is related to CO_2 diffusion across the permeable polymeric shell prior to capturing CO_2 chemically into the core liquid. The regeneration occurs by heating. The selection of materials is crucial to maintain the cyclic capacity of the solvent and the mechanical integrity of the spherical-shaped nanoparticles [47].

12.4.2.3 Porous Liquid Solvents

Porous liquids are neat liquids that dispose permanent cavities. They are classified into three types – types I, II, and III. Type I porous liquids exhibit the best performance for CO_2/N_2 separation such as large pore size, lower viscosity, and volatility. In particular, the combination of PDMS with amino-functionalized imidazole-based zeolites reports promising results for CCS in terms of low viscosity – around 50 mPa s – and elevated CO_2 absorption capacity. In addition, their properties can be adjusted at desirable values by varying the size of the pore, the molecular weight of the PDMS external layers covering the zeolite, and the number of amino groups included in the zeolite framework [50].

12.5 COMMERCIAL SOLVENTS USED AT INDUSTRIAL SCALE

The current solvents commercially developed are summarized in Table 12.10. They were tested using flue gases from coal-fired power plant units close to depleted oil fields for enhanced oil recovery (EOR). They can provide CO_2 purities over 99%. Although operational issues such as foaming, fouling, and plugging have been reported, the facilities ensured capacity factors above 75% [41].

The energy requirements reported for solvent regeneration range between 2.3 and 3 GJ/t CO_2. KoSol-5 solvent can reduce further its energy demand and reach values below 2.3 GJ/t CO_2 combined advanced layout configuration and heat integration [41]. Results from Petra Nova facility state that KM CDR process based on the advanced sterically hindered amine solvent, namely, KS-1, can capture CO_2 at 55–60 €/t CO_2, near to the reference cost – 40 €/t CO_2 – indicated by DOE for

TABLE 12.10

Summary of Commercial-Scale Solvents Tested at Large-Scale Facilities

Solvent	KS-1	Cansolv	KoSol's	Econoamine FG+
Maximum CO_2 capture capacity (Mt/y)	1.4	1	0.12	0.6
TRL	Commercial (10)	Commercial (10)	Large Demo (9)	Commercial (10)
Facility	Petra Nova	Boundary Dam	Shidongkou	-
Power (MWe)	240	115	-	-
Company	NRG Energy	Shell	China Huaneng Group	Fluor
CO_2 source	Coal-fired PC	Lignite-fired PC	Coal-fired PC	Hard coal PC
Regeneration energy (GJ/t CO_2)	2.5–3	2.3	KoSol-4 3	2.9
			KoSol-5 2.5–2.6	
CO_2 utilization	EOR	EOR	Food industry	-
Reference	-	[51]	[52]	[41]

economic feasibility of CCS applied to energy production from fossil fuels [30,36]. In addition, further reduction up to 30% and 20% may be achieved on investment and operating cost, respectively, from experiences carried out on this facility in operation [51].

12.6 CONCLUSIONS

Concerns about climate change are promoting mitigation actions against climate change. In this sense, post-combustion capture based on chemical absorption is a mature technology that can be applied at industrial processes that use fossil fuels as an energy source. Currently, the application of this CCS technology is more focused on energy-intensive industrial processes such as cement production, oil refining, and iron and steel instead of its prime application field – energy production from fossil fuels.

In the last decades, much efforts have been done to strengthen chemical absorption in order to be ready for commercial-scale implementation. Several commercial units can be found under operation – Boundary Dam and Petra Nova – avoiding the emission of more than 1 Mt CO_2 per year in each unit. Investment should be returned back based on the utilization of the captured CO_2. At this moment, EOR is the most economically attractive option but it seems that a CO_2 by-product market should enhance the implementation of CO_2 capture technologies.

Chemical absorption technology has its main drawbacks: energy penalty and amine emissions and the new development of solvents play a key role to overcome these barriers. Conventional amines show a linear behavior in terms of kinetics and regeneration energy. Those amine-based solvents that exhibit high absorption rates have higher energy requirements during the regeneration step, and vice versa. Novel blend formulations can partially overcome this issue but the amine emission associated with solvent degradation and amine volatility keep limiting their use at industrial scale.

This chapter summarizes the cutting edge of solvents currently under research for CCS applications. Based on the state of art in the field of CO_2 absorption, IL, phase-change solvents, and polymeric liquid solvents are most promising. IL shows extremely low volatility, and then, it should be neglected. However, the mass transfer limitation due to high viscosity significantly reduces the absorption rates, and hence, the investment costs of absorber units are prohibitive. Phase-change solvents can potentially reduce the energy penalty during the solvent regeneration by means of lowering the amount of solvents to be regenerated but they show elevated volatility during operation. The most novel family of solvents is porous liquid. They exhibit promising results in terms of CO_2 absorption capacity and CO_2 desorption but they are only evaluated at a laboratory scale.

REFERENCES

[1] F. Vega, S. Camino, L. M. Gallego, M. Cano, and B. Navarrete, Experimental study on partial oxy-combustion technology in a bench-scale CO_2 capture unit, *Chemical Engineering Journal*, 362 (2019), pp. 71–80.

[2] G. T. Rochelle, Thermal degradation of amines for CO_2 capture, *Current Opinion in Chemical Engineering*, 1 (2012), pp. 183–190.

[3] J. Oexmann and A. Kather, Minimising the regeneration heat duty of post-combustion CO_2 capture by wet chemical absorption: The misguided focus on low heat of absorption solvents, *International Journal of Greenhouse Gas Control*, 4 (2010), pp. 36–43.

[4] M. K. Mondal, H. K. Balsora and P. Varshney, Progress and trends in CO_2 capture/separation technologies: A review, *Energy*, 46 (2012), pp. 431–441.

[5] L. Kohl and R. B. Nielsen, *Gas Purification*, 5th ed. Gulf Publisher Company, Houston, 1997.

[6] T. Sema, A. Naami, K. Fu, M. Edali, H. Liu, H. Shi, Z. Liang, R. Idem and P. Tontiwachwuthikul, Comprehensive mass transfer and reaction kinetics studies of CO_2 absorption into aqueous solutions of blended MDEA-MEA, *Chemical Engineering Journal*, 209 (2012), pp. 501–512.

[7] Y. Artanto, J. Jansen, P. Pearson, G. Puxty, A. Cottrell, E. Meuleman and P. Feron, Pilot-scale evaluation of AMP/PZ to capture CO_2 from flue gas of an Australian brown coal-fired power station, *International Journal of Greenhouse Gas Control*, 20 (2014), pp. 189–195.

[8] P. D. Vaidya and E. Y. Kenig, CO_2-alkanolamine reaction kinetics: A review of recent studies, *Chemical Engineering and Technology*, 30 (2007), pp. 1467–1474.

[9] A. Aboudheir, P. Tontiwachwuthikul, A. Chakma and R. Idem, Kinetics of the reactive absorption of carbon dioxide in high CO_2-loaded, concentrated aqueous monoethanolamine solutions, *Chemical Engineering Science*, 58 (2003), pp. 5195–5210.

[10] P. Spigarelli and S. K. Kawatra, Opportunities and challenges in carbon dioxide capture, *Journal of CO_2 Utilization*, 1 (2013), pp. 69–87.

[11] H. P. Huang, Y. Shi, W. Li and S. G. Chang, Dual alkali approaches for the capture and separation of CO_2, *Energy Fuels*, 15 (2001), pp. 263–268.

[12] A. Olajire, CO_2 capture and separation technologies for end-of-pipe applications – A review, *Energy*, 35 (2010), pp. 2610–2628.

[13] H. Knuutila, H. F. Svendsen and M. Anttila, CO_2 capture from coal-fired power plants based on sodium carbonate slurry; a systems feasibility and sensitivity study, *International Journal of Greenhouse Gas Control*, 3 (2009), pp. 143–151.

[14] G. Barbieri, A. Brunetti, F. Scura and E. Drioli, CO_2 separation by membrane technologies: Applications and potentialities, *Chemical Engineering Transactions*, 24 (2011), pp. 775–780.

[15] M. Aghaie, N. Rezaei and S. Zendehboudi, A systematic review on CO_2 capture with ionic liquids: Current status and future prospects, *Renewable and Sustainable Energy Reviews*, 96 (2017), pp. 502–525.

[16] D. Hospital-Benito, J. Lemus, C. Moya, R. Santiago, V. R. Ferro and J. Palomar, Techno-economic feasibility of ionic liquids-based CO_2 chemical capture processes, *Chemical Engineering Journal*, 407 (2021), p. 127196.

[17] M. Ramdin, T. W. De Loos and T. J. H. Vlugt, State-of-the-art of CO_2 capture with ionic liquids, *Industrial and Engineering Chemistry Research*, 51 (2012), pp. 8149–8177.

[18] O. Y. Orhan, Effects of various anions and cations in ionic liquids on CO_2 capture, *Journal of Molecular Liquids*, 333 (2021), p. 115981.

[19] D. Hospital-Benito, J. Lemus, C. Moya, R. Santiago and J. Palomar, Process analysis overview of ionic liquids on CO_2 chemical capture, *Chemical Engineering Journal*, 390 (2020), p. 124509.

[20] B. Kazmi, F. Raza, S. A. A. Taqvi, Z. ul H. Awan, S. I. Ali and H. Suleman, Energy, exergy and economic (3E) evaluation of CO_2 capture from natural gas using pyridinium functionalized ionic liquids: A simulation study, *Journal of Natural Gas Science and Engineering*, 90 (2021), p. 103951.

[21] F. Vega, M. Cano, S. Camino, L. M. G. Fernández, E. Portillo and B. Navarrete, Solvents for carbon dioxide capture, in *Carbon Dioxide Chemistry, Capture and Oil Recovery*, 1st ed., edited by I. Karamé, J. Shaya and H. Srour, Intech Open, London, 2018.

[22] A. Hafizi, M. Rajabzadeh, M. H. Mokari and R. Khalifeh, Synthesis, property analysis and absorption efficiency of newly prepared tricationic ionic liquids for CO_2 capture, *Journal of Molecular Liquids*, 324 (2021), p. 115108.

[23] X. Luo and C. Wang, The development of carbon capture by functionalized ionic liquids, *Current Opinion in Green and Sustainable Chemistry*, 3 (2017), pp. 33–38.

[24] C. Wang, H. Luo, H. Li, X. Zhu, B. Yu, and S. Dai, Tuning the physicochemical properties of diverse phenolic ionic liquids for Equimolar CO_2 capture by the substituent on the anion, *Chemistry— A European Journal*, 18 (2012), pp. 2153–2160.

[25] R. Wang, Y. Yuying, W. Mengfan, L. Jinshan, Z. Shihan, A. Shanlong and W. Lidong, Energy efficient diethylenetriamine–1-propanol biphasic solvent for CO_2 capture: Experimental and theoretical study, *Applied Energy* 290 (2021). https://doi.org/10.1016/j.apenergy.2021.116768.

[26] L. M. Galán Sánchez, G. W. Meindersma and A. B. de Haan, Kinetics of absorption of CO_2 in amino-functionalized ionic liquids, *Chemical Engineering Journal*, 166 (2011), pp. 1104–1115.

[27] C. D. Freitas, L. P. Cunico, M. Aznar and R. Guirardello, Modeling vapor liquid equilibrium of ionic liquids + gas binary systems at high pressure with cubic equations of state, *Brazilian Journal of Chemical Engineering*, 30 (2013), pp. 63–73.

[28] S. Zeng, X. Zhang, L. Bai, X. Zhang, H. Wang, J. Wang, D. Bao, M. Li, X. Liu and S. Zhang, Ionic-liquid-based CO_2 capture systems: Structure, Interaction and Process, *Chemical Reviews*, 117 (2017), pp. 9625–9673.

[29] H. Zhai and E. S. Rubin, Systems analysis of ionic liquids for post-combustion CO_2 capture at coal-fired power plants, *Energy Procedia*, 63 (2014), pp. 1321–1328.

[30] J. de Riva, J. Suarez-Reyes, D. Moreno, I. Díaz, V. Ferro and J. Palomar, Ionic liquids for post-combustion CO_2 capture by physical absorption: Thermodynamic, kinetic and process analysis, *International Journal of Greenhouse Gas Control*, 61 (2017), pp. 61–70.

[31] M. T. Mota-Martinez, P. Brandl, J. P. Hallett and N. MacDowell, Challenges and opportunities for the utilisation of ionic liquids as solvents for CO_2 capture, *Molecular System Design and Engineering*, 3 (2018), pp. 560–571.

[32] M. B. Shiflett, D. W. Drew, R. A. Cantini and A. Yokozeki, Carbon dioxide capture using ionic liquid 1-butyl-3-methylimidazolium acetate, *Energy Fuels*, 24 (2010), pp. 5781–5789.

[33] J. Zhang, Y. Qiao and D. W. Agar, Intensification of low temperature thermomorphic biphasic amine solvent regeneration for CO_2 capture, *Chemical Engineering Research and Design*, 90 (2012), pp. 743–749.

[34] H. Zhou, X. Xu, X. Chen and G. Yu, Novel ionic liquids phase change solvents for CO_2 capture, *International Journal of Greenhouse Gas Control*, 98 (2020), p. 103068.

[35] R. Wang, Y. Yang, M. Wang, J. Lin, S. Zhang, S. An and L. Wang, Energy efficient diethylenetriamine–1-propanol biphasic solvent for CO_2 capture: Experimental and theoretical study, *Applied Energy*, 290 (2021), p. 116768.

[36] S. Zhang, Y. Shen, L. Wang, J. Chen and Y. Lu, Phase change solvents for post-combustion CO_2 capture: Principle, advances, and challenges, *Applied Energy*, 239 (2019), pp. 876–897.

[37] M. Rabensteiner, G. Kinger, M. Koller and C. Hochenauer, Pilot plant study of aqueous solution of piperazine activated 2-amino-2-methyl-1-propanol for post combustion carbon dioxide capture, *International Journal of Greenhouse Gas Control*, 51 (2016), pp. 106–117.

[38] X. Zhou, X. Li, J. Wei, Y. Fan, L. Liao and H. Wang, Novel nonaqueous liquid-liquid biphasic solvent for energy-efficient carbon dioxide capture with low corrosivity, *Environmental Science and Technology*, 54 (2020), pp. 16138–16146.

[39] P. Kazepidis, A. I. Papadopoulos, F. Tzirakis and P. Seferlis, Optimum design of industrial post-combustion CO_2 capture processes using phase-change solvents, *Chemical Engineering Research and Design*, 175 (2021), pp. 209–222.

[40] M. S. Alivand, O. Mazaheri, Y. Wu, G. W. Stevens, C. A. Scholes and K. A. Mumford, Development of aqueous-based phase change amino acid solvents for energy-efficient CO_2 capture: The role of antisolvent, *Applied Energy*, 256 (2019), p. 113911.

[41] E. Nessi, A. I. Papadopoulos and P. Seferlis, A review of research facilities, pilot and commercial plants for solvent-based post-combustion CO_2 capture: Packed bed, phase-change and rotating processes, *International Journal of Greenhouse Gas Control*, 111 (2021), p. 103474.

[42] U. E. Aronu, R. Pellegrin, K. Hjarbo, A. Chikukwa, I. Kim, A. Tobiesen, B. Hargrove and T. Mejdell, Integrated phase change solvent-contactor process for CO_2 scrubbing from industrial exhaust gases: Pilot plant demonstration. In *14th Greenhouse Gas Control Technologies (GHGT-14) Conference, Melbourne, 21-26 October*, (2018), pp. 2–4.

[43] M. Hasib-Ur-Rahman, H. Bouteldja, P. Fongarland, M. Siaj and F. Larachi, Corrosion behavior of carbon steel in alkanolamine/room-temperature ionic liquid based CO_2 capture systems, *International Journal of Greenhouse Gas Control*, 51 (2012), pp. 8711–8718.

[44] W. Jiang, F. Wu, G. Gao, X. Li, L. Zhang and C. Luo, Absorption performance and reaction mechanism study on a novel anhydrous phase change absorbent for CO_2 capture, *Chemical Engineering Journal*, 420 (2021), p. 129897.

[45] Q. Huang, G. Jing, X. Zhou, B. Lv and Z. Zhou, A novel biphasic solvent of amino-functionalized ionic liquid for CO_2 capture: High efficiency and regenerability, *Journal of CO_2 Utilization*, 25 (2018), pp. 22–30.

[46] J. Zhang, O. Nwani, Y. Tan and D. W. Agar, Carbon dioxide absorption into biphasic amine solvent with solvent loss reduction, *Chemical Engineering Research and Design*, 89 (2011), pp. 1190–1196.

[47] A. Sattari, A. Ramazani, H. Aghahosseini and M. K. Aroua, The application of polymer containing materials in CO_2 capturing via absorption and adsorption methods, *Journal of CO_2 Utilization*, 48 (2021), p. 101526.

[48] X. Xu, C. Heath, B. Pejcic and C. D. Wood, CO_2 capture by amine infused hydrogels (AIHs), *Journal of Materials Chemistry A*, 6 (2018), pp. 4829–4838.

[49] M. Sadeghpour, R. Yusoff and M. K. Aroua, Polymeric ionic liquids (PILs) for CO_2 capture, *Reviews in Chemical Engineering*, 33 (2017), pp. 183–200.

[50] X. Li, D. Yao, D. Wang, Z. He, X. Tian, Y. Xin, F. Su, H. Wang, J. Zhang, X. Li, M. Li and Y. Zheng, Amino-functionalized ZIFs-based porous liquids with low viscosity for efficient low-pressure CO_2 capture and CO_2/N_2 separation, *Chemical Engineering Journal*, 429 (2022), p. 132296.

[51] M. G. Plaza and C. Pevida, Current status of CO_2 capture from coal facilities, in *New Trends in Coal Conversion*, edited by I. Suárez-Ruiz, M. A. Díez, and F. Rubiera, Woodhead Publishing, 2019, pp. 31–58.

[52] J. H. Lee, N. Kwak, H. Niu, J. Wang, S. Wang, H. Shang and S. Gao, KEPCO-China Huaneng post-combustion CO_2 capture pilot test and cost evaluation, *Korean Chemical Engineering Research*, 58 (2020), pp. 150–162.

13 Metal Oxide-Based Nanocomposites for Photocatalytic Reduction of CO_2

Rashi Gusain, Neeraj Kumar, and Suprakas Sinha Ray
University of Johannesburg
Council for Scientific and Industrial Research

CONTENTS

13.1 INTRODUCTION

Continuous emission of greenhouse gases into the environment is causing global warming and alarming the survival on the planet Earth [1]. CO_2 is one of the primary gases of greenhouse gases and contributes significantly to global climate changes. The National Oceanic and Atmospheric Administration's Mauna Loa observatory noticed that the CO_2 concentration had surpassed 415 ppm in the atmosphere, which is the maximum level recorded yet [2]. International Panel on Climate Change (IPCC) anticipated that by 2,100, the level of CO_2 will reach up to 590 ppm, and the mean temperature of the globe will rise by 1.90°C [3]. It will affect globally and cause several severe issues on the planet, e.g., melting the ice at Earth's pole, which can also increase the sea level [4]. Fast industrialization, combustion of fossil fuels, decay of plants and animals carcasses by microorganisms, and respiration are the major causes of CO_2 emission into the environment. Figure 13.1 displays the various sources of CO_2 emission in the environment. Therefore, there is serious demand to mitigate CO_2 emissions and find renewable energy resources, which can alleviate the global warming effect and meet the growing energy demand.

In general, CO_2 can be reduced in the atmosphere by lowering the direct emission of CO_2 in the atmosphere, CO_2 capture and storage (CCS), or CO_2 utilization by converting it into valuable chemicals or fuels [5–8]. Capturing CO_2 and artificial conversion of CO_2 into useful chemicals

DOI: 10.1201/9781003206385-13

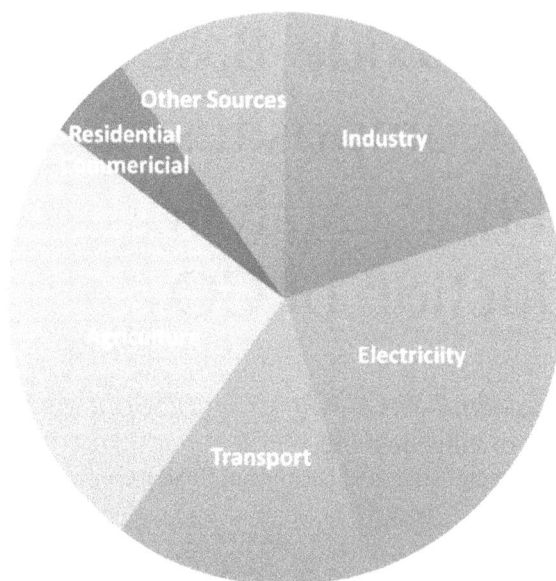

FIGURE 13.1 Various sources of CO_2 emission in the atmosphere.

and fuels are gaining the attention to develop various strategies through biological, photochemical, thermochemical, or electrochemical approaches [9,10]. Photocatalytic reduction of CO_2 is an environmentally friendly, clean, and sustainable approach to convert CO_2 into valuable products such as methane (CH_4), methanol (CH_3OH), carbon monoxide (CO), formic acid (HCOOH), acetic acid (CH_3COOH), etc. [1]. Artificial photocatalytic conversion of CO_2 not only mitigates the CO_2, but the reaction products (chemical fuels) can also meet the future energy demand. Inoue et al. were the first to propose the photo-electrochemical reduction of CO_2 using several semiconductor electrodes (TiO_2, ZnO, CdS, GaP, SiC, and WO_3) in 1979 [11]. Later, the photocatalytic reduction of CO_2 into CH_4 was proposed by Thampi et al. using Ru-loaded TiO_2 [12]. Afterward, considerable interest has been arisen to develop strategies for the enhanced photocatalytic reduction of CO_2.

Photocatalytic remediation of CO_2 is a well-accepted advanced oxidation process (AOP) to reduce CO_2 using a semiconductor as a photocatalyst. Semiconductor materials with appropriate bandgaps play a vital role in photocatalysis. Various semiconductor materials such as metal oxide (e.g., TiO_2, ZnO, etc.) [13,14], metal dichalcogenides (e.g., MoS_2, WS_2, etc.) [15,16], carbon-based materials (graphene derivatives, g-C_3N_4, carbon quantum dots, etc.) [17,18], MXenes [19], layered dihydroxides [20], and polymer-based nanocomposite [21] materials have been studied for the photocatalytic conversion of CO_2. Metal oxides are abundantly available, economically approachable, exhibit large active surface area, controlled morphology, and are easy to modify and are investigated for various applications, including photocatalytic CO_2 reduction. TiO_2 was the first catalytic material used for the CO_2 catalytic conversion into chemicals [11]. Since then, several metal oxides and metal oxide-based nanocomposites such as magnetic metal oxides [22], graphene-metal oxides [23], metal-metal oxides [24], and porous material-supported metal oxides [25] have also been extensively explored for the photocatalytic CO_2 conversion into valuable chemicals. For efficient photocatalysis, the designing of an appropriate photocatalyst is also an important step. There are several reviews on the photo-reduction of CO_2 using various semiconductor materials. But the role of photocatalysts based on several metal oxides for photocatalytic CO_2 reduction at one place is still scarce. Therefore, this chapter is providing the insights of the application of various metal oxides and metal oxide nanocomposites for the CO_2 photocatalytic reduction. The following section will explain the fundamentals of the CO_2 photocatalytic reduction mechanism and the role of metal oxides as photocatalyst in the photo-reduction of CO_2.

13.2 PHOTOCATALYTIC REDUCTION OF CO_2

Heterogeneous photocatalysis is an advanced oxidation process (AOP), which has also popularly been investigated for the photocatalytic reduction of CO_2 into useful solar fuels and chemicals [26–28]. Photocatalysis is an attractive alternative to various thermos-catalytic reactions performed at high temperatures and pressure. It can easily be operated at ambient conditions using a renewal source of energy supply. Semiconductor materials are used as photocatalysts due to the presence of a suitable bandgap. In a typical photocatalysis reaction, under solar or simulated solar light irradiation, it involves the absorption of a photon by the semiconducting material (photocatalyst) to excite the e^- in the conduction band (CB) and leaving h^+ in behind the valence band (VB) (Figure 13.2). To generate the charge pairs, the energy of the photon (E_{hv}) should be equal to or greater than the bandgap energy (E_g) of the semiconductor material. These charge pairs (e^- and h^+) are responsible for initiating the various photocatalytic redox reactions. The general mechanism of the photocatalytic reaction is shown in Figure 13.2. This process generally includes the following steps: (i) adsorption of the CO_2 molecules on the surface of semiconducting photocatalyst; (ii) absorption of the light photon by the semiconductor photocatalyst; (iii) generation, separation, and transport of the charge pairs on the surface of photocatalyst; (iv) generation of reaction product via the photochemical redox reaction at the surface of photocatalyst; and (v) desorption of the reaction product from the photocatalyst surface for further usage.

CO_2 is a highly thermodynamically stable molecule and usually necessitates high energy inputs to further transform C=O into other units such as C-O, C-H, etc. It has been observed that the appropriate CB and VB reduction potentials of the semiconductor photocatalyst materials regulate the reactivity of the photogenerated e^- and h^+ pairs, which drives to reduce the CO_2 molecule into various other value-added chemicals. The precise mechanism of CO_2 photo-reduction is still unclear and can be explained via several elementary reaction pathways [29]. As shown in Figure 13.2, the photocatalytic reduction of CO_2 can be directed using one electron via the formation of $CO_2^{\bullet-}$ as an intermediate and follow the subsequent steps. However, it is challenging as it requires the intensively negative redox potential (–1.9 eV) of the photocatalyst.

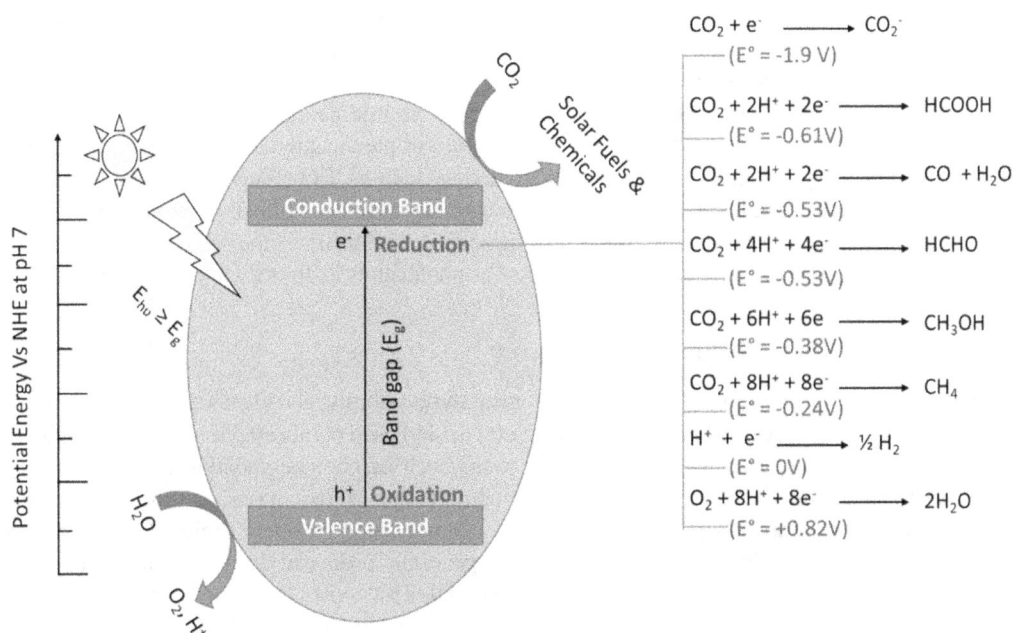

FIGURE 13.2 General mechanism of CO_2 photo-reduction.

Multistep reduction of CO_2 using multielectrons and protons is an alternative and approachable pathway with significantly lower redox potential. Therefore, the artificial photocatalytic reduction of CO_2 is a multistep reaction, which requires 2, 4, 6, 8, or 12 electrons and protons [30]. As a result of CO_2 photo-reduction, a few diversified products can be generated, such as CO, CH_4, CH_3OH, HCHO, HCOOH, and so on. These products generations via CO_2 photo-reduction are associated with the semiconductor materials' reduction potentials as shown in Figure 13.2 to normal hydrogen electrode (NHE) at pH 7. Therefore, the CB and VB potentials always play a vital role in determining the reactivity of photo-induced charge pairs and the product generation. During photocatalysis exercises, photogenerated electrons actively participate in the CO_2 reduction and holes in the water oxidation. Furthermore, doping of semiconducting materials with co-catalyst such as metal or metal oxide and introduction of defects delay the recombination of photogenerated charge pairs and enhance the photocatalytic efficiency [31,32].

There are several challenges to developing an efficient heterogeneous photocatalyst for CO_2 reduction. To perform the highly efficient photocatalytic reduction, the photocatalyst should exhibit the following characteristics: (i) active response for the visible and near-infrared spectrum of the solar light; (ii) high surface area; (iii) semiconductor materials should generate multiple electrons and protons easily; (iv) appropriate bandgap and the CB potential edge should be more negative than the required theoretical values; (v) longer life span of photo-induced charge carriers; (vi) beneficial defects, edges, and dangling bonds; and (vii) ease to functionalize [33]. An appropriate photocatalyst should actively respond to visible and near-infrared light irradiation, as these are the major portion of the sunlight spectrum. Various metal oxide nanoparticles and nanocomposites are promising heterogeneous photocatalysts to reduce CO_2 into various value-added chemicals and solar fuels. The following section will be detailed about the metal oxides and metal oxide nanocomposites for the CO_2 photo-reduction.

13.3 METAL OXIDE NANOCOMPOSITES FOR PHOTOCATALYTIC REDUCTION OF CO_2

Metal oxide nanoparticles have been extensively investigated in various applications such as lubrication, electrochemistry, catalysis, energy storage, sensors, biomedical and environmental remediation [34–43]. The high active surface area and surface-to-volume ratio of metal oxide nanoparticles make them suitable candidates for various surface-associated applications [44]. Metal oxide-nanostructured materials are crystalline, controlled structural, nontoxic, environmental, and thermal stability and exhibit appropriate bandgap, which shows their feasibility in photocatalytic environmental remediation applications [45]. Several metal oxide nanomaterials such as TiO_2, ZnO, CuO, WO_3, and their nanocomposites with organic and/or inorganic nanomaterials have been broadly explored for the photocatalytic reduction of CO_2 [46–49]. The following section demonstrates the presentation of various metal oxide nanoparticles and their nanocomposites as photocatalysts for the CO_2 photo-reduction.

13.3.1 TITANIA (TiO_2)-BASED NANOCOMPOSITES

Owing to low cost, nontoxic, rich availability, photocatalytic stability, and chemical inertness, TiO_2 is one of the most widely accepted and investigated metal oxide-based photocatalyst materials [46,50,51]. It exhibits a wide range of photocatalysis applications such as photodegradation of contaminants, water desalination, water splitting, dye bleaching, and reduction of toxic gases. It is the first material used for CO_2 photo-reduction [52], and since then, it has been extensively explored for its photocatalytic characteristics. Methane and methanol were generally produced due to photocatalytic treatment of CO_2 using TiO_2 due to its high reduction potential. TiO_2 exists in five polymorphs in nature, i.e., anatase, brookite, rutile, monoclinic, and orthorhombic. Among all the polymorphs of TiO_2, anatase and rutile forms are most stable and display exceptionally promising photocatalytic activity. The physical and chemical characteristics of the material change with the change in phase structure, hence determining the photocatalytic performance. For advanced applications, several TiO_2 nanostructures have been engineered for photocatalytic purposes, such as nanorods [53], nanotubes [54], nanofibers [55], and TiO_2 nanocomposites [46] etc., and their different morphologies significantly impact the photocatalytic air purification [56]. Among these all nano-architectures, 1-dimensional TiO_2 has been implicated as highly efficient photocatalyst due to high surface-to-volume ratio and vectorial electron percolation pathways [57,58]. For example, Figure 13.3 represents the various nanostructures of TiO_2 (solid microspheres (TSMS), mesoporous microspheres (TMMS), hollow spheres (THS), nanosheets (TNS), nanotubes (TNT), sea-urchin (TSU)), and their effect on methyl ethyl ketone (MEK) photo-oxidation [56]. For practical applications, researchers have been employed several strategies to reduce the cost and enhance the photocatalytic potential of TiO_2-based photocatalysts.

Facets engineering of material is a promising strategy and significantly impacts the photocatalytic potential of TiO_2 nanoparticles [59–61]. The fluorinated anatase TiO_2 (F-TiO_2) nanosheets exhibit a high percentage of exposed [001] facets with high surface energy compared to [101] facets and hence favored the photocatalytic CO_2 reduction [62]. Functionalization of fluorine on TiO_2 was performed by treating TiO_2 with various amounts of hydrofluoric acid (HF) (2, 4, 6, 8, and 12 mL). The fluorination of TiO_2 altered the surface into electronegative and attracted the electron-deficient CO_2, while the optical characteristics of TiO_2 remained unaffected. The photocatalytic CO_2 reduction potential of F-TiO_2 nanosheets was also compared with TNPs, TNTs, and TNSs. Among all TiO_2 nanocomposites, 6HF-TiO_2 with [001]:[101] facets ratio of 72:28 shows the highest potential of CO_2 photoreduction under UV light illumination for 20 hours. Furthermore, the fluorine presence on the surface of F-TiO_2 prolongs the life span of charge carriers and improves photocatalytic activity. The amount of HF can easily adjust the growth of [101] and [001] facets during TiO_2 nanosheets synthesis. Han et al. also supported the higher reactivity of [001] facets for photocatalytic CO_2 reduction than low energy [101] facets due to the high density of exposed Ti and O atoms [60].

TiO_2 exhibits a wide bandgap (TiO_2 (Rutile): 3.03 eV & TiO_2 (Anatase): 3.2 eV) and allows only the absorption of UV-range photons, which is only 4% of the solar spectrum [63]. To allow the absorption of visible and near-infrared photons, the bandgap of TiO_2 nanomaterials is engineered

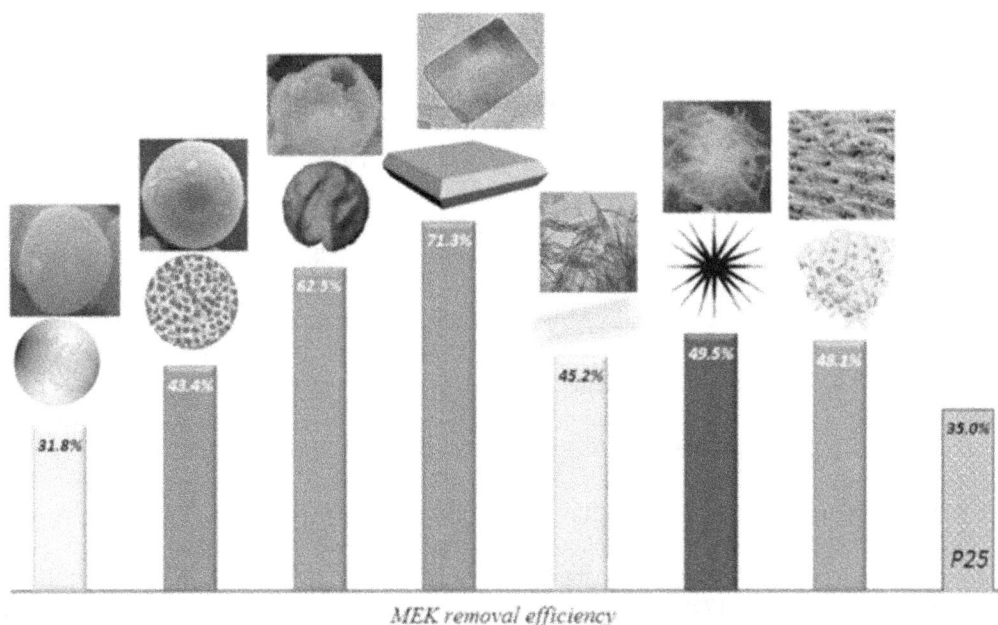

MEK removal efficiency

FIGURE 13.3 FESEM images and photocatalytic efficiency of various TiO_2 nanoparticles with different morphologies. (Reproduced with permission from [56]. Copyright 2020 Elsevier Science Ltd.)

by introducing defects [64], doping [65], coupling with semiconductors [66], surface modification, and functionalizing with some organic and inorganic foreign materials. The visible light response of TiO_2 can also be enhanced by introducing the color centers [67]. This can be done either by the heat treatment in the inert environment or by small cations (e.g., H^+, Li^+) intercalation. Fabrication of TiO_2 with metals, nonmetals, and metal oxides semiconductors as co-catalyst has been popularly accepted to improve photocatalytic performance. The coupling of TiO_2 with other semiconductor materials with different energy levels facilitates advancement in the photocatalytic approach by enhancing the charge separation, delaying recombination of charge carriers, and improving interfacial charge transferring ability. For example, CuO-TiO_2 hollow microspheres were also reported as a photocatalytic system with improved light-harvesting efficiency for the photo-reduction of CO_2 into CO and CH_4 under UV light illumination [68]. Unlike TiO_2 hollow microsphere, the enhanced photo-reduction of CO_2 in the presence of CuO-integrated TiO_2 hollow microsphere photocatalyst can be attributed to the improved e^- trapping, delayed recombination of photogenerated e^- and h^+, and visible light harvesting. Incorporation of TiO_2 surface with noble metals (e.g., Ag, Au, Pd) is the most favorable approach to improve the visible light harvesting and charge separation efficiency of the material by localized surface plasmon resonance (LSPR). Doping of TiO_2 with metals causes overlap of Ti 3d orbitals with metals d orbitals, resulting in the shift in the absorption spectrum and favoring the absorption of visible light photons. The rate of CH_4 generation via CO_2 photo-reduction using noble metal-doped TiO_2 was found in the following order for different noble metals $Ag < Rh < Au < Pd < Pt$ [69]. Generally, Ag and Au nanoparticles are loaded on the TiO_2 surface and investigated for CO_2 photo-reduction due to their comparable stability in air. Ag-incorporated TiO_2 nanomaterials usually promote the CH_4 and CO production from CO_2 photo-reduction under visible light [70,71]. Ag nanoparticles on TiO_2 surface enhance photocatalysis by plasmonic effect with hot electron evolution [71]. The synergetic effect of interfacial charge transfer and plasmonic hot electron injection in the Ag-TiO_2 heterostructures leads to enhancement in the photo-reduction of CO_2. Au is also known to exhibit the LSPR effect on TiO_2 nanostructural materials. Montmorillonite (MMT)-dispersed Au/TiO_2 nanomaterial was prepared using the sol-gel method and used for solar-simulated CO_2 photo-reduction to generate CO [72]. The synergetic effect of delayed recombination of photo-induced charge carriers by MMT and LSPR by Au leads to faster adsorption and catalytic photo-reduction of CO_2 and desorption of CO. The photo-reduction efficiency of MMT-supported Au/TiO_2 was six times more than that of bare TiO_2.

Nonmetal doping (e.g., N, F, C, S) of TiO_2 at the O sites has also revealed great success in attaining visible light harvesting during photocatalysis [73,74]. Mixing O 2p orbitals with the p orbitals of doped nonmetals results in narrowing the bandgap of TiO_2 and visible light absorption. Among all nonmetals, N and C are the most favorable dopants for enhancing the photocatalytic efficiency of TiO_2 under visible light [75,76]. Due to the comparable atomic size of N with O, it can easily be doped in TiO_2 structure. Considerable efforts have been put to dope N into TiO_2 bulk or on the surface using dry and wet preparation methods. The excellent photocatalytic activity of N-doped TiO_2 is due to the strong visible light harvesting, good crystallinity, and quick photo-induced charge carrier separation [77]. The fast charge separation is accredited to developing paramagnetic [O-Ti^{4+}-N^{2-}-Ti^{4+}-VO] clusters in N-TiO_2 nanomaterial. During the doping process, the created oxygen vacancy trapped the e^- and stimulated to produce super-oxygen anion radical ($O_2^{\bullet-}$), which acts as a necessary reactive species in photocatalytic treatment. Other than N, O atoms from the TiO_2 matrix can easily be substituted by the F atoms due to similar ionic radius [78]. It was also observed that the hydrogen and fluorine in hydrogenated F-doped TiO_2 enables the adsorption in UV, visible, and infrared light irradiation along with fast photo-induced charge separation [78]. The doping of S into TiO_2 lattice can narrow the bandgap of material from 3.2 to 1.7 eV and performs with higher photocatalytic reduction potential [79]. As a result of nonmetal doping in the TiO_2 matrix, it changes the lattice parameters of the materials, and created the trap states within the band edges from electronic perturbations, which results in bandgap narrowing [80]. This not only allows visible light harvesting but also increases the life span of photo-induced charge carriers. Several researchers have also done on the co-doping of nonmetals in TiO_2 for improved photocatalytic purposes under visible light irradiation [81]. This provides the advantage of the presence of atom N by enhancing the visible light response and atom F by enhanced charge separation.

The photocatalytic efficiency of TiO_2 is also affected by the various supports/immobilization materials. The fixation of TiO_2 material on solid supports, such as zeolites [82], ceramic foam [83], non-woven fabric [84], and porous metal [85], reduces the active surface area and limits the mass transfer, hence declining its photocatalytic performance. However, the dispersed TiO_2 fixed on zeolites and other mesoporous support displays advanced potential [86]. TiO_2/Y-zeolite photocatalyst with little quantity of Ti content has shown high photocatalytic conversion rate of CO_2 into CH_4 (12.5 µmol/gcat/h) than powder TiO_2 (0.3 µmol/gcat/h) [87]. In TiO_2/Y-zeolite photocatalyst, zeolite frameworks offer exceptional pore channels, ion-exchange capacities, and remarkable internal surface topologies, promoting CO_2 absorption and selective photo-reduction. The functionalization of TiO_2 with graphene derivatives (e.g., graphene oxide (GO), reduced GO (rGO), graphene quantum dots (GQDs)) is also an attractive approach to improve the photocatalytic performance. Tan et al. reported the photocatalytic potential of GO-doped-O-rich TiO_2 (GO–OTiO_2) for catalytic photo-reduction of CO_2 under low power energy saving daylight bulbs irradiation [88]. On introducing GO into the photo-stability of oxygen-rich TiO_2 has significantly improved. Also, the photo-activity of GO–OTiO_2 was found to be 1.6 and 14.0 times higher than that of oxygen-rich TiO_2 and commercial Degussa P25, respectively. This was due to the synergistic effects of visible light harvesting by oxygen-rich TiO_2 and quick separation and transfer of charge carriers by GO nanosheets. In another study, Lin et al. reduced the CO_2 in the presence of H_2O vapor under Xe lamp irradiation using TiO_2/N-doped rGO (TiO_2/NrGO) nanocomposites [89]. The quick e^- transfer and enhanced CO_2 adsorption on the rGO surface promote the CO_2 photo-reduction using TiO_2/NrGO. Other than graphene derivatives, other nanocarbon families have also been fabricated with TiO_2 to enhance photocatalytic performance. Graphitic carbon nitride (g-C_3N_4), with an increased concentration of N on the surface of C exhibits a narrow bandgap (2.7 eV) and has been incorporated with TiO_2 for CO_2 photo-activation under visible light. In one study, g-C_3N_4 was dispersed with Cu-TiO_2 and showed excellent CO_2 photo-conversion into CH_4 [90]. Similarly, Z-scheme P-O linked g-C_3N_4/TiO_2 nanocomposite was also fabricated for CO_2 photo-reduction with an enhanced visible light response and separation and transport of charge carriers [91]. As a result of CO_2 photocatalytic reduction, several products formed, such as acetic acid (46.9 mg/L/h), methanol (38.2 mg/L/h), and formic acid (28.8 mg/L/h), which was found to be 3.3-, 3.5-, and 3.8-fold higher than the bare TiO_2 nanotubes, respectively.

13.3.2 ZINC OXIDE (ZnO)-BASED NANOCOMPOSITES

ZnO compromises tetrahedral configuration and exhibits three forms: (i) hexagonal wurtzite, (ii) rocksalt, and (iii) cubic zinc blende (Figure 13.4) [92–94]. Among all forms, wurtzite is the most thermodynamically stable. However, a cubic zinc blend can be grown on cubic substrates to enhance stability. ZnO is also one of the most popularly investigated metal oxide photocatalysts for CO_2 photo-reduction. It has been proposed as an alternative to TiO_2 due to its comparable small bandgap and potential to absorb a larger range of the solar spectrum than TiO_2.

Furthermore, the low cost, nontoxicity, high quantum efficiency, and photo-stability promote its application in photocatalysis [33,95]. The mechanism of ZnO for photocatalysis is quite similar to TiO_2 [96]. The photocatalytic efficiency of ZnO was also compared with other semiconductor materials together with TiO_2 for dye degradation under sunlight [97]. ZnO performed the best photocatalytic activity among all semiconductors, which is associated with absorbing a large fraction of the solar spectrum.

Although ZnO exhibits the visible light response toward the solar spectrum, whereas the quick recombination of charge carriers is the barrier for improved photocatalytic activity, this step drops the quantum yield of the product and causes energy wasting. Therefore, to inhibit the recombination of photo-induced charge pairs, ZnO can be fabricated or doped with several other metals, metal

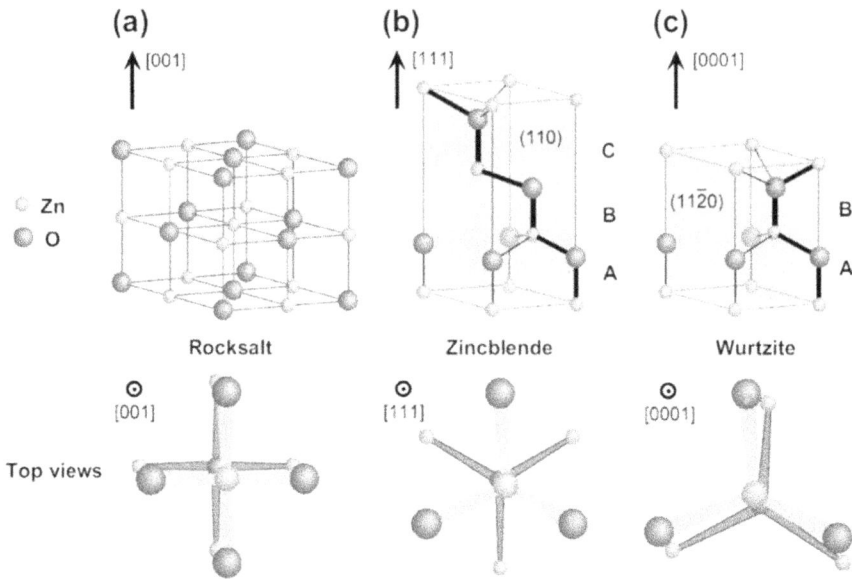

FIGURE 13.4 Ball and stick representation of ZnO crystalline structures and projections of planes of the (a) cubic rocksalt, (b) zinc blende, and (c) wurtzite. (Reproduced with permission from [93]. Copyright 2013 Elsevier Science Ltd.)

oxides, metal chalcogenides, nonmetals, carbon derivatives, etc. [14]. Metal doping can obstruct the e^--h^+ pair recombination step by providing efficient charge separation. Also, metal dopants may act as electron sinks and trap electrons, avoiding recombination [98]. Furthermore, with the enhancement in charge separation, reactive species such as hydroxyl radical and superoxide anions generate, which improves the photocatalytic reaction.

Graphene is one of the most efficient carbon family members used to couple with ZnO to improve the solar spectrum photon absorption, CO_2 adsorption, and photocatalytic activity enhancement. Numerous reports have been published on the designing and synthesis of ZnO and graphene derivative-based photocatalyst [99–101]. In a report, a series of ZnO/reduced GO nanohybrids were prepared by varying the amount of GO (1%, 5%, 10%, and 20%) and employed for the visible light photo-reduction of CO_2 [100]. TEM image (Figure 13.5a) shows the distribution of ZnO nanoparticles on the GO nanosheets. Also, the surface area of ZnO/rGO nanocomposite was observed higher than that of pure ZnO, which can benefit CO_2 adsorption and reduction. The CO_2-to-methanol formation was noticed five times higher for ZnO/rGO than for pure ZnO. The proposed mechanism for CO_2 photo-reduction using ZnO-rGO is shown in Figure 13.5a. Under UV-visible light irradiation, the charge carriers generated, and excited electrons jumped to the CB, leaving holes in the VB. rGO nanosheets provide enough active surface area for CO_2 adsorption and trap the electrons from the CB for CO_2 photo-reduction. Holes are responsible for the oxidation of water into O_2 and H^+. CO_2 reacts with the electron and H^+ and is reduced to methanol. Also, the amount of GO significantly influences the photocatalytic activity. Figure 13.5a shows the methanol production using ZnO/rGO nanocomposite with various amounts of rGO. 10% rGO content in ZnO/rGO nanocomposites shows the highest methanol production, which might be due to material's high active surface area. Further increasing the rGO amount prevents the photon from reaching the ZnO surface and reduces the photocatalytic activity by reducing charge carrier generation.

ZnO-g-C_3N_4 heterostructures were also studied for the CO_2 photo-reduction [102]. The coupling of ZnO with g-C_3N_4 improves the stability of the material. The amount of g-C_3N_4 content in the ZnO matrix also makes a significant impact on photocatalytic efficiency. Figure 13.5b shows the CO_2 photo-reduction product formation (CO, CH_4 and C_2H_4) using different ratios of g-C_3N_4 in ZnO.

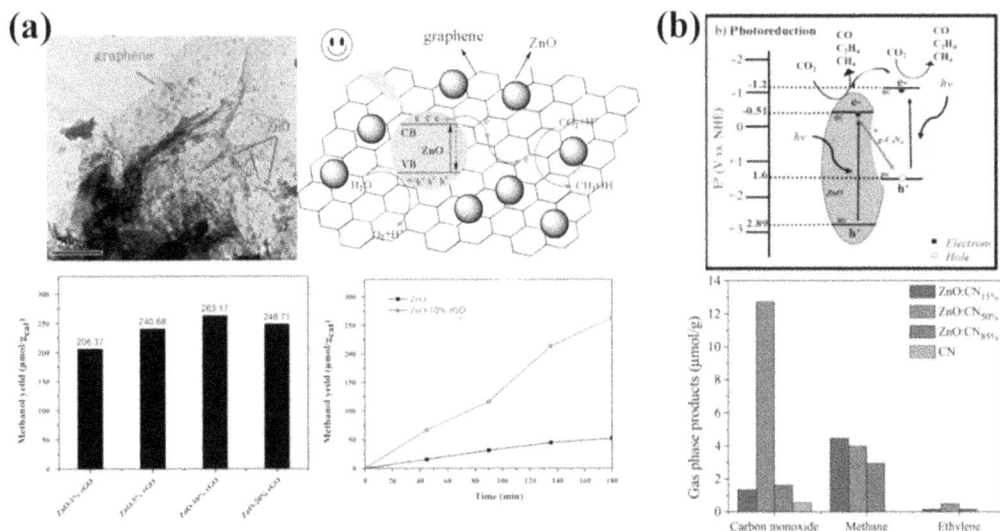

FIGURE 13.5 (a) TEM image of ZnO/rGO nanocomposite; mechanism of CO_2 photo-reduction using ZnO/rGO as photocatalyst; methanol yield using various ZnO/rGO nanocomposites with different amounts of rGO; comparison of ZnO and ZnO/rGO nanocomposites for methanol production by CO_2 reduction under visible light irradiation. (Reproduced with permission from [100]. Copyright 2015 Elsevier Science Ltd.) (b) Bandgap energy and reduction potentials of ZnO-g-C_3N_4 nanocomposite for the CO_2 photo-reduction; Generation of CO, CH_4, and C_2H_4 using g-C_3N_4 and various ZnO-g-C_3N_4 photocatalysts via CO_2 photo-reduction. (Reproduced with permission from [102]. Copyright 2021 Elsevier Science Ltd.)

Pristine g-C_3N_4 could only produce CO in a small amount. However, introducing little amount of ZnO (ZnO-CN85%, 85% of g-C_3N_4 in ZnO) improves the photocatalytic efficiency and produces CO and CH_4. Further increasing the amount of ZnO (ZnO-CN50%) shows the highest production of CO, CH_4, and C_2H_4. However, with a small amount of g-C_3N_4 (ZnO-CN15%), the CO_2 photo-reduction drops, which might be due to the agglomeration of ZnO rods and decreasing life span of charge carriers. Also, the band edge potentials (negative CB edges of g-C_3N_4) of heterostructure support the photo-reduction of CO_2 (Figure 13.5b). Besides carbon derivatives, ZnO's low charge separation efficiency can also be improved by doping with metal and nonmetals. It has been noticed that nonmetal dopants such as C, N, S, and F introduce the oxygen vacancy defects and shift the bandgap of ZnO. Dopants also act as electron scavengers and free the positive holes, which inhibit the recombination of charge carriers and allow the efficient reduction of CO_2 [14]. Oliveira et al. doped the ZnO with N and studied for the photocatalytic degradation of Rhodamine dye (RhB) and CO_2 photo-reduction [103]. N atom present on defect sites enhances the CO_2 adsorption, which improves the interaction of CO_2 with charge carriers and hence photo-reduction [104]. Besides nonmetals, metals have also been fabricated on the defect sites of ZnO to avoid recombination of charge carriers by trapping the electrons and promote photocatalytic efficiency. For example, Ismail et al. prepared Pt-doped mesoporous ZnO for CO_2 photo-conversion into methanol [105]. Pt-doped ZnO exhibits a large surface area and high visible light response than pristine ZnO. 1.5% Pt-doped ZnO nanoparticles show 18.5-fold high efficiency for CO_2-to-methanol conversion.

13.3.3 TUNGSTEN OXIDE (WO₃)-BASED NANOCOMPOSITES

WO_3 is a nontoxic and photo-chemically stable semiconductor material with a narrow bandgap (2.7–2.8 eV) [106]. It is a suitable photocatalyst material for visible light irradiation and absorbs a large range of the solar spectrum. Nonetheless, the high rate of photogenerated charge carriers recombination limits its photocatalytic potential. Also, the CB edge potential of WO_3 (~0.5 eV vs. NHE) is highly positive than the required reduction potential of O_2 (−0.33 V vs. NHE). Hence, the photo-induced electrons in the CB of WO_3 are not sufficient to reduce the oxygen molecule and could not generate the reactive oxygen species for photocatalytic degradation [107]. The electronic properties of WO_3 are also affected by the crystal structure of the molecule. WO_3 exhibits the four crystalline structures: (i) monoclinic (γ-WO_3), (ii) triclinic (δ-WO_3), (iii) orthorhombic (β-WO_3), and (iv) tetragonal (α-WO_3). Among all the most stable phases is the monoclinic phase, followed by the triclinic phase. The orthorhombic phase is stable at higher temperatures [108].

Different strategies have been projected for the improvement in the photocatalytic efficiency of WO_3, i.e., structural modifications [109,110], doping of WO_3 with metals and nonmetals [111,112], and coupling of WO_3 with other semiconductors [113,114]. Morphological modifications such as shapes, size, exposed facets, and crystalline nature of nanostructural material significantly affect photocatalytic performance [110,115]. For example, [002], [020], and [200] exposed facets of WO_3 have been observed with high reactivity [115–117]. Xie et al. have prepared two different WO_3 nanostructural materials (quasi-cubic and sheet-like structures) with exposed facets followed by hydrothermal routes but in two different conditions [115]. The quasi-cubic structure was observed on treating WO_3 with HF, whereas sheet-like crystals were obtained using HNO_3. The bandgap of quasi-cubic WO_3 was noticed to be narrower (2.71 eV) than that of sheet-like WO_3 (2.79 eV). Also, the band edge potentials were also influenced by the fraction of exposed facets in both WO_3 crystals. Sheet-like WO_3 exhibit mainly (002)-exposed facets with small percentage of (200)- and (002)-exposed facets with CB potential −0.3 eV, which allows the photo-conversion of CO_2 into CH_4. On the other hand, quasi-cubic, WO_3 composed of [001]-, [010]-, and [100]-exposed facets and could not reduce the CO_2.

Additionally, ultrathin and single-crystal WO_3 nanosheets also alter the bandgap of the materials by size quantization effect and hence affect the photocatalytic reduction of CO_2. The CB potential of the single-crystal WO_3 nanosheet with ~4–5 nm thickness was observed around −0.42 eV [118], enough to reduce the CO_2 to CH_4. However, the calculated CB potential of commercial WO_3 is positive and cannot reduce CO_2. Coupling WO_3 with other semiconductor materials is also a positive approach to enhancing photocatalytic efficiency by synergistic effects of improved light absorption ability, more surface-active sites, and delayed recombination of charge carriers. The coupling of In_2O_3 with WO_3 forms the type II heterojunction due to band alignments in the material. In_2O_3-WO_3 nanocomposite shows a high affinity towards the CO_2 photo-reduction [119]. In the composite, the In_2O_3 is present in cubic form and WO_3 is present in the monoclinic phase. The obtained band potentials of the coupled semiconductor materials were found to be feasible for CO_2-to-CH_3OH conversion. Although the coupling with a semiconductor material, in type II heterojunction, improves the charge separation and transport, the redox capability of the system is compromised [120]. This limitation can be overcome by the preparation of Z-scheme photocatalyst, in which the coupled semiconductors are not in physical contact. In contrast, the redox pair plays the part of a bridge between them. WO_3 is generally used as an oxidation photosystem (OPS) in the Z-scheme photocatalyst due to the presence of highly positive VB potential (~2.83–3.22 eV) [121]. Ho et al. prepared the hierarchal Z-scheme CdS-WO_3 photocatalyst to reduce the CO_2 into CH_4 under visible light illumination [122]. The bandgap of the WO_3 and CdS in the Z-scheme catalyst was observed as 2.5 and 2.3 eV, which is appropriate to absorb the visible spectrum of solar light. Also, the VB edges for WO_3 and CB edge for CdS were noticed at ~3.0 and −0.6 eV, respectively, which are capable to oxidize the water molecule and reduce the CO_2 molecule into CH_4.

Doping of noble metal with the WO_3 causes the plasmonic energy transfer at the interface between semiconductor material and metal, allowing the broad solar spectrum photon absorption and trapping the CB electrons that can be employed for the multielectron CO_2 photo-reduction. Wu et al. incorporated Pd and Au in the WO_3 and prepared Pd-Au/TiO_2-WO_3 photocatalyst for CO_2 photo-reduction [123]. The incorporation of metals in the TiO_2-WO_3 matrix significantly improves photocatalytic performance by hindering the charge carriers recombination. Besides doping and coupling of metals with metallic semiconductors, WO_3 was also combined with metal-free semiconductors such as g-C_3N_4, graphene, and graphene derivatives to enhance photocatalytic performance [124,125].

Oxygen vacancy-modified Z-scheme WO_{3-x} photocatalyst was prepared with g-C_3N_4 and employed in CO_2 photo-conversion into CO [124]. The photocatalytic potential of the designed photocatalyst was found to be higher than that of WO_3 and g-C_3N_4. Oxygen vacancies in the photocatalyst alter the bandgap and extend the visible light activation to promote the charge carrier parting and prolong the life span of electrons, which results in improved CO_2 photo-conversion.

Similarly, one-dimensional (1D) WO_3 nanobelt/graphene nanocomposite was synthesized following an in situ hydrothermal route for the photo-conversion of CO_2 into fuels [125]. The composite results in elevated CB potential of WO_3 by graphene, which could effectively reduce the CO_2 into hydrocarbons. Therefore, with the modification in WO_3 either by morphological modification, by doping, or by functionalization, the photocatalytic potential of WO_3 can be improved.

13.3.4 COPPER OXIDE (CUO & CU₂O)-BASED NANOCOMPOSITES

Copper oxide nanoparticles exist in two forms, i.e., CuO and Cu_2O. CuO is widely investigated as p-type semiconductor material employed for photocatalysis due to its outstanding characteristics such as low cost, abundant availability, and constricted energy bandgap (1.2–2.0 eV) [126,127]. Monoclinic CuO exhibits remarkable physical and chemical properties such as good redox potential, large active surface area, excellent stability, and superthermal conductivity, making it attractive for photocatalysis. CuO exhibits the small gap to allow the visible light response and the appropriate CB (~0.46 eV) and VB (~2.16 eV) potentials that are enough to release the active species such as ˙OH and ˙O_2^- radicals, which plays a vital role in the photodegradation [128]. However, the small bandgap of CuO also promotes the recombination of photo-induced charge carriers, which compromises the rate of photodegradation. Therefore, CuO has been functionalized or doped with several organic and inorganic materials such as other metal oxides (TiO_2, ZnO, MgO, etc.), metals, nonmetals, graphene derivatives, and ionic liquids to delay the recombination and enhance the photocatalytic rate by altering the electronic structure, charge transport characteristics, and light absorption properties [23,37,129,130]. Furthermore, the size, shape, and composition of the CuO also significantly influence the photocatalytic characteristics. Ribeiro et al. have prepared CuO using different preparation methods, e.g., solvothermal (CuO-Solv), co-precipitation (CuO-Prec), direct calcination (CuO-cal), and commercial CuO (CuO-Com), and demonstrated the role of CuO in CO_2 photo-reduction under UV irradiation [48]. Figure 13.6a reveals the FESEM images of CuO before and after CO_2 photo-reduction into CH_4. The generation of CH_4 via CO_2 photo-reduction using various CuO was found to be in the following order: CuO-Solv > CuO-Prec > CuO-Cal > CuO-Com. The high CO_2 photo-reduction rate of CuO-Solv might be due to the increased surface area of the material among all CuO. On UV illumination, the CuO surface reduces to metallic copper (Cu^0), promoting CO_2 reduction to formate and acetate as an intermediate for CH_4 generation. This process eventually further oxidizes Cu^0 to Cu^{2+} with the aid of holes (h^+) and its leaching to the solution. The Cu^{2+} might react with the dissolved CO_2 and form $Cu_2(OH)_2CO_3$ nanorods, as shown in Figure 13.6a. These nanorods remain stable in CO_2 photo-reduction process, which significantly influences the photo-reduction rate. In another article, Riberio et al. also explained the role of $Cu_2(OH)_2CO_3$ in CO_2 photo-reduction [131].

Gusain et al. studied the rGO-CuO nanocomposites, comprising CuO nanorods of various breadths, incorporated with rGO for CO_2 photodegradation [23]. They have concluded several factors regarding the photocatalytic potential of rGO-CuO: (i) breadth of CuO nanorods plays a significant role in controlling the photocatalytic characteristics of materials, (ii) rGO-CuO performs sevenfold higher than the pristine CuO, and (c) rGO-CuO (Cu^{2+}) shows superior photocatalytic performance than rGO-Cu_2O (Cu^{1+}) under similar conditions for CO_2 photo-reduction (Figure 13.6b). The enhanced photocatalytic activity of rGO-CuO than CuO is due to the slow recombination and fast transport of photo-induced charge carriers due to the rGO skeleton. Figure 13.6b reveals the proposed mechanism for CO_2 reduction under visible light illumination. First, CuO absorbs the visible light photon and excites the e^- and h^+ pairs. rGO in the rGO-CuO acts as an electron sink and promotes the quick transfer of electrons at catalytic sites for CO_2 reduction into methanol. rGO

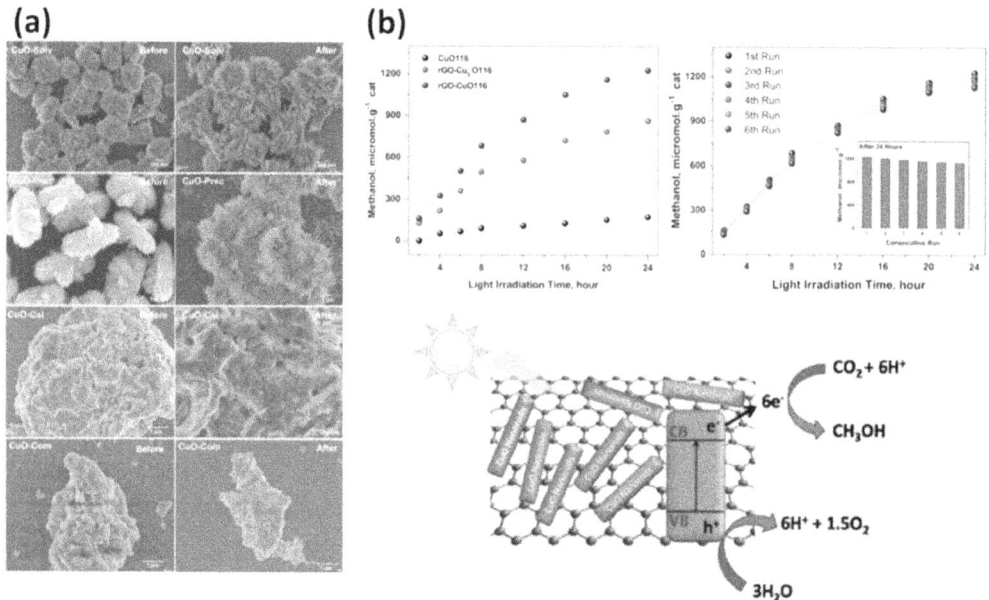

FIGURE 13.6 (a) FESEM images of CuO nanoparticles before and after CO_2 photocatalytic reduction. Reproduced with permission from [48]. Copyright 2020 Elsevier Science Ltd. (b) Comparison of photocatalytic efficiency of CuO, rGO-CuO, and rGO-Cu_2O and recyclability tests of rGO-CuO for CO_2 photo-reduction and mechanism of CO_2 photo-reduction using rGO-CuO as photocatalyst under visible light irradiation. (Reproduced with permission from [23]. Copyright 2016 Elsevier Science Ltd.)

also provides a large surface area for CO_2 adsorption. The photo-induced holes oxidize the water molecules into protons, and one mole of CO_2 reacts with six moles of protons to generate CH_3OH. In addition, the material showed high stability and excellent recyclability for five cycles (Figure 13.6b). CuO has also been decorated with several other metal oxides and evaluated for photocatalysis applications [132,133]. The incorporation of metal oxides affects the structural design and introduces the new band edges, promoting the charge transfer and increasing their life span for photocatalysis. Figure 13.7a shows the CuO/ZnO nanocomposite band edge potentials, which eases the visible light photon absorption and CO_2 photo-reduction [134]. The [111] plane of CuO allows epitaxial growth of [101] plane of ZnO, which creates defect-free interface and eases the photo-excited electron transfer from CuO to ZnO and hole transfer from ZnO to CuO. CuO is also fabricated with wide bandgap semiconductors to allow visible light absorption. Therefore, heterostructures of CuO/TiO_2 have also been investigated for the CO_2 photo-reduction, as the incorporation of CuO into TiO_2 improves the material's visible light response and enhances photocatalytic activity [135].

Cu_2O has also been explored as the photocatalyst for CO_2 reduction. Although Cu_2O also exhibits an appropriate bandgap for the visible light response, the VB of Cu_2O is not positive enough to oxidize the water and generate ˙OH radical required for CO_2 photo-reduction [137]. However, the CB edges of Cu_2O are negative enough to reduce the CO_2. Therefore, it has always been functionalized with other organic/inorganic material to form heterostructure and used as a photocatalyst to reduce CO_2 [138]. The heterostructure of Cu_2O with selecting semiconductor material that exhibits strong oxidation potential can be profitable for the CO_2 photo-reduction. Cu_2O was incorporated with reduced titania (RT) to form Z-scheme heterostructure and efficiently reduce the CO_2 into CH_4 [136]. Figure 13.7b shows the potential edges for the Z-scheme RT-Cu_2O photocatalyst and the mechanism of CO_2 photo-reduction. The photocatalyst material showed excellent recyclability. It has also been explored that the photocatalytic efficiency of Cu_2O is facet dependent, and [110] facets are active for the CO_2 photo-reduction, while [100] facets are inactive [139]. Other than metals, nonmetals copper oxides have also been functionalized with carbon derivatives (e.g., graphene

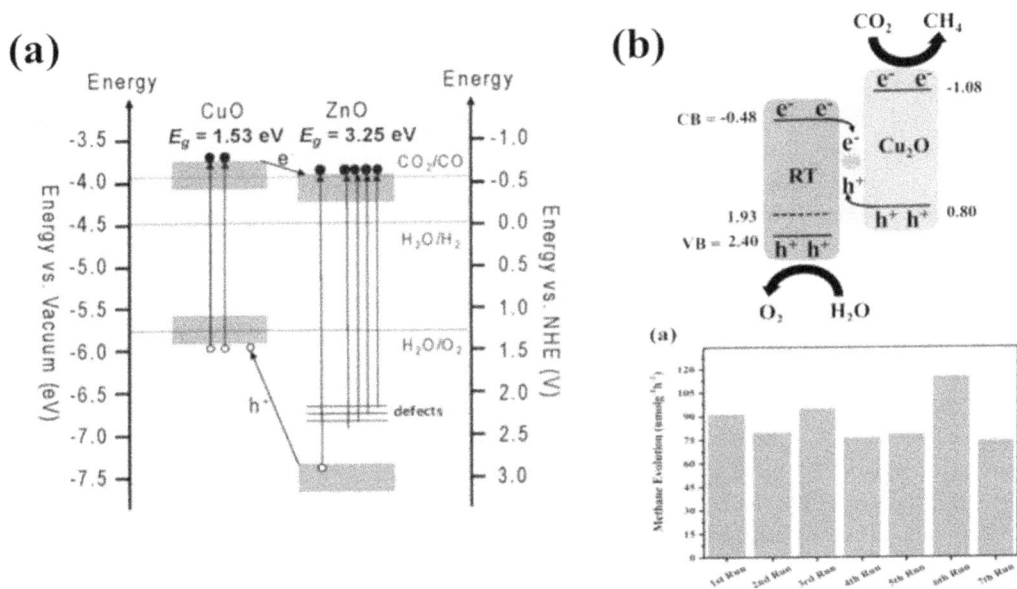

FIGURE 13.7 (a) Bandgap energy and band edge potentials of CuO/ZnO photocatalyst (vs. NHE at pH 7) for CO_2 photo-reduction. (Reproduced with permission from [134]. Copyright 2015 American Chemical Society.) (b) Band edge potential of reduced titania (RT) and Cu_2O for CO_2 photo-reduction under visible light illumination and recyclability experiment. (Reproduced with permission from [136]. Copyright 2020 Elsevier Science Ltd.)

derivatives, CNTs, g-C_3N_4, carbon quantum dots (QDs), etc.) for enhanced CO_2 photo-reduction. Liu et al. encapsulated the CuO QDs with the metal-organic framework (MIL-125(Ti)) and coupled with g-C_3N_4 for CO_2 photo-reduction (CuO QDs/g-C_3N_4/MIL-125(Ti)) [140]. CuO QDs/g-C_3N_4/MIL-125(Ti) significantly improves the charge transfer and reduces CO_2 into CO, CH_3OH, CH_3CHO, and CH_3CH_2OH. In CuO QDs/g-C_3N_4/MIL-125(Ti) heterostructure, CuO QDs enhance the adsorption and activation of CO_2, which results in a rapid generation of active intermediate radicals and thus the resultant reaction products.

13.3.5 Cerium Oxide (CeO₂)-Based Nanocomposites

CeO_2 is a rare earth metal oxide widely studied in catalysis, including photocatalysis, due to its nontoxicity, specific chemical reactivity, high thermal stability, and rigidity [141,142]. Ceria (CeO_2) exhibits the wide energy bandgap (3.1 eV) and high dielectric constant ($\varepsilon = 24.5$) [143]. This makes CeO_2 a potent catalyst, but it can only absorb the UV spectrum of solar light, limiting its application. Therefore, it has been modified following several strategies to enhance the solar spectrum photon absorption and photocatalysis performance. CeO_2 was designed with g-C_3N_4 to promote the CO_2 photo-reduction, and CO and CH_4 were observed as the main reaction products [143]. To check the effect of g-C_3N_4 on CeO_2, Han et al. synthesized a series of CeO_2-modified g-C_3N_4 and carefully studied the band structures supported by theoretical and experimental results [144]. The incorporation of g-C_3N_4 in CeO_2 improves the visible light response of the photocatalyst and promotes the separation of charge carriers, which eventually enhances the photocatalytic activity. The doping of metals on the oxygen vacancies of the CeO_2 also promotes photocatalytic performance [145,146]. Wang et al. doped the CeO_2 at oxygen vacancies and observed the enhanced CO yield by CO_2 photo-reduction [145]. These oxygen vacancies help in the broad solar spectrum photon absorption and quick parting and transfer of the charge pairs. Furthermore, CeO_2 is also immobilized with other metal oxide semiconductors to maximize potential band edges and stability. Seeharaj et al. prepared the TiO_2/rGo/CeO_2 heterojunction as a photocatalyst to reduce CO_2 to methanol and ethanol [147]. The high interfacial contact area between all the materials (TiO_2 nanosheets, rGO nanosheets, and CeO_2 nanoparticles) promotes the photo-induced charge carriers to react with adsorbed CO_2 molecules. The prepared heterojunction also hindered the electron-hole recombination effectively. Spindle-structured CeO_2 was also modified with the attapulgite and applied for CO_2 photo-reduction [148]. Attapulgite effectively inhibits the agglomeration of CeO_2 nanoparticles and improves the surface area for CO_2 adsorption and conversion. Therefore, to improve the photocatalytic potential of CeO_2, doping or fabrication has been proven an effective approach.

13.3.6 ZIRCONIUM DIOXIDE (ZrO_2)-BASED NANOCOMPOSITES

ZrO_2 is another metal oxide semiconductor material, which has been explored for heterogeneous photocatalysis reactions due to its high CB edge position (–1.0 eV vs NHE) [149–151]. Its exciting photo-induced electrons exhibit a large driving force for photocatalytic reduction of CO_2. But the wide bandgap (5.0 eV) limits its photo-excitation under a wide solar spectrum. It can be used as a photocatalyst under UV light irradiation. Sayama et al. reported the photo-excitation of ZrO_2 under UV light illumination in the aqueous medium [151,152]. They have also investigated that Cu (1 wt %)-ZrO_2 photocatalyst suspended in $NaHCO_3$ aqueous solution can be used for the photocatalytic reduction of CO_2 and photo-decomposition of H_2O molecule [151]. Kohno et al. studied the application of ZrO_2 for the photo-conversion of CO_2 in the presence of H_2 under UV light irradiation [153,154]. During the photocatalytic reaction in the presence of H_2, using ZrO_2-based photocatalyst, CO_2 selectively reduced to CO as the main reaction product at room temperature. They have also studied the reaction mechanism and found that ZrO_2 reduces CO_2 into CO_2^- anion radical, which reacts with the H_2 and generates formate anions ($HCOO^-$) as reactive species [149]. These formate anions further react with another CO_2 molecule present in the reaction medium and selectively generate the CO molecules. In another study, the photocatalytic efficiency of ZrO_2 for the reduction of CO_2 was performed in the presence of CH_4 instead of H_2 and CO [155]. In the proposed mechanism, first, gaseous CO_2 was adsorbed on the ZrO_2 surface, which converts into CO_2^- anion radical under light irradiation. The active CO_2^- anion radical reacts with CH_4 present in the reaction medium and yields surface acetate and formate. The acetate did not react further and stayed as adsorbed on the surface, while the formate acted as a reductant and reduced the CO_2 present in the reaction medium to yield CO. The formate itself further oxidizes into the surface-adsorbed CO_2.

The selective photo-conversion of CO_2 into CO on using ZrO_2 as photocatalyst further motivates the researchers to explore the ZrO_2. Zhang et al. uniformly disperse various weight ratios of Ni single atom on defective ZrO_2 to enhance the UV-visible light response for CO_2 photo-reduction [156]. Ni-immobilized ZrO_2 increases the selectively CO production rate by 6 and 40 times higher than defective and perfect ZrO_2, respectively. DFT calculations also show that the Ni immobilization on ZrO_2 decreases the energy barriers on the O sites of ZrO_2 (010) essential for CO production.

13.3.7 OTHER METAL OXIDE-BASED NANOCOMPOSITES

Besides TiO_2, ZnO, WO_3, CuO, ZrO_2, and CeO_2, other metal oxide-based nanocomposites have also been explored for the photocatalytic reduction of CO_2. SnO_2 is an n-type semiconductor photoactive material with a wide bandgap (~3.6 eV) at room temperature [157,158]. It exhibits excellent electrical and optical properties such as high theoretical specific capacity, low resistivity, excellent electron mobility, and optical transparency. Also, it shows no adverse health effects on the human body. The reduction potential of SnO_2 is negative enough (–0.41 eV vs. NHE at pH7) to reduce the CO_2 under the solar spectrum [159]. Under light irradiation, SnO_2 nanoparticles get excited and reduce the CO_2 into reaction products (e.g., formic acid) and return to the ground state to accept the electron from water and regenerate it as SnO_2 [158]. However, it exhibits limited light absorption capacity and fast photo-induced charge pairs recombination, limiting its photocatalytic potential. The fabrication of SnS_2 to SnO_2 enhanced the photocatalytic performance of the resultant SnS_2-SnO_2 nanocomposite by improving the light-absorbing capacity, retarding the recombination of charge carriers, and introducing more available active spots for CO_2 adsorption and photo-conversion [160]. The nanocomposite successfully reduces the CO_2 into CH_4, CO, and C_2H_4 under visible light illumination. Under visible light exposure, e^-–h^+ pairs generated in the SnS_2 and electrons from the CB of SnS_2 jumped to the SnO_2 CB; this delayed the recombination process. As a result, the charge carriers dissociate the water molecule into H^+ and O_2 and reduce the CO_2 into H_2O and CO. The nanocomposite also shows excellent recyclability. Z-scheme SnO_{2-x}/g-C_3N_4 nanocomposite also reduces the CO_2 efficiently under visible light illumination to produce CH_4, CH_3OH, and CO [161]. The heterojunction formation between SnO_2 and g-C_3N_4 facilitates the photo-conversion of CO_2 by promoting visible light absorption and separation and transfer of the photo-induced charge carriers.

Iron oxide is also an n-type semiconductor, which is plentifully available and also investigated as a photocatalyst. Moreover, the recovery of the catalyst using an external magnet makes it a promising material for the application. Iron oxide exhibits in various forms, such as α-FeOOH, β-FeOOH, γ-FeOOH, α-Fe_2O_3, and γ-Fe_2O_3, which have been studied for photocatalytic activity. Iron oxide in the α-Fe_2O_3 phase exhibits excellent thermal stability and low bandgap (2.2 eV), making it feasible for visible light absorption and hence photocatalytic application [163]. However, the low CB edges of α-Fe_2O_3 generated electrons with low energy, which are insufficient to carry out the CO_2 reduction reactions. Thus, it is desirable to fabricate the α-Fe_2O_3 in suitable ways to promote the CO_2 photocatalytic reduction reaction. In this regard, Wong et al. reported Z-scheme α-Fe_2O_3/g-C_3N_4 nanocomposite for CO_2 photo-reduction under visible light irradiation [164]. The incorporation of α-Fe_2O_3 in g-C_3N_4 not only improves the CO_2 adsorption and visible light harvesting but also enhances the photocatalytic reduction of CO_2 into CO 2.2 times higher than that of pristine g-C_3N_4.

Similarly, α-Fe_2O_3 was also introduced with Cu_2O to fabricate Z-scheme α-Fe_2O_3/Cu_2O (n-p junction) heterostructures to reduce the CO_2 into CO [162]. The superior photocatalytic efficiency of the Z-scheme α-Fe_2O_3/Cu_2O photocatalyst was due to the effective separation of the photo-induced charge carriers in the constructed nanocomposite. Figure 13.8 shows the most probable photo-reduction mechanisms following the double charge transfer mechanism and Z-scheme mechanism. If the photocatalytic reaction follows double charge transfer, then the accumulated electrons in the α-Fe_2O_3 CB would

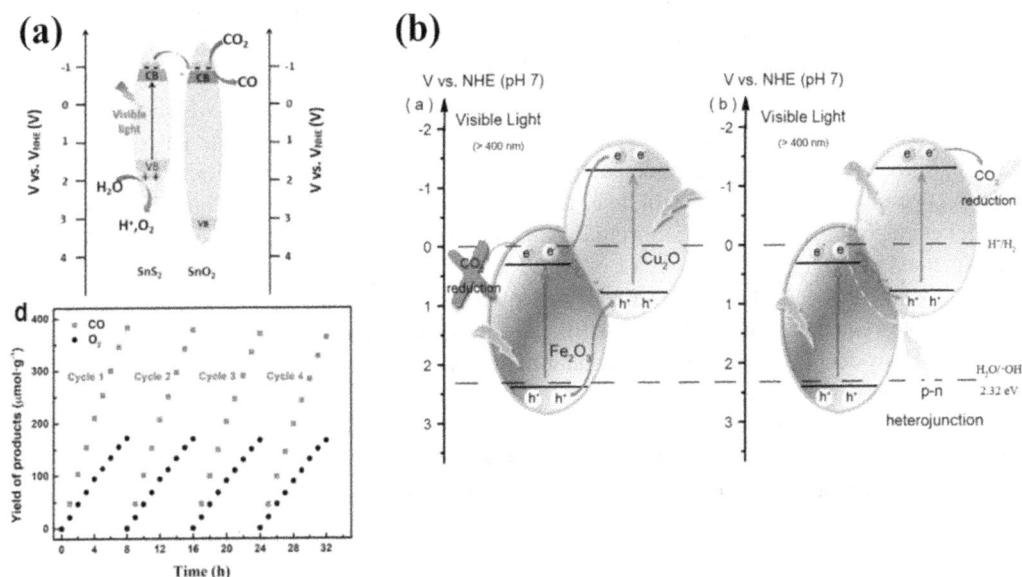

FIGURE 13.8 Schematic representation of photo-induced charge pairs separation via double charge transfer mechanism and Z-scheme mechanism. (Reproduced with permission from [162]. Copyright 2015 American Chemical Society.)

not be able to generate CO by CO_2 reduction due to the more positive potential level of the α-Fe_2O_3 CB (0.037 eV vs. NHE). Additionally, the holes in the VB of Cu_2O cannot also oxidize the water for an active radical generation. Therefore, the double charge transfer mechanism cannot be applied to photocatalytic reduction of CO_2 using α-Fe_2O_3/Cu_2O. However, in the Z-scheme mechanism, through the p-n junction in the heterostructure, excited e$^-$ from the CB of α-Fe_2O_3 transfers to the VB of Cu_2O and recombines with the hole. This results in the generation of abundant e$^-$ in the Cu_2O CB and h$^+$ in the α-Fe_2O_3 VB, which participate in the reduction and oxidation of CO_2 and H_2O, respectively.

Bismuth oxide (Bi_2O_3) is another nontoxic p-type semiconductor with a narrow bandgap (2.8 eV) and higher oxidation potential of the valence hole that attracts its application in photocatalysis [165]. However, it has not been much explored for the CO_2 photo-reduction. The low migration and fast recombination of charge carriers make its poor performance in photocatalysis. Therefore, it is fabricated with other heterostructures to improve the photocatalytic ability. For example, Bi_2O_3/g-C_3N_4 Z-scheme photocatalyst was designed for CO_2 photo-reduction into CO [166]. The performance of the nanocomposite was found to be 1.8 times higher than that of pristine g-C_3N_4, which is mainly due to the successful separation and transfer of charge carriers through the interface.

13.4   CONCLUSION

This chapter provides an overview of the recent research and development of metal oxide-based photocatalysts for CO_2 reduction under solar light illumination. A wide range of metal oxides such as TiO_2-, ZnO-, WO_3-, CeO_2-, CuO-, and ZrO_2-based nanocomposites having different morphological features and crystalline phases have been discussed with the recent development, opportunities, and challenges for the photocatalytic reduction of CO_2. The narrow bandgap of the metal oxide nanoparticles allows the absorption of solar spectrum photons for photo-excitation of the electrons, which is necessary to initiate the photocatalytic reaction. However, one of the major drawbacks of using metal oxide nanoparticles as photocatalysts is the insufficient absorption of the large range of solar spectrum photons and fast recombination of photogenerated charge pairs. Therefore, material characteristics such as a large active surface area for CO_2 adsorption, improved light harvesting and fast carrier charge generation, separation, and transport can be manipulated through the morphological and structural control during the fabrication step, which leads to advancements in the photocatalysis. In general, the doping of metal oxides with metal, nonmetal, co-doping, and heterojunction systems with other semiconductors significantly improves the visible light response, delays the recombination of charge carriers, and enhances the electron transport, which consequently enhances photocatalysis. Enhancing the separation and transport of charge carriers at the excited state significantly upgrades the photocatalytic activity. Trapping of photo-induced electrons is required to increase the availability of holes to oxidize the organic molecules. Thus, the factors such as facet engineering, crystallinity, morphology, and fabrication needs of metal oxide nanoparticles are explained in detail. The presence of oxygen vacancies in the metal oxide matrices also plays a crucial role in enhanced CO_2 photo-reduction. Although considerable research has been carried out on CO_2 photo-reduction, some challenges still persist: (i) the efficient selectivity of the generation of one reaction product by the CO_2 photo-reduction; (ii) the selective adsorption and reduction of CO_2 by photocatalyst in the mixture of gases; and (iii) the large-scale implantation of the photocatalysis process for CO_2 reduction with an economical approach. Therefore, more efforts are required in effective nanomaterials development and reaction advancement. Additionally, advanced research should be boosted at the molecular level through in situ characterization to enhance CO_2 photo-reduction efficiency.

ACKNOWLEDGMENT

The authors would like to thank the Department of Science and Innovation, the Council for Scientific and Industrial Research, and the University of Johannesburg for financial support.

REFERENCES

1. Roy SC, Varghese OK, Paulose M, Grimes CA. Toward solar fuels: Photocatalytic conversion of carbon dioxide to hydrocarbons. *ACS Nano.*, 2010, 4, 1259–1278.
2. Zhang T, Zhang W, Yang R, Liu Y, Jafari M. CO_2 capture and storage monitoring based on remote sensing techniques: A review. *J. Clean. Prod.*, 2021, 281, 124409.
3. Netz B, Davidson O, Bosch P, Dave R, Meyer L. Climate change 2007: Mitigation. Contribution of Working Group III to the Fourth Assessment Report of the Intergovernmental Panel on Climate Change. Summary for Policymakers. Climate change 2007: Mitigation Contribution of Working Group III to the Fourth Assessment Report of the Intergovernmental Panel on Climate Change Summary for Policymakers. 2007.
4. Khatib H. IEA world energy outlook 2011—A comment. *Energy Policy*, 2012, 48, 737–743.
5. Mardani A, Streimikiene D, Cavallaro F, Loganathan N, Khoshnoudi M. Carbon dioxide (CO_2) emissions and economic growth: A systematic review of two decades of research from 1995 to 2017. *Sci. Total*, 2019, 649, 31–49.
6. De Coninck H, Stephens JC, Metz B. Global learning on carbon capture and storage: A call for strong international cooperation on CCS demonstration. *Energy Policy*, 2009, 37, 2161–2165.
7. Alper E, Orhan OY. CO_2 utilization: Developments in conversion processes. *Petroleum*, 2017, 3, 109–126.
8. Centi G, Perathoner S, Salladini A, Iaquaniello G. Economics of CO_2 utilization: A critical analysis. *Front. Energy Res.*, 2020, 8, 567986.
9. Sharma T, Sharma A, Sharma S, Giri A, Kumar A, Pant D. Recent developments in CO_2-capture and conversion technologies. In: *Chemo-Biological Systems for CO_2 Utilization*. CRC Press, Boca Raton, 1st Edition, 2020, pp. 1–14.
10. Yaashikaa P, Kumar PS, Varjani SJ, Saravanan A. A review on photochemical, biochemical and electrochemical transformation of CO_2 into value-added products. *J. CO_2 Util.*, 2019, 33, 131–147.
11. Inoue T, Fujishima A, Konishi S, Honda K. Photoelectrocatalytic reduction of carbon dioxide in aqueous suspensions of semiconductor powders. *Nature*, 1979, 277, 637–368.
12. Thampi KR, Kiwi J, Graetzel M. Methanation and photo-methanation of carbon dioxide at room temperature and atmospheric pressure. *Nature*, 1987, 327, 506–508.
13. Shehzad N, Tahir M, Johari K, Murugesan T, Hussain M. A critical review on TiO_2 based photocatalytic CO_2 reduction system: Strategies to improve efficiency. *J. CO_2 Util.*, 2018, 26, 98–122.
14. Ong CB, Ng LY, Mohammad AW. A review of ZnO nanoparticles as solar photocatalysts: Synthesis, mechanisms and applications. *Renew. Sust. Energ. Rev.*, 2018, 81, 536–551.
15. Kumar N, Kumar S, Gusain R, Manyala N, Eslava S, Ray SS. Polypyrrole-promoted rGO–MoS_2 nanocomposites for enhanced photocatalytic conversion of CO_2 and H_2O to CO, CH_4, and H_2 products. *ACS Appl. Energy Mater.*, 2020, 3, 9897–9909.
16. Dai W, Yu J, Luo S, Hu X, Yang L, Zhang S, et al. WS_2 quantum dots seeding in Bi_2S_3 nanotubes: A novel Vis-NIR light sensitive photocatalyst with low-resistance junction interface for CO_2 reduction. *Chem. Eng. J.*, 2020, 389, 123430.
17. Sun Z, Wang H, Wu Z, Wang L. g-C_3N_4 based composite photocatalysts for photocatalytic CO_2 reduction. *Catal. Today*, 2018, 300, 160–172.
18. Darkwah WK, Teye GK, Ao Y. Graphene nanocrystals in CO_2 photoreduction with H_2O for fuel production. *Nanoscale Adv.*, 2020, 2, 991–1006.
19. Tahir M, Ali Khan A, Tasleem S, Mansoor R, Fan WK. Titanium carbide (Ti_3C_2) MXene as a promising co-catalyst for photocatalytic CO_2 conversion to energy-efficient fuels: A review. *Energy Fuels*, 2021, 35, 10374–10404.
20. Yang Z-Z, Wei J-J, Zeng G-M, Zhang H-Q, Tan X-F, Ma C, Li X-C, Li Z-H, Zhang C. A review on strategies to LDH-based materials to improve adsorption capacity and photoreduction efficiency for CO_2. *Coord. Chem. Rev.* 2019, 386, 154–182.
21. Tran VV, Nu TTV, Jung H-R, Chang M. Advanced photocatalysts based on conducting polymer/metal oxide composites for environmental applications. *Polymers*, 2021, 13, 3031.
22. Pradipta AR, Mauludi K, Kartini I, Kunarti ES. Synthesis of Fe_3O_4/TiO_2 nanocomposite as photocatalyst in photoreduction reaction of CO_2 conversion to methanol. *Key Eng. Mater.*, 2020. 454–458.
23. Gusain R, Kumar P, Sharma OP, Jain SL, Khatri OP. Reduced graphene oxide–CuO nanocomposites for photocatalytic conversion of CO_2 into methanol under visible light irradiation. *Appl. Catal. B*, 2016, 181, 352–362.
24. Sayed M, Xu F, Kuang P, Low J, Wang S, Zhang L, et al. Sustained CO_2-photoreduction activity and high selectivity over Mn, C-codoped ZnO core-triple shell hollow spheres. *Nat. Commun.*, 2021, 12, 1–10.

25. He L, Zhang W, Liu S, Zhao Y. Three-dimensional porous N-doped graphitic carbon framework with embedded CoO for photocatalytic CO_2 reduction. *Appl. Catal. B*, 2021, 298, 120546.

26. Ikreedeegh RR, Tahir M. A critical review in recent developments of metal-organic-frameworks (MOFs) with band engineering alteration for photocatalytic CO_2 reduction to solar fuels. *J. CO_2 Util.*, 2021, 43, 101381.

27. Kovačič Za, Likozar B, Huš M. Photocatalytic CO_2 reduction: A review of Ab initio mechanism, kinetics, and multiscale modeling simulations. *ACS Catal.*, 2020, 10, 14984–15007.

28. He Y, Lei Q, Li C, Han Y, Shi Z, Feng S. Defect engineering of photocatalysts for solar-driven conversion of CO_2 into valuable fuels. *Mater. Today*, 2021, 50, 358–384.

29. Xiang X, Pan F, Li Y. A review on adsorption-enhanced photoreduction of carbon dioxide by nanocomposite materials. *Adv. Compos. Hybrid Mater.*, 2018, 1, 6–31.

30. Shen H, Peppel T, Strunk J, Sun Z. Photocatalytic reduction of CO_2 by metal-free-based materials: Recent advances and future perspective. *Solar RRL*, 2020, 4, 1900546.

31. Bai S, Zhang N, Gao C, Xiong Y. Defect engineering in photocatalytic materials. *Nano Energy*, 2018, 53, 296–336.

32. Kannan K, Radhika D, Sadasivuni KK, Reddy KR, Raghu AV. Nanostructured metal oxides and its hybrids for photocatalytic and biomedical applications. *Adv. Colloid Interface Sci.*, 2020, 281, 102178.

33. Mukwevho N, Gusain R, Fosso-Kankeu E, Kumar N, Waanders F, Ray SS. Removal of naphthalene from simulated wastewater through adsorption-photodegradation by ZnO/Ag/GO nanocomposite. *J. Ind. Eng. Chem.*, 2020, 81, 393–404.

34. Chavali MS, Nikolova MP. Metal oxide nanoparticles and their applications in nanotechnology. *SN Appl. Sci.*, 2019, 1, 607.

35. Gusain R, Khatri OP. Ultrasound assisted shape regulation of CuO nanorods in ionic liquids and their use as energy efficient lubricant additives. *J. Mater. Chem. A.*, 2013, 1, 5612–5619.

36. Gusain R, Gupta K, Joshi P, Khatri OP. Adsorptive removal and photocatalytic degradation of organic pollutants using metal oxides and their composites: A comprehensive review. *Adv. Colloid Interface Sci.*, 2019, 272, 102009.

37. Gusain R, Singhal N, Singh R, Kumar U, Khatri OP. Ionic-liquid-functionalized copper oxide nanorods for photocatalytic splitting of water. *ChemPlusChem.* 2016, 81, 489.

38. Mukwevho N, Fosso-Kankeu E, Waanders F, Kumar N, Ray SS, Yangkou Mbianda X. Photocatalytic activity of $Gd_2O_2CO_3$·ZnO·CuO nanocomposite used for the degradation of phenanthrene. *SN Appl. Sci.*, 2018, 1, 10.

39. Kumar N, Mittal H, Reddy L, Nair P, Ngila JC, Parashar V. Morphogenesis of ZnO nanostructures: Role of acetate (COOH–) and nitrate (NO_3^-) ligand donors from zinc salt precursors in synthesis and morphology dependent photocatalytic properties. *RSC Adv.*, 2015, 5, 38801–38809.

40. Kumar N, George BPA, Abrahamse H, Parashar V, Ray SS, Ngila JC. A novel approach to low-temperature synthesis of cubic HfO_2 nanostructures and their cytotoxicity. *Sci. Rep.*, 2017, 7, 9351.

41. Mukwevho N, Kumar N, Fosso-Kankeu E, Waanders F, Bunt J, Ray SS. Visible light-excitable ZnO/2D graphitic-C_3N_4 heterostructure for the photodegradation of naphthalene. *Desalin. Water Treat.*, 2019, 163, 286–296.

42. Ray SS, Gusain R, Kumar N. *Carbon Nanomaterial-Based Adsorbents for Water Purification: Fundamentals and Applications.* Elsevier; 1st Edition, 2020.

43. Kumar N, George BPA, Abrahamse H, Parashar V, Ngila JC. Sustainable one-step synthesis of hierarchical microspheres of PEGylated MoS_2 nanosheets and MoO_3 nanorods: Their cytotoxicity towards lung and breast cancer cells. *Appl. Surf. Sci.*, 2017, 396, 8–18.

44. Kumar N, Sinha Ray S. Synthesis and functionalization of nanomaterials. In: Sinha Ray S, editor. *Processing of Polymer-based Nanocomposites: Introduction.* Cham: Springer International Publishing; 2018. pp. 15–55.

45. Umukoro EH, Kumar N, Ngila JC, Arotiba OA. Expanded graphite supported p-n MoS_2-SnO_2 heterojunction nanocomposite electrode for enhanced photo-electrocatalytic degradation of a pharmaceutical pollutant. *J. Electroanal. Chem.*, 2018, 827, 193–203.

46. Li K, Teng C, Wang S, Min Q. Recent advances in TiO_2-based heterojunctions for photocatalytic CO_2 reduction with water oxidation: A review. *Front. Chem.*, 2021, 9, 637501.

47. Hegazy I, Geioushy R, El-Sheikh S, Shawky A, El-Sherbiny S, Kandil A-HT. Influence of oxygen vacancies on the performance of ZnO nanoparticles towards CO_2 photoreduction in different aqueous solutions. *J. Environ. Chem. Eng.*, 2020, 8, 103887.

48. Nogueira AE, da Silva GT, Oliveira JA, Torres JA, da Silva MG, Carmo M, et al. Unveiling CuO role in CO_2 photoreduction process–Catalyst or reactant? *Catal. Commun.*, 2020, 137, 105929.

49. Murillo-Sierra J, Hernández-Ramírez A, Hinojosa-Reyes L, Guzmán-Mar J. A review on the development of visible light-responsive WO_3-based photocatalysts for environmental applications. *Chem. Eng. J. Adv.*, 2020, 5, 100070.

50. Al Zoubi W, Al-Hamdani AAS, Sunghun B, Ko YG. A review on TiO_2-based composites for superior photocatalytic activity. *Rev. Inorg. Chem.*, 2021, doi: 10.1515/revic-2020-0025.

51. Kumar N, Mittal H, Alhassan SM, Ray SS. Bionanocomposite hydrogel for the adsorption of dye and reusability of generated waste for the photodegradation of ciprofloxacin: A demonstration of the circularity concept for water purification. *ACS Sustain. Chem. Eng.*, 2018, 6, 17011–17125.

52. Thampi KR, Kiwi J, Grätzel M. Methanation and photo-methanation of carbon dioxide at room temperature and atmospheric pressure. *Nature.* 1987, 327, 506–508.

53. Gupta T, Cho J, Prakash J. Hydrothermal synthesis of TiO_2 nanorods: Formation chemistry, growth mechanism, and tailoring of surface properties for photocatalytic activities. *Mater. Today Chem.*, 2021, 20, 100428.

54. Alberoni C, Barroso-Martín I, Infantes-Molina A, Rodríguez-Castellón E, Talon A, Zhao H, et al. Ceria doping boosts methylene blue photodegradation in titania nanostructures. *Mater. Chem. Front.*, 2021, 5, 4138–4152.

55. Kang S, Khan H, Lee C, Kwon K, Lee CS. Investigation of hydrophobic $MoSe_2$ grown at edge sites on TiO_2 nanofibers for photocatalytic CO_2 reduction. *Chem. Eng. J.*, 2021, 420, 130496.

56. Mamaghani AH, Haghighat F, Lee C-S. Role of titanium dioxide (TiO_2) structural design/morphology in photocatalytic air purification. *Appl. Catal. B*, 2020, 269, 118735.

57. Zhang Q-H, Han W-D, Hong Y-J, Yu J-G. Photocatalytic reduction of CO_2 with H_2O on Pt-loaded TiO_2 catalyst. *Catal. Today*, 2009, 148, 335–340.

58. Qu J, Zhang X, Wang Y, Xie C. Electrochemical reduction of CO_2 on RuO_2/TiO_2 nanotubes composite modified Pt electrode. *Electrochim. Acta.* 2005, 50, 3576–3580.

59. He H, Zapol P, Curtiss LA. A theoretical study of CO_2 anions on anatase (101) surface. *J. Phys. Chem. C*, 2010, 114, 21474–21481.

60. Han X, Kuang Q, Jin M, Xie Z, Zheng L. Synthesis of titania nanosheets with a high percentage of exposed (001) facets and related photocatalytic properties. *J. Am. Chem. Soc.*, 2009, 131, 3152–3153.

61. Yu J, Low J, Xiao W, Zhou P, Jaroniec M. Enhanced photocatalytic CO_2-reduction activity of anatase TiO_2 by coexposed {001} and {101} facets. *J. Am. Chem. Soc.*, 2014, 136, 8839–8842.

62. He Z, Wen L, Wang D, Xue Y, Lu Q, Wu C, Chen J., Song S. Photocatalytic reduction of CO_2 in aqueous solution on surface-fluorinated anatase TiO_2 nanosheets with exposed {001} facets. *Energy Fuels*, 2014, 28, 3982–3993.

63. Kim H-S, Lee J-W, Yantara N, Boix PP, Kulkarni SA, Mhaisalkar S, Gratzel M., Park N-G. High efficiency solid-state sensitized solar cell-based on submicrometer rutile TiO_2 nanorod and $CH_3NH_3PbI_3$ perovskite sensitizer. *Nano Lett.*, 2013, 13, 2412–2417.

64. Zhao H, Pan F, Li Y. A review on the effects of TiO_2 surface point defects on CO_2 photoreduction with H_2O. *J. Materiomics*, 2017, 3, 17–32.

65. Sasan K, Zuo F, Wang Y, Feng P. Self-doped $Ti^{3+}–TiO_2$ as a photocatalyst for the reduction of CO_2 into a hydrocarbon fuel under visible light irradiation. *Nanoscale.* 2015, 7, 13369–13372.

66. Zhao H, Liu L, Andino JM, Li Y. Bicrystalline TiO_2 with controllable anatase–brookite phase content for enhanced CO_2 photoreduction to fuels. *J. Mater. Chem. A*, 2013, 1, 8209–8216.

67. Nah YC, Paramasivam I, Schmuki P. Doped TiO_2 and TiO_2 nanotubes: Synthesis and applications. *ChemPhysChem.* 2010, 11, 2698–2713.

68. Fang B, Xing Y, Bonakdarpour A, Zhang S, Wilkinson DP. Hierarchical $CuO–TiO_2$ hollow microspheres for highly efficient photodriven reduction of CO_2 to CH_4. *ACS Sustain. Chem. Eng.*, 2015, 3, 2381–2388.

69. Xie S, Wang Y, Zhang Q, Deng W, Wang Y. MgO-and Pt-promoted TiO_2 as an efficient photocatalyst for the preferential reduction of carbon dioxide in the presence of water. *ACS Catal.*, 2014, 4, 3644–3653.

70. Tan D, Zhang J, Shi J, Li S, Zhang B, Tan X, et al. Photocatalytic CO_2 transformation to CH_4 by Ag/Pd bimetals supported on N-doped TiO_2 nanosheet. *ACS Appl. Mater. Interfaces*, 2018, 10, 24516–24522.

71. Hong D, Lyu L-M, Koga K, Shimoyama Y, Kon Y. Plasmonic Ag@TiO_2 core–shell nanoparticles for enhanced CO_2 photoconversion to CH_4. *ACS Sustain. Chem. Eng.*, 2019, 7, 18955–18964.

72. Tahir M. Synergistic effect in MMT-dispersed Au/TiO_2 monolithic nanocatalyst for plasmon-absorption and metallic interband transitions dynamic CO_2 photo-reduction to CO. *Appl. Catal. B*, 2017, 219, 329–343.

73. Irie H, Watanabe Y, Hashimoto K. Carbon-doped anatase TiO_2 powders as a visible-light sensitive photocatalyst. *Chem. Lett.*, 2003, 32, 772–773.

74. Phongamwong T, Chareonpanich M, Limtrakul J. Role of chlorophyll in *Spirulina* on photocatalytic activity of CO_2 reduction under visible light over modified N-doped TiO_2 photocatalysts. *Appl. Catal. B*, 2015, 168, 114–124.

75. Fujishima A, Zhang X, Tryk DA. TiO_2 photocatalysis and related surface phenomena. *Surf. Sci. Rep.*, 2008, 63, 515–582.

76. Emeline AV, Kuznetsov VN, Rybchuk VK, Serpone N. Visible-light-active titania photocatalysts: The case of N-doped s—properties and some fundamental issues. *Int. J. Photoenergy*, 2008, 2008, 258394.

77. Zeng L, Lu Z, Li M, Yang J, Song W, Zeng D, et al. A modular calcination method to prepare modified N-doped TiO_2 nanoparticle with high photocatalytic activity. *Appl. Catal. B*, 2016, 183, 308–316.

78. Samsudin EM, Abd Hamid SB, Juan JC, Basirun WJ, Centi G. Synergetic effects in novel hydrogenated F-doped TiO_2 photocatalysts. *Appl. Surf. Sci.*, 2016, 370, 380–393.

79. McManamon C, O'Connell J, Delaney P, Rasappa S, Holmes JD, Morris MA. A facile route to synthesis of S-doped TiO_2 nanoparticles for photocatalytic activity. *J Mol Catal A Chem.*, 2015, 406, 51–57.

80. Hamal DB, Klabunde KJ. Synthesis, characterization, and visible light activity of new nanoparticle photocatalysts based on silver, carbon, and sulfur-doped TiO_2. *J. Colloid Interface Sci.*, 2007, 311, 514–522.

81. Xie Y, Li Y, Zhao X. Low-temperature preparation and visible-light-induced catalytic activity of anatase F–N-codoped TiO_2. *J Mol. Catal. A Chem.*, 2007, 277, 119–126.

82. Sun Q, Hu X, Zheng S, Sun Z, Liu S, Li H. Influence of calcination temperature on the structural, adsorption and photocatalytic properties of TiO_2 nanoparticles supported on natural zeolite. *Powder Technol.*, 2015, 274, 88–97.

83. Yao Y, Ochiai T, Ishiguro H, Nakano R, Kubota Y. Antibacterial performance of a novel photocatalytic-coated cordierite foam for use in air cleaners. *Appl. Catal. B*, 2011, 106, 592–599.

84. Khataee AR, Fathinia M, Aber S, Zarei M. Optimization of photocatalytic treatment of dye solution on supported TiO_2 nanoparticles by central composite design: Intermediates identification. *J. Hazard. Mater.*, 2010, 181, 886–897.

85. Hu H, Xiao W-J, Yuan J, Shi J-W, Chen M-X, Shang Guan W-F. Preparations of TiO_2 film coated on foam nickel substrate by sol-gel processes and its photocatalytic activity for degradation of acetaldehyde. *J. Environ. Sci.*, 2007, 19, 80–85.

86. Anandan S, Yoon M. Photocatalytic activities of the nano-sized TiO_2-supported Y-zeolites. *J. Photochem. Photobiol. C: Photochem. Rev.*, 2003, 4, 5–18.

87. Anpo M, Yamashita H, Ichihashi Y, Fujii Y, Honda M. Photocatalytic reduction of CO_2 with H_2O on titanium oxides anchored within micropores of zeolites: Effects of the structure of the active sites and the addition of Pt. *J. Phys. Chem. B*, 1997, 101, 2632–2636.

88. Tan L-L, Ong W-J, Chai S-P, Goh BT, Mohamed AR. Visible-light-active oxygen-rich TiO_2 decorated 2D graphene oxide with enhanced photocatalytic activity toward carbon dioxide reduction. *Appl. Catal. B*, 2015, 179, 160–170.

89. Lin L-Y, Nie Y, Kavadiya S, Soundappan T, Biswas P. N-doped reduced graphene oxide promoted nano TiO_2 as a bifunctional adsorbent/photocatalyst for CO_2 photoreduction: Effect of N species. *Chem. Eng. J.*, 2017, 316, 449–460.

90. Jin B, Yao G, Jin F, Hu YH. Photocatalytic conversion of CO_2 over C_3N_4-based catalysts. *Catal. Today*, 2018, 316, 149–154.

91. Wu J, Feng Y, Li D, Han X, Liu J. Efficient photocatalytic CO_2 reduction by P–O linked g-C_3N_4/TiO_2-nanotubes Z-scheme composites. *Energy*, 2019, 178, 168–175.

92. Morkoç H, Özgür Ü. *Zinc Oxide: Fundamentals, Materials and Device Technology.* John Wiley & Sons; 2008.

93. Özgür Ü, Avrutin V, Morkoç H. Zinc oxide materials and devices grown by MBE. In: M Henini, editor. *Molecular Beam Epitaxy.* Oxford: Elsevier; 2013, pp. 369–416.

94. Özgür Ü, Avrutin V, Morkoç H. Zinc oxide materials and devices grown by molecular beam epitaxy. In: M Henini, editor. *Molecular Beam Epitaxy.* Elsevier; 2018, pp. 343–75.

95. Lee KM, Lai CW, Ngai KS, Juan JC. Recent developments of zinc oxide based photocatalyst in water treatment technology: A review. *Water Res.*, 2016, 88, 4284–48.

96. Daneshvar N, Salari D, Khataee A. Photocatalytic degradation of azo dye acid red 14 in water on ZnO as an alternative catalyst to TiO_2. *J. Photochem. Photobiol. A: Chem.*, 2004, 162, 317–322.

97. Sakthivel S, Neppolian B, Shankar M, Arabindoo B, Palanichamy M, Murugesan V. Solar photocatalytic degradation of azo dye: Comparison of photocatalytic efficiency of ZnO and TiO_2. *Sol. Energy Mater Sol. Cells*, 2003, 77, 65–82.

98. Wang C, Astruc D. Recent developments of metallic nanoparticle-graphene nanocatalysts. *Prog. Mater. Sci.*, 2018, 94, 306–383.

99. Li B, Liu T, Wang Y, Wang Z. ZnO/graphene-oxide nanocomposite with remarkably enhanced visible-light-driven photocatalytic performance. *J. Colloid Interface Sci.*, 2012, 377, 114–121.

100. Zhang L, Li N, Jiu H, Qi G, Huang Y. ZnO-reduced graphene oxide nanocomposites as efficient photocatalysts for photocatalytic reduction of CO_2. *Ceram. Int.*, 2015, 41, 6256–6262.

101. Heo JN, Kim J, Do JY, Park N-K, Kang M. Self-assembled electron-rich interface in defected ZnO: rGO-Cu: Cu_2O, and effective visible light-induced carbon dioxide photoreduction. *Appl. Catal. B*, 2020, 266, 118648.

102. de Jesus Martins N, Gomes IC, da Silva GT, Torres JA, Avansi Jr W, Ribeiro C, et al. Facile preparation of ZnO: g-C_3N_4 heterostructures and their application in amiloride photodegradation and CO_2 photoreduction. *J. Alloys Compd.*, 2021, 856, 156798.

103. Oliveira JA, Nogueira AE, Goncalves MC, Paris EC, Ribeiro C, Poirier GY, et al. Photoactivity of N-doped ZnO nanoparticles in oxidative and reductive reactions. *Appl. Surf. Sci.*, 2018, 433, 879–886.

104. Zhang J, Shao S, Zhou D, Xu Q, Wang T. ZnO nanowire arrays decorated 3D N-doped reduced graphene oxide nanotube framework for enhanced photocatalytic CO_2 reduction performance. *J. CO_2 Util.*, 2021, 50, 101584.

105. Albukhari SM, Ismail AA. Highly dispersed Pt nanoparticle-doped mesoporous ZnO photocatalysts for promoting photoconversion of CO_2 to methanol. *ACS Omega.* 2021, 6, 23378–23388.

106. Nagarjuna R, Challagulla S, Sahu P, Roy S, Ganesan R. Polymerizable sol–gel synthesis of nano-crystalline WO_3 and its photocatalytic Cr (VI) reduction under visible light. *Adv. Powder Technol.*, 2017, 28, 3265–3273.

107. Mu W, Xie X, Li X, Zhang R, Yu Q, Lv K, et al. Characterizations of Nb-doped WO_3 nanomaterials and their enhanced photocatalytic performance. *RSC Adv.*, 2014, 4, 36064–36070.

108. Tahir MB, Nabi G, Rafique M, Khalid N. Nanostructured-based WO_3 photocatalysts: Recent development, activity enhancement, perspectives and applications for wastewater treatment. *Int. J. Environ. Sci. Technol.*, 2017, 14, 2519–2542.

109. Ahmed B, Kumar S, Ojha AK, Donfack P, Materny A. Facile and controlled synthesis of aligned WO_3 nanorods and nanosheets as an efficient photocatalyst material. *Spectrochim. Acta A Mol. Biomol. Spectrosc.*, 2017, 175, 250–261.

110. Farhadian M, Sangpour P, Hosseinzadeh G. Morphology dependent photocatalytic activity of WO_3 nanostructures. *J. Energy Chem.*, 2015, 24, 171–177.

111. Liu Y, Li Y, Li W, Han S, Liu C. Photoelectrochemical properties and photocatalytic activity of nitrogen-doped nanoporous WO_3 photoelectrodes under visible light. *Appl. Surf. Sci.*, 2012, 258, 5038–5045.

112. Mehmood F, Iqbal J, Jan T, Gul A, Mansoor Q, Faryal R. Structural, photoluminescence, electrical, anti cancer and visible light driven photocatalytic characteristics of Co doped WO_3 nanoplates. *Vib. Spectrosc.*, 2017, 93, 78–89.

113. Song C, Wang X, Zhang J, Chen X, Li C. Enhanced performance of direct Z-scheme CuS-WO_3 system towards photocatalytic decomposition of organic pollutants under visible light. *Appl. Surf. Sci.*, 2017, 425, 788–795.

114. Praus P, Svoboda L, Dvorský R, Reli M, Kormunda M, Mančík P. Synthesis and properties of nanocomposites of WO_3 and exfoliated g-C_3N_4. *Ceram. Int.*, 2017, 43, 13581–13591.

115. Xie YP, Liu G, Yin L, Cheng H-M. Crystal facet-dependent photocatalytic oxidation and reduction reactivity of monoclinic WO_3 for solar energy conversion. *J. Mater. Chem.*, 2012, 22, 6746–6751.

116. Li Y, Tang Z, Zhang J, Zhang Z. Exposed facet and crystal phase tuning of hierarchical tungsten oxide nanostructures and their enhanced visible-light-driven photocatalytic performance. *Cryst. Eng. Comm.*, 2015, 17, 9102–9110.

117. Wang X, Fan H, Ren P. Effects of exposed facets on photocatalytic properties of WO_3. *Adv Powder Technol.*, 2017, 28, 2549–2555.

118. Chen X, Zhou Y, Liu Q, Li Z, Liu J, Zou Z. Ultrathin, single-crystal WO_3 nanosheets by two-dimensional oriented attachment toward enhanced photocatalystic reduction of CO_2 into hydrocarbon fuels under visible light. *ACS Appl. Mater. Interfaces*, 2012, 4, 3372–3377.

119. Gondal MA, Dastageer MA, Oloore LE, Baig U. Laser induced selective photo-catalytic reduction of CO_2 into methanol using In_2O_3-WO_3 nano-composite. *J. Photochem. Photobiol. A: Chem.*, 2017, 343, 40–50.

120. Yuan X, Jiang L, Chen X, Leng L, Wang H, Wu Z, et al. Highly efficient visible-light-induced photoactivity of Z-scheme Ag_2CO_3/Ag/WO_3 photocatalysts for organic pollutant degradation. *Environ. Sci. Nano*, 2017, 4, 2175–2185.

121. Murillo-Sierra JC, Hernández-Ramírez A, Hinojosa-Reyes L, Guzmán-Mar JL. A review on the development of visible light-responsive WO_3-based photocatalysts for environmental applications. *Chem. Eng. J. Adv.*, 2021, 5, 100070.

122. Jin J, Yu J, Guo D, Cui C, Ho W. A hierarchical Z-scheme CdS–WO₃ photocatalyst with enhanced CO₂ reduction activity. *Small.* 2015, 11, 5262–5271.
123. Zhu Z, Huang W-R, Chen C-Y, Wu R-J. Preparation of Pd–Au/TiO₂–WO₃ to enhance photoreduction of CO₂ to CH₄ and CO. *J. CO₂ Util.*, 2018, 28, 247–254.
124. Huang S, Long Y, Ruan S, Zeng Y-J. Enhanced photocatalytic CO₂ reduction in defect engineered Z-scheme WO₃₋ₓ/g-C₃N₄ heterostructures. *ACS Omega.* 2019, 4, 15593–15599.
125. Wang P-Q, Bai Y, Luo P-Y, Liu J-Y. Graphene–WO₃ nanobelt composite: Elevated conduction band toward photocatalytic reduction of CO₂ into hydrocarbon fuels. *Catal. Commun.*, 2013, 38, 82–85.
126. Anandan S, Yang S. Emergent methods to synthesize and characterize semiconductor CuO nanoparticles with various morphologies – an overview. *J. Exp. Nanosci.*, 2007, 2, 23–56.
127. Raizada P, Sudhaik A, Patial S, Hasija V, Parwaz Khan AA, Singh P, et al. Engineering nanostructures of CuO-based photocatalysts for water treatment: Current progress and future challenges. *Arab. J. Chem.*, 2020, 13, 8424–8457.
128. Kumar KY, Muralidhara H, Nayaka Y, Hanumanthappa H, Veena M, Kumar SK. Hydrothermal synthesis of hierarchical copper oxide nanoparticles and its potential application as adsorbent for Pb (II) with high removal capacity. *Sep. Sci. Technol.*, 2014, 49, 2389–2399.
129. Malwal D, Gopinath P. Enhanced photocatalytic activity of hierarchical three dimensional metal oxide@CuO nanostructures towards the degradation of Congo red dye under solar radiation. *Catal. Sci. Technol.*, 2016, 6, 4458–4472.
130. Jung S, Yong K. Fabrication of CuO–ZnO nanowires on a stainless steel mesh for highly efficient photocatalytic applications. *Chem. Commun.*, 2011, 47, 2643–2645.
131. Nogueira AE, Oliveira JA, da Silva GTST, Ribeiro C. Insights into the role of CuO in the CO₂ photoreduction process. *Sci. Rep.*, 2019, 9, 1316.
132. Luo K, Li J, Hu W, Li H, Zhang Q, Yuan H, et al. Synthesizing CuO/CeO₂/ZnO ternary nano-photocatalyst with highly effective utilization of photo-excited carriers under sunlight. *Nanomaterials*, 2020, 10, 1946.
133. Yendrapati TP, Gautam A, Jain SL, Bojja S, Pal U. Controlled addition of Cu/Zn in hierarchical CuO/ZnO p-n heterojunction photocatalyst for high photoreduction of CO₂ to MeOH. *J. CO₂ Util.*, 2019, 31, 207–214.
134. Wang W-N, Wu F, Myung Y, Niedzwiedzki DM, Im HS, Park J, et al. Surface engineered CuO nanowires with ZnO islands for CO₂ photoreduction. *ACS Appl. Mater. Interfaces*, 2015, 7, 5685–5692.
135. Edelmannová M, Lin K-Y, Wu JC, Troppová I, Čapek L, Kočí K. Photocatalytic hydrogenation and reduction of CO₂ over CuO/TiO₂ photocatalysts. *Appl. Surf. Sci.*, 2018, 454, 313–318.
136. Ali S, Lee J, Kim H, Hwang Y, Razzaq A, Jung J-W, et al. Sustained, photocatalytic CO₂ reduction to CH₄ in a continuous flow reactor by earth-abundant materials: Reduced titania-Cu₂O Z-scheme heterostructures. *Appl. Catal. B*, 2020, 279, 119344.
137. Zhang F, Li Y-H, Qi M-Y, Tang Z-R, Xu Y-J. Boosting the activity and stability of Ag-Cu₂O/ZnO nanorods for photocatalytic CO₂ reduction. *Appl. Catal. B*, 2020, 268, 118380.
138. Aguirre ME, Zhou R, Eugene AJ, Guzman MI, Grela MA. Cu₂O/TiO₂ heterostructures for CO₂ reduction through a direct Z-scheme: Protecting Cu₂O from photocorrosion. *Appl. Catal. B*, 2017, 217, 485–493.
139. Wu YA, McNulty I, Liu C, Lau KC, Liu Q, Paulikas AP, et al. Facet-dependent active sites of a single Cu₂O particle photocatalyst for CO₂ reduction to methanol. *Nat. Energy*, 2019, 4, 957–968.
140. Li N, Liu X, Zhou J, Chen W, Liu M. Encapsulating CuO quantum dots in MIL-125 (Ti) coupled with g-C₃N₄ for efficient photocatalytic CO₂ reduction. *Chem. Eng. J.*, 2020, 399, 125782.
141. Li Q, Song L, Liang Z, Sun M, Wu T, Huang B, et al. A review on CeO₂-based electrocatalyst and photocatalyst in energy conversion. *Adv. Energy Sustain. Res.*, 2021, 2, 2000063.
142. Wang M, Shen M, Jin X, Tian J, Zhou Y, Shao Y, et al. Mild generation of surface oxygen vacancies on CeO₂ for improved CO₂ photoreduction activity. *Nanoscale*, 2020, 12, 12374–12382.
143. Li H, Wang G, Zhang F, Cai Y, Wang Y, Djerdj I. Surfactant-assisted synthesis of CeO₂ nanoparticles and their application in wastewater treatment. *RSC Adv.*, 2012, 2, 12413–12423.
144. Han Z, Yu Y, Zheng W, Cao Y. The band structure and photocatalytic mechanism for a CeO₂-modified C₃N₄ photocatalyst. *New J Chem.*, 2017, 41, 9724–9730.
145. Wang M, Shen M, Jin X, Tian J, Li M, Zhou Y, et al. Oxygen vacancy generation and stabilization in CeO₂₋ₓ by Cu introduction with improved CO₂ photocatalytic reduction activity. *ACS Catal.*, 2019, 9, 4573–4581.
146. Wang H, Guan J, Li J, Li X, Ma C, Huo P, et al. Fabricated g-C₃N₄/Ag/m-CeO₂ composite photocatalyst for enhanced photoconversion of CO₂. *Appl. Surf. Sci.*, 2020, 506, 144931.

147. Seeharaj P, Kongmun P, Paiplod P, Prakobmit S, Sriwong C, Kim-Lohsoontorn P, et al. Ultrasonically-assisted surface modified TiO_2/rGO/CeO_2 heterojunction photocatalysts for conversion of CO_2 to methanol and ethanol. *Ultrason. Sonochem.* 2019, 58, 104657.

148. Zheng J, Zhu Z, Gao G, Liu Z, Wang Q, Yan Y. Construction of spindle structured CeO_2 modified with rod-like attapulgite as a high-performance photocatalyst for CO_2 reduction. *Catal. Sci. Technol.*, 2019, 9, 3788–3799.

149. Kohno Y, Tanaka T, Funabiki T, Yoshida S. Identification and reactivity of a surface intermediate in the photoreduction of CO_2 with H_2 over ZrO_2. *J. Chem. Soc. Faraday Trans.*, 1998, 94, 1875–1880.

150. Kohno Y, Hayashi H, Takenaka S, Tanaka T, Funabiki T, Yoshida S. Photo-enhanced reduction of carbon dioxide with hydrogen over Rh/TiO_2. *J. Photochem. Photobiol. A: Chem.*, 1999, 126, 117–123.

151. Sayama K, Arakawa H. Photocatalytic decomposition of water and photocatalytic reduction of carbon dioxide over zirconia catalyst. *J. Phys. Chem.*, 1993, 97, 531–533.

152. Sayama K, Arakawa H. Effect of carbonate addition on the photocatalytic decomposition of liquid water over a ZrO_2 catalyst. *J. Photochem. Photobiol. A: Chem.*, 1996, 94, 67–76.

153. Kohno Y, Tanaka T, Funabiki T, Yoshida S. Photoreduction of carbon dioxide with hydrogen over ZrO_2. *Chem. Commun.*, 1997, 9, 841–842.

154. Kohno Y, Tanaka T, Funabiki T, Yoshida S. Photoreduction of CO_2 with H_2 over ZrO_2. A study on interaction of hydrogen with photoexcited CO_2. *Phys. Chem. Chem. Phys.*, 2000, 2, 2635–2639.

155. Kohno Y, Tanaka T, Funabiki T, Yoshida S. Reaction mechanism in the photoreduction of CO_2 with CH_4 over ZrO_2. *Phys. Chem. Chem. Phys.*, 2000, 2, 5302–5307.

156. Xiong X, Mao C, Yang Z, Zhang Q, Waterhouse GIN, Gu L, et al. Photocatalytic CO_2 reduction to CO over Ni single atoms supported on defect-rich zirconia. *Adv. Energy Mater.*, 2020, 10, 2002928.

157. Long J, Xue W, Xie X, Gu Q, Zhou Y, Chi Y, et al. Sn^{2+} dopant induced visible-light activity of SnO_2 nanoparticles for H_2 production. *Catal. Commun.*, 2011, 16, 215–219.

158. Chowdhury AH, Das A, Riyajuddin S, Ghosh K, Islam SM. Reduction of carbon dioxide with mesoporous SnO_2 nanoparticles as active photocatalysts under visible light in water. *Catal. Sci. Technol.*, 2019, 9, 6566–6569.

159. de Mendonça VR, Lopes OF, Nogueira AE, da Silva GT, Ribeiro C. Challenges of synthesis and environmental applications of metal-free nano-heterojunctions. In: *Nanophotocatalysis and Environmental Applications*. Springer; 2019, pp. 107–138.

160. You F, Wan J, Qi J, Mao D, Yang N, Zhang Q, et al. Lattice distortion in hollow multi shelled structures for efficient visible-light CO_2 reduction with a SnS_2/SnO_2 junction. *Angew. Chem.*, 2020, 132, 731–734.

161. He Y, Zhang L, Fan M, Wang X, Walbridge ML, Nong Q, et al. Z-scheme SnO_{2-x}/g-C_3N_4 composite as an efficient photocatalyst for dye degradation and photocatalytic CO_2 reduction. *Sol. Energy Mater Sol. Cells*, 2015, 137, 175–184.

162. Wang J-C, Zhang L, Fang W-X, Ren J, Li Y-Y, Yao H-C, et al. Enhanced photoreduction CO_2 activity over direct Z-scheme α-Fe_2O_3/Cu_2O heterostructures under visible light irradiation. *ACS Appl. Mater. Interfaces*, 2015, 7, 8631–8639.

163. Akhavan O, Azimirad R. Photocatalytic property of Fe_2O_3 nanograin chains coated by TiO_2 nanolayer in visible light irradiation. *Appl. Catal. A Gen.*, 2009, 369, 77–82.

164. Jiang Z, Wan W, Li H, Yuan S, Zhao H, Wong PK. A hierarchical Z-scheme α-Fe_2O_3/g C_3N_4 hybrid for enhanced photocatalytic CO_2 reduction. *Adv. Mater.*, 2018, 30, 1706108.

165. Xie T, Liu C, Xu L, Yang J, Zhou W. Novel heterojunction Bi_2O_3/$SrFe_{12}O_{19}$ magnetic photocatalyst with highly enhanced photocatalytic activity. *J. Phys. Chem. C*, 2013, 117, 24601–24610.

166. Peng H, Guo R-T, Lin H, Liu X-Y. Synthesis of Bi_2O_3/g-C_3N_4 for enhanced photocatalytic CO_2 reduction with a Z-scheme mechanism. *RSC Adv.*, 2019, 9, 37162–37170.

Index

Note: **Bold** page numbers refer to tables; *italic* page numbers refer to figures.

For Product Safety Concerns and Information please contact our EU
representative GPSR@taylorandfrancis.com
Taylor & Francis Verlag GmbH, Kaufingerstraße 24, 80331 München, Germany

www.ingramcontent.com/pod-product-compliance
Lightning Source LLC
Chambersburg PA
CBHW082106220326
41598CB00066BA/5639